STRATEGIC DEFENSES

Ballistic Missile Defense Technologies

Anti-Satellite Weapons, Countermeasures, and Arms Control

TWO REPORTS BY THE OFFICE OF TECHNOLOGY ASSESSMENT

OTA Reports are the principal documentation of formal assessment projects. These projects are approved in advance by the Technology Assessment Board. At the conclusion of a project, the Board has the opportunity to review the report, but its release does not necessarily imply endorsement of the results by the Board or its individual members.

Princeton University Press, Princeton, New Jersey

Published by Princeton University Press, 41 William Street, Princeton, New Jersey 08540
In the United Kingdom: Princeton University Press, Guildford, Surrey

First Princeton University Press edition, 1986
LCC 85-43466
ISBN 0-691-07711-8 (alk. paper)   ISBN 0-691-02252-6 (pb.)

Reprinted by arrangement with the Office of Technology Assessment, Congress of the United States, Washington, D.C. 20510

Publisher's Note: Undertaken by the Office of Technology Assessment at the request of the House Armed Services Committee and the Senate Foreign Relations Committee, and made public in September 1985, the two reports included in *Strategic Defenses* were published by the Government Printing Office in a limited quantity. It is our purpose in republishing these reports in one volume to make them more readily available to students, scholars, and the general public.

Clothbound editions of Princeton University Press books are printed on acid-free paper, and binding materials are chosen for strength and durability. Paperbacks, while satisfactory for personal collections, are not usually suitable for library rebinding.

Printed in the United States of America by Princeton University Press, Princeton, New Jersey

Ballistic Missile Defense Technologies

Advisory Panel on New Ballistic Missile Defense Technologies

Guyford Stever, *Chairman*
President, Universities Research Associates

Solomon Buchsbaum
Executive Vice President
AT&T Bell Labs

Ashton Carter
Kennedy School of Government
Harvard University

Robert Clem
Director of Systems Sciences
Sandia National Laboratories

Sidney D. Drell
Deputy Director
Stanford Linear Accelerator Center

Daniel J. Fink
President
D. J. Fink Associates, Inc.

Richard Garwin
IBM Fellow
Thomas J. Watson Research Center

Noel Gayler, Admiral, USN (Ret.)
American Committee on East-West Accord

Colin Gray
President
National Institute for Public Policy

George Jeffs
President
North American Space Operations,
Rockwell International

General David Jones, USAF (Ret.)
Former Chairman
Joint Chiefs of Staff

Robert S. McNamara
Former President of the World Bank

Michael M. May
Associate Director-at-Large
Lawrence Livermore National Laboratory

H. Alan Pike
Program Manager, Space Stations
Lockheed Missiles & Space Co.

Frederick Seitz
President Emeritus
The Rockefeller University

Robert Selden
Associate Director for Theoretical and
Computational Physics
Los Alamos National Laboratory

Marshall D. Shulman
Director
Harriman Institute for Advanced Study of
the Soviet Union
Columbia University

Ambassador Gerard C. Smith
President
Consultants International Group, Inc.

Sayre Stevens
Vice President
System Planning Corp.

Maj. General John Toomay, USAF (Ret.)
Consultant

Seymour Zeiberg
Vice President
Research and Engineering Operations
Martin Marietta Aerospace

NOTE: OTA appreciates and is grateful for the valuable assistance and thoughtful critiques provided by the advisory panel members. The views expressed in this OTA report, however, are the sole responsibility of the Office of Technology Assessment. Participation on the Advisory Panel does not imply endorsement of the report.

OTA Project Staff on New Ballistic Missile Defense Technologies

Lionel S. Johns, *Assistant Director, OTA*
Energy, Materials, and International Security Division

Peter Sharfman, *International Security and Commerce Program Manager*

Thomas H. Karas, *Project Director*

Michael Callaham

Richard DalBello

Gerald Epstein

Anthony Fainberg[1]

Robert Rochlin[2]

Alan Shaw

Contractors

Fredrick Drugan William Green Brian McCue

Administrative Staff

Jannie Coles Dorothy Richroath Jackie Robinson

---

[1]Since February 1985.

[2]On detail from U.S. Arms Control and Disarmament Agency.

Workshop on Soviet Military Doctrine and BMD (December 1984)

Raymond Garthoff
The Brookings Institution

Daniel Goure
Science Applications International Corp.

David Holloway
Center for International Security and Arms Control
Stanford University

Stephen Meyer
Center for International Studies
Massachusetts Institute of Technology

Harriet Fast Scott

Edward Warner
The RAND Corp.

Participating Members of Advisory Panel

Marshall D. Shulman
Sayre Stevens
H. Guyford Stever

Foreword

President Reagan's Strategic Defense Initiative has kindled a national debate over the roles of strategic offensive nuclear weapons, ballistic missile defenses, and arms control in U.S. national security policy. It has also underscored the important ramifications of U.S. military space policy.

At the requests of the House Armed Services Committee and the Senate Foreign Relations Committee, OTA undertook an assessment of the opportunities and risks involved in an accelerated program of research on new ballistic missile defense technologies, including those that might lead to deployment of weapons in space. Debate over the relevant political, military, and technical issues has been hotly contested by participants with widely varying assumptions and points of view. OTA has not attempted to resolve the debate, but rather to try to clarify the issues and enhance the level of discourse.

This report examines both the "why" and the "what" of ballistic missile defenses. **Why** would we want ballistic missile defense weapons if we could have them? Would the advantages of deploying them outweigh the disadvantages? **What** technologies are under investigation for BMD applications? How might those applications serve our strategic goals? These policy and technology questions interact with one another in complex ways: what seems technologically possible conditions perceptions of policy options, while policy choices shape technological pursuits.

Closely related to BMD technology issues are questions about the development and deployment of anti-satellite weapons. Whether or not the United States decides to deploy BMD systems in space, the other military uses of space will continue to be of national importance. How might the United States deal with the potential threat of current and future Soviet anti-satellite weapons to U.S. military space activities? After consultation with the staffs of the requesting committees, OTA decided to prepare a companion report, *Anti-Satellite Weapons, Countermeasures, and Arms Control*. The relative role each of those elements (the weapons, the countermeasures, and arms control) plays will be strongly affected by the course followed in the development and deployment of space-based BMD systems.

OTA gratefully acknowledges the contributions of the many individuals, firms, laboratories, and government agencies who assisted its research and writing for this report.

JOHN H. GIBBONS
Director

Contents

Office of Technology Assessment

The Office of Technology Assessment (OTA) was created in 1972 as an analytical arm of Congress. OTA's basic function is to help legislative policymakers anticipate and plan for the consequences of technological changes and to examine the many ways, expected and unexpected, in which technology affects people's lives. The assessment of technology calls for exploration of the physical, biological, economic, social, and political impacts that can result from applications of scientific knowledge. OTA provides Congress with independent and timely information about the potential effects—both beneficial and harmful—of technological applications.

Requests for studies are made by chairmen of standing committees of the House of Representatives or Senate; by the Technology Assessment Board, the governing body of OTA; or by the Director of OTA in consultation with the Board.

The Technology Assessment Board is composed of six members of the House, six members of the Senate, and the OTA Director, who is a nonvoting member.

OTA has studies under way in nine program areas: energy and materials; industry, technology, and employment; international security and commerce; biological applications; food and renewable resources; health; communication and information technologies; oceans and environment; and science, transportation, and innovation.

Chapter 1

Executive Summary

Contents

Table

Chapter 1
Executive Summary

THE PRESIDENTIAL CHALLENGE

President Reagan's speech of March 23, 1983, renewed a national debate that had been intense in the late 1960s but much subdued since 1972. Wouldn't the United States be more secure attempting to defend its national territory against ballistic missiles while the Soviet Union did the same? Or would it be more secure attempting to keep such defenses largely banned by agreement with the Soviet Union?

The President posed the question,

> What if free people could live secure in the knowledge that their security did not rest upon the threat of instant retaliation to deter a Soviet attack, that we could intercept and destroy strategic ballistic missiles before they reached our own soil or that of our allies?[1]

Calling upon the U.S. scientific community ". . . to give us the means of rendering these nuclear weapons impotent and obsolete," he announced that he was

> . . . directing a comprehensive and intensive effort to define a long-term research and development program to begin to achieve our ultimate goal of eliminating the threat posed by strategic nuclear missiles. This could pave the way for arms control measures to eliminate the weapons themselves.

After that speech the President ordered studies to explore further the promise of ballistic missile defense, and in 1984 the Department of Defense established an organization to expand and accelerate research in ballistic missile defense technologies. This research program was called the "Strategic Defense Initiative" (SDI).

If there were a national consensus on the role, if any, ballistic missile defense (BMD) should play in our national strategy, assessing the likelihood of attaining the necessary capabilities at an acceptable cost would be difficult enough. There is extensive controversy over the potential of various BMD technologies and the possibilities for applying them in affordable weapons systems that would be effective against a Soviet offensive threat which includes countermeasures to our defenses. But there is also extensive controversy over whether various levels of ballistic missile defense capability, if attainable, would be desirable. A fair assessment of the technological possibilities must weigh them against a range of strategic criteria which are themselves matters of controversy.

This report is intended to illuminate, rather than adjudicate, the BMD debate. It provides more questions than answers. But the questions will remain relevant in the years to come, because their answers will affect national policies with or without ballistic missile defense. For the short term, the important questions have to do with what kind of research the United States should conduct on BMD and with how future BMD technical possibilities affect current offensive force planning and diplomatic activities. For the longer term, the important questions have to do with what kind of BMD we could reasonably expect to deploy, whether we would want to, and what the consequences might be.

[1]Transcript of televised speech, Mar. 23, 1983. For text of relevant passages, see app. H.

THE BMD R&D DEBATE

The near-term debate over BMD research and development (as opposed to deployment) has focused on the following issues in particular:

1. What are (or should be) the central goals of the U.S. BMD research and development program;
2. The feasibility of reaching those goals;
3. The relationship between this research and arms control negotiations with the Soviet Union.

Participants in the debate over ballistic missile defense hold differing views on:

- Soviet motivations, intentions, and capabilities;
- Whether current U.S. nuclear strategy and nuclear forces are now, and will continue to be, adequate to deter Soviet threats and aggression;
- The past role and future prospects of arms control in contributing to U.S. national security;
- How optimistic or pessimistic one should be about the technical feasibility of rendering nuclear ballistic missiles "impotent and obsolete."

These differing views have shaped the debates both about BMD research and about BMD deployment.

Goals

Strategic Defense Initiative Goals

Few are comfortable with a situation in which U.S. security depends heavily on our threatening mass destruction with nuclear weapons. Fewer still are comfortable with the vulnerability of the U.S. population to Soviet nuclear attack. President Reagan's speech appeared to offer a way of eventually escaping this condition. **Although some people have interpreted some of President Reagan's statements to mean that he envisions development of a virtually perfect defense of the U.S. population against all types of nuclear attack, pursuit of defenses able to protect the U.S. population and that of its allies in the face of a determined Soviet effort to overcome them does not appear to be a goal of the Strategic Defense Initiative program.**[2]

Rather, some of the President's language and many subsequent policy statements indicate that the Administration envisions a more complex scenario that might eventually lead to deep reductions in the nuclear arsenals with which the United States and the Soviet Union now threaten one another. The steps in this scenario are:

1. A research program to seek ballistic missile defenses that would be cheaper to deploy than the offensive weapons needed to penetrate them.
2. A decision in the early or mid-1990s to develop such defenses for deployment near the end of the century.
3. Negotiations with the Soviet Union for agreed mutual deployment of defenses coupled with reductions in offensive weapons. In this transition stage, the threat of nuclear retaliation would play a still important, but presumably declining, role in deterring Soviet threats and aggression.
4. An ultimate stage in which ballistic missile defenses, air defenses, and negotiated reductions of offensive weapons to extremely low levels have eliminated the ability of the United States and the Soviet Union to destroy one another's societies with nuclear weapons.

Administration officials have stated, however, that negotiating with the Soviets does not mean giving the Soviets a veto over a U.S. decision to deploy BMD. In their view, if defenses become cheaper than the weapons they must intercept, the Soviets ought to see the rationality of the U.S. negotiating scenario. But if the Soviets refuse to negotiate,

[2] According to the Department of Defense "Report to the Congress on the Strategic Defense Initiative, 1985":

> The goal of the SDI is to conduct a program of vigorous research focused on advanced defensive technologies that may lead to strategic defense options that could:
> - support a better basis for deterring aggression;
> - strengthen strategic stability;
> - increase the security of the United States and its allies; and
> - eliminate the threat posed by ballistic missiles.
>
> The SDI seeks, therefore, to exploit emerging technologies that may provide options for a broader-based deterrence by turning to a greater reliance on defensive systems.

U.S. security would increase anyway because (a) Soviet ballistic missiles would be less capable of achieving military objectives than they had been in the past; and (b) if the Soviets and the United States spent equal amounts on strategic forces, the assumed cost advantage of the defense would lead to a continuing decline in ability of the Soviet offensive forces to penetrate U.S. defenses.

Although the pursuit of this scenario appears to be the central purpose of the Strategic Defense Initiative, other goals have also been ascribed to it. These include:

- maintaining an ability to deploy U.S. ballistic missile defenses promptly in case the Soviets should "break out" of the ABM Treaty;
- hedging against Soviet unilateral development and deployment of advanced ballistic missile defense technologies by gaining an understanding of what is feasible (U.S. responses could include comparable defenses, more offensive weapons, offensive countermeasures, or all three);
- developing new technologies which may or may not be applied ultimately to BMD, but which could have other military and civilian applications.

Other Perspectives on Goals

The differing views of BMD debate participants cited above lead to support for differing research goals or different placements of research emphasis. Some approve of the SDI long-term goals but believe that there should be greater emphasis on moving toward near-term deployment of land-based and space-based BMD systems. Others question the SDI goals on strategic or technical grounds. They suggest that the United States should emphasize technology development and hedging against Soviet BMD potentials and that moving toward a deployment decision in the foreseeable future should not be a goal. Those who stress maintaining a base for quickly deploying BMD to deter or respond to a Soviet ABM Treaty break-out tend to favor research emphasis on "terminal" defenses, designed primarily (or, in some cases, exclusively) to protect U.S. ICBM silos and probably using nuclear warheads. A description of how various BMD research goals might present congressional choices for alternate research and development programs is presented in a later section.

Technical Feasibility

A second major focus of the debate over BMD is technical feasibility—the likelihood that the research will lead to the development of BMD systems that could achieve Administration goals. There are at least two layers of technical issues involved in this part of the debate. One is whether particular technology performance levels (for example, those of sensors, pointing and tracking systems, computers, chemical lasers or electromagnetic rail guns) could be scaled up and integrated into effective weapons systems. The second layer of technical issues is whether the weapons systems could operate effectively against determined Soviet efforts to counter them. Proponents of the SDI believe that the technologies are sufficiently promising to be worth intensive research. In addition, they point out that for many years the Soviets have been conducting research in advanced BMD-related technologies (such as lasers) and that the SDI as a research program would be justified if on no other grounds than hedging against possible Soviet progress in these areas.

Skeptics argue that offensive nuclear weapons are so likely—unless offenses are tightly constrained in number and quality—to continue to dominate defensive weapons that pursuing the SDI goals is not justifiable. They question whether Soviet research into advanced BMD-related technologies is likely to lead to actual defensive systems that U.S. missiles could not penetrate. They believe that the best hedge against such Soviet programs is continuing or accelerating work on U.S. offensive penetration aids. They may support continued U.S. research on BMD, but they are concerned about the potential consequences of certain SDI demonstration experiments.

Arms Control

Most BMD systems based on advanced technologies could not be developed, tested, or deployed under the ABM Treaty regime.[3] One issue is whether or not our program of BMD research will be compatible with the ABM Treaty. A more fundamental issue, however, is whether or not the ABM Treaty continues to be compatible with our national interest.

Differing views on the nature of the United States-Soviet strategic relationship come to the fore most strongly in debates over the interplay between the Strategic Defense Initiative and arms control.

Supporters of the SDI tend to argue from the following perspective:

- The Soviet Union has been relentless—and at least partly successful—in its pursuit of strategic nuclear superiority over the United States. In particular, the Soviets have obtained a "first strike" capability against U.S. land-based ICBMs. In the future, the Soviets might conceivably find means of detecting and destroying U.S. missile-launching submarines as well. The Soviets can be expected to exploit such advantages by attempting to intimidate the United States and its allies.
- Past arms control agreements have not successfully limited the Soviet offensive buildup. In particular, the ABM Treaty and the companion Interim Offensive Agreement, contrary to U.S. hopes, led to no significant Soviet offensive restraint. Instead, behaving as if nuclear war would be like other wars, only bigger, the Soviets have deployed far more weapons than they need for deterrence.
- The SDI has already caused the Soviets to return to arms control negotiations which they had previously walked out of.[4] The best prospect for future arms control agreements lies in persuading the Soviets that their "first strike" ICBMs will become obsolete in the face of U.S. defenses, and that the most promising way of adding to Soviet security is negotiating the reduction of both U.S. and Soviet offensive weapons while both sides emphasize defenses. Failing such persuasion, a competition in which defensive weapons had an economic advantage over offensive weapons would be more in the U.S. interest than the current situation because in the long run it should reduce net Soviet offensive capabilities.
- Given the asymmetries between the societies and the strategic objectives of the United States and Soviet Union, the arms control process as it has been conducted to date may never be to the net benefit of the United States. On the other hand, BMD may permit pursuit of a common interest in the "assured survival" of each society.

Many critics of the SDI have another perspective:

- Given the continuing mutual abilities of the United States and the Soviet Union to destroy one another's societies with several kinds of nuclear delivery vehicle (ICBM, SLBM, cruise missile, bomber), the Soviets do not have and cannot reasonably hope to obtain an exploitable strategic nuclear advantage. Even the narrower possibility of destroying most U.S. land-based ICBMs in their silos is so fraught with uncertainties that the Soviets would be irrational to try it. Moreover, there are other potential means,

[3]While laboratory research into any type of BMD system is permitted under the treaty, there are severe limitations on field testing and development of ABM systems. Only fixed, land-based systems can be developed or tested, and only one specified fixed, land-based system can be deployed. See app. A.

[4]The official position of both the United States and the Soviet Union is that the ongoing Geneva talks are new negotiations and do not represent a resumption of the previous ones.

such as mobile basing, to increase the survivability of the ICBM leg of the nuclear triad.

- While certain issues of Soviet compliance with past arms control agreements need to be resolved, by and large those agreements have kept Soviet offenses below the levels they might otherwise have reached. The ABM Treaty successfully limited Soviet deployment of anti-ballistic missile launchers and spared the United States the need to build countering offensive and defensive weapons. Abandonment of the Treaty could lead to a more costly and more dangerous arms race.
- Rather than having driven the Soviets back to the negotiating table, the SDI might instead have merely provided them a face-saving way to reverse their previous decision—which they now regret—to stay out of arms control talks until newly deployed nuclear weapons were removed from Europe. Even though negotiations have resumed, we should believe the Soviets when they say that U.S. BMD research and deployment would lead them to seek and deploy more offensive weapons and countermeasures rather than to agree to offensive reductions. Negotiations offer a better chance of reducing the net Soviet offensive threat to the United States than does ballistic missile defense. Whatever value SDI does have in encouraging arms control can best be realized if we agree to constraints on BMD technology development, for example by clarifying or extending provisions in the ABM Treaty, in exchange for Soviet agreement to deep cuts in offensive forces.
- Over the longer term, the best hope for avoiding nuclear war lies not in new kinds of military strategy or technology, but rather in maintaining a stable balance of invulnerable retaliatory forces until the political relationship between the two superpowers can be considerably improved.

ALTERNATIVE BMD RESEARCH PROGRAMS

The issues facing Congress in the near term concern the U.S. research program on technologies for strategic defense. **There is general agreement that these technologies merit investigation. Support for BMD research, however, does not necessarily imply support for the Strategic Defense Initiative. Possible BMD research programs can differ greatly from the SDI in emphasis, direction, and level of effort. Moreover, research programs having different perceived and intended purposes—even if they have similar technical content—can have very different consequences.**

Decisions to be made by Congress this year and in the years to come will have a major impact in either ratifying or re-directing major changes which have been initiated in the U.S. BMD research program and in U.S. arms control policy by President Reagan's Strategic Defense Initiative:

Urgency.—Research under the SDI is intended to proceed at a "technology-limited" pace to permit a decision to be made at the earliest possible date on whether to enter full-scale engineering development; entering such development would clearly be inconsistent with ABM Treaty constraints. The pre-SDI program had no such mandate for an early decision on maintaining or abandoning the ABM Treaty.

Visibility.—The SDI has much higher visibility and a much higher level of Presidential attention than the previous program of research in BMD-relevant technologies. The decision to spotlight BMD has already been made, and its consequences are already being felt. These consequences certainly include a decision by the Soviets to at least explore their options to respond to the increased probability of a U.S. BMD deployment.

Direction.—Under the SDI, emphasis has shifted away from fairly well-understood, or "mature," technologies, which generally include use of nuclear-armed interceptors, towards non-nuclear defenses which would use much more speculative but potentially more powerful technologies.

Budget.—Over the next decade, much more is to be spent on BMD research than would have been allocated in the absence of the SDI. In the proposed FY 1986 budget, the BMD funding level was more than twice its projected FY 1986 level under the pre-SDI program. Subsequent increases under SDI are to be even greater, and by FY 1990 are projected to be over eight times the FY 1984 funding level.

Arms Control Policy.—Instead of the pre-SDI approach of seeking deep reductions of offensive forces along with maintenance of the ABM Treaty ban on defenses against ballistic missiles, current arms control policy seeks "greatly reduced levels of nuclear arms *and an enhanced ability to deter war based on an increasing contribution of non-nuclear defenses against offensive nuclear arms.*"[5]

Different approaches that can be taken towards ballistic missile defense research proceed from different sets of basic assumptions about the value and feasibility of BMD, and from differing assessments of the consequences of pursuing BMD research. Three such approaches can be distinguished and are presented below. These approaches differ primarily in emphasis and urgency, rather than in which technologies are to be studied. Most BMD-relevant technologies would be investigated, at some level, in all three.

The first approach is the SDI as proposed by the Reagan Administration. The second approach would proceed to BMD deployment faster than the SDI would be able to, and the third approach would conduct BMD research and development at a slower rate than the SDI. Each of the last two approaches is further broken down into two suboptions which differ in the emphasis given to existing versus near-term technologies (in the second approach) or near-term versus far-term technologies (in the third). The five research suboptions are defined as follows:

1. SDI approach.—Vigorously investigate advanced BMD technologies with the intent to decide in the 1990s on whether or not to enter full-scale engineering development and subsequent deployment. This approach assumes that while technology now within the state of the art is not good enough to be worth deploying today, the potential of advanced BMD technologies is sufficiently promising that a technology-limited effort (i.e. a program limited by what is technologically feasible rather than by funding constraints) is warranted to develop that potential. It also assumes that if successfully developed, such technologies could make possible a national security regime (weapons systems and arms control) preferable to the current one.

2a. Early deployment approach.—Emphasize early and incremental deployment of currently available BMD technology. This approach places high strategic value on the modest levels of defensive capability which could probably be obtained today. Although the ABM Treaty permits the United States to defend some ICBMs with a single, highly constrained defensive deployment, most early deployment proposals go beyond these constraints and could not be pursued under the existing treaty regime.

2b. Intermediate deployment approach.—Emphasize research on BMD technologies advanced beyond those available today but which, un-

[5]Quoted from "The U.S. Strategic Concept," enunciated by Ambassador Paul H. Nitze in an address before the International Institute for Strategic Studies, London, Mar. 28, 1985. (Emphasis added.)

like many SDI technologies, might be applicable to deployments in the early to mid-1990s. This approach assumes investigations of longer-run technologies should not delay deployments in the nearer term.

3a. Funding-limited approach.—Investigate advanced BMD technologies at a funding level well below that requested for the SDI and with a much-reduced sense of urgency. Like the SDI, this approach would focus on advanced technologies that may eventually make a highly capable defense possible. Unlike the SDI, however, it does not assume that we will know in a few years whether we can achieve that goal. The program would not aim towards facilitating a development decision at a particular time, nor would it include tests or demonstrations which might raise questions of compliance with the ABM Treaty.

3b. Combination approach.—Balance research in advanced BMD technologies with the development of near-term deployment options which would include "traditional" BMD technologies (nuclear-armed, radar-guided interceptors) of the sort specifically mentioned in the ABM Treaty. This program, conducted at a funding level well below that requested for the SDI, would aim to deter Soviet abandonment of the ABM treaty; to hedge against future Soviet BMD developments; to prevent technological surprise; and to investiage the long-term potential of advanced BMD technologies. Like the funding-limited approach, it would not include demonstrations or development work which might raise questions of compliance with the ABM Treaty.

Important issues that will be relevant to a decision among these alternative research approaches are discussed below. First, however, note is taken of Soviet BMD research.

SOVIET BMD RESEARCH AND COMPARISON WITH U.S. RESEARCH

Both the United States and the Soviet Union have conducted research and development activities in BMD since before the ABM Treaty was signed. Both have acquired considerable experience with the "traditional" BMD technologies, such as nuclear-armed, radar-guided interceptors, of the sort specifically mentioned in the ABM Treaty. **However, although the level of Soviet "traditional" BMD technology probably does not exceed our own, the Soviets, with a working BMD production base, are almost certainly better equipped in the near-term to deploy a large-scale, "traditonal" BMD system than we are.**

The Soviets have deployed and maintained an ABM system around Moscow utilizing "traditional" BMD technologies. They have also extensively upgraded and modernized that system. Ever since the United States decided that its own, similar system was not effective enough to justify maintaining it, the Moscow ABM has been the world's only operational ABM system. According to the Department of Defense publication *Soviet Military Power, 1985*, the Soviets are "developing a rapidly deployable ABM system to protect important target areas in the U.S.S.R.." That report concludes that "the aggregate of [their] ABM and ABM-related activities suggests that the U.S.S.R. may be preparing an ABM defense of its national territory."[6] Officials of the CIA, however, have said that they do not judge it likely that the Soviets would in fact move to such a deployment in the near term.[7] These officials point out that while the Soviets could expand their presently limited ABM system by the early 1990s,

> In contemplating such a deployment . . . [they] will have to weigh the military advan-

[6] Both quotations from Department of Defense, *Soviet Military Power, 1985*, p. 48.

[7] Unclassified testimony of National Intelligence Officer Lawrence K. Gershwin before a joint session of the Subcommittee on Strategic and Theater Nuclear forces of the Senate Armed Services Committee and the Defense Subcommittee of the Senate Committee on Appropriations, June 26, 1985.

Moscow Ballistic Missile Defense

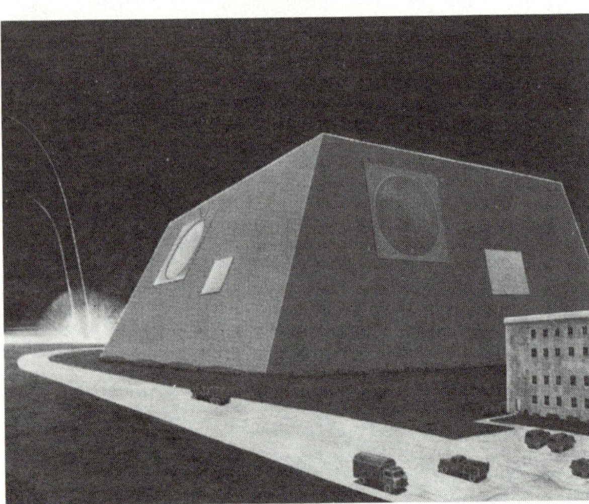

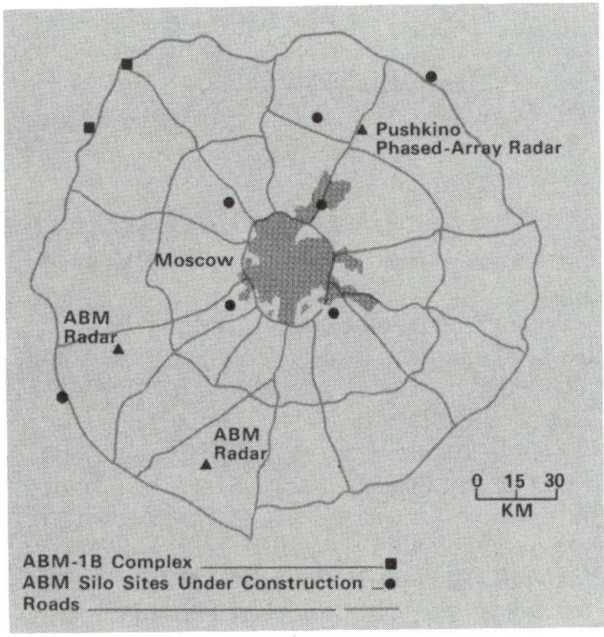

Photo credit: U.S. Department of Defense

The Moscow ballistic missile defenses identified in the map at right include the Pushkino ABM radar, above, Galosh anti-ballistic missile interceptors, top left, and new silo-based high-acceleration interceptors, top right.

tages they would see in such defenses against the disadvantages they would see in such a move, particularly the responses by the United States and its allies.

One of the functions of a U.S. BMD research program is to deter or respond to a near-term Soviet ABM Treaty breakout. A U.S. response to such a situation would most likely consist of deployment of a near-term U.S. defense, deployment of offensive countermeasures that would ensure that our strategic forces could penetrate Soviet defenses, or some combination of the two. Should a defensive response be desired, a research approach which prepared options for near-term deployment would be needed.

Photo credit: U.S. Army

The Missile Site Radar (background) of the Safeguard ABM System was designed to refine the data received from the long-range Perimeter Acquisition Radar, track the attacking ICBM reentry vehicles, and fire Sprint and Spartan interceptor missiles (in cells, foreground), to intercept them. Though this site was permitted under the 1972 ABM Treaty and its 1974 protocol, the United States decided that its limited capabilities did not justify the cost and deactivated the system in 1976.

Offensive countermeasures intended to penetrate, counter, or evade the Soviet defense are at least as important in deterring or responding to a Soviet defensive deployment as U.S. defensive options are. Offensive countermeasure research would accompany any of the BMD research options above. In addition to providing capability against Soviet defenses, an offensive countermeasures research program must be an integral portion of any research and technology development program studying BMD so that possible counters to U.S. defenses can be anticipated.

The Soviets are also vigorously developing advanced technologies potentially applicable to BMD,[8] in addition to modernizing the "traditional" system they have deployed around Moscow. However, the **quality** of that work is difficult to determine, and its **significance** is therefore highly controversial. It has been estimated that the total Soviet effort in directed energy research is larger than that in the United States. **However, in terms of basic technological capabilities, the United States**

[8] *Soviet Military Power, 1985,* op. cit., pp. 43-44.

clearly remains ahead of the Soviet Union in key areas required for advanced BMD systems, including sensors, signal processing, optics, microelectronics, computers and software. The United States is roughly equivalent to the Soviets in other relevant areas such as directed energy and power sources. The Soviet Union does not surpass the United States in any of the 20 "basic technologies that have the greatest potential for significantly improving military capabilities in the next 10 to 20 years" which were surveyed by the Under Secretary of Defense for Research and Engineering.[9]

[9]*The FY 1986 Department of Defense Program for Research, Development, and Acquisition,* Statement by the Under Secretary of Defense, Research, and Engineering, 99th Cong., 1st sess., 1985, p. II-4.

ISSUES FOR R&D PROGRAMS

1. Maintenance of the ABM Treaty

The five research options cited above each have different implications for the ABM Treaty. Administration policy is that the **SDI** approach is intended to remain within Treaty bounds until a decision is made to develop BMD systems for deployment. However, proposed technology experiments raise technical questions concerning compliance with Treaty constraints on BMD development and testing.[10] Moreover, the sense of urgency and the high visibility imparted to the SDI also raise political questions concerning the degree to which the United States is committed to maintaining the ABM Treaty regime. **Early** or **Intermediate deployment** would probably imply abandonment of the Treaty, though intermediate deployment might allow time for attempts at reaching agreement with the Soviet Union on Treaty revisions to permit limited deployments. The **funding-limited** and **combination** approaches would relax the urgency of BMD research, easing the political questions; to the extent that technology demonstrations were de-emphasized, the technical questions of treaty compliance would be relaxed as well. Advocates of these approaches would strive not to damage the Treaty regime before we had identified a preferable alternative which we had confidence could be attained.

[10]In app. A of this report, OTA points out that if one accepts the Defense Department's current interpretation of key terms of the ABM Treaty, one may also reasonably accept the conclusion that the current SDI program will be Treaty compliant; however, applying a different interpretation to key Treaty terms could lead to the opposite conclusion.

The United States can plan for revision of or withdrawal from the ABM Treaty, or it can attempt to make the Treaty more effective. The middle course—trying to bolster the effectiveness of the ABM Treaty in the short run (thereby preventing short-term Soviet BMD testing and deployment) while explicitly and publicly preparing to decide whether to abandon it in the future—may be the most difficult to implement. If we choose to maintain the Treaty in the near term, **an important issue for Congress to consider is how we can carry out our BMD research program so that it does not either prematurely compromise the ABM Treaty by encouraging Soviet exploitation of technical ambiguities, or stimulate the Soviets to begin deploying BMD and enhanced offensive forces at a time more advantageous to them than to us.** If we were to allow the ABM Treaty regime to erode, and then find at the end of our BMD research program that the new BMD technologies did not fulfill expectations, we could end up with the worst of both worlds: no arms control to limit BMD, Soviet BMD deployment, no effective U.S. BMD, and, quite possibly, augmented Soviet offensive forces intended to overcome an anticipated U.S. BMD.

At the same time, current issues of Soviet non-compliance with the Treaty must be addressed as well. If they cannot be satisfactorily resolved, the United States in effect would have adopted stricter standards of compliance than those observed by the Soviets, which might put us at a competitive disadvantage.

Congress may wish to review the standards and the procedures by which U.S. activities are judged to comply with existing treaty commitments—perhaps by establishing an independent and nonpartisan commission to review Soviet BMD activities and to advise Congress and the President on compliance questions associated with BMD activities proposed by the U.S. Department of Defense.

2. Requirements for Arms Control

In addition to their differing effects on the ABM Treaty, the alternate BMD research approaches pose different requirements for arms control.

The role of arms control under the **SDI** approach would be to facilitate a safe transition to a state of highly constrained offenses coupled with highly effective defenses. Such a transition agreement would have to be negotiated before actual deployments began. And it might need to take effect during the research and development stages, in order to regulate offensive and defensive developments. The negotiability of any such agreement is very much in question. Nobody has yet suggested how the problems of measuring, comparing, and monitoring disparate strategic forces—problems which have plagued past arms control negotiations—could be satisfactorily resolved in the far more difficult situation where both offensive and defensive forces must be included.

By deploying BMD in excess of ABM Treaty limits without waiting for the establishment of a replacement arms control regime, most **early deployment** approaches imply abandonment not only of the ABM Treaty but of the entire arms control process. Not content with the condition of strategic parity prerequisite to arms control (or, alternatively, believing that the Soviets are not willing to settle for such a state) supporters of these approaches would instead attempt to attain and maintain a lead over the Soviets in strategic forces.

Supporters of the **intermediate deployment** approach might see the possibility of negotiating with the Soviets over a transition not to "defense dominance," but to agreed force postures with an increased role for defenses relative to offenses. On balance, however, if such an agreement could not be reached, they would probably see uncoordinated deployments by the two sides as being more in the U.S. interest than the current ABM Treaty regime.

Under the **funding-limited** and **combination** approaches, negotiations with the Soviets which attempted to establish the boundaries between permitted and proscribed BMD research would be desirable for the purpose of clarifying activities on both sides. If the prospect of the United States' developing advanced technologies under the SDI approach sufficiently concerns the Soviets, Soviet desires for limitations which would have the effect of constraining U.S. research and technology development might give the United States considerable bargaining leverage. Such an agreement would almost certainly have to permit laboratory research, which would be extremely difficult to ban verifiably, but it might constrain more observable activities such as demonstrations of ABM "sub-components" and other field experiments which the Department of Defense argues are currently not prohibited by the ABM Treaty. Although it might be difficult to construct a verifiable and equitable agreement of this sort, the task might be easier than reaching agreement on the mutual introduction of strategic defenses.

3. Anti-Satellite Weapon Arms Control

At the Spring 1985 U.S.-Soviet arms control negotiations in Geneva, the Soviets emphasized the importance they attach to limiting weapons deployed in or directed at space. As the companion OTA report, *Anti-Satellite Weapons, Countermeasures, and Arms Control*, indicates, anti-satellite weapon technologies and BMD weapon technologies are closely related. Therefore, those favoring uninhibited research on BMD would find arms control measures limiting antisatellite weapon testing highly constrictive. Indeed, to attempt to remain compliant with the ABM Treaty, some technology demonstrations now planned under the SDI would be conducted as anti-satellite tests. On the other hand, those interested in strengthening the testing limitations in the ABM Treaty would find anti-satellite weapon test restrictions a useful tool in further constraining BMD development.

4. R&D/Deployment Coupling

There is an inherent conflict between seeking the ability to make deployment decisions in the near term and seeking to keep control over whether and when such a deployment might be made. Vigorous U.S. R&D programs could lead the Soviets to infer an intent to deploy, and might stimulate them to preempt such a deployment. Therefore, proposals for a vigorous R&D program should demonstrate the ability to cope with a Soviet defensive breakout and associated Soviet offensive actions in a timely way. Offensive countermeasures would probably contribute more than defensive actions towards our ability to respond to Soviet defensive breakout.

If our research program is not to be presumed to be a prelude to deployment, there must be a clearly perceived threshold which requires a positive decision—not merely the lack of a negative one—to cross. The limitations posed by the ABM Treaty provide such a threshold.

Also required, however, is a clear set of decision criteria that must be met before BMD development continues past the point requiring ABM Treaty renegotiation or abrogation. As the level of effort devoted to BMD research increases, a momentum or constituency will be created that will favor continuing and enlarging the research effort and then moving from research to demonstrations to deployment. For this reason, it would be easier to establish decision criteria before a few more years of BMD research growth had occurred and before the time comes to make the actual decision.

5. Technology Experiments

Technology demonstration experiments are the most expensive and one of the most controversial aspects of a BMD research program.

Photo credit: U.S. Department of Defense

Homing Overlay Experiment—Launch: Launch from Kwajalein Missile Range of modified Minuteman missile with device to home in on ICBM target.

Demonstrations may be useful to gauge technical progress or to provide public evidence that the technology effort in general is succeeding. Moreover, demonstrations will be needed sooner or later to determine whether some system components are feasible. On the other hand, advancing our understanding of basic principles and technologies may be preferable to demonstrating the existing state of the art. There is a risk that demonstrations may "lock in" sub-optimal levels of technology and divert resources which would otherwise go towards developing improved options.

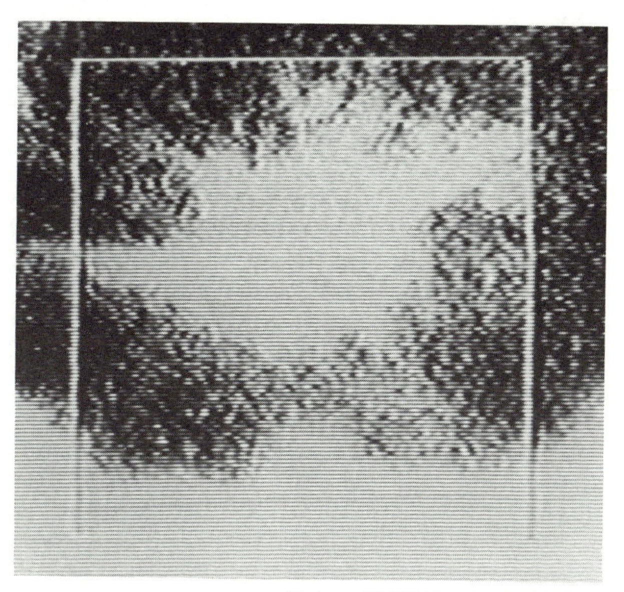

Photo credit: U.S. Department of Defense

Homing Overlay Experiment—Kinetic Kill: Video recording of telescopic view of impact of homing vehicle on reentry vehicle target. Debris resulting from collision is spreading out from center of rectangle.

Demonstrations of BMD technology are also complicated by ABM Treaty constraints on developing and testing ABM components or systems. Experiments that raise treaty compliance questions run the risk of provoking a Soviet reaction that could eliminate the option of deferring BMD deployment until technology had advanced further. One possible way to assess whether this risk is worth taking might be to require that before such experiments are approved there should be developed both (1) a plausible system architecture which would use the particular technologies to be demonstrated and (2) a corresponding arms control approach. Congress may wish to satisfy itself beforehand that, if the technologies are proven feasible, such an architecture and arms control regime appear likely to meet satisfactorily whatever criteria are established for proceeding with BMD.

6. Research and Development of Offensive Forces

In the absence of an agreement to forgo or drastically reduce them, there will be a role for U.S. strategic offensive nuclear forces for the foreseeable future. To ensure their effectiveness in the event that the Soviets deploy defenses, the United States will need to continue its development of penetration aids and other offensive countermeasures. By minimizing the potential effectiveness of Soviet defense, the existence of such countermeasures would help deter the Soviets from abrogating the ABM Treaty or any subsequent agreement limiting defenses.

However, prudence dictates we assume that any offensive countermeasure that can be developed by the United States could also be available to the Soviets, and we therefore must consider what such countermeasures would do if deployed against *our* defenses. Development by either side of powerful offensive countermeasures conflicts with the long-term goal of minimizing the role for offenses—a problem which will be exacerbated if *defensive* technologies have applications in *offensive* roles (e.g. attacking satellites or aircraft, or, particularly, attacking enemy defenses).

7. Relations With Allies

Beyond its effects on the ABM Treaty, the U.S. BMD research program can have other foreign policy consequences which should be taken into account in evaluating options. Most of our allies support United States BMD research as a counter to Soviet research, and some have inquired how they can participate in this research. However, for the most part they generally have deep reservations about the wisdom of deploying a strategic defense. **Whether the U.S. BMD research program now, and any deployment in the future, can be conducted so as to avoid endangering the cohesion of our alliances is an important issue.**

8. Technology Transfer

The ABM Treaty prohibits the "transfer to other states" of "ABM systems or their components," or of "technical descriptions or blue prints" worked out for their construction. These provisions prohibit the signatory nations from using their allies to circumvent ABM Treaty constraints. As a result, allied participation in a treaty-compliant research program would have to be limited to research which had not reached the "system" or "component" level. More of a problem for research at this stage would be restrictions which the United States itself might impose, as it does now, on the transfer of military technology to its allies for fear that such technologies may eventually reach the Soviet Union.

In some discussions of BMD research or deployment approaches it has been suggested that the United States might intentionally transfer BMD technologies to the Soviet Union to prove that the United States did not seek military superiority. **Any such transfer would raise two very significant issues: If BMD plans or devices are transferred, potential adversaries might be able to study them to discover vulnerabilities, enabling them to circumvent or destroy our own such components. If technological capability is transferred, rather than specific devices, the American advantage which had enabled us to develop that technology first would necessarily be compromised. Furthermore, many BMD-relevant technologies have applications in other military areas that we may not want to help the Soviets develop. Approaches towards BMD which assume that we can and should maintain technological supremacy over the Soviets would not be consistent with transfer of U.S. BMD technology to them.**

DEPLOYMENT ISSUES

Decisions about BMD should be made in the light of their overall impact on U.S. national security. National security depends on more than military capability. It is also affected by such factors as Soviet perceptions and actions, arms control, the cohesion of our alliances, national unity and resolve, and economic strength. It is beyond the scope of this report to attempt to define or measure national security or to explore the merits of alternative approaches to enhancing it. **Instead, we address the narrower question of how a decision to deploy BMD might affect our national security. One way to approach this question is to establish a set of criteria that BMD deployment would have to meet to some degree in order to produce net benefits for national security.**

Most participants in the current debate would probably agree on what criteria to apply in making a U.S. decision about whether to deploy BMD. But there is considerable disagreement over how stringent each of these criteria should be and what relative weights should be assigned to each. There are also strong disagreements about both the strategic and the technical prospects for satisfying them.

We label the criteria:

1. Potential Role in U.S Nuclear Strategy
2. Crisis Stability Effects
3. Arms Race Stability Effects
4. Diplomatic Stability Effects
5. Feasibility
6. Cost

The national debate over BMD in the years to come is likely to center on the application of these criteria. It is possible that the Soviets, who maintain a vigorous BMD research and development program, will choose to "break out" or "creep out" from the ABM Treaty before the United States has decided whether to deploy BMD itself. Soviet judgments about the relation of their BMD deployments to the above criteria could be quite different from U.S. judgments. But a U.S. decision about how to respond to Soviet BMD deployments would still need to take these criteria into account.

Criterion 1: Potential Role in U.S Nuclear Strategy

BMD should enhance the effectiveness of current U.S. national strategy or permit the adoption of a new and better one. And things should not get much worse before they get better: the transition to an improved strategy should not make the world significantly more dangerous than it is now. On the other hand, if one believes that current strategy in the absence of BMD is likely to lead to a worse situation for the United States, then one might settle for a BMD deployment which simply kept things from getting as bad as they might otherwise have. In addition, the BMD deployment should be no more costly or risky than alternative means, if they exist, of achieving the same strategic goals.

For analytic purposes, OTA has postulated five levels of ballistic missile defense capabilities (including none at all) and, where appropriate, of air defenses. In this part of the discussion we do not consider the feasibility or cost of obtaining and sustaining those levels of protection against Soviet offensive weapons (these are discussed in this summary under Criteria 5 and 6). Rather, we simply attempt to explore some of the strategic implications of those levels if we and the Soviets could both achieve and sustain them. (Analysis in chapter 5 of this report also addresses asymmetrical situations—cases in which one side or the other has a higher level of defense capability. For the most part, however, we assume that, in accordance with stated U.S. policy, the United States does not seek superiority over the Soviets and will not permit Soviet superiority over the United States.)

These postulated levels are:

No Additional Strategic Defense.—No defenses against ballistic missiles beyond those permitted by the ABM Treaty; passive and air defenses comparable to current levels.

Level 1: Protection of Some ICBMs.—Defenses able to assure the survival of a useful fraction of U.S. land-based ICBMs, but which would offer little or no protection to cities.

Level 2: Either/Or.—Defenses—including BMD and air defense—able to ensure the survival of most land-based ICBMs **or** a high degree of urban survival against a follow-on (or simultaneous attack), but not both.

Level 3: Most ICBMs, Some Cities.—Defenses that could intercept enough Soviet missile and and air-breathing weapons to deny the Soviets the ability to destroy most U.S. land-based ICBMs in their silos but could not deny them the ability to destroy many U.S. cities if all their offenses were concentrated on cities.

Level 4: Extremely Capable.—Defenses which would permit the Soviets to destroy, by any means, few if any targets in the United States. They could not be confident that they could destroy any U.S. cities.

The Soviets might, of course, have comparable levels of defensive capabilities. The combination of U.S. and Soviet capabilities would determine what nuclear strategies were available to the United States.

The stated goal of the Administration's Strategic Defense Initiative is that the early stages of BMD deployment should make the existing U.S, deterrent strategy more effective, while later stages would allow us to move to a different strategy.

The following discussion summarizes the potential implications for U.S. nuclear strategy of each of the five BMD capability levels postulated above.

No Additional Strategic Defense

Some argue that in the absence of ballistic missile defenses, U.S. nuclear forces will be less and less able to support current U.S. nuclear strategy, which has been called a "countervailing strategy." It attempts to deter the Soviet Union from nuclear attack or threat of attack on the United States or its allies by persuading the Soviets that U.S. nuclear counterattacks would, primarily, lead to unacceptable damage to valued Soviet assets (*punishment*), and, secondarily, would cause such Soviet attacks to fail in their geo-political objectives (*denial*). Although the United States would first attempt to repel a Soviet non-nuclear attack with conventional forces, it holds out the possibility to the Soviets that their aggression could escalate into a nuclear conflict. Thus the United States seeks not only to deter a nuclear attack on itself, but to obtain "extended deterrence" by making the Soviets fear that other kinds of aggression could lead to nuclear escalation.

In 1985, the Department of Defense estimated that the Soviet SS-18 ICBMs could destroy 80 percent of the U.S. Minuteman ICBMs in their protective silos.[11] Some argue that this vulnerability of the U.S. land-based missile force could, under some conditions, offer the Soviets an incentive to launch a preemptive nuclear strike against the United States. Even if the Soviets did not wish to launch a preemptive attack, however, they might believe that their ability to destroy a high proportion of U.S. land-based intercontinental missiles gave them a basis for nuclear blackmail against the United States or its allies.

Others argue that the Soviets are highly likely to remain deterred from such strikes by the threat of retaliation from thousands of U.S. nuclear weapons deployed on alerted bombers and submarine-launched ballistic missiles. Moreover, they argue, the probability of success of a necessarily high-precision attack on the U.S. Minuteman force is subject to many uncertainties, including the possibility that the U.S. might launch its missiles before Soviet missiles could reach them.

Level 1: Defense of Some ICBMs

To the extent that the Soviets have confidence in the success of an attack on U.S. land-based missiles, on U.S. command and control facilities, and on some other military targets, even relatively modest levels of BMD performance could reduce that confidence and thereby enhance U.S. deterrence.

At the same time, given the kinds of offensive missile forces the Soviet Union and the United States now have, modest levels of Soviet BMD performance would reduce the net number of ballistic missile warheads reaching the Soviet Union in a U.S. retaliatory attack. (See chapter 5 of this report for a detailed explanation.) This reduction in the total effective size of the U.S. retaliatory force may be thought worthwhile when weighed against the advantages of preserving at least some land-based ICBMs in addition to the SLBMs and bomber weapons and of having a partly protected command and control system.

On the other hand, in the view of those who think that the probability of Soviet success in a disarming first strike is already sufficiently uncertain, the uncertainty that BMD could add to Soviet military planning would do little or nothing to enhance deterrence. In addition, if the survivability of land-based ICBMs and command and control facilities were the only goal of U.S. defenses, then there might be other, less costly, means of achieving that goal, such as making the ICBMs mobile (and thus difficult for the Soviets to target) and increasing the redundancy and mobility of communications nodes and command centers.

Another goal might be to deter Soviet limited nuclear strikes against various kinds of military targets in the United States or a theater of war abroad, particularly in Europe. Depending on how they were configured, low to moderate levels of BMD might offer some protection to such targets. But, unlike multiple, relatively hard-to-destroy targets like missile silos, these other military targets would, as a whole set, still be highly vulnerable to a determined Soviet attack using many hundreds of nuclear weapons. The presence of BMD would force the Soviets to use more missiles than otherwise would be necessary, thus raising the threshold of violence and perhaps increasing the Soviet perception of risk of large-scale nuclear retaliation from the United States. At the same time, the presence of Soviet BMD would similarly narrow the range of U.S. limited options for using nuclear ballistic missiles.

[11] The Soviets have other accurate missiles that might raise that estimate, and as they add still more accurate missiles to their arsenal, the estimate could go up further. U.S. Department of Defense, *Soviet Military Power, 1985*, p. 30.

Modest levels of nationwide ballistic missile defense might protect the United States and the Soviet Union against relatively small missile attacks from other nuclear missile powers. On the other hand, small nuclear powers interested in nuclear weapons as instruments of terrorism may not rely on ballistic missiles as delivery vehicles, but might use, for example, small aircraft or boats to smuggle their weapons into superpower territory. For the near future, however, it is the Soviets, not we, who face Chinese, French, and British nuclear missiles. **From the point of view of U.S. allies Britain and France, Soviet BMD could degrade the credibility of their own nuclear deterrent forces.** If, on the other hand, the United States were able to provide them with effective missile defenses of their own, they might or might not consider that to be a fair trade.

Yet another benefit of even limited nationwide BMD capabilities would be the probable interception of accidental or unauthorized launches of very few ballistic missiles.

Level 2: Either/Or

(Defenses—including BMD and air defense —able to ensure the survival of most land-based ICBMs **or** a high degree of urban survival against a follow-on or simultaneous attack, but not both.) If the United States and the Soviet Union both had this level of defense capability, both Soviet and U.S. strategic planners would face still greater uncertainties. For example, in planning a first strike on the United States, the Soviets would have to consider not only how the U.S. ICBM silos might be defended, but also how U.S. defenses might be allocated between silo (and other military target) defense and city defense. They would have to be careful to retain sufficient offensive capability to threaten many U.S. cities after they had attacked U.S. silos.

A Soviet first strike followed by a U.S. retaliation could have a wide range of outcomes. The Soviets would have little ability to determine in advance what the actual outcome would be. Depending on how the United States had allocated its defenses and how the Soviets allocated their offenses, the Soviet strike might destroy nearly all, some, or virtually none of the U.S land-based ICBM force. As a result, a U.S. retaliatory attack, on the other hand, might or might not be large and well-coordinated enough to penetrate Soviet defenses. Depending on how one measures "success" in a nuclear strike, the Soviets might emerge "better" or "worse" off than they would have been if neither side had had defenses. **However, at this level, a significant and extremely dangerous possibility is that the Soviets might calculate that a first strike against U.S. retaliatory forces combined with Soviet defenses could keep damage from a U.S. retaliatory strike to a relatively low level. If the Soviets similarly calculated that the United States could strike first and defend successfully against their retaliation, that would be an additional incentive for the Soviets to attack preemptively.**

The very wide range of possible outcomes of a strategic nuclear war under these circumstances, and the difficulty of predicting which might occur, should reinforce military conservatism on both sides. But, particularly during a crisis, as the uncertainties of striking first go up, so do the potential gains, in terms of reducing the other side's ability to retaliate. And, perhaps more important, the potential risks of waiting for the other side to go first also increase.

Level 3: Defense of Most ICBMs, Plus Some Cities

(Ballistic missile and air defenses that could unconditionally deny the Soviets the ability to destroy most U.S. land-based ICBMs in their silos but could not deny them the ability to destroy many U.S. cities if all their offenses were concentrated on cities). At this level of defense capability on both sides, neither the Soviets nor the United States could have confidence in almost any plausible plan for attacking military targets, no matter how they allocated their warheads.[12] Both sides

[12]Some might argue, though, that several hundred or a thousand nuclear weapons reaching key strategic command and con-

would still be able to do considerable damage to many "soft" civilian or military targets (though perhaps markedly less than if neither had defenses[13]), but each would have to expect comparable retaliatory destruction imposed by the other. Soviet decisions to challenge U.S. interests would not be reinforced by any possibility that the Soviets could improve their military position by a preemptive strike on U.S. offensive forces.

At this high level of defense capability, both sides would also want very capable air defense systems, in order to deny any attempts to accomplish significant military ends with nuclear weapons delivered by bomber or cruise missile. Substantial civil defense capabilities might further reduce the level of casualties predicted.

At a such high level of BMD capability on both sides, the Soviets might also perceive a reduced risk that conventional or tactical nuclear war would escalate to strategic nuclear war. Insofar as the risk of escalation to nuclear war had discouraged the Soviets from aggressive acts, they might now be more tempted to use or threaten to use military force. On the other hand, U.S. leaders might be more willing to commit conventional or tactical nuclear forces to block Soviet aggression if they believed that escalation to a war that would damage U.S. territory were unlikely.[14]

trol nodes would be a plausible military accomplishment. But if sufficient survivability measures had been incorporated in the command and control system, weapons penetrating the BMD system might not accomplish much.

[13]Some might argue, though, that several hundred nuclear weapons reaching cities would be comparable in horror to several thousand.

[14]It should be noted that some policy analysts believe that the United States relies too much on the extended deterrence thought to be provided by nuclear weapons. In their view, more emphasis should be placed on conventional forces adequate to repel aggression as the primary means of deterring threats or aggression. Some holding this view believe that nuclear weapons should be used only to deter the use of nuclear weapons by threatening punishment. In this latter view, any nuclear first-strike attack by either the Soviet Union or the United State, assuming either current levels of nuclear delivery capability or much lower levels, would be irrational because the cost would be out of proportion to any conceivable gain. In other words—again, in this view—there is already little or no military utility to nuclear weapons, and ideas of "nuclear war-fighting" are unrealistic. Thus BMD deployment at "Level 3" would have no significant value for U.S. national security.

Level 4: Extremely Capable

(Defenses that would permit the Soviets to destroy few or no U.S. targets; they might be able destroy some U.S. cities, but their military planners could not have confidence in their ability to do so). The previous hypothesized levels of defense capability all retain a key element—many say *the* key element—of today's situation: the threat of massive nuclear destruction should the Soviets attack us. At this high level of capability, however, denial of Soviet ability to inflict damage on the United States would supplant retaliation as the key element of U.S. security. The survival of U.S. society as a whole would no longer depend on the *rationality* of Soviet decisions, but on the *inability* of the Soviet Union to inflict mortal damage upon us. If we believed that the Soviets had virtually no chance of delivering any nuclear weapons at all on U.S. cities or those of our allies, we might do away altogether with threats of retaliation. If, on the other hand, we believed that there was at least some risk of their being able and willing to do so, we might want to retain some residual (albeit low-confidence) retaliatory capability.

In either case, the threat of nuclear retaliation would play a much smaller role in U.S. security policy than it does today. As the Administration's long-range scenario for the Strategic Defense Initiative implies, **this level of protection could probably only be reached by a combination of defense deployments and negotiated deep reductions of offenses.** The principle of "extended deterrence" would have been abandoned, but in an international climate in which the superpowers had negotiated vast reductions in their nuclear offensive capabilities toward one another, they might also be able to negotiate reductions in the conventional and nuclear threats to U.S. allies. We return below to the question of BMD and arms control negotiations, as well as to questions of technical feasibility and cost.

Criterion 2: Crisis Stability

The deployment of BMD should not increase incentives to launch a strategic nuclear first

strike in a crisis situation. **Preferably, such incentives should be decreased.** The motive for a Soviet decision to escalate a crisis to a central nuclear war might not be to gain a clear political or military objective: instead, it may be to reduce what they fear could be a severe loss.[15] In time of crisis we would not want the Soviet leadership to calculate that its least bad option was to start a nuclear war. We would not want our own force posture to lead them to believe either that they could gain in some way by striking first or that the United States would be likely to preempt. (The issue is not whether U.S. policy would actually allow a preemptive U.S. attack, but whether the Soviets might fear that possibility.)

No Additional Strategic Defense.—Those who believe that vulnerability of the U.S. land-based ICBM force and U.S. command and control facilities might offer the Soviets an incentive to launch a preemptive nuclear attack see this vulnerability as crisis-destabilizing. It is also possible that the growing accuracy of U.S. missile warheads that could attack Soviet missiles might induce the Soviets to believe they must "use or lose" their vulnerable weapons under some circumstances.

Others argue that current crisis stability is relatively high and likely to remain so as long as both sides continue to maintain thousands of relatively survivable warheads on submarines and bombers.

Level 1: Some ICBMs.—Insofar as ICBM vulnerability is a destabilzing factor, the ability on both sides to defend some ICBMs should be crisis-stabilizing. Again, if protection of retaliatory capabilities were the only goal for a BMD system, it should be compared in cost-effectiveness to other means of achieving the same end. On the other hand, those who see crisis stability as high and likely to remain so are likely to view defense of ICBMs as unnecessary for that purpose.

[15]Many factors would go into a decision to escalate a crisis to strategic nuclear war. Calculations about the likelihood that the other side might launch a preemptive attack and about the disadvantages of waiting for it to do so would be only one set of such decision factors. In this report we treat only the even more limited question of how BMD (or its absence) might affect such calculations.

Level 2: Either/Or.—(Defenses—including BMD and air defense—able to ensure the survival of most land-based ICBMs **or** a high degree of urban survival against a follow-on (or simultaneous attack), but not both.)

As indicated above, there would be a far more serious potential for crisis instability if both sides had a "Level 2" strategic defense capability. That the Soviets would be less certain that an attack on U.S. ICBMs would succeed ought to be a stabilizing factor. On the other hand, at "Level 2" there would be at least the possibility—not previously available—that a first strike combined with defenses could keep damage from a retaliatory strike to a relatively low level. Worst of all, it is possible that both sides could arrive at a highly unstable situation in which each might perceive a chance of assuring its own survival by striking first, and only by striking first.

Level 3: Most ICBMs, Some Cities.—If both sides had ballistic missile and air defenses that could unconditionally deny the other side the ability to destroy most U.S. land-based ICBMs in their silos, but could not deny them the ability to destroy many of one's cities if all the offenses were concentrated on cities, crisis stability should be quite high. The advantages of attacking first should be marginal, the threat of retaliatory destruction still substantial.

Level 4: Extremely Capable.—At a level of defense at which few or no military targets and few or no cities could be destroyed, a strategic nuclear crisis would seem to be out of the question. An aggressor calculating that he might in some way deliver a few weapons on enemy territory would have to contend with the risk that the victim could retaliate on a similar level. Nor could a first strike do anything to reduce such residual retaliatory capabilities.

Importance to Crisis Stability of BMD System Survivability

One criterion for a BMD system which many Administration officials have cited is system survivability—the ability of the system to perform at desired levels despite direct attack on its components. We may take it for granted that neither side would bother to deploy a BMD system which could obviously be rendered ineffective by enemy attack. Rather, the question would be about the degrees of confidence on each side regarding the continuing survivability of its own and the other side's defensive systems.

If one side or the other had a BMD system that was itself vulnerable, preemption would leave the attacked side defenseless and the attacker at least partially defended against retaliation—even if the victim of attack launched ICBMs before they could be destroyed.

If both sides had vulnerable BMD systems, the net result of simultaneous successful attacks on both systems could be to leave the two sides in an offensive stand-off similar to the one existing now. **However, an extremely unstable situation would arise if each sides's space-based BMD system were vulnerable to attack, but only from the other's BMD system.** Each would have powerful incentives to "use or lose" his system, to attack before the other side did. The one that struck first might substantially disarm the other side.

Criterion 3: Arms Race Stability and Arms Control

Related to, but separate from, the issue of crisis stability is the issue of arms race stability. What incentives would BMD deployment by one side offer the other to agree to negotiate arms control measures limiting or reducing those forces? On the other hand, what incentives would BMD deployment by one side offer to the other side to increase its offensive or defensive forces in a way which would induce the first side to further increase its own forces?

There is a degree of paradox associated with the uncertainties that BMD deployment could introduce in the calculations of the two sides. On the one hand, increased uncertainty about the likelihood of successful attacks could increase crisis stability by making the aggressor less willing to gamble on a favorable outcome from a first strike. On the other hand, in the face of growing uncertainty about the effectiveness of its military forces, each side will have an incentive to try to reduce that uncertainty by deploying additional offensive and defensive weapons and countermeasures.

BMD deployments at any level would be much less likely to destabilize the strategic nuclear competition if they could be coordinated in advance by explicit agreement between the United States and the Soviet Union. If the Soviets could be persuaded that U.S. defenses hold the potential for rendering offenses obsolescent by making them less and less able to reach their targets, then the Soviets might have an increased incentive to try to negotiate mutual reduction of U.S. and Soviet offenses. Moreover, if both sides could agree for other reasons on the desirability of reducing offenses and increasing defenses, then the incentive of a favorable "cost-exchange ratio" of defenses over offenses would not be necessary. Or, to put it another way, a favorable ratio could be negotiated: decreasing offenses would make defenses more effective. A "race" between offenses and defenses would be circumvented.

An arms control agreement for phasing in BMD would have to establish acceptable levels and types of offensive and defensive capabilities for each side and means for verifying them adequately. It would have to specify offensive system limitations that prevented either side from obtaining a superior capability to penetrate the other's defenses. It would have to specify the BMD system designs for each side that would not exceed the BMD capabilities agreed to. **It is important to note, however, that no one has as yet specified in any detail just how such an arms control agreement could be formulated.**

Without such an agreement, as the United States and the Soviet Union began to deploy BMD, each might easily suspect the other of attempting to gain military advantage by seeking the ability to destroy most of the opponent's

land-based missiles and then use defenses to keep retaliatory damage to a very low level. If either side feared that its retaliatory capabilities were about to be lost or greatly reduced relative to those of the other side, there would be an incentive to add offensive capabilities and defensive capabilities at the same time. Those additions, in turn, could look to the other side like the pursuit of a "first-strike capability" and stimulate further reactive offensive and defensive deployments. The potential interactions could be extremely complex, depending as they would on the actual deployments made by each side, the effectiveness of those deployments as perceived by the other side, and the future deployments each side anticipated that the other would make. Land-based ICBMs, sea-based SCBMs, bombers, cruise missiles, and air defenses would all affect strategic stability—positively or negatively.

We do not yet know at what point the Soviets might decide that their best chance of avoiding military inferiority was to abandon their offense and stress defense. Would they do so after calculating *in advance* that offensive responses would be economically futile, or only after a considerable acceleration of the strategic arms competition had *proved* the fact through experience? Thus far, they have repeatedly declared that their reaction to the SDI will be to augment their offensive forces and pursue countermeasures. Most observers seem to believe that these (along with Soviet BMD deployments) are the most likely initial Soviet reaction to possible U.S. BMD deployments.[16]

But there is much disagreement about when, if ever, the Soviets might reverse their decision and agree to deep offensive reductions. Some argue, for example, that even if increments of offense were more expensive than corresponding increments of defense, the Soviets would still add offenses. In the long run, of course, if the United States stayed in such an arms race, the Soviets would find themselves with declining offensive capabilities. But, for the near term before any BMD deployment, if the Soviets perceive the likelihood of U.S. BMD deployment later on, then they are likely to remain unwilling to agree to offensive arms reductions.

Criterion 4: Diplomatic Stability

Relations with other nations benefit from a degree of mutual understanding of each other's intentions and from some predictability of action. While it is clear that many kinds of military deployments will affect our international relationships, we would do well to try to introduce changes in ways that minimize adverse effects on our overall relations with foreign nations.

The deployment of BMD would have significant effects, positive or negative, on our relations with our allies, with our adversaries, and possibly with other countries. Moreover, the *manner* in which we carried out a deployment decision could also affect our diplomatic relations. As Presidential arms control advisor Paul Nitze pointed out in a speech in London in March 1985,

> President Reagan has made clear that any future decision to deploy new defenses against ballistic missiles would be a matter for negotiation.
>
> This does not mean a Soviet veto over our defense programs; rather, our commitment to negotiation reflects a recognition that we should seek to move forward in a cooperative manner with the Soviets . . .
>
> Before negotiating such a cooperative transition with the Soviet Union, we would consult fully with our allies.[17]

[16]Some believe that the Soviets would actually deploy countermeasures to BMD such as penetration aids before increasing already planned levels of offensive weapons. Others believe that in any case the rate of growth of Soviet offensive forces is already so high that prospective U.S. BMD deployments would have little effect on that rate. Yet another argument is that since the Soviets have heavily emphasized other forms of strategic defense (e.g., air defense and passive defenses), they may be more willing than they admit to shift to an emphasis on ballistic missile defenses.

[17]Speech before International Institute for Strategic Studies, op. cit.

The Administration has made a frequently restated commitment to develop defenses that would defend U.S. allies as well as U.S. territory.

Proponents of BMD argue that deployment could enhance U.S. diplomatic relations in at least two ways. First, if U.S. deterrence of a Soviet nuclear attack on the United States were enhanced, our allies should feel more secure about our commitment to fight if they are attacked since it might be less dangerous for us to do so. Second, if their territory were also protected against Soviet missiles, our allies should feel directly more secure from the Soviet threat. This might be especially true if the deployment of BMD led to mutual deep reductions of offensive nuclear forces. Some would argue, however, that even defense of NATO *military* targets against Soviet missiles would strengthen allied feelings of security by enhancing deterrence of Soviet attack. Third, a new strategic relationship with the Soviets, in which we had negotiated a transition to a "defense dominant" world, might lead to a healthier relationship both between the United States and its allies and between the Western allies and the Soviet bloc. As Paul Nitze put it,

> Clearly, were we able to move cooperatively with the Soviet Union toward a nuclear-free world, that would presuppose a more cooperative overall relationship than exists at present—one in which efforts to establish a conventional balance at lower levels should be fruitful.[18]

On the other hand, skeptics about the diplomatic promise of BMD make the following sorts of predictions:

- The likely Soviet response to U.S. BMD deployment, prospective and actual, would be to add offense rather than negotiate offensive reductions; thus, the idea of a negotiated transition to a safer world or indeed of any offensive arms reductions at all in the context of BMD deployments is probably unrealistic.
- If a "defense dominant" situation had made escalation of conflict to nuclear war between the United States and the Soviet Union much less likely, most of the effects of U.S. "extended deterrence" would be lost. Either the likelihood of Soviet conventional aggression might increase, or large additions to Western conventional forces might be necessary, or both.
- European allies of the United States may correctly believe that BMD cannot protect them as well as it could the United States, particularly in view of their proximity to Warsaw Pact territory and the variety of shorter-range nuclear delivery means available to the Soviets; thus, they may see U.S. and Soviet BMD as tending to decouple their defense from that of the United States and, conceivably, make Europe a "safer" sphere of conflict for the Soviets.
- Soviet BMD might render the British and French nuclear deterrent forces ineffective, thus leading those allies to oppose the U.S. initiative in upsetting the strategic equation.
- Many national leaders see the ABM Treaty as the keystone of East-West arms control; if the United States leads the way to abandonment of that Treaty regime, U.S. allies may question whether the United States is serious about arms control and may seek to distance themselves from the United States. In addition, signatories to the 1970 Nuclear Non-Proliferation Treaty may also see U.S. and Soviet abandonment of the ABM Treaty as abandonment of the arms control process (a process the nuclear powers committed themselves to in the 1970 treaty) and be more inclined to develop their own nuclear weapons.

Criterion 5: Feasibility

There are two important levels of technical feasibility. First, it must be feasible to apply the technologies under consideration in working *components* of a BMD system. Second, it must be feasible to make the components work together effectively as an *operational system* in the face of attempts of the adversary to overcome that system.

---

[18] Ibid., p. 6.

General Issues

Whether new ballistic missile defense technologies could lead to the kind of defense we would want depends both on the potential of the technologies and on the kind of defense we would want.

- **Levels of BMD performance intended to enhance deterrence by increasing the uncertainty of the Soviets as they calculate the risks and benefits of a strike on U.S. ICBM silos and command and control facilities might be attained with technologies now fairly well-understood.**
- **Levels of BMD performance intended to assure complete denial of military objectives (such as destruction of most U.S. missile silos) to Soviet ballistic missiles would require major technological advances.**
- **Levels of BMD performance intended to offer substantial protection to U.S. cities and other "soft" targets against nuclear attack would require still more extensive advances.** These higher levels of BMD capability (such as clearly denying military utility to ballistic missiles or substantially protecting cities) will almost surely require a multi-layered, multi-weapon BMD system. Therefore, lower levels of BMD capability might be attained if a few technical developments prove fruitful, while higher levels will require that more key technologies become available together.
- **A strategic defense which could assure the survival of all or nearly all U.S. cities in the face of unconstrained Soviet nuclear offensive forces (missiles, bombers, cruise missiles, other means of attack) does not appear feasible.** As we have seen, current Administration policy envisages pursing the goal of assured survival through a **combination** of defensive weapons and negotiated deep reduction of Soviet and U.S. offensive weapons.

A wide variety of technologies could, in principle, be developed to produce components for a multi-tiered ballistic missile defense system. Candidate technologies for kill mechanisms include various types of lasers, kinetic energy vehicles (self-propelled or projectile), and particle beams. **No known physical law stands in the way of developing such components and assembling them into a layered system intended to intercept ballistic missiles and their warheads in the boost-phase, the post-boost (reentry vehicle separation) phase, the midcourse, and the reentry phase.** Physical laws do limit the potential performance of some kinds of components, however. For example, neutral particle beams cannot penetrate the atmosphere, and thus could not intercept missiles while they were still in the atmosphere.

For most of the new ballistic missile defense technologies, much research is still necessary to determine whether the physical principles involved can be affordably applied in working weapon systems. **Many of the technologies being considered for BMD still require improvements in performance of orders of magnitude (factors of ten) before they can be used in weapons.** Systems for boost phase, post-boost phase, and midcourse BMD are likely to require many satellites in orbit, satellites which must be highly reliable while relatively inaccessible to maintenance.[19]

Massive improvements in computer speed, reliability, durability in a hostile environment, and software capabilities would be required. Current research gives cause for some optimism about meeting the hardware requirements, though **most analysts agree that generating the necessary software would be a monumental task.**

Space-based BMD systems would require a much more capable space transportation system than the United States now has, and would probably require a substantial lowering of space launch costs. This requirement would be less stringent, however, in the case of systems employing ground-based lasers.

Another issue is the susceptibility of sensors to defeat by various countermeasures. Their

[19]However, some system designs might require fewer than others. For example, those using ground-based lasers, or those using weapons "popped up" from the ground, would require fewer satellites than those using entirely space-based weapons.

sensitive nature, necessary for long distance detection, makes them vulnerable to various kinds of temporary or permanent blinding. They would have to be designed to operate against a background of nuclear explosions. Making decoys look like targets and making targets look like decoys may spoof sensor systems. Space-basing makes the the sensors potentially vulnerable to antisatellite weapon attack.

For all space-based BMD system components, survivability against directed energy, nuclear, or kinetic energy weapon attack is a major issue. For example, space mines might be planted to tail sensor satellites or battle stations. As the companion OTA report, *Anti-Satellite Weapons, Countermeasures, and Arms Control,* indicates, there are many potentially effective means of interfering with or destroying space systems, as well as many potential countermeasures for dealing with those means. Whether the means of protecting satellites will be adequate to ensure the survivability of particular space-based BMD systems will depend in part on the kind of system deployed[20] and in part on future Soviet anti-satellite capabilities. Insufficient information is now available to resolve the survivability question.

The Soviet Union will have about as long to develop offensive countermeasures to defensive systems as the United States does to develop the defensive systems, and vice-versa. No one can confidently predict today whether defensive technologies will dominate offensive delivery technologies in the future. It is clear that a U.S. BMD research program should devote considerable effort to exploring BMD countermeasures, both to determine whether defense at the desired level of effectiveness is feasible and affordable, and to hedge against Soviet BMD advances.

[20]For example, many sensors, redundantly distributed among numerous satellites in a variety of orbits, could increase overall system survivability.

Components and Systems

In the absence of officially proposed BMD architectures (system designs) and in the absence of specific weapon designs, it is impossible to estimate BMD feasibility and costs. The Strategic Defense Initiative Program is charged with doing research to find out whether current technologies can be scaled up to the necessary performance levels and then whether they can be applied in engineering effective, reliable, and affordable weapons. How much Congress will choose to invest in this research program will depend in part on its judgment about the benefits and risks considered in this report and on its beliefs in the premises and predictions of differing policy advocates.

It is possible, however, to give a general conception of the likely ultimate feasibility and costs by conveying a feeling for the requisite scale of a highly capable, multi-layered BMD system. OTA has postulated a BMD system architecture purely as an illustration of the kinds of tasks involved in deploying a very ambitious BMD system. **We definitely do not predict that the example we have hypothesized will ever be proposed or built, nor do we assert that the technologies assumed for it are more or less promising than any others.** The example is given (see table 1-1) to illustrate that deploying a large-scale, multi-tiered BMD system would be a formidable, complex, and expensive job.

A highly capable BMD system designed along the lines of our postulated example would pose the following challenging requirements:

- **A boost-phase defensive layer effective in the face of proliferation and countermeasures.** The boost phase interception of ballistic missiles must be highly effective to keep the tasks of the succeeding defensive layers manageable. Soviet deployment of many additional rocket boosters appears possible in the very near term. Rocket boosters which finish burning very quickly and upper stages which dispense their separate reentry vehicles very rapidly appear feasible to deploy in significantly less

Table 1-1.—Hypothetical Multi-Layered BMD System

System level	System elements	Description	Comments
Level 1 **Terminal Defense** (defense of hardened sites using endoatmospheric rockets to intercept reentry vehicles (RVs) as they approach their targets)	Early warning satellites; ground-based radar; airborne optical sensors; ground-based battle management computers; fast endoatmospheric interceptors.	Warning of launch provided by high-orbit satellites: RVs detected and tracked in region of ground targets by ground radar and airborne sensors; ground computers assign interceptors to RVs; kill assessment[a] permits reassignment of defense interceptors; atmospheric interception used; air effects used to discriminate between RVs and decoys.	Homing either infrared (IR) or radar; interceptors should be relatively inexpensive, since many needed; may be nuclear or nonnuclear.
Level 2 **Light Midcourse and Terminal Defense** (additional layer added with some interception capability in midcourse and some ability to discriminate RVs from decoys in space to reduce burden on terminal layer; some area defense)	Level 1 plus: exoatmospheric homing interceptors, range hundreds of km; pop-up[b] IR sensors (possibly satellite-based instead); self-defense capability for space assets.	As in level 1 for terminal defenses; longer range interceptors added which can intercept some RVs above atmosphere, providing some area defense; this requires some discrimination capability, furnished by passive IR pop-up sensors, launched towards cloud of decoys and attacking RVs; the new layer reduces the burden on the terminal layer.	Passive IR sensors used for crude discrimination and possibly kill assessment; data base of Soviet RV and decoy signatures needed; sensors must be able to function in a hostile nuclear environment.
Level 3 **Heavier Midcourse Layer** (effective midcourse layer added, giving realistic two-layer system, with each layer highly effective)	Level 2 plus: ultraviolet laser radar (ladar) imaging on satellites; highly capable space-based battle management system; space-based kinetic energy weapons; effective self-defense in space; significant space-based power.	Satellite-based ultraviolet laser radar (ladar) used to image objects; discrimination provided by comparing images with data base of Soviet RV and decoy characteristics; RVs attacked by in-orbit kinetic-energy weapons, which also defend all space-based components of system; this level has fully developed terminal and midcourse layers, but no boost or post-boost phase defense.	Ladar imaging rapid with resolution good to 1 meter or less for adequate discrimination and birth-to-death tracking of RVs; kinetic weapon homing capability good to less than a meter.
Level 4 **Boost-Phase Plus Previous Layers** (boost-phase intercept added to kill boosters or post-boost vehicles before RVs and decoys dispersed)	Level 3 plus: ground-based high-intensity lasers (either excimer or free electron); space-based mirrors for relay and aim; high resolution tracking and imaging in boost phase; self-defense for all phases.	This level adds a boost- and post-boost-phase layer, consisting of very bright ground-based laser beams directed to their targets by orbiting mirrors; sensing by infrared sensors, imaging by ultra-violet ladar; battle management to handle all layers doing discrimination, kill assessment, and target assignments and reassignments. Boost- and post-boost-phase layers may be combined, since post-boost phase could be shortened to 10 seconds or so.	Extremely capable battle management system needed; kill assessment required for boost phase as well as midcourse.
Level 5 **Extremely Effective Layer** (Level 4 with better capability; meant to permit only minimal penetration to targets by enemy RVs)	Level 4 plus: more terminal and exoatmospheric interceptors; electromagnetic launchers for midcourse and boost-phase intercepts; large capacity space-based power; all systems extremely reliable.	More interceptors are added in terminal and midcourse layers; electromagnetic launchers used for boost, post-boost and midcourse intercepts; high capacity space power needed; all systems, including battle management must be extremely reliable.	Essentially same as Level 4, but more of it and higher reliability; newer technologies used as they become available.

[a]Kill assessment refers to the process of determining whether a struck target has been effectively disabled.
[b]Pop-up components are ground-based assets which are launched into space for action upon warning of an enemy attack.

SOURCE: Office of Technology Assessment.

Photo credit: U.S. Department of Defense

Artist's conception of possible space-based mirror for redirecting ground-based laser beam to target in space

time than the defensive systems they would be designed to counter. Boost-phase defenses need to be effective against both.

- **Sensors and computers able to discriminate rapidly between decoys and reentry vehicles in the midcourse phase (as the objects separate from the post-boost vehicles, and before they reenter the atmosphere).** Techniques now fairly well understood for making reentry vehicles and decoys "look" like one another to various sensor systems will make target discrimination one of the most challenging tasks for midcourse interception systems.
- **Sensors that can function nearly continuously under attack and against a background of nuclear detonations.**
- **A system of battle management computers and software of very high complexity.** A control system will be required to be able to track thousands—possibly hundreds of thousands—of objects simultaneously, assign weapons to attack the correct targets, account for targets destroyed, and assign other weapons to missed targets. This task will require extremely large, complex computer programs of very high reliability.
- **Communications links among sensors, battle management centers, and weapons that can function reliably in the face of jamming attempts, attack, and interference from nuclear detonations.**
- **Space-based power supply systems, each of ten or more megawatts, with high reliability, quick response, and affordable maintainability.**
- **Means of protecting space-based BMD assets from a wide range of possible means of attack.**
- **Ground-based exoatmospheric (late midcourse layer) interceptors that are inexpensive because they must be numerous.**
- **Ground-based interceptors for the final (terminal) layer that are inexpensive because they must be numerous.**

Not only do issues of technological feasibility need to be resolved, but so do issues of **operational** feasibility. That is, the developed components must be combined into an **integrated, reliable system** that could operate effectively and maintain that effectiveness over time as new countermeasures appeared. Such a system could never be fully tested operationally—as indeed strategic offensive nuclear systems have never been. But we would want to have high confidence in the effectiveness of a defensive system to consider steep reductions in our offensive retaliatory forces.

Criterion 6: Cost

Another part of the decision about investment in BMD research depends on a weighing of potential benefits and risks against ultimate costs. If some of the research can lead only to the demonstration of the technical feasibility of systems so costly that the nation would never want or be able to pay for them, then the decision on whether to do the research would be different than if the expected costs

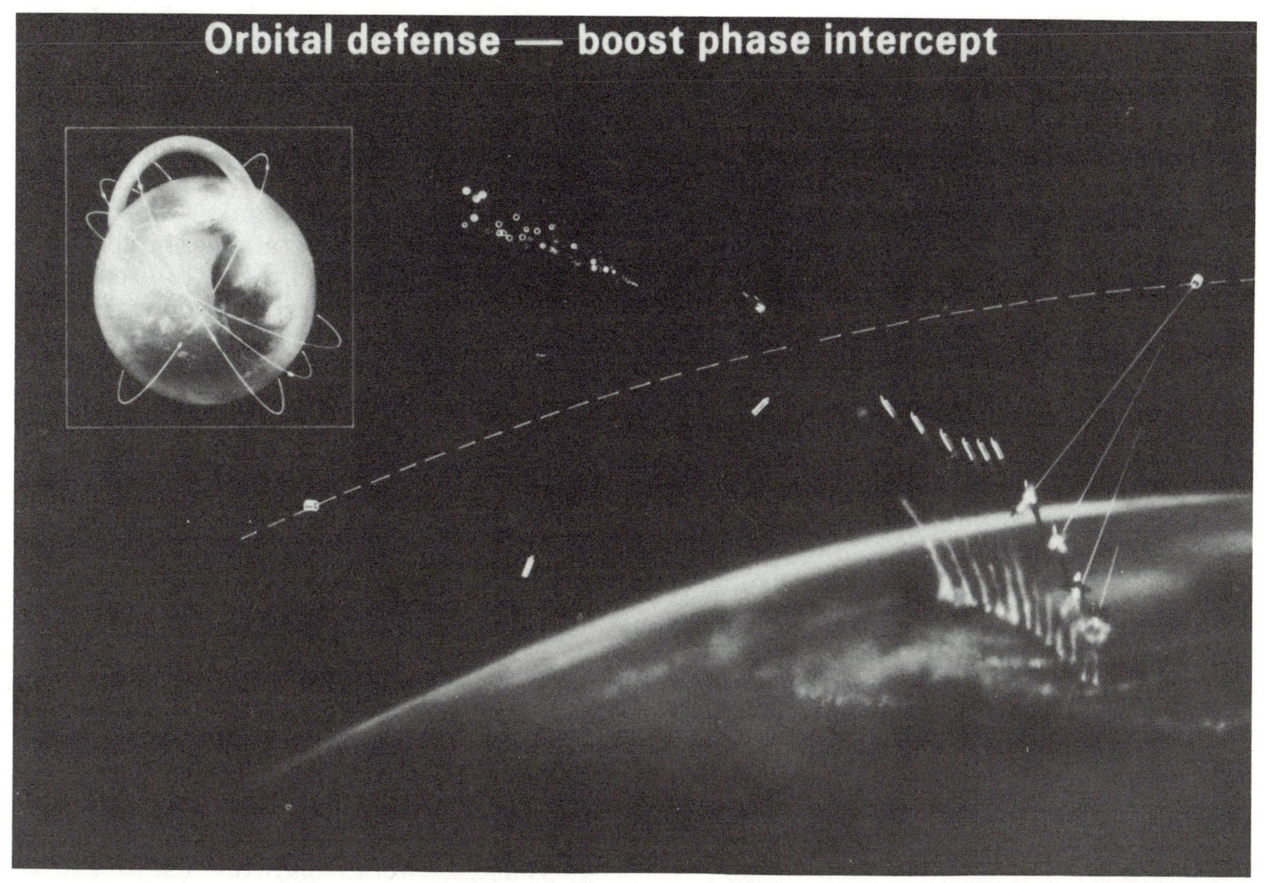

Photo credit: U.S. Department of Defense

Boost-phase intercept would provide great leverage, since each booster may contain many targets and decoys. In the illustration, some boosters, penetrating this layer of the defense, have deployed RVs and decoys. Inset shows "threat tube" of Soviet missiles and orbits of some of the satellites in the defense constellation. The number of battle stations required depends on weapon range (related to brightness for directed-energy weapons), retarget time, number of targets, and time available for attacking the targets.

were commensurate with the expected benefits. Everyone can agree that a multi-tiered BMD system with significant space basing would be very expensive, but how expensive depends on many unknowns.

Besides illustrating the need for the kinds of technical developments described under Criterion 5, examination of our hypothetical system also indicates how the presence of so many unknown factors makes realistic cost estimates impossible now. It does not demonstrate that deploying a large-scale BMD system would be either affordable or too costly. **However, the burden of proof is on those who maintain that BMD can be affordable and display a favorable cost/exchange ratio with offenses to provide credible estimates of eventual system costs.**

Important issues of cost include the following:

- **Allocation of Defense Department research funds.** The first six years of the Strategic Defense Initiative are scheduled to consume a total of approximately $33 billion in defense research funds. Succeeding years before a development decision may bring yet higher annual costs. These should be weighed against opportunity costs in other areas of defense research.
- **Allocation of national technical research re-**

Photo credit: U.S. Department of Defense

Artist's conception of space-based sensors on a surveillance and tracking satellite.

sources. The supply of specialized scientists, engineers, and research facilities is not highly elastic in the short term. BMD research would divert some of these national resources from other important tasks. On the other hand, BMD research might produce substantial "spin-off" results which could lead to advances in technologies applicable to other civilian or military purposes. Over the long term, the research might also stimulate training of additional scientists and engineers.

- **Allocation of military procurement funds.** BMD procurement could absorb funds needed for other military programs, such as ground, naval, or air forces. On the other hand, under some scenarios, BMD might reduce the need for expenditures on offensive nuclear forces and even on conventional forces.
- **Allocation of industrial resources.** In the procurement stage, BMD deployment might divert engineering and manufacturing resources from other production. On the other hand, it might contribute to an industrial base for other activities, such as commercial development of space.
- **Total costs.** The total system costs for BMD will remain difficult to predict for some time. Also difficult to determine are the potential effects of BMD on other U.S. military needs. For example, a BMD deployment which led to negotiation of deep reductions in offensive forces would eventually allow shifting expenditures away from strategic offenses. In the short run, however, until the cost-exchange ratios in the offense-defense competition on both sides became clear, *increased* expenditures might be required to maintain offensive forces on a par with those of the Soviet Union.

Defenses intended to protect substantial parts of the United States' and its allies' populations would also require a highly capable air defense system, since making ballistic missiles obsolete would not in itself suffice to assure population survival. Effective population defense might also be judged to require large civil defense expenditures as a complement to the active missile and air defenses.

A change in U.S. strategy which placed greater emphasis on non-nuclear capabilities for deterring aggression against U.S. allies might require costly enhancements of our air, sea, and land conventional forces and those of our allies. An alternative, however, which could reduce rather than increase the cost of such forces might be substantial *conventional* arms control, particularly in the NATO-Warsaw Pact arena.

We referred earlier to the concept of the "cost-exchange ratio" between defense and offense. That is, increments of defense should cost less than the corresponding increments of offense that they must neutralize. If so, then the offense would have a strong disincentive to try to keep up with the defense. We also need to estimate at what point the Soviets would decide to concede such a reality and stop trying to maintain offensive capabilities. However, a favorable cost-exchange ratio would not suffice if the defense system as a whole were too expensive to deploy. **One goal of research should be to identify a BMD system whose base cost and cost-exchange ratio with offenses was such that the combined cost of overcoming existing Soviet offenses and countering their response to our defenses was affordable.** Just what the cost-exchange ratio needs to be would depend on how willing the Soviets might be to try to outspend us to maintain their offensive capabilities.

CONCLUDING REMARKS

Debated Issues

The question of the role of ballistic missile defense in U.S. national security is complex. However, national debate has tended to polarize between support of and opposition to the SDI.

Both proponents and opponents agree on two major points:

- **The United States should adopt whatever BMD posture will be most likely to minimize the risk of nuclear war.**
- **The United States should be carrying out some research on BMD technology.**

The strongest disagreements regarding SDI center on two related issues:

- **How likely is it that technology will reach a point where it would be desirable to deploy BMD?** This disagreement partly reflects differing guesses about the future cost and rate of technological progress. More significantly, however, it reflects differing views about how valuable BMD would be for our national security, and how effective a BMD system must be for the benefits of deployment to outweigh the risks.
- **Should the research program be carried out with the vigorous commitment that characterizes the SDI?** The central idea of the SDI seems to be an ardent belief that a program of urgent, centrally directed, and generously funded research and development would have a good probability of bringing us within a few years to the point where we would be justified in deciding to deploy a high-technology ballistic missile defense system. The central concern of SDI opponents, apart from skepticism that such a system could be effective and affordable, is that the technology development may be much more likely to destabilize the superpower strategic balance and set off an arms race than to justify a decision to deploy. For this reason they favor a less urgent, less expensive, and less prominent research program, mainly to hedge against unexpected technological breakthroughs and as a means of deterring the Soviets from abandoning the ABM Treaty by providing the United States with an adequate response if they do.

Proponents of SDI are not all of one mind. However, they stress some or all of the following:

1. *The most important national goal we can have is assured survival; that is, President Reagan's goal of a world in which "free people could live secure in the knowlege that their security did not depend on the threat of instant retaliation to deter a Soviet attack, that we could intercept and destroy strategic ballistic missiles before they reach our soil or that of our allies."* This goal may be attainable, particularly if the development of defenses induces the Soviets to agree on reductions of offensive forces, and therefore it is worth pursuing vigorously.
2. *Even if we cannot achieve assured survival, a strategic policy that relies to a significant extent on strategic defenses would be better for the United States*

than our existing policy of deterring aggression only by the threat of retaliation by our offensive nuclear forces.

3. *The strategic balance has gradually been shifting against the United States, and developing and deploying ballistic missile defenses (with or without accompanying arms control measures limiting strategic offensive forces) offers the best opportunity to reverse this trend.*
4. *Many of the new ideas proposed for ballistic missile defense are now ripe for intensive research and development.* If the United States develops these technologies vigorously, we can expect major improvements in potential BMD capabilities. There are grounds for believing that defensive technologies may improve so much faster than offensive technologies that it will become cheaper to deploy defenses than to deploy offensive countermeasures to overcome the defenses. This would give the Soviets a powerful incentive to agree to reduce offensive arms and concentrate on building their own defensive systems. If the Soviets exploited these BMD technologies and we did not, our security might be severely jeopardized.

Opponents of SDI argue some or all of the following:

1. *Assured survival is so extremely improbable in the forseeable future as to be irrelevant as a national goal.* If it could be attained at all, it would require drastic reductions and stringent limitations of all offensive nuclear arms even if very effective defenses could be deployed. But since the vigorous pursuit of defensive capabilities now would make such offensive arms control much less likely to be attained, we should pursue offensive arms control first and defensive deployments afterwards, if at all.
2. *Ballistic missile defenses that are highly effective, but not adequate to provide assured survival, could create dangerous instabilities.* Developing them would set off an offensive/defensive arms race. Furthermore, defensive deployments could provide great incentives to preempt in a crisis by holding out the possibility of "victory" to the side launching a massive first strike and defending against the presumably less effective retaliatory second strike. If deployed BMD systems were themselves vulnerable to attack, the incentive to strike first could be even greater.
3. *The buildup of Soviet strategic forces in recent years, while certainly undesirable, has not reduced the U.S. ability to deter a Soviet attack.* The continuing Soviet buildup does not pose a serious threat to the credibility of our deterrent. A U.S. strategic defense would not improve the strategic balance. Modernization of our strategic forces and vigorous efforts to make the arms control process effective would be far more likely than BMD to improve U.S. security.
4. *While nobody can predict with certainty the results of future research, it is highly unlikely that we could develop BMD systems which could not be overcome by affordable Soviet countermeasures.* Therefore, the SDI is not the most fruitful area in which to concentrate our limited resources for military R & D. While research on BMD is necessary, an overly vigorous U.S. BMD program would be likely to stimulate a buildup of Soviet offensive forces, which would preclude meaningful offensive arms control measures and make it harder to maintain the survivability of our retaliatory forces.

OTA Findings

1.—Both the capability of a BMD system to defend the United States, and the strategic value to the United States of any given BMD capability, depend on the interaction of all the kinds of the defenses actually deployed with all the kinds of offensive threat against which they must actually defend. In the past, the enormous destructive power of nuclear weapons has meant that offensive strategic technologies have had a large and fundamental advantage over defensive technologies. Unless this imbalance between

the offense and defense disappears, strategic defenses might be plausible for limited purposes, such as defense of ICBM silos or complication of enemy attack plans, but not for the more ambitious goal of assuring the survival of U.S. society. This imbalance might be changed either by political decisions of both superpowers to reduce the kinds and levels of offensive deployments to capabilities much less than available technology permits, or by development and deployment of defensive systems able to overcome whatever offenses could be developed and deployed in the same period. While it is certainly possible that defensive technological development could outpace the development of offensive weapons and countermeasures to defenses, this does not appear very likely.

2.—**Assured survival of the U.S. population appears impossible to achieve if the Soviets are determined to deny it to us.** This is because the technical difficulties of protecting cities against an all-out attack can be overcome only if the attack is limited by restraints on the quantity and quality of the attacking forces. The Reagan Administration currently appears to share this assessment.

3.—**If the Soviets chose to cooperate in a transition to mutual assured survival, it would probably be necessary to negotiate adequately verifiable arms control agreements on reducing present and restricting future offensive forces and on the manner, effectiveness, and timing of defensive deployments.** OTA was unable to find anyone who could propose a plausible agreement for offensive arms reductions and a cooperative transition that could be reached before both the Soviets and the United States learn more about the likely effectiveness and costs of advanced BMD technologies. Indeed, such a transition could hardly be planned until engineering development was well advanced on the actual defensive systems to be deployed. Even then, adequate verification would be difficult. Without such agreement on the nature and timing of a buildup of defensive forces, it would be a radical departure from previous polcies for either side to make massive reductions in its offensive forces in the face of the risk that the other side's defenses might become highly effective against the reduced offenses before one's own defenses were ready. Such a transition would be more appealing to both sides if BMD technologies could be developed which cost less to deploy than the offensive countermeasures needed to overcome them than it would be if the historic and current advantages of offense over defense persist. In essence, **the question is whether a vigorous U.S. program to develop BMD, and the prospect that both sides might deploy effective BMD, will make the Soviets more willing than they have been in the past (or now say they are) to agree to deep reductions of strategic offensive forces on terms acceptable to the United States.**

4.—**There is great uncertainty about the strategic situation that would arise if BMD deployment took place without agreement between the United States and Soviet Union to reduce offensive forces as defensive forces grew.** Until the actual offensive systems (including ICBMs, SLBMs, bombers, and cruise missiles) and defensive systems (including BMD and air defenses) were specified and well understood, no one could know with confidence whether a situation of acute crisis instability (i.e. striking first could appear to lead to "victory") could be avoided. A fear on either side that the other could obtain such a first strike capability could lead both sides to build up both their offenses and their defenses. Such build-ups would make it even more difficult to negotiate a cooperative transition from offense dominance to defense dominance.

5.—**The technology is reasonably well in hand to build a BMD system that could raise significantly the price in nuclear warheads of a Soviet attack on hardened targets in the United States; such a system, if combined with a re-basing of U.S. ICBMs, could protect a substantial fraction of those U.S. land-based missiles against a Soviet first strike.** However, it is not clear whether BMD would be the best way to provide missile survivability, nor is it clear whether the combination of a U.S. program protecting ICBMs and the Soviet response—perhaps expansion of their Moscow defense to other Soviet cities—would on balance strengthen or weaken our deterrent.

6.—**It is impossible to say at this time how effective an affordable BMD system could be.** To answer this question requires extensive research on sensor, command and control, and weapons technologies; on system architecture (including survivability and computer software); and on counter-countermeasures. Credible cost estimates based on this research will also be necessary.

7.—**The decision whether to push ahead vigorously with the SDI or to scale back the Adminsistration proposal involves a balancing of opportunities against risks, in the face of considerable uncertainty.** The SDI offers an *opportunity* to substantially increase our nation's safety *if* we obtain great technical success and a substantial degree of Soviet cooperation. The argument that sufficiently great U.S. technical success would force the Soviets to cooperate in their own security interests is logically compelling, but there can be no assurance that the Soviets would actually behave as we think they should. The SDI carries a *risk* that a vigorous BMD research program could bring on an offensive and defensive arms race, and a further risk that BMD deployment, if it took place without Soviet cooperation, could create severe instabilities. Whether BMD deployed in the face of intense Soviet efforts to counter it would enhance U.S. security depends on a judgment that decreased Soviet confidence that they could destroy targets in the United States or on allied territory would, in Soviet minds, outweigh their increased confidence that targets in the Soviet Union would survive because of their own BMD.

8.—**Whatever type of BMD research program the United States decides to pursue, it would be prudent to carry out that research in such a way as to minimize Soviet incentives to decide to deploy their own BMD beyond the limits set by the ABM Treaty before the United States has completed the research necessary to make our decision.** This might be done by unilaterally restraining our BMD research. We would have greater influence over Soviet actions, however, if we reached agreement with the Soviets regarding disputed interpretations of the ABM Treaty—including the boundaries of permitted research—and regarding the conditions under which future BMD deployments would be desirable. Such an agreement would also reduce Soviet incentives to build up their offensive forces in order to overcome anticipated U.S. defenses. However, it must be recognized that acting to deter a Soviet decision to deploy BMD may require limiting and slowing our own BMD research.

Chapter 2
Introduction

Contents

Chapter 2
Introduction

THE STRATEGIC DEFENSE INITIATIVE

President Reagan's speech of March 23, 1983, proposed a major shift in U.S. nuclear strategy. For at least 25 years, since the earliest Soviet deployments of Intercontinental Ballistic Missiles, the United States has relied on the threat of retaliation to deter Soviet nuclear attack on the United States. During the 1960s both sides worked on developing weapons that were intended to defend against ICBMs. In the United States, a debate also arose over whether such defenses were feasible and desirable. Would the United States be more secure attempting to defend its national territory against ballistic missiles while the Soviet Union did the same? Or would it be more secure attempting to keep such defenses largely banned by agreement with the Soviet Union? In 1972 President Nixon chose the latter by signing the SALT I ABM (Anti-Ballistic Missile) Treaty, and the Senate consented by ratifying it.

In his speech President Reagan said that even if current arms reduction negotiations with the Soviets were to succeed,

> . . . it will still be necessary to rely on the specter of retaliation—on mutual threat . . . Wouldn't it be better to save lives than to avenge them? . . . What if free people could live secure in the knowledge that their security did not rest upon the threat of instant U.S. retaliation to deter a Soviet attack; that we could intercept and destroy strategic ballistic missiles before they reached our own soil or that of our allies?[1]

He held out the prospect, then, for a substantial change in U.S. nuclear strategic policy. With this change, the United States would move away from its current deterrent posture against the Soviet Union, which stresses offensive counter-threats to deter potential Soviet aggression. Instead, deterrence would emphasize preventing Soviet ballistic missiles from reaching their targets at all.

The President called upon

> . . . the scientific community in our country, those who gave us nuclear weapons, to turn their great talents now to the cause of mankind and world peace: to give us the means of rendering these nuclear weapons impotent and obsolete.[2]

He did add a caution to his proposal:

> . . . defense systems have limitations and raise certain problems and ambiguities. If paired with offense systems, they can be viewed as fostering an aggressive policy, and no one wants that.

He nevertheless announced that he was

> . . . directing a comprehensive and intensive effort to develop a long-term research and development program to begin to achieve our ultimate goal of eliminating the threat posed by strategic nuclear missiles. This could pave the way for arms control measures to eliminate the weapons themselves.[3]

Studies Following the President's Speech

Presidential National Security Study Directive 6-83 (NSSD 6-83) called for the Defense Department to study and report on how such a research and development program might best be shaped. The Defense Department established two groups of consultants to study ballistic missile defense (BMD). The most prominent of these, a "Defensive Technologies Study Team" prepared a study on "Eliminating the Threat Posed by Nuclear Ballistic Missiles." That committee of 50 defense scientists and engineers was chaired by Dr. James C. Fletcher, former NASA administrator, and be-

[1]Ronald Reagan, televised speech of Mar. 23, 1983.

[2]Ibid.

[3]Ibid.

came known as the "Fletcher Panel." The Fletcher Panel produced a technology research and development plan (with "fiscally constrained" and "technology-limited" alternatives), the aim of which was to

> . . . allow knowledgeable decisions on whether, several years from now, to begin an an engineering validation phase that, in turn, could lead to an effective defensive capability in the 21st century.[4]

The Department of Defense also created a second panel to carry out NSSD 6-83: the Future Security Strategy Study Team, headed by Fred S. Hoffman, which produced a report entitled "Ballistic Missile Defense and U.S. National Security." Saying that "A combination of technical and strategic uncertainties makes it impossible to say when or whether we can reach the ultimate goal" of fully defending our people against nuclear ballistic missiles, the Hoffman Panel paid particular attention to how "defenses might also reinforce deterrence" by increasing the uncertainties faced by nuclear attack planners.[5]

The Strategic Defense Initiative Organization

Following these studies and the acceptance of their major findings by the Secretary of Defense and the President, early in 1984 the Defense Department began to establish the BMD research program under the rubric "Strategic Defense Initiative Program." In March, Secretary of Defense Caspar Weinberger appointed Air Force Lieutenant General James A. Abrahamson to head this program. In April, the Secretary chartered the Strategic Defense Initiative Organization and appointed Lt. General Abrahamson as its Director. This Organization was charged with undertaking

> . . . a comprehensive program to develop key technologies associated with concepts for defense against ballistic missiles. The technology plan identified by the Defensive Technologies Study and the policy approach outlined in the Future Security Strategy Study will serve as general guides for initiating this program . . . The SDIP will place principal emphasis on technologies involving non-nuclear intercept and destruction concepts. The basic approach will be to consider layered systems that can be deployed in such a way as to increase the contribution of defenses to deterrence and move the United States toward its ultimate goal of a thoroughly reliable defense . . . The program shall protect U.S. options for near-term deployment of limited ballistic missile defenses.[6]

[4]As paraphrased in "The Strategic Defense Initiative: Defensive Technologies Study," Department of Defense, March 1984, p. 4.

[5]A third group, an Interagency Working Group headed by Franklin C. Miller of the Defense Department, also produced a BMD-related report on "Future Security Strategies." (The executive branch has denied Congress access to this report.)

[6]Caspar Weinberger, Memorandum on "Strategic Defense Initiative Organization (SDIO) Charter," Apr. 24, 1984.

ORGANIZATION OF THE OTA STUDY

The national debate about ballistic missile defense technologies will take place in the context of larger issues of national security strategy. On the one hand, BMD development and deployment would be carried out to fulfill the requirements of a U.S. national strategy. The answer to the question of whether we can build a BMD system depends on how good a BMD system we need. How good a system we need depends on what our national strategy would require the the system to do. On the other hand, the emergence of new BMD-related technologies has suggested to many that new strategies, once infeasible, may become available. President Reagan's call for a Strategic Defense Initiative stemmed both from a dissatisfaction with our existing national strategy and from the belief that changes in strategy might be made technically feasible. Thus the issue of what is technologically possible is embedded in a debate about what is strategically desirable and practical.

The absence of a national consensus about what our strategy ought to be makes difficult the question of what kind of BMD capabilities, if any, we should pursue. Differing strategic perspectives lead to disagreements over whether particular levels of BMD capability, integrated into an appropriate, U.S. nuclear strategy, would:

- make nuclear war less likely or more likely;
- ameliorate the effects of a nuclear war should it occur or not;
- lead to more effective international agreements to limit offensive arms or to a greatly accelerated arms race.

Estimates of which of these results BMD deployments might produce depend in part on difficult judgments about what kind of strategic relationship the United States should try to sustain with the Soviet Union.

But those strategic judgments depend at least in part on technical estimates of the potential effectiveness of strategic defenses. Such technical estimates will be based partly on projections of levels of technological achievement (what kinds of system could we build?) and partly on projections of potential Soviet strategic and technological responses.

Thus the questions, "What kind of ballistic missile defense, if any, would we want if we could have it?" and "What kind of ballistic missile defense can we have?" feed back upon one another. Since we cannot afford to carry out research on every kind of weapon that may be technically possible, our research on BMD should be guided by our strategic objectives. But decisions about our strategic objectives should be informed by what is technically possible, so research may lead to new strategic objectives.

This study tries to bring light to the debate by clarifying both the strategic and the technological issues. It begins by reviewing current U.S. nuclear strategy and the reasons for the absence of a role for ballistic missile defense in that strategy. It outlines some strategic ideas that various advocates have offered for altering the current strategy, but does not attempt to choose among those ideas. That choice is left to the reader.

Second, the report assumes, for the sake of discussion, that various levels of BMD capability might be available to the United States and the Soviet Union, and examines how one would go about analyzing the ways such capabilities might serve various strategic goals. Third, it explores some of the possible consequences for crisis stability, arms race stability, and arms control that BMD might have. Fourth, it reviews the technologies being researched for their applicability to BMD tasks. Fifth, it reviews some of the alternative overall BMD research program objectives that Congress may wish to consider. The approach of this study, then, is to try to assist Congress in understanding the potential implications, both long- and short-term, of the new BMD technologies.

Chapter 3 of this volume briefly reviews some historical background to the current BMD debate, recalling the nature of the earlier technologies and the strategic assumptions behind the national decision in 1972 to agree by treaty with the Soviet Union to forgo their deployment. It also reviews the debates since the ratification of that treaty over whether the decision was, in retrospect, a wise one or not. Finally, it attempts to delineate the differences in politics and technology between the current era and the one in which the earlier decisions about BMD were taken. **The information in this chapter should be useful for understanding how it is that U.S. nuclear strategy today does not contain a role for BMD and why some proponents now argue that it should.**

This study first analyzes not the question, "Could we build a BMD system?," but the related questions "Why would we want one?" and "How capable would it have to be?" To set the stage for these questions, **we start in chapter 4 with a review of the principles of current U.S. nuclear strategy and of some proposals for altering that strategy that have appeared in public debate.** The chapter also explores some of the implications of such changes in strategy, particularly for our commitments to allies. The United States might want BMD to enhance its current nuclear strategic posture, which consists of trying to deter Soviet aggression through a mix of threatening retaliatory punishment and being able to deny the Soviets the goals of such aggression. For reasons explained in chapter 3, current U.S. strategy relies on nuclear offensive forces, protected only by passive means, and not on active defenses against ballistic missiles.

Successfully building and sustaining relatively low levels of BMD capability might, in various ways, strengthen the current nuclear strategic policy. Reliance on considerably higher levels of strategic defense, however, would amount to a substantial alteration in existing policy. We would come to rely much more on simply denying the Soviets the damage they might intend with their nuclear ballistic missiles and rely much less on our threat of retaliation to deter them. With extremely high levels of defense against all forms of Soviet nuclear delivery vehicle, we could even consider largely abandoning the threat of nuclear retaliation against the Soviets. (If we could, on the other hand, build a highly effective defense while retaining a highly effective offense against Soviet territory, we might regain the strategic superiority over the Soviets which we possessed for the first 15 years or so of the nuclear age.)

Chapter 5 tries to indicate what must go into a persuasive analysis of how various postulated levels of BMD performance might either enhance the current nuclear deterrent posture of the United States or promote movement to a different strategy. Because any move to a new strategy will necessarily start with modifications of our current strategy, the chapter devotes the majority of its discussion to the question of how the additions of BMD to that strategy might be expected to work. Deterrence, whether relying on the threat to deny military successes or on the threat of punishment, rests on the perceptions and calculations of the one deterred on the outcome of a conflict that he might consider starting. Calculations about the outcome of a nuclear war would be affected by the presence of ballistic missile defense on both sides. The chapter examines the strategic implications of several levels of defense capability, ranging from none at all to extremely high.

Whether BMD can make a satisfactory contribution to U.S. strategic goals depends on a great deal more than whether certain levels of technical performance can be achieved against postulated offensive threats. If certain kinds of BMD looked technically feasible, there would still be several important questions we would want answers to before we decided on deployment. In particular, we would want the addition of BMD to enhance international stability in a crisis, not increase the incentives presented to either side to initiate nuclear conflict. **Chapter 6 describes some conditions for crisis stability and looks at some ways in which BMD might either enhance it or weaken it.**

For BMD to be effective in serving our national strategy, it must not stimulate offensive responses on the part of the Soviets that leave the United States exposed to a more severe nuclear threat than it was before. Nor should BMD deployment lead to an arms race of offense against defense on both sides that was so costly that we could not or would not want to sustain it. Instead, we would want to see BMD contribute to arms race stability. In some hypothesized cases, BMD leads not just to arms race stability, but to new possibilities for arms control. Some argue, on the other hand, that moving toward BMD could erode the current strategic arms control regime while lessening the prospects for future agreements. **Chapter 6, then, also discusses arms race stability and arms control in relation to BMD.**

Chapter 7 introduces the technologies which might form the basis for new ballistic missile defense systems in the coming two or three decades. Potential countermeasures to weapons using these technologies are also identified. The interplay of defenses, countermeasures, and counter-countermeasures cannot be discussed in detail, because many concepts are classified. But the chapter does attempt to give an idea of the nature of the problem. Because most of these technologies are in a relatively undeveloped state, Congress will not likely be faced in the near future with full-scale BMD deployment decisions. Rather, it will have to judge how public money should be spent on BMD research in the next few years.

Chapter 8 describes an imaginary design for a multi-layered BMD system. The purpose of this hypothetical construct is **not** to predict what kind of BMD system the United States might actually choose to deploy after the current research program is completed. Rather, **it is used as a means of illustrating the kinds of technological problems that must be solved, the kinds of feasibility issues that will arise, and the kinds of cost factors that will have to be considered if the decision to build a large-scale ballistic missile defense is to be taken.**

Once we had defined the future strategic condition we would like to be in, and once we had chosen the technologies we believe should be applied to BMD, we would have to see a plausible path from our present condition to the future one. And we would like to have some assurance that there were feasible ways of **maintaining** that condition once it was reached. We would want to have some confidence that the transition to the new situation, as well as the new situation, would make nuclear war less likely, not more likely.

Chapter 9 presents alternate descriptions of how the transition from our present strategic nuclear posture to one incorporating significant strategic defenses might take place—or might be avoided. Beginning with the strategic evolution envisaged by Administration proponents of the President's Strategic Defense Initiative, it examines a variety of cases, looking at different imaginable outcomes both of BMD development and deployment and of non-deployment. It attempts to present the premises, values, and conclusions of those advancing such viewpoints. This exercise should serve as useful background to the current debate over BMD research and development.

Chapter 10 examines the general goals and shape of the current BMD research program, its implications, and possible alternatives to it. The chapter does not attempt to define the details of such alternate programs, but the differences in purpose and shape that might underlie them. It attempts to relate such alternatives to the strategic context established in the earlier chapters. Even though no deployment decisions are now before Congress, eventual goals must at least be considered at the time research and development programs are undertaken. The decision to find out what is feasible implies some ultimate goals. How the research is carried out and at what levels will be affected by those goals. Moreover, even a research program can have important national and international consequences.

Since many of the BMD-relevant technologies could lead to space-based weapons systems and components, issues concerning anti-satellite (ASAT) weapons are closely related to ballistic missile defense issues. Because of special Congressional interest in some of the nearer term issues around ASAT, and in consultation with the staffs of the requesting committees, OTA undertook a subsidiary study of ASAT issues. In the resulting companion report, *Anti-satellite Weapons, Countermeasures, and Arms Control*, OTA has attempted to make clear the implications of ASAT and BMD for one another. Decisions about one cannot be rationally made without considering implications for the other.

Chapter 3

Ballistic Missile Defense Then and Now

Contents

Chapter 3

Ballistic Missile Defense Then and Now

INTRODUCTION

This chapter briefly reviews events and decisions of the 1960s and early 1970s which explain why the United States does not now have ballistic missile defense. It pays particular attention to the rationale of the Johnson and Nixon Administrations for ultimately declining to deploy large-scale ballistic missile defense and instead agreeing with the Soviets to severely limit it. The chapter also describes the positions of those who subsequently supported or questioned the desirability of U.S. adherence to that agreement.

With that debate over values and premises as background, the chapter then recounts some of the factors that produced the renewal of the public debate over what is now generally called "BMD," for "Ballistic Missile Defense."

THE U.S. ABM PROGRAM TO 1969

In the late 1950s, the U.S. Army repeatedly sought authorization to begin producing an anti-ballistic missile (ABM) system called the Nike-Zeus.[1] The Army's goal was a nationwide defense against Soviet ICBMs. Derived from the air defense missile, the Nike-Hercules, the Nike-Zeus interceptor would have been directed by ground-based radars toward incoming Soviet missile reentry vehicles (RVs). When within range of the reentry vehicle, the nuclear weapon aboard the interceptor would explode, destroying the RV. The Eisenhower Administration resisted Army urgings of Nike-Zeus deployment, though the Army continued to win substantial support in Congress for BMD deployment.

The Kennedy Administration was unconvinced that the Nike-Zeus system—with its relatively slow rocket booster, mechanically steered radar, and limited computational capacity—would perform adequately against foreseeable Soviet ICBM threats. Moreover, Secretary of Defense McNamara's systems analysts concluded that it would cost the United States considerably more to offset Soviet missiles than it would cost the Soviets to deploy them. In addition, trying to limit damage to the U.S. population with ABM made even less sense without an extensive civil defense program, which seemed an unlikely prospect.[2] The 1963 Defense budget authorized research on a new BMD system, to be called the Nike-X. The new system would employ faster burning rockets (later called Sprint), electronically steered phased-array radars, and new computers, and would intercept incoming reentry vehicles just after they entered the atmosphere (making it easier to sort out genuine warheads from decoys).

In 1965 the U.S. Army began to develop another interceptor, the Spartan, which would detonate a nuclear warhead above the atmosphere, where it would generate intense X-rays that might be expected to knock out several incoming reentry vehicles at once. While the Sprint rocket had a limited range of about 25 miles, the Spartan had one of several hundred miles.

[1]The following survey of early BMD developments drawn from Alain C. Enthoven and K. Wayne Smith, *How Much is Enough? Shaping the Defense Program, 1961-1969* (New York: Harper Colophon, 1972), pp. 184-196; David N. Schwartz, "Past and Present: The Historical Legacy," *Ballistic Missile Defense*, Ashton B. Carter and David N. Schwartz (eds.) (Washington, DC: The Brookings Institution, 1984), pp. 330-349; and J. P. Ruina, "The U.S. and Soviet Strategic Arsenals," *SALT: The Moscow Agreements and Beyond*, Mason Willrich and John B. Rhinelander (eds.) (New York: The Free Press, 1974), pp. 34-65.

[2]See Fred Kaplan, *The Wizards of Armageddon* (New York: Simon & Schuster, 1983), pp. 321-324.

Photo credit: U.S. Army

Army Nike-Zeus ABM interceptor in test firing. Derived from the Nike-Hercules air defense missile, the Nike-Zeus with its nuclear warhead was designed to intercept incoming ballistic missile reentry vehicles at altitudes of about 100 nautical miles. The Eisenhower and Kennedy Administrations, doubting the systems likely performance against foreseeable Soviet ICBM threats, did not support its deployment.

By the end of 1966, pressures on the Johnson Administration to deploy the Nike-X had grown strong. Evidence that the Soviets were deploying an ABM system had become unambiguous. Over Administration objections, Congress had voted money to begin U.S. deployment. The Joint Chiefs of Staff recommended to the President that the United States deploy, as a first step, the Spartan as an area defense of the whole United States and the Sprint to defend 25 cities with later expansion to cover 52 cities. This system was intended to reduce casualties in the event of full-scale nuclear war with the Soviet Union.

After hearing arguments for and against deployment in December 1966, President Johnson requested money in the fiscal year 1968 budget to permit deployment in January 1967, but postponed an actual decision pending attempts to interest the Soviets in limiting ABMs. The Secretary of Defense continued to believe that although the Nike-X might be somewhat effective against current Soviet missiles, that effectiveness would be shortlived. McNamara explained to Congress in March 1967:

> . . . the Soviets have it within their technical and economic capacity to offset any further damage limiting measures we might undertake, provided they are determined to maintain their deterrent against us. It is the virtual certainty that the Soviets will act to maintain their deterrent which casts such grave doubts on the advisability of our deploying the NIKE X system for the protection of our cities against the kind of heavy, sophisticated missile attack they could launch in the 1970s. In all probability, all we would accomplish would be to increase greatly both their defense expenditures and ours without any gain in real security to either side.[3]

The Joint Chiefs of Staff were recommending deployment of a system that at least promised to be effective against *current* Soviet

[3]U.S. Congress, House Committee on Armed Services, *Hearings on Military Posture* 90th Cong., 1st sess., 1967, p. 874.

Photo credits: U.S. Army

Interceptor missiles deployed as part of the Safeguard ABM System (deactivated in 1976) defending Minuteman ICBM silos near Grand Forks, North Dakota (see photo, p. 51). The Sprint (on right) was designed as part of the Nike-X ABM program. The nuclear-armed Sprint accelerated rapidly to intercept incoming reentry vehicles after they had entered the atmosphere, making it easier to discriminate them from decoys. The Spartan (on left) was to operate above the atmosphere, where intense X-rays from its nuclear warhead were intended to knock out several reentry vehicles at once.

ICBMs, but McNamara proposed only to pursue the development, test, and evaluation of Nike-X. He also proposed that the United States initiate negotiations with the Soviet Union designed to limit the deployment of an anti-ballistic missile system. During the first half of 1967, the State Department and the White House attempted without great success to interest the Soviet Union in such negotiations.

On September 18, 1967, Secretary McNamara gave a speech in which he first explained his reasons for opposing ABM deployment, then announced that the United States would deploy a partial ABM system.[4] The rationale he offered for deployment, however, was intended to lessen congressional pressures for a large-scale system. The proposed U.S. ABM would not attempt to protect U.S. cities against a large Soviet missile attack, but instead would offer a shield against the much smaller threats of a potential Chinese ICBM fleet or an accidental Soviet attack. Even so, the Nike-X system to be deployed—called "Sentinel"—closely resembled the first stages of a system designed to defend against Soviet missiles.

As the United States prepared to deploy its ABM system, it also continued to attempt to engage the Soviets in negotiations to limit ABMs as well as offensive strategic arms. In the summer of 1968 the two countries agreed in principle to begin such negotiations, but the Soviet invasion of Czechoslovakia in August made them politically impossible. The Strategic Arms Limitation Talks (SALT) finally began under the Nixon Administration in November 1969.

Meanwhile, during 1968, senatorial and public opposition to the ABM deployment began to develop. To the surprise of ABM advocates, who had expected people to welcome deployment of a system to defend them and who had expected opposition from cities not included on the initial deployment list, ABM opponents were able to mobilize opposition from groups living near the proposed deployment areas. When the Nixon Administration took office in January 1969, Secretary of Defense Melvin Laird suspended the Sentinel deployment and ordered a review of the ABM program. In March 1969, President Nixon announced plans to deploy a somewhat different system, to be called "Safeguard." The announced purpose of the Safeguard system was to defend not cities, but ICBM silos. Nixon had accepted the McNamara reasoning, explaining:

> Although every instinct motivates me to provide the American people with complete protection against a major nuclear attack, it is not now within our power to do so. The heaviest defense system we considered, one designed to protect our major cities, still could not prevent a catastrophic level of U.S. fatalities from a deliberate all-out Soviet attack. And it might look to an opponent like the prelude to an offensive strategy threatening the Soviet deterrent.[5]

Although the Spartan (exoatmospheric) missiles were no longer to be located near large cities as with Sentinel, the Safeguard system would still offer a thin area defense as well as

[4]For an explanation of the apparent paradox, see Morton Halperin, *Bureaucratic Politics and Foreign Policy* (Washington, DC: The Brookings Institution, 1974), pp. 1-7 and 297-310.

[5]U.S. Arms Control and Disarmament Agency, *Documents on Disarmament, 1969* (Washington, DC: U.S. Government Printing Office, 1970), p. 103.

Photo credit: U.S. Army

The 12-story Perimeter Acquisition Radar (PAR) was built in northeastern North Dakota as part of the Safeguard ABM System. It was to detect and track attacking ballistic missile reentry vehicles at long-range until they were close enough to be handed over to the shorter range Missile Site Radar pictured below, p. 51. When the Grand Forks ABM site was deactivated in 1976, this PAR became part of the NORAD missile early warning system.

a site defense of ICBMs. The Safeguard proposals set off rounds of hearings in Congress and considerable public debate.[6] The Safeguard program narrowly missed being held up by Congress when the Senate defeated a delaying amendment in a 50-50 tie.[7]

[6]For the contrast between the Sentinel and Safeguard proposals, see Herbert F. York, "Military Technology and National Security," *Progress in Arms Control? Readings From Scientific American* (San Francisco: W.H. Freeman, 1979), pp. 45-56.

[7]Stanford Arms Control Group, *International Arms Control: Issues and Agreements*, 2d ed., Coit D. Blacker and Gloria Duffy (eds.) (Stanford, CA: Stanford University Press, 1984), p. 225.

SOVIET ABM PROGRAM TO 1970

The pressures—political and strategic—on the Johnson Administration in 1967 to begin deployment of an ABM system were strengthened by reports of Soviet ABM deployments.[8] Some argued that the Soviet Talinn air defense system, with its SA-5 interceptors, might be "upgraded" to ABM capability (earlier, it had been argued that the Talinn system was designed as an ABM system).[9] In 1964, during their annual May Day military display, the Soviets had paraded a larger interceptor missile, the Galosh, through Moscow. They had also begun to deploy the necessary radar systems (the so-called "Hen House" early warning radar and the "Dog House" battle management radar) and a ring of Galosh launch sites around Moscow.

As late as 1967, it may have appeared that the Galosh system, with its long-range, nuclear-armed interceptors, would be extended to other cities as well. During that year, however, only six of eight prepared sites around Moscow were under active construction. By 1969 the Soviets had halted construction of two more sites. In 1969 and 1970 they installed missiles at four sites with 16 launchers each. The Galosh system deployment stopped at 64 launchers, and even for the defense of Moscow the number was clearly inadequate to deal with the impending deployment of U.S. ballistic missiles with multiple, independently targetable reentry vehicles (MIRVs), or even to deal with a determined attack with single-warhead missiles.

[8]For more detailed descriptions of Soviet ABM programs, see Sidney Graybeal and Daniel Goure, "Soviet Ballistic Missile Defense (BMD) Objectives: Past, Present and Future," *Ballistic Missile Defense Advanced Technology Center*, contract No. DASG-60-79-C-0132, *U.S. Arms Control Objectives and the Implications for Ballistic Missile Defense*, proceedings of a symposium held at the Center for Science and International Affairs, Harvard University, Nov. 1-2, 1979, pp. 69-101; Sayre Stevens, "The Soviet BMD Program," *Ballistic Missile Defense*, Carter and Schwartz (eds.), op. cit., pp. 330-349; and John Prados, *The Soviet Estimate* (New York: The Dial Press, 1982), pp. 151-171.

[9]See Prados, op. cit., pp. 160-166.

SALT I: THE ABM TREATY AND THE INTERIM AGREEMENT ON OFFENSIVE STRATEGIC ARMS

The controversies over the deployment of the U.S. Safeguard ABM system and over the degree of progress in Soviet ABM developments took place as the Nixon Administration prepared its positions for entering strategic arms limitation talks with the Soviet Union. After its own review of the issues, the Nixon Administration ended up agreeing with the Johnson Administration that it was highly desirable to attempt to limit ballistic missile defenses. By the time the negotiations began, the Soviets had apparently come to the same conclusion (after having resisted the idea in early talks with the Johnson Administration).

Provisions of the SALT I Agreements

ABM Treaty

The texts of the SALT I agreements between the United States and the Soviet Union were completed at Helsinki in May 1972.

The centerpiece of those agreements was the treaty on "... The Limitation of Anti-Ballistic Missile Systems."[10] Each side agreed "... not to deploy ABM systems for a defense of the territory of its country and not to provide a base for such a defense, and not to deploy ABM systems for defense of an individual region..." with certain very limited exceptions. The exceptions were that each side could deploy 100 ABM launchers within a 150-kilometer radius of its national capital and another 100 within a 150-kilometer radius of an area containing ICBM launchers. These provisions allowed the Soviets to keep the system they were building around Moscow and it allowed the United States to keep its first Safeguard installation in North Dakota.[11]

In 1974 the two sides agreed in a protocol to the treaty that each would be limited at any one time to one of the two areas provided for in the treaty. In practical terms, that meant that the Soviets would retain the system around Moscow and the United States would keep its system in North Dakota. The United States judged that the minimal effectiveness of its North Dakota installation did not justify the cost, and deactivated it in 1976. The Soviets, though allowed 100 ABM launchers around Moscow, at first kept the system at 64 and later reduced it to 32. More recently, they have begun to upgrade and expand it, possibly to the full 100 allowed launchers.

The ABM Treaty was to be of unlimited duration: the parties agreed that the defense of most of their national territories against strategic (long-range) ballistic missiles would be banned until one or both decided to abrogate or seek to amend the treaty. In order to keep the treaty up to date, a review was provided for every 5 years. In addition, the treaty created a Standing Consultative Commission where the two sides could discuss not only matters of compliance with the treaty, but "possible proposals for further increasing the viability" of the treaty, "including proposals for amendments." It also stated that each side had the right to withdraw from the treaty, with 6 months' notice, "... if it decides that extraordinary events related to the subject matter of this Treaty have jeopardized its supreme interests."

The two parties agreed that "in order to insure fulfillment of the obligation not to deploy ABM systems ... in the event ABM systems based on other physical principles ... are created in the future, specific limitations on such systems and their components would be subject to discussion in accordance" with the provisions for the Standing Consultative Commission and for amendments.

Interim Agreement

When they signed the ABM Treaty, President Nixon and Secretary Brezhnev also signed an "Interim Agreement ... on Certain Measures With Respect to the Limitation of Strategic Offensive Arms." This agreement froze the number of land-based ICBM launchers on each side and set ceilings on the numbers of SLBM launchers each could deploy (up to the limits, land-based ICBM launchers could be "traded in" for SLBM launchers). The Interim Agreement on offensive forces expired in 5 years, although the two sides continued to observe it as SALT II negotiations extended on for 7 years.

Implications and Aftermath of SALT I

Points of view on the orginal desirability and subsequent success of the ABM Treaty vary widely. Supporters of the treaty believe that the treaty enhanced U.S. security, though they differ in the degree of dissatisfaction they feel with the offensive limitations agreed upon in SALT I and SALT II. Some critics of continued adherence to the ABM Treaty do not quarrel with the original idea of the agreement, but

[10] For the full text of the treaty and associated agreed and unilateral statements, see app. B.

[11] For a detailed analysis of the ABM Treaty Provisions, see George Schneiter, "The ABM Treaty Today," *Ballistic Missile Defense*, Carter and Schwartz (eds.), op. cit., pp. 221-250; John B. Rhinelander, "The SALT I Agreements," *SALT: The Moscow Agreements and Beyond*, Willrich and Rhinelander (eds.), op. cit., pp. 125-159; and U.S. Congress, Office of Technology Assessment, *Arms Control in Space: Workshop Proceedings*, OTA-BP-ISC-28 (Washington, DC: U.S. Government Printing Office, May 1984), pp. 33-34.

Photo credit: U.S. Army

The Missile Site Radar (background) of the Safeguard ABM System was designed to refine the data received from the long-range Perimeter Acquisition Radar, track the attacking ICBM reentry vehicles, and fire Sprint and Spartan interceptor missiles (in cells, foreground), to intercept them. Though this site was permitted under the 1972 ABM Treaty and its 1974 protocol, the United States decided that its limited capabilities did not justify the cost and deactivated the system in 1976.

believe that subsequent U.S. policy overly neglected U.S. BMD research while too gently tolerating possible Soviet violations of the agreement. Other critics tend to believe that the very premises under which the treaty was entered into were erroneous.

Supporters

Supporters of the ABM Treaty believe that the agreement was basically "stabilizing," in the senses both of "arms race stability" and "crisis stability." Proponents of limiting BMD have argued that anti-missile defenses would "destabilize" the offensive arms race by stimulating the opponents to build up their offensive forces in order to try to overcome the enemy defenses. Recall Secretary McNamara's belief in the ". . . virtual certainty that the Soviets will act to maintain their deterrent . . ." and President Nixon's conclusion that BMD ". . . might look to an opponent like the prelude to an offensive strategy threatening the Soviet deterrent." This reasoning led, conversely, to the idea stated in the ABM Treaty that limiting ABMs would be a "substantial factor" in curbing the offensive arms race.

Ballistic missile defense, said ABM Treaty supporters, might also induce "crisis instability" by affecting the structure of incentives before the two sides in a confrontation. Nobody seriously believed in 1972 that a BMD system could limit the damage from a nuclear war to "acceptable" levels, and thus make the possessor of a BMD system less afraid of nuclear war. However, even a less capable BMD system might offer an incentive to attack first if its owner believed that nuclear war had become inevitable and that damage could be kept acceptably low only if the other side's forces had first been substantially weakened by a "counterforce" blow. An even more subtle destabilizing effect of owning a BMD system might be to induce in the *other* side the *expectation* that one intended to strike first, and therefore gave him an incentive to preempt that first strike by going first himself. Such reasoning that ABMs might increase the risk of nuclear war, then, led to the premise in the ABM Treaty preamble that limiting ABMs would decrease it.

Some supporters of the treaty agree that its effects on limiting the offensive arms race are difficult to discern. As one observed in 1974,

> To the great disappointment of many of the strongest supporters of the ABM Treaty, its conclusion has not resulted in the noticeable slowdown in strategic offensive weapons programs that would have been expected according to the action-reaction theory. Even U.S. MIRV programs, which had been specifically rationalized as being required to penetrate possible Soviet ABM defenses, are proceeding without change. It has become increasingly clear that strategic weapons programs have the bases for their support in a multiplicity of interests and that, once underway, expedient and changing rationales will be used to sell them.[12]

It is difficult to identify an offensive strategic weapons program on either side which was stopped by any provision of either the SALT I or the SALT II agreements. During the early and mid-1970s the United States more than doubled the deliverable strategic nuclear warheads in its arsenal (though the total nuclear weapons inventory and the size of individual strategic warheads dropped from the 1960s). Much of the numerical increase came in the form of submarine-launched ballistic missile (SLBM) warheads, which were too inaccurate to threaten Soviet ICBM silos, but which were also invulnerable to a Soviet preemptive first strike. The Soviets, meanwhile, had built a lead in numbers of SLBM and ICBM launchers and in the carrying capacity of the missiles in those launchers. In the mid to late 1970s, the addition of multiple reentry vehicles to their large ICBMs multiplied their strategic warhead count severalfold. That large force, coupled with increased accuracy of the reentry vehicles, appeared to threaten a substantial portion of the U.S. ICBM silos (see figure 3-1). By most static measures of strategic nuclear force, the Soviets were taking a lead.[13]

On the other hand, we have no way of knowing whether the offensive competition might not have been even more vigorous than it was if each side had been attempting to guarantee the penetration of its forces against substantial ballistic missile defenses on the other side. The only way of testing that proposition would have been to forgo the treaty. In any case, we have at least avoided the costly deployment of BMD systems which, many would argue, would have provoked offensive countermeasures and would have been technically ineffective at the same time.

Supporters of the ABM Treaty also see it as a significant step in a larger process of arms control negotiation between the United States and the Soviet Union. SALT I led to SALT II, SALT II was to lead to SALT III, and so on. The SALT process seemed to be one sign of a recognition by both sides that cooperative action to reduce the likelihood of nuclear war is desirable. Abandonment of the ABM Treaty, would, conversely, signify to some a retreat from that recognition.

Supporters of the ABM Treaty agree that Soviet violations of the treaty must be dealt with firmly if the treaty and the arms limita-

[12] George W. Rathjens, "Future Limitations of Strategic Arms," Willrich and Rhinelander (eds.), op. cit., p. 228.

[13] Cf. U.S. Department of Defense, *Soviet Military Power, 1985*, pp. 25-41.

Figure 3-1.—U.S. and Soviet Strategic Forces, 1970-1984

Delivery systems

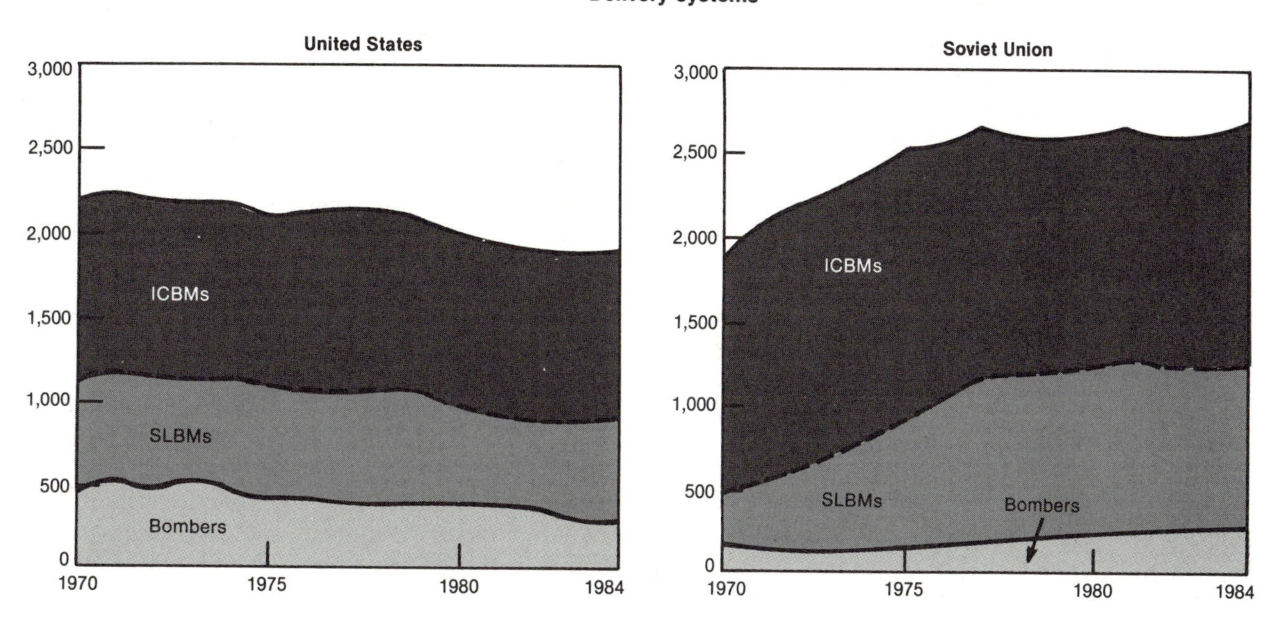

Nuclear weapons

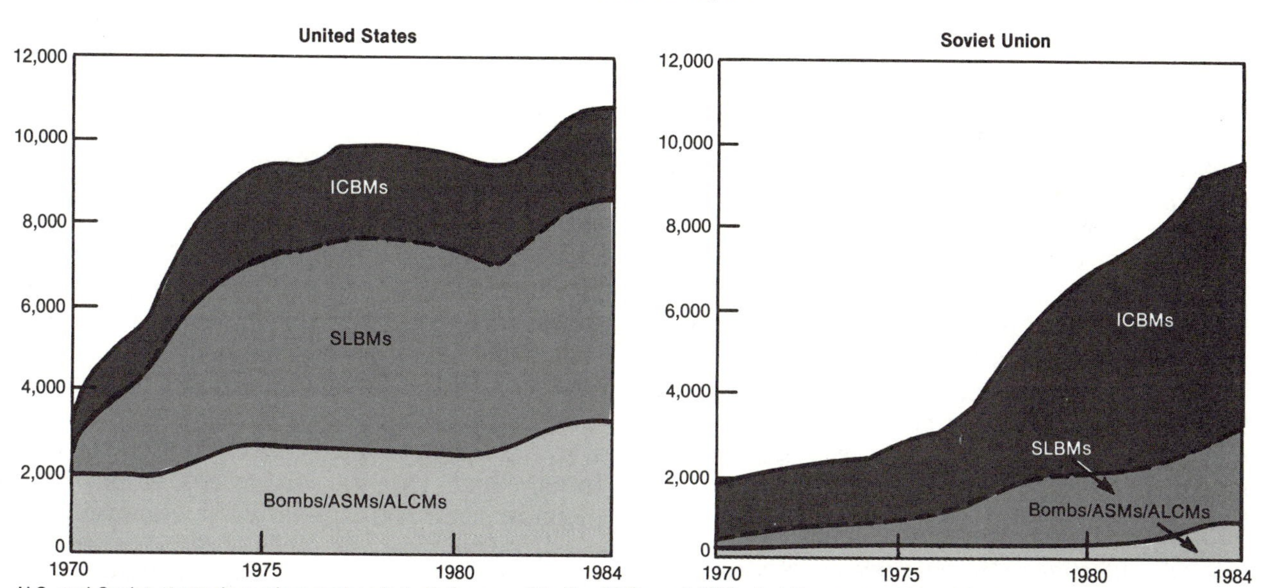

U.S. and Soviet strategic nuclear warheads both increased in the 1970s and 1980s, but the composition of forces on the two sides differed. The United States maintained a substantial fraction of its warheads as bombers and air-to-surface missiles while it added many multiple, independently targetable reentry vehicles (MIRVs) to its SLBM force. The Soviets added SLBM warheads, but concentrated on deploying many, relatively large, warheads on increasingly accurate ICBMs.

SOURCE: Congressional Research Service.

tion process are to survive. Some argue that such firmness should have been exercised much sooner. But they argue that the chances of successfully enforcing Soviet compliance would be much better in the context of a clear U.S. intent to strengthen arms control rather than in a context of public threats to abandon the arms control process.[14]

The policy of every Administration since 1972, including the present one, has been that adherence to the ABM Treaty has been, on balance, in the national security interest of the United States. The Reagan Administration has stated that Strategic Defense Initiative research will be carried on within the limits of the treaty. There are, however, critics of the treaty within the Administration. The following section offers a range of critical views of the ABM Treaty, but does *not* describe current Administration policy.

Critics

Various critics of the ABM Treaty disagree with the proponents on almost every count. Some believe that SALT I slowed the pace of the U.S. strategic force modernization programs since 1972 while the deployments of Soviet ICBMs and SLBMs in the same period could hardly have been higher even in the presence of U.S. BMD and the absence of the modest offensive limitations in SALT I and SALT II. They point out that many in the United States had hoped in vain that SALT I would prevent the Soviets from acquiring the ability to threaten destruction of substantial numbers of U.S. land-based ICBMs in a preemptive nuclear strike. In a unilateral statement attached to the ABM Treaty, the United States declared its belief that:

> . . . an objective of the follow-on negotiations should be to constrain and reduce on a long-term basis threats to the survivability of our respective strategic retaliatory forces.

Moreover, the Jackson Amendment to the joint resolution of approval of the SALT I Interim Agreement (H.J.R. 1227, Sept. 30, 1972) stated that Congress considered that:

> . . . the success of the interim agreement and the attainment of more permanent and comprehensive agreements are dependent upon the preservation of longstanding United States policy that neither the Soviet Union nor the United States should seek unilateral advantage by developing a first strike potential.[15]

Critics of the ABM Treaty argue that seeking first strike potential is exactly what the Soviets have been doing. In their view, the Soviets have always rejected the notion that they should be deterred by the U.S. retaliatory capability. Instead, the Soviets believe they should actively pursue the capability to fight and win a nuclear war with acceptable losses to the Soviet economy, society, political regime, and military forces. In this view, Soviet war plans call for a preemptive first-strike against U.S. land-based ICBMs, bombers on the ground, and submarines in port. Extensive Soviet air defenses and civil defense would protect key Soviet industrial, political, and military targets from a weakened U.S. retaliatory strike.

Critics of the ABM Treaty believe that, although Soviet ballistic missile defenses could limit damage to the Soviet Union even further, the Soviets nevertheless decided to forgo them because they feared that U.S. technology would produce a greatly superior BMD system. Now that they have used the treaty to slow U.S. BMD developments while pushing ahead with their own, they may soon be ready clandestinely or openly to deploy BMD systems, which, though not perfect, would complement their damage-limiting strategy. Indeed, in this view, evidence of Soviet cheating on the ABM Treaty (as well as other arms control agreements) suggests that the Soviets are already set on that course.[16] Many treaty critics believe that, lulled into a false sense of

[14]Cf. Michael Krepon, "Both Sides Are Hedging," *Foreign Policy*, No. 56, fall 1984, pp. 153-172.

[15]Stanford Arms Control Group, op. cit., 249.

[16]For a discussion of Soviet cheating and recommended U.S. responses to it, see Colin Gray, "Moscow Is Cheating," *Foreign Policy*, fall 1984, pp. 141-152.

security by the SALT agreements, the United States failed to make necessary efforts in air defense, civil defense, ballistic missile defense, and offensive force modernization. Some go so far as to conclude that the resulting asymmetry in U.S. and Soviet strategic capabilities

> . . . virtually guarantees that in case of a nuclear war the U.S. will suffer defeat and probably suffer annihilation as a functioning society while the Soviet Union and its system will survive and with sufficient power intact to establish the world hegemony that its leadership has always considered its ultimate due.[17]

In these circumstances, far from enhancing crisis stability, the ABM Treaty has contributed to increasing Soviet incentives to nuclear risk-taking:

> . . . the U.S. lack of strategic defense considerably reduces the credibility of U.S. deterrence in Soviet eyes and may facilitate a Soviet belief in safe expansion. As a result, in crisis situations the Soviets may consider themselves less restrained than the United States and act accordingly.[18]

[17]Michael J. Deane, *Strategic Defense in Soviet Strategy* (Miami, FL: Advanced International Studies Institute, 1980), p. 114.
[18]Ibid.

In this view, then, arms control has led to naive inaction on the part of the United States and the attainment of strategic superiority by the Soviet Union. What the United States should do is pursue nuclear war-fighting capabilities, including offensive counterforce capabilities, air defense, civil defense, and ballistic missile defense, that will give it a credible "theory of victory" with which to deter Soviet aggression.[19]

Others argue that while limiting ballistic missile defense may have been a reasonable policy when the available technology was more primitive, new technologies call for new policies. They say that at the very least, the addition of ballistic missile defenses could enhance the current U.S. deterrent posture. And some suggest that defenses might permit a dramatic change in strategy from offensive to defensive emphasis.

We return in chapter 4 to the question of what it might take to deter the Soviets.

[19]Cf. Colin Gray, "Nuclear Stategy: The Case for a Theory of Victory," *International Security*, vol. 4, No. 1, summer 1979, pp. 54-87.

THE CURRENT BALLISTIC MISSILE DEFENSE DEBATE

Strategic Nuclear Forces: The ICBM Vulnerability Issue

As the Soviets added MIRVed missiles to their ICBM force during the late 1970s, the Defense Department was predicting that the growing numbers of more accurate Soviet ICBM warheads would place the U.S. land-based ICBMs at increasing risk of destruction in a preemptive strike. By the early to mid-1980s, some argued, the United States would have entered a "window of vulnerability" in which 90 percent or more of its land-based ICBMs could be destroyed within minutes.

There has been considerable debate, though, over how significant this problem is and what to do about it. Some argued that the Soviets would have open to them the possibility that they could launch a preemptive strike on U.S. ICBMs and on U.S. bomber bases (as well as on many missile-carrying submarines in port), leaving the U.S. President with only the less accurate SLBM weapons to retaliate, perhaps mainly against Soviet cities. Since this choice would then bring about the destruction of U.S. cities in counter-retaliation, the argument went, the President would have a strong incentive to withhold retaliation and capitulate to whatever Soviet demands followed the Soviet strike. Given this theoretical first-strike capability, the Soviets would be inclined to attempt nuclear intimidation of the United States and might succeed without ever having to fire a missile.

Critics of this point of view argued that:

1. the Soviets could, for various reasons, have little confidence that they could execute this partially disarming first strike successfully;[20]
2. a "surgical" strike against U.S. missile, bomber, and submarine bases is not possible—millions would be killed and the Soviets could not count on U.S. restraint in retaliation; and
3. U.S. SLBMs and bombers would be capable of damaging a great variety of Soviet military, political, and economic targets—the President would not be limited to retaliating against urban populations.

The Carter and Reagan Administrations took positions which implied that the ICBM vulnerability issue was important but not urgent. The Carter Administration proposed deploying a new ICBM, the MX, which would be based deceptively among multiple protective structures so as to raise the price in warheads of a Soviet attack to unacceptable levels. The fully deployed system was scheduled to restore relative invulnerability to land-based ICBMs in about 1989.

Rejecting the Carter Administration's multiple protective structure basing mode, the Reagan Administration first explored alternate "survivable" basing modes, then referred the ICBM issue to a "President's Commission on Strategic Forces," chaired by Brent Scowcroft. The Scowcroft Commission recommended the deployment of 100 MX missiles in fixed, presumably vulnerable, silos now occupied by Minuteman missiles. It also recommended development of a small, possibly mobile, ICBM that might reduce the ICBM vulnerability problem sometime in the early 1990s.[21] At various times during the course of debate over ICBM vulnerability, ballistic missile defense had been suggested as a measure for protecting the missiles.[22] The Scowcroft Commission

[20]Some argue that imputations of the required degrees of accuracy to Soviet reentry vehicles are, for various technical reasons, not justifiable and that the Soviets would be foolish to have confidence in a theoretical, basically untestable, capability. In addition, Soviet ICBMs could not attack U.S. ICBMs without giving U.S. bombers enough warning to become airborne. Furthermore, the Soviets must take into account the possibility that U.S. ICBMs would be launched on warning, escaping before the Soviet ICBMs arrived. For many years it has been U.S. policy to have a capability to launch on warning. Although we have no declared policy to do so, the possibility that we might is a part of our deterrent posture. See U.S. Congress, Office of Technology Assessment, *MX Missile Basing*, OTA-ISC-140 (Washington, DC: U.S. Government Printing Office, September 1981), for a discussion of the technical requirements for launch-on-warning. (On the other hand, some argue that an attack of shorter range, submarine-launched missiles producing nuclear detonations above the U.S. missile fields could pin the missiles down until the Soviet ICBMs arrived.)

[21]*Report of the President's Commission on Strategic Forces* (April 1983), reprinted in U.S. Congress, House Committee on Armed Services, *Defense Department Authorization and Oversight, Hearings on H.R. 2287, Department of Defense Authorization for Appropriations for Fiscal Year 1984 and Oversight of Previously Authorized Programs, Part 2 of 8 Parts, Strategic Programs*, 98th Cong., 1st sess., 1983, pp. 33-62.

[22]For a discussion of the technical issues, see U.S. Congress, Office of Technology Assessment, *MX Missile Basing*, op. cit., pp. 109-143.

Photo credit: U.S. Air Force

Artist's concept of a new small intercontinental ballistic missile (SICBM) now under research and development by the U.S. Air Force. In this design, the missile would be about 46 feet long and weigh about 30,000 lbs. It would deliver a 1,000 lb payload at ranges in excess of 6,000 miles. The President's Commission on Strategic Forces, appointed by President Reagan and chaired by Brent Scowcroft, recommended deployment of a small, possibly mobile ICBM that might alleviate the ICBM vulnerability problem in the early 1990s.

concluded, however, that the vulnerability of the Minuteman and MX silos

> . . . in the near term, viewed in isolation, is not a sufficiently dominant part of the overall problem of ICBM modernization to warrant other immediate steps being taken such as closely spacing new silos or ABM defense of those silos.[23]

Some proponents of ballistic missile defense, however, disagree. They say that the immediate goal of pursuing ballistic missile defense should be to reduce or eliminate the "military utility" of Soviet ICBMs, which, presumably, means their ability to destroy a large number of ICBM silos as well as other hardened military targets. Indeed, the defense of ICBMs became the major focus of U.S. BMD research through the late 1970s and early 1980s.

Technological Developments

"Conventional" BMD

In the years since the signing of the ABM Treaty, the United States has continued research on ballistic missile defense technology. Although some work was conducted on "exotic" technologies with possible long-term application, the major focus was on systems that might be deployed within a few years in response to a Soviet "breakout" from the ABM Treaty. The systems to which most attention was paid were designed primarily to partially defend hard targets, such as ICBM shelters, against "counterforce" attacks. The goal would not be to protect every single shelter perfectly but to try to assure the survival of adequate retaliatory forces after a Soviet first strike by raising the "price" of successful attack on U.S. ICBMs to levels the Soviets would not want to pay. ("Price" here is measured as either a percentage of available Soviet missile forces or the financial and political cost of deploying additional forces.)

The Army Ballistic Missile Defense Program Office developed some subsystems for a successor to the Sprint missile component of the old Safeguard system: the Low Altitude Defense System, or LoADS. The LoADS would

[23]Ibid., p. 51. For further discussion of the Scowcroft Commission findings, see ch. 6 of this report.

Photo credit: U.S. Army

In the 1970s and early 1980s, the Army developed the Low Altitude Defense System (LoADS) as a successor to the Sprint ABM interceptor and its associated Missile Site Radar (see photos above, pp. 47 and 51). It would have used small, possibly mobile, short-range phased-array radars and computers to direct small, nuclear-armed missiles to intercept incoming reentry vehicles after they had entered the atmosphere. Because nuclear explosions would occur so close to the ground, these weapons would have been suitable for protecting only hardened targets such as ICBM shelters.

use small, possibly mobile, short-range phased-array radars and computers to direct small missiles carrying nuclear warheads to incoming enemy reentry vehicles after they had entered the atmosphere. Because nuclear explosions would occur so close to the ground, these weapons would have been suitable for protecting only hardened targets such as ICBM shelters.[24] The Army has also worked on an endoatmospheric (inside the atmosphere) nonnuclear kill missile for the LoADS, but has not established its feasibility.[25]

The Army has also been developing a nonnuclear exoatmospheric (above the atmosphere) interceptor. A sensor and kill vehicle (which collides with the incoming reentry vehicle) were demonstrated in a test of the Homing Overlay Experiment held in the summer of 1984. A full-blown system with many such interceptors would probably use missile-borne or airborne optical (long-wave infrared) sensors to detect and track the numerous incoming reentry vehicles, and would need to be able to discriminate between warheads and decoys.

The Army has argued that while either the LoADS or the high-altitude system could work well standing alone, they could be even more effective if deployed together in a "layered" defense.[26]

Newer Technologies Potentially Applicable to BMD[27]

Those who advocate greater efforts now in BMD research and development argue that technical advances since the early 1970s point the way to solving many or all of the technical problems that made BMD less attractive when the ABM Treaty was signed.

Technological limits of the late 1960s and early 1970s made it seem that BMD systems could be of only limited effectiveness and that it would likely be less costly to improve the ability of offenses to penetrate defenses than it would be to build the defenses in the first place. The systems under development were limited to ground-based interceptors that would operate during the last few minutes of the offensive trajectory, in the terminal phase, and in late midcourse. Guidance would have been provided by large radars located at or near the interceptor launchers. The radars, vulnerable to attack, would themselves have been prime targets for the offense. Proliferation of the radars would have been difficult because they had to be large and expensive.

The speed and capacity of available computers limited the ability of the radars to operate successfully in a complicated environment and of automated battle management systems to handle large attacks. It would have been very difficult to discriminate targets from decoys or other penetration aids. This problem would have forced either the commitment of very large numbers of interceptors to kill comparatively few targets, or the delay of any attempt at intercept until the incoming warheads entered the atmosphere, where discrimination becomes easier. Either would have put a substantial strain on data handling and weapon resources. This situation limited the range at intercept, and therefore the area each site could protect, forcing up requirements for numbers of interceptors.

Guidance and warhead technology had not yet made it feasible to consider trying to use nonnuclear warheads to destroy missile reentry vehicles. Nuclear explosions threatened collateral damage problems which further limited the region over which the intercept could take place.[28] They also posed the risk of blackout of the radars once the first intercept was

[24]Such systems are apparently not being considered under the Strategic Defense Initiative.

[25]Such low-altitude nonnuclear interception could still not assure protection of soft targets like cities because the incoming warheads could be salvage-fused—i.e., designed to detonate at the moment of impact with the interceptor.

[26]See William A. Davis, Jr., "Current Technical Status of U.S. BMD Programs," *U.S. Arms Control Objectives and the Implications for Ballistic Missile Defense,* op. cit., pp. 37-40.

[27]Much of the basic material comes from the unclassified version of the DOD Defensive Technologies Study, submitted to Congress March 1984, and from a recent paper by James Fletcher, the leader of that study (James C. Fletcher, "The Technologies for Ballistic Missile Defense," *Issues in Science and Technology*, fall 1984).

[28]Again, if the incoming warheads were salvage-fused, the collateral damage might be greater from the the intercepted warhead than from a nuclear interceptor.

made. Analyses showed that the cost of relatively easy countermeasures—e.g., adding to the offensive forces or even just adding crude decoys—would be less than the cost of building the BMD systems.

For some, technical advances of recent years suggest solutions to the problems previously limiting the promise of ballistic missile defense. The advances are, for the most part, more embryonic than mature. They will have to be further proven, and in many cases vastly scaled up from present performance levels, before they can be designed and engineered into working BMD weapon components. Nevertheless, advocates of greater investment in the development of these technologies believe they offer the promise of building weapons with nonnuclear kill mechanisms; weapons that could attack missiles in their boost and midcourse phases; sensors, computers, and, especially, software for high-speed, high-volume target tracking and discrimination; and computers and software for high-capacity battle management. The new technologies are discussed in detail in chapters 7 and 8.

Soviet BMD Activities

Meanwhile, in the early 1980s, Soviet BMD developments were giving U.S. officials some cause for concern. According to the Department of Defense document, *Soviet Military Power, 1985*, since 1980 the Soviets have been upgrading the Moscow ABM system from 64 launchers to the 100 allowed by the ABM Treaty:

> When completed, the new system will be a two-layer defense composed of silo-based long-range modified GALOSH interceptors designed to engage targets outside the atmosphere; silo-based high acceleration interceptors designed to engage targets within the atmosphere; associated engagement and guidance radars; and a new large radar at Pushkino designed to control ABM engagements. The silo-based launchers may be reloadable. The first new launchers are likely to be operational this year [1985], and the new defenses could be fully operational by 1987.[29]

In addition, "the Soviets have developed a rapidly deployable ABM system for which sites could be built in months rather than years." Soviet early warning and tracking radars, including one site under construction which violates the ABM Treaty, could support an "ABM deployment to protect important target areas in the U.S.S.R." in the next 10 years. Another hypothesized addition to such a system would be the SA-10 (under deployment) and SA-X-12 (under development) surface-to-air missiles which "may have the potential to intercept some types of U.S. strategic ballistic missiles."[30]

According to the Department of Defense report, then, the Soviets are "developing a rapidly deployable ABM system to protect important target areas in the U.S.S.R." The report concludes that "the aggregate of [their] ABM and ABM-related activities suggests that the U.S.S.R. may be preparing an ABM defense of its national territory."[31] Officials of the CIA, however, have said that they do not judge it likely that the Soviets would in fact move to such a deployment in the near term.[32] These officials point out that, while the Soviets could expand their presently limited ABM system by the early 1990s,

> In contemplating such a deployment . . . [they] will have to weigh the military advantages they would see in such defenses against the disadvantages they would see in such a move, particularly the responses by the United States and its allies.[33]

[29]*Soviet Military Power 1985* (Washington, DC: U.S. Department of Defense, 1984), p. 48.

[30]Ibid.

[31]Ibid.

[32]Unclassified testimony of National Intelligence Officer Lawrence K. Gershwin before a joint session of the Subcommittee on Strategic and Theater Nuclear forces of the Senate Armed Services Committee and the Defense Subcommittee of the Senate Committee on Appropriations, June 26, 1985.

[33]Prepared testimony of Robert M. Gates and Lawrence K. Gershwin, ibid.

Moscow Ballistic Missile Defense

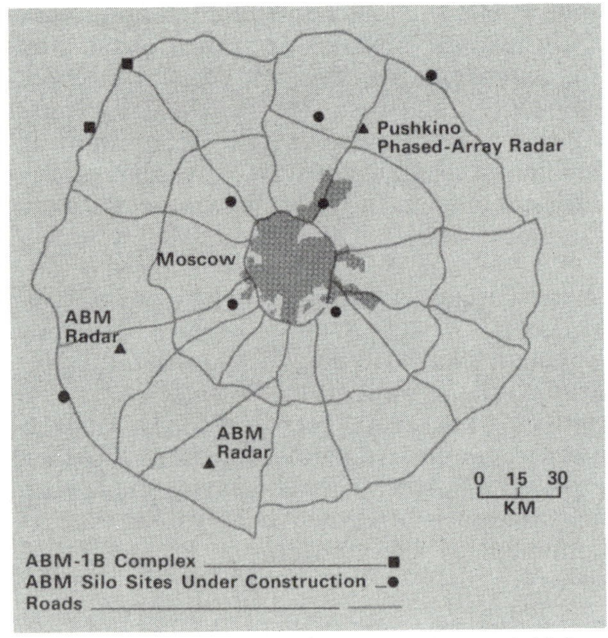

Photo credit: U.S. Department of Defense

The Moscow ballistic missile defenses identified in the map at right include the Pushkino ABM radar, above, Galosh anti-ballistic missile interceptors, top left, and new silo-based high-acceleration interceptors, top right.

The Defense Department also reports that the Soviets are working on ground-based lasers for ballistic missile defense, although "initial operational deployment is not likely in this century."[34] They also have a "vigorous program underway for [ground- and space-based] particle beam development and could have a prototype space-based weapon ready for testing in the late 1990s."[35] The CIA, on

[34]The CIA says:

We are concerned about a large Soviet program to develop ground-based laser weapons for terminal defense against reentry vehicles. There are major uncertainties, however, concerning the feasibility and practicality of using ground-based lasers for BMD.

Testimony of Gates and Gershwin, prepared testimony, ibid.

[35]Ibid., p. 44.

Photo credit: U.S. Department of Defense

Artist's concept of large phased-array early warning and missile tracking radar under construction at Krasnoyarsk in the Soviet Union. The U.S. Government has judged this radar to violate the ABM Treaty because of its siting, orientation, and capability.

the other hand, estimates that the "technical requirements are so severe" that there is a "low probability" that the Soviets will test such a prototype before the year 2000.[36] The Soviets are also reported to be working on particle beam weapons.

Soviet BMD developments, then, lead some to project either of two threatening possibilities. One is that the Soviets might decide formally (or at least overtly) to abandon the ABM Treaty and rapidly deploy ballistic missile defenses that gave them a strategic advantage over the United States before it could respond adequately. This possibility is sometimes referred to as a "break out" from the treaty. The second threatening possibility is that the Soviets might "creep out" of the treaty. That is, they might feign adherence to the ABM Treaty but gain a significant unilateral ballistic missile defense capability through treaty violations and through technical advances in systems (e.g., theater ballistic missile defenses) nominally permitted by the treaty.

[36]Testimony of Gates and Gershwin, op. cit.

Political Developments

Decline of Detente

In 1972, when the Nixon Administration signed the SALT I agreements, U.S. policy toward the Soviet Union was one of *detente*, in which the United States was attempting to ameliorate its adversarial relationship with the Soviet Union through various cooperative arrangements, including but not limited to arms control. According to one of the architects of this policy, a "network of agreements" was meant to provide "incentives and penalties" that might "moderate Soviet behavior."[37] Although arms control negotiations between the two superpowers continued through the 1970s, during the same period U.S.-Soviet cooperation declined and conflict increased. The So-

[37]Henry A. Kissinger, *Years of Upheaval* (Boston: Little, Brown & Co., 1982), p. 246.

viet invasion of Afghanistan in late 1979 led the Carter Administration to withdraw the signed SALT II treaty from senatorial consideration. The Reagan Administration came to office with a stated intent of correcting what it saw as the undue softness of previous Administrations toward the Soviets and serious neglect of U.S. military strength. Harsh Soviet reaction to political liberalization in Poland and Soviet destruction of a Korean airliner that had strayed over Soviet air space did not improve Soviet standing in U.S. eyes.

Decline of Arms Control

During his 1980 election campaign President Reagan emphasized that the SALT II accords signed by the Carter Administration were "fatally flawed." In office, he decided not to request Senate confirmation of them (while promising not to violate them.) He declined to pursue ratification of two other previously negotiated agreements, the Threshold Test Ban Treaty and the Peaceful Nuclear Explosions Treaty. A view that seemed to be widely held within the Reagan Administration was that previous arms control agreements had resulted in substantial net advantages to the Soviet Union, and that only a determined U.S. program of "strategic modernization" would persuade the Soviets to agree to equitable limitations.

Although discussions were begun with the Soviets on strategic and intermediate-range nuclear force limitations, no progress was made. Some argued that the Soviets actually had no wish to reach an equitable agreement, but wished only to score propaganda points against the United States, to divide the NATO alliance, and to prevent deployment of Pershing II and ground-launched cruise missiles in Western Europe. Others argued that while Soviet bargaining intentions and tactics were certainly open to question, the Reagan Administration, manned in key positions by people hostile to arms control, did not negotiate seriously.[38]

With the debatable, or at least ambiguous, success of previous arms control arrangements and the lack of apparent progress toward new limitations, there has been a growing public concern about the eventual outcome of the strategic arms race and a general desire for nuclear arms reduction agreements.[39]

By the time of President Reagan's speech of March 23, 1983, several conditions held:

- the competition in strategic offensive nuclear weapons continued;
- there was considerable skepticism in the Administration and in Congress that arms control could do much to contain the Soviet military threat to the United States;
- the near-term potential for mutually beneficial negotiations with the Soviets seemed slim;
- there was deep suspicion toward the Soviet Union inside the Administration and widely shared by the U.S. public;
- advocates of ballistic missile defense for the United States were arguing that new technologies had put effective defenses within sight;
- the Department of Defense was concerned about Soviet BMD developments; and
- there was strong public feeling that something should be done to curb the nuclear arms race.

[38]Cf. Strobe Talbot, *Deadly Gambits: The Reagan Administration and the Stalemate in Nuclear Arms Control* (New York: Knopf, 1984).

[39]See Jamie Kalven, "A Talk With Louis Harris," *The Bulletin of the Atomic Scientists*, August/September 1982, pp. 3-5. See also, Daniel Yankelovich and John Doble, "The Public Moood," *Foreign Affairs*, fall 1984, pp. 33-46.

WHAT IS NEW?

If President Reagan meant to set a bold precedent with his March 23d speech, he succeeded. The Strategic Defense Initiative Program is probably the first major national weapons research program which was *begun* with a public Presidential appeal for a national commitment. The Initiative has made BMD once again a central issue of national debate over defense policy. But there are striking technical and political differences between the new debate and the old one.

In the late 1960s, the Nixon Administration policy was (until the ABM Treaty was negotiated) to propose immediate deployment of fully developed, currently available systems. The costs and capabilities of these systems were understood reasonably well. The likely countermeasures (multiple reentry vehicles and other penetration aids) had also been invented and, by the time of the signing of the ABM Treaty, tested in the United States. There was wide (though certainly not complete) agreement that when the Soviets adopted these countermeasures the proposed U.S. BMD system would be substantially reduced in effectiveness.

Today, while there are those who advocate early BMD deployment using near-term technology, the SDI focus is on BMD systems which are still only conceptual, based on technologies that are yet to be developed or matured. Similarly, the likely countermeasures are mostly conceptual, and their effectiveness and cost remain speculative.

Some experts consider these technologies to be promising, not only for "enhancing deterrence," but perhaps ultimately for protecting most U.S. cities and population from the threat of nuclear destruction. Most experts agree that at least some research should be done on them. Although some argue for early deployment of BMD based on currently available technology, the debate now centers mainly on what kind of research to pursue, at what funding level, and for what ends.

Nevertheless, the SDI cannot be adequately characterized as "just a research program to find out what is possible." The President has called for a national commitment of scientific and technological resources to find effective defenses against ballistic missiles. The proposed research program envisages a steadily rising level of expenditures and a series of "experiments" to demonstrate capabilities that could lead to engineering development decisions in the early 1990s and deployment decisions in the late 1990s. Just where the lines are between research on the one hand and development on the other is not entirely clear: if the research is highly successful, there will be pressures for moving to the early stages of development. Then, if early development is highly successful, there will be pressures for deployment. And whether or not decisions to deploy BMD systems are ever made, a U.S. research program may affect Soviet weapons decisions and U.S.-Soviet political relations.

The political environment today differs from what it was when the United States decided to exclude ballistic missile defense from its strategic posture. Although the country was still in a bitter war with a Soviet ally (the Democratic Republic of Vietnam) in 1972, the Nixon Administration had embarked on a policy of detente with the Soviet Union, acting on the assumption that a judicious mixture of competition and conflict was possible. Arms control was seen as a possible tool both for reducing the risk of fatal conflict between the two sides and for establishing political bonds which might ameliorate the causes of conflict. Detente, seen by many as a failed policy, has been discarded by the United States and arms control has come under increasing suspicion and criticism. At the same time, public fears of the consequences of unrestrained arms competition have grown. Although in 1985 the United States and the Soviet Union embarked on a new series of arms control talks, no one expected early progress.

Another significant difference between the BMD debate before 1972 and the one now is that then the ABM Treaty did not exist, and today it does. In the late 1960s, the United States entered into negotiations with the Soviets intending to persuade them that forgoing BMD would be mutually advantageous; in 1985, the announced U.S. intent is to persuade them that *having* BMD would be mutually advantageous. Although the Secretary of Defense and other Administration officials have expressed dissatisfaction with the treaty, the Administration has not yet chosen to seek revision of it, let alone abandon it. It has stated that SDI research will be conducted within treaty constraints. The ABM Treaty is widely seen in the United States, among the NATO allies,[40] and perhaps in the Soviet Union, as the most significant arms control agreement between the superpowers. Its abrogation by either side would symbolize to many abandonment of the serious pursuit of arms control and resignation to a largely unconstrained nuclear arms race.

An important consideration in pursuing BMD, even at the research and development level, is when and how the ABM Treaty would have to be modified or abrogated and what the consequences of such changes would be for U.S. national, NATO alliance, and U.S.-Soviet politics. On the other hand, those who see mainly disadvantages to the treaty believe that any risks in its abrogation or attempted modification are far outweighed by the risks (e.g., as militarily significant Soviet violations) of continued U.S. adherence to it.

Protagonists in the U.S. debate over BMD disagree about how central the ABM Treaty should be in the debate. But most can probably agree that the question of the survival of that particular treaty is subsidiary to the primary issue of whether BMD deployed by the United States and the Soviet Union would lead to a safer world.

[40]See David Yost, "Ballistic Missile Defense and the Atlantic Alliance," *International Security*, vol. 7, No. 2, fall 1982, pp. 143-174.

Chapter 4

Deterrence, U.S. Nuclear Strategy, and BMD

Contents

Chapter 4

Deterrence, U.S. Nuclear Strategy, and BMD

OVERVIEW

Depending on how the policies and forces of the United States and the Soviet Union changed to accommodate it, the introduction of ballistic missile defenses into our military posture could well represent a major shift in national strategy. Alternatively, it might only be an incremental adjustment. To understand the role that BMD can play in national strategy, we must first understand what our present strategy is. We can then ask whether or how ballistic missile defenses might address some of the problems that have so far been identified with our strategy—or whether it might enable adoption of a strategy significantly better than the present one.

This chapter provides that background. After a brief summary of current U.S. nuclear strategy, it discusses what possible Soviet actions that strategy seeks to deter. The chapter goes on to describe our current strategy in greater depth and presents a discussion of some of the problems with our strategy that critics have identified. The chapter concludes by identifying possible evolution of or replacements to our strategy, paying particular attention to the roles that ballistic missile defenses might play.

INTRODUCTION

The overall strategic objective of our current nuclear strategy is, and consistently has been, to avoid nuclear attack on this nation while preserving other national interests. To accomplish this, our strategy has attempted to achieve three major goals:

- deter the Soviets from nuclear attack on the United States by convincing them that the outcome would be unacceptable to them;
- convince the Soviets that we will attempt to preserve our national interests by means short of nuclear war, but that attacks on those interests might well lead to nuclear war; and
- terminate nuclear war, if it cannot be avoided, at the lowest possible level of violence and on terms most favorable to us.

We strive to deter nuclear attack by fostering a perception among the Soviet leadership that they would suffer unacceptable losses in a nuclear war, and that under no circumstances would such a war leave them better off in terms of achieving their geopolitical objectives than they otherwise would have been. For this strategy to be credible, we must also foster the perception among the Soviets that we are not only willing to fight a nuclear war if necessary, but that nothing they could do could make us incapable of doing so. However, we also do not want our forces to be structured in such a way as to give the Soviets increased incentive to strike first in a crisis. We therefore strive to balance potential war-fighting capability against crisis stability.

In the event of attack, U.S. strategy incorporates two broad elements. We would seek to deny the Soviets success in achieving the goals motivating such an attack, and we would threaten retaliation. The *perception* of these capabilities contributes to deterring attack; the *possession* of these capabilities is intended to make possible the termination of hostilities on favorable terms if they cannot be avoided. These elements apply both to deterring a Soviet first strike and to deterring and responding to subsequent Soviet actions. This discussion stresses "intending" to terminate hostilities, rather than successfully doing so, because it is by no means obvious that any plan for initiating even limited use of nuclear weapons can avoid the destruction of the societies of both parties to the conflict.

We would accomplish these elements, denial of success and retaliation, with offensive and passive defensive means. We deny the Soviets success in attacking military installations by means of a variety of passive measures such as hardening them and making them redundant (e.g., ICBM silos), dispersing them (e.g., air and naval forces), and hiding them (e.g., ballistic missile submarines). We do not attempt to deny success to attacks on our cities, on economic targets, or on "soft" military targets. We threaten retaliation by maintaining survivable offensive forces that are capable on balance of riding out attack and then reaching and destroying Soviet military and civilian assets. In short:

- The survival of the United States depends on rational behavior of the Soviet leadership. We seek to deter them from attacking, but if they intend to destroy the United States and suffer the consequences, we cannot prevent them from doing so.
- Deterrence rests primarily on offensive forces. We rely more heavily on the threat of retaliation than we do on denial of success.
- We rely on the use of passive defenses, not active ones, for the survivability of our offensive forces.

DETERRING THE SOVIETS

The principal target of U.S. nuclear strategy—the Soviet Union—is obvious. The mechanism by which that strategy works, however, is not simple. From what actions do we want to deter the Soviet Union? How does the deterrent mechanism operate? These questions are the subjects of a vast literature;[1] the problem can only be outlined here. In general, there are three broad, and to some extent overlapping, categories of Soviet behavior the United States would like to deter:

- A surprise, "bolt from the blue," strategic nuclear attack intended to disarm the United States and, conceivably, remove it as an international competitor to the Soviet Union.
- Initiation or threatened initiation of nuclear war against the United States as an escalation of an ongoing crisis, conventional war, or theater nuclear war.
- Threats of or acts of military aggression against U.S. allies or against countries whose conquest the United States would see as challenging vital U.S. interests.

Exactly what one believes the United States must do to deter the Soviet Union from the kinds of behavior listed above depends on one's perceptions of Soviet motivations, strategy, and military capabilities. However, determining Soviet intentions is a controversial procedure, and U.S. Sovietologists offer a wide range of interpretations of Soviet views. Before examining current U.S. strategy in any detail, we will first explore this diversity of opinion.

It arises, in part, from apparent contradictions in Soviet statements and writings on the subject. Examination of actual Soviet nuclear force deployments helps narrow the controversy somewhat, but still does not persuasively resolve the debated questions to everyone's satisfaction. A recent OTA workshop[2]

[1]See app. M for references to a representative sampling.

[2]OTA workshop of Soviet military strategy and policy, held Dec. 12, 1984; summary to be made available separately from this report.

suggested that the conflicting statements of Soviet strategic doctrine emanate from two overlapping, but distinguishable spheres: the "sociopolitical" and the "military-technical."

The former consists of propositions of the following kinds, which are often heard emanating from the highest levels of Soviet political leadership, and often from high military leaders as well:

- The Soviet Union will not be the first to use nuclear weapons.
- Nuclear war with the United States would be mutual suicide.
- It is impossible to keep a nuclear war limited.
- Soviet nuclear policy is defensive and retaliatory in nature.
- A rough parity of nuclear forces now exists between the Soviet Union and the United States.
- The Soviet Union does not seek nuclear superiority.

On the other hand, many contemporaneous Soviet military writings on operational levels, the "military-technical" arena, stress such strategic principles as:

- It is important to seize the offensive at all levels of warfare.
- Getting in the first blow (preemptive attack) can decide the outcome of a nuclear war.
- Nuclear warfare might be contained within a particular theater of operations.
- A combination of offensive attack and strategic defense (e.g., air defense and civil defense) could limit damage to the Soviet Union from a nuclear war.
- The Soviet Union would prevail in a nuclear war.

Analysts of Soviet military policy agree that both of these bodies of doctrine co-exist in Soviet writings, as indeed they do in U.S. writings. However, there is disagreement on which would take precedence under what circumstances. When it actually comes to running risks of engaging in nuclear war with the United States, which precepts are Soviet decisionmakers most likely to follow?

Examining actual Soviet nuclear force deployments seems to some analysts to support the notion that the "military-technical" set of doctrines has been given considerable operational application.[3] The Soviets have built a large land-based ICBM force which appears capable of destroying the bulk of the U.S. land-based ICBM force in a first strike. Their anti-submarine warfare programs seek the ability to threaten our sea-based deterrent. They have a massive air defense system and a large civil defense program. Although they have deployed no nationwide ballistic missile defense capability (which would be prohibited by the 1972 ABM Treaty), they appear to continue preparations to be able to do so.

On the other hand, other characteristics of the Soviet strategic posture, especially when viewed in the light of the relevant U.S. strategic capabilties, suggest that Soviet leaders should and do give some credence to the "sociopolitical" set of propositions above. Even in a bolt-from-the-blue surprise attack, the Soviet Union cannot expect to escape a devastating retaliatory blow against a wide range of military, economic, and political targets. This follows for a number of reasons:

- Only one-quarter of U.S. strategic nuclear warheads are deployed on land-based ICBMs which are thought to be at risk to Soviet preemption; the rest are on bombers and submarines. In normal times, half the submarines are invulnerable at sea and many bombers are poised for rapid take-off. If nuclear forces were in a "generated" posture, such as in a crisis when a preemptive strike could be anticipated, even more submarines would be at sea and more bombers would be on alert, widely dispersed, and ready for quick take-off.
- Although the Soviet air defense system is impressive, the U.S. Defense Department believes that structural and electronic upgrades to current U.S. bombers,

[3]One such analysis is done by Stephen Meyer, "Insight From Mathematical Modeling in Soviet Mission Analysis, Part II," a report done under contract MDA-903-82-K-0107 with the Defense Advanced Research Projects Agency.

Figure 4-1.—U.S.S.R. ICBMs

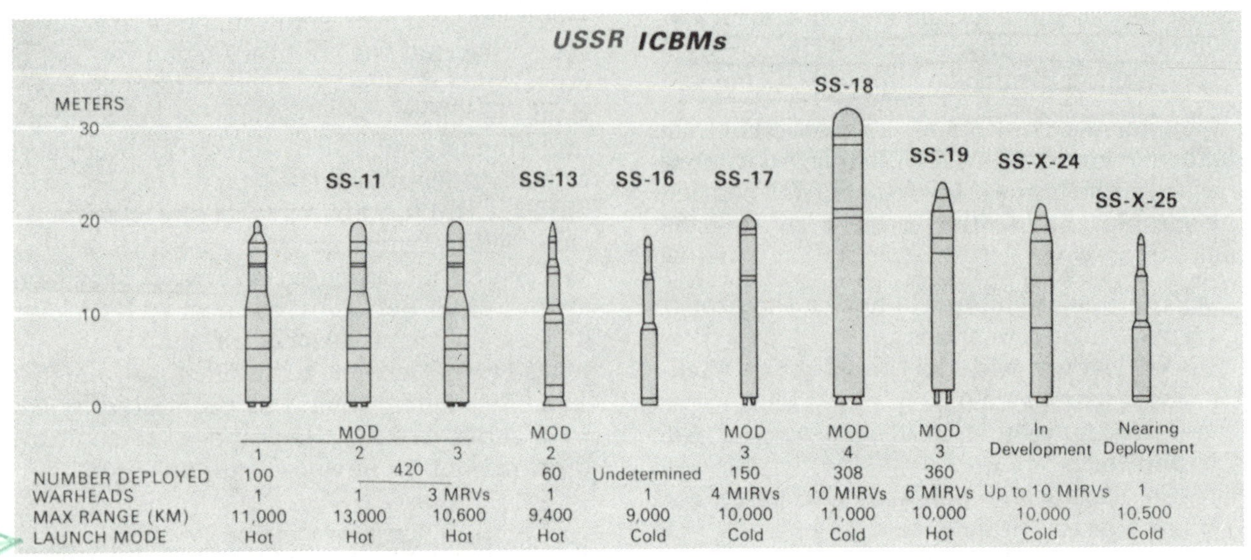

The Soviet ICBM arsenal. The Soviets have built a large land-based ICBM force which appears capable of destroying the bulk of the U.S. land-based ICBM force in a first strike.

SOURCE: U.S. Department of Defense.

Photo credit: U.S. Navy

The U.S.S. Ohio, first of the Trident ballistic missile submarines. About half of U.S. strategic nuclear warheads are deployed on submarine-launched ballistic missiles aboard Poseidon and Trident submarines. In normal times, about half of these are hidden under water; during a "generated" alert, still more would be sent to sea.

current and advanced cruise missiles, the B-1B bomber, and use of new bomber technology will continue to assure penetration of those defenses for the foreseeable future.

- Although the Soviet civil defense program is large, it remains a matter of controversy in the United States as to how well it could actually protect Soviet political and economic assets against U.S. strategic forces.[4]

The Soviets have also taken measures to protect their own nuclear forces from a nuclear strike. They have hardened their ICBM silos to withstand high overpressures; they have a large submarine-launched missile force; they appear to be developing a mobile land-based ICBM. Such survivability measures can be interpreted as maintenance of a secure "third-strike" reserve force which would be protected from a U.S. retaliatory attack and which therefore could be used to deter the United States from retaliating.[5] However, the survivability measures can also be viewed as providing an

[4]Two contrasting views are given by Leon Goure, "War Survival in Soviet Strategy" (Washington, DC: Advanced International Studies Institute, 1976); and U.S. Central Intelligence Agency, "Soviet Civil Defense," Director of Central Intelligence, N178-10003, July 1978. See also note 8, below.

[5]Some argue that if the United States calculated that after it retaliated the Soviets would be left with a larger reserve of nuclear weapons, a U.S. President would be even more hesitant about retaliating.

Photo credits: U.S. Air Force

"Air-breathing," means of delivering U.S. strategic nuclear weapons. Top left, air-launched cruise missile being launched from Air Force B-52 bomb bay. Top right, B1-B bomber. Advanced Technology Bomber (ATB) design, right, still highly classified. According to the Organization of the Joint Chiefs of Staff, "The B1-B is designed to penetrate Soviet defenses well into the 1990s. The strategic modernization program also calls for the development of an ATB with stealth characteristics. Plans call for the ATB to deploy in the 1990s to neutralize an increasingly sophisticated Soviet air defense system." [*United States Military Posture for FY 1986,* p. 25.]

invulnerable retaliatory force capable of inflicting unacceptable damage on an attacker, which would support a strategy of deterrence.

Many U.S. Sovietologists believe that the concept of strategic preemption to limit damage (if not to completely decide the outcome of the conflict) is an important element in Soviet military doctrine and force deployments. There is less agreement about the conditions under which the Soviets might choose to exercise a preemptive option.[6]

Returning to the three general categories of Soviet actions the United States would like to deter, we can see how differing interpretations of Soviet nuclear strategy lead to differing assessments of what deterrence requires of U.S. nuclear forces.

Bolt-From-the-Blue First Strike

The requirement to absorb such an attack and still retain the capability to deliver an un-

[6]Sovietologist Raymond Garthoff argues from Soviet military and political writings that: 1) Soviet doctrines of strategic preemption cover only certain narrow cases, with a launch on warning or launch under attack being more likely; and 2) a decision to preempt, and therefore to start a nuclear war, would be suicidal because such a war would be disastrous for both sides. (c.f. "Mutual Deterrence, Parity, and Strategic Arms Limitation in Soviet Policy," Chapter 5 of *Soviet Military Thinking,* Derek Leebaert (ed.) (London: George Allen & Unwin, 1981), pp. 92-124. Another analyst concludes:

> The Soviet leaders have been forced to recognize that their relationship with the United States is in reality one of mutual vulnerability to devastating nuclear strikes, and that there is no immediate prospect of escaping from this relationship. Within the constraints of this mutual vulnerability they have tried to prepare for nuclear war, and they would try to win such a war if it came to that. But there is little evidence to suggest that they think victory in a global nuclear war would be anything other than catastrophic.

(David Holloway, *The Soviet Union and the Arms Race* (London and New Haven: Yale University Press, 1984), p. 179.)

On the other hand, still other analysts argue that:

> There is a regrettable tendency in the West to view the Soviet Union almsot entirely in "mirror image" terms . . . The simple fact of the matter is that U.S. and Soviet concepts of the benefits of "victory" and its relative costs reflect philosophical and societal parameters that are in no way symmetrical . . . The data available suggest, in fact, that the Soviet leadership, in the pursuit of its hegemonial objectives, may be prepared to incur losses in societal and human values that would be "unthinkable," at least in cold blood, within Western polities, but which in Soviet eyes are bearable, viewed, for instance, in relation to the total Soviet military and civilian casualties in World War II.

Jacquelyn K. Davis, Robert L. Pfaltzgraff, Jr., and Uri Ra'anan, "Soviet Strategic-Military Thought and Force Levels: Implications for American Security," in Jacquelyn K. Davis, et al., *The Soviet Union and Ballistic Missile Defense* (Cambridge, MA: The Institute for Foreign Policy Analysis, Inc., 1980), p. 25.

Yet another analyst argues that emphases in Soviet strategic doctrine have varied over tiem but always according to the dictates of Soviet political leadership when it takes a stand. Currently, he argues, " . . . the primary rationale for all Soviet nuclear options is now retaliation—the inhibition of American escalation." James M. McConnell, "Shifts in Soviet Views on the Proper Focus of Military Development," *World Politics,* April 1985, p. 337.

acceptable retaliatory strike to the Soviet Union has been a fundamental determinant of the U.S. strategic posture. Although this is the scenario most analysts agree is the least likely, it is also one of the most stressing; the chances of its occurring could well increase if we were to ignore this case and, as a result, become dangerously vulnerable to it.

In this scenario, the Kremlin leaders sit down one day and decide that a world without the United States as a major power would be a more comfortable one for the Soviet Union, and that they have the means of bringing that world about at acceptable cost. Alternatively, they decide that they face some intolerable trends (perhaps the disintegration of their position in Eastern Europe, or economic collapse at home) and that only a victory over the United States can rescue them. They would presumably estimate that a surprise attack on the United States could disarm it sufficiently so that it might prefer negotiations to retaliation and that, at worst, whatever retaliatory damage the United States could inflict would be an acceptable price for the defeat of the United States.

Some have argued that the Soviets might attempt a more or less surgical strike on U.S. land-based missiles (and perhaps bombers), leaving only the less accurate submarine-launched U.S. missiles available for retaliation. Since the present generation of these missiles is not accurate enough to destroy Soviet reserve ICBM silos or other very hard targets (e.g., command bunkers or shelters for the Soviet political and military leadership), the United States might be deterred from retaliating at all, hoping to spare its cities from a Soviet "third strike"; instead, U.S. leaders would be forced to sue for peace on Soviet terms.

Aside from the operational uncertainties Soviet military planners would face, this scenario minimizes several considerations.[7] First, an attack on U.S. land-based strategic forces would inevitably lead to the deaths of millions of Americans. The Soviets would be imprudent, to say the least, to believe that the United States would fail to retaliate. They also need to consider that the United States maintains the option of launching its land-based ICBMs on warning of attack, leaving only empty silos to await the Soviet first strike. In addition, although currently less accurate than land-based missiles, U.S. SLBMs are aimed at a wide variety of military, political, and economic targets—targets presumably chosen to be those the Soviet leadership would least like

Photo credit: U.S. Navy

U.S. Navy Trident C-4 SLBM in test launch. The C-4 is more accurate than its predecessor, the Poseidon SLBM, but not as accurate as the D-5 (Trident II) SLBM, which, when it becomes operational in the late 1980s, is expected to be nearly as accurate as land-based ICBMs.

[7]On Soviet planning uncertainties, see Benjamin S. Lambeth, "Uncertainties for the Soviet War Planner," *International Security*, vol. 7, winter 1983, pp. 139-166. See also Stanley Sienkiewicz, "Observations on the Impact of Uncertainty in Strategic Analysis," *World Politics*, vol. XXXII, October 1979, pp. 90-110.

to lose. Finally, surviving U.S. bombers and cruise missiles could accurately attack Soviet hard targets.

Moreover, there is no consensus in the United States about just what levels of national loss the Soviet leadership might be prepared to suffer to obtain various objectives.[8] The question becomes even more complicated when one moves from a scenario in which the Soviets deliberately choose to begin a nuclear war (as would result from bolt-from-the-blue attack) to scenarios in which the Soviets are only running a *risk* of nuclear war.

Escalatory Confrontation

What might deter the Soviet Union from *risking* nuclear confrontation remains extremely controversial. But suppose the risk has been taken. The Soviets may have miscalculated U.S. willingness to escalate a conflict, thus miscalculating the risk of war involved in some act of aggression. It is also possible to imagine scenarios in which the Soviets either do not believe the United States to have made a deterrent commitment, or do not believe that their own actions constitute what others would see as aggression. Varying perceptions of Soviet motivations lead to varying degrees of willingness to postulate such scenarios. In any case, the situation postulated here is not that the Soviets have decided in advance that victory at a reasonable price is achievable, nor that they believe the consequences of a nuclear war to be acceptable. Rather, it is that they believe (and think that the United States probably shares the belief) that past miscalculations and the pressure of events have made central nuclear war imminent and quite possibly inevitable. The question is whether they would launch a preemptive strike on the United States or whether they would wait and take the chance either that the United States would not strike, or that they could launch their own forces upon detection of U.S. attack.

[8]"A Garthoff-Pipes Debate on Soviet Strategic Doctrine," *Strategic Review*, vol. 10, fall 1982, pp. 36-63.

In this case, compared to the "bolt-from-the-blue" scenario, the Soviet calculus of risk is different. They do not necessarily believe that a nuclear war will be to their strategic advantage. They assume that the United States will retaliate devastatingly if struck; moreover, they have some doubt as to whether the United States, expecting a Soviet strike, will wait for it. If the Soviets were absolutely certain that strategic nuclear war was inevitable, they would presumably see no choice but to launch a preemptive strike. Although the Soviet Union might suffer grievous damage from a U.S. retaliation, that damage might be reduced at least marginally by the combination of Soviet counterforce strikes and defensive measures.

Suppose, however, that the situation remained at least somewhat ambiguous. The Soviets would be confronting a U.S. strategic force in a high state of alert: many submarines in port might have been sent to sea; additional alerted bombers might have been dispersed to many airfields; in the expectation of imminent attack, U.S. land-based ICBMs might be prepared to be launched under warning of attack so as to escape a Soviet disarming strike. Attacked by this augmented retaliatory force, Soviet civil defense and air defense capabilities might not go far in preventing damage.[9] The

[9]It is possible that *greater* immediate damage could be done to the Soviet Union as a result of having civil defenses. Evacuation of Soviet cities could be interpreted as a signal that the Soviets were considering a preemptive strike, so the United States might respond by "generating" its strategic forces, or putting them in a high state of alert. In a generated posture, U.S. forces would be less vulnerable to a Soviet first strike. Therefore, it is conceivable that more Soviets would be killed by U.S. retaliation which had been bolstered as a result of Soviet evacuation than would die in U.S. retaliation for a "bolt-from-the-blue" attack that the Soviets mounted without evacuating their cities and therefore without warning the United States of their plan.

One study estimated that a retaliatory second-strike attack by U.S. forces in their "day-to-day" posture against Soviet nuclear forces, other military targets, and industry would kill 60 to 64 million Soviets, over the short term, if they did not evacuate cities but instead took protection in the "best available" shelter. If the Soviets successfully evacuated 80 percent of their urban population and caused the United States to generate its forces, a U.S. retaliatory strike would kill 23 to 34 million Soviets. However, that total would rise to to 54 to 65 million if the evacuated population were targeted. The study did not consider long-term effects, and no analysis was made to determine the feasibility of implementing such an evacuation suc-

Soviets would face the difficult choice of launching an attack which assured great damage to themselves, or taking a chance that a strategic exchange could still be avoided but at the same time risking even greater damage if the gamble on U.S. restraint failed. It is widely believed that this is the kind of scenario which is the most likely to lead to a nuclear war.

Threats of Aggression and Aggression Against U.S. Allies and Interests

The extension of U.S. nuclear forces to deter against attack on NATO allies as well as against attack on the United States proper, called "extended deterrence," provides the greatest challenge to U.S. nuclear strategy. This commitment to United States allies is central to the North Atlantic Treaty Organization:

> The Parties agree that an armed attack against one or more of them in Europe or North America shall be considered an attack against them all; and consequently, they agree that, if such armed attack occurs, each of them, in exercise of the right of individual or collective self-defense recognized by Article 51 of the Charter of the United Nations, will assist the Party or Parties so attacked by taking forthwith, individually and in concert with the other Parties, such action as it deems necessary, including the use of armed force, to restore and maintain the security of the North Atlantic area.[10]

In the 1950s, when the Soviet Union had very little capability for a nuclear attack on the United States, it was more or less plausible for the United States to threaten nuclear punishment for aggression against its allies. (The Soviets attempted to compensate somewhat for this asymmetry in nuclear deterrence in the late 1950s and early 1960s by deploying hundreds of medium- and intermediate-range nuclear missiles that could at least reach Western Europe.)

But during the 1960s, the Soviets began to acquire their own ICBMs, stationed in silos that would be hard for the United States to knock out in a quick first strike. By that time, the United States had added thousands of tactical and "theater" nuclear weapons to NATO forces. In a strategy of "flexible response," the United States would answer Soviet aggression at whatever level seemed necessary—including first use of nuclear weapons—to repel the attack. Given the Warsaw Pact numerical superiority in many categories of conventional force, it was widely assumed that NATO would have to resort to nuclear counterattack at a fairly early stage. Such nuclear counterattacks might lead to termination of the conflict before it escalated to central nuclear war—*but then again they might not.* Thus, U.S. strategy in NATO held out the ultimate prospect, if not the immediate threat, that the U.S. assured destruction capability might still be called into play.

At the same time, the United States would have to reckon with the risk that escalation of a European war might lead to assured Soviet-inflicted destruction of the United States. As Henry Kissinger told a European audience in 1979:

> The European allies should not keep asking us to multiply strategic assurances that we cannot possibly mean, or if we do mean, we should not want to execute because if we execute, we risk the destruction of civilization.*
>
> The Soviets might not believe that the United States would really run such a risk in order to defend Europe.

Extended deterrence, therefore, poses an inherent dilemma which U.S. nuclear strategy has not fully solved: to the extent that U.S. strategic nuclear forces are believable as "NATO's ultimate deterrent," their use in that role risks the United States' own destruc-

---

cessfully. "In fact," the study noted, "it is highly questionable whether the United States or the Soviet Union could effectively achieve this [civil defense] posture." U.S. Arms Control and Disarmament Agency, "An Analysis of Civil Defense in Nuclear War," December 1978, figure 13, pp. 11 and 12.

[10]The North Atlantic Treaty, Article 5, signed on Apr. 4, 1949 in Washington, DC, *The Signing of the North Atlantic Treaty—Proceedings*, Department of State Publication 3497, (Washington, DC: U.S. Government Printing Office, June 1949).

*Henry A. Kissenger, "NATO Defense and the Soviet Threat," *Survival*, November/December 1979, p. 266.

tion. To the extent that such use is not believable, those forces cannot effectively deter attack on NATO. The prospect of inviting Soviet retaliation directly against the United States for use of nuclear weapons in defense of Europe looks as repugnant to some Americans as its converse of confining a superpower-initiated nuclear war to European soil looks to some Europeans.

The reasoning behind "flexible response" is that it is credible to threaten the **possibility, but not the certainty**, of escalation to general nuclear war. In a situation where escalation might or might not occur, with that possibility not necessarily under the direct control of either side, what would otherwise have been an unbelievable threat might acquire credence. Strategist Thomas Schelling describes the role of uncertainty, "the threat that leaves something to chance," in this situation:

> The brink is not, in this view, the sharp edge of a cliff where one can stand firmly, look down, and decide whether or not to plunge. The brink is a curved slope that one can stand on with some risk of slipping, the slope [getting] steeper and the risk of slipping greater as one moves toward the chasm . . . One does not, in brinksmanship, frighten the adversary who is roped to him by getting so close to the edge that if one *decides* to jump one can do so before anyone can stop him. Brinksmanship involves getting onto the slope where one may fall in spite of his own best efforts to save himself, dragging his adversary with him.[11]

The threat of first use of nuclear weapons by NATO depends either on the assumption that any nuclear use may lead to uncontrolled escalation, making the threat of a NATO nuclear response an effective deterrent to both conventional and nuclear aggression, or on the assumption that NATO can maintain "escalation dominance" on the Warsaw Pact, preventing the use of nuclear weapons beyond the level that NATO chooses to use them. Soviet deployments of tactical, theater, and intermediate-range nuclear forces, in conjunction with their central strategic forces, are sufficient to deny NATO high confidence in imposing escalation dominance on the Warsaw Pact. Therefore, measures have been taken to tighten the perceived "coupling" between Europe and U.S. central strategic forces to bolster the United States' "extended deterrent."[12]

For example, the recent deployment by the U.S. of intermediate-range Pershing II missiles and ground-launched cruise missiles (GLCMs) in Europe was undertaken in part

[11]Thomas C. Schelling, *The Strategy of Conflict* (London: Oxford University Press, 1960), p. 199. "Children," says Schelling, "understand this perfectly."

[12]This and the following paragraph draw on Robert S. McNamara's arguments in "The Military Role of Nuclear Weapons: Perceptions and Misperceptions," *Foreign Affairs*, fall 1983, p. 59.

Photo credit: U.S. Air Force

Test firing of U.S. Air Force Ground-Launched Cruise Missile (GLCM). This mobile missile is being deployed in Europe partly to respond to NATO fears that the United States might be unwilling to use nuclear forces based at sea or in its own homeland in defense of Europe.

to respond to NATO fears that the United States would be unwilling to use nuclear forces based at sea or in its own homeland in defense of Europe. These European-based systems, in addition to the submarine-launched ballistic missiles which the United States had already assigned for NATO use, were intended to strengthen the connection between conventional and tactical nuclear forces, on one hand, and American central strategic forces, on the other. In striking Soviet territory, they might precipitate a Soviet retaliatory strike on American territory which in turn might generate a U.S. central strategic attack. The Soviets, no longer perceiving a "firebreak" between conventional aggression (which might result in NATO first use of tactical nuclear weapons) and central strategic exchange, would be deterred from making the initial conventional attack.

In calculating whether to challenge the United States in areas where the United States appears to have a military commitment, the Soviets must weigh the gains they hope to achieve (or losses they hope to avoid) against a calculated risk of nuclear war. They cannot know with certainty what the United States' responses will be. Therefore, they must estimate the probability of various U.S. reactions. It might be that what they predict to be the most likely outcome is the one they seek. Alternatively, there might be a less likely result which was nevertheless so desirable that the Soviets would judge their overall risks to be tolerable, considering the possible gains. They might even act to minimize the most favorable outcome obtainable by the United States rather than to maximize their own benefit (should those cases differ).

The United States, similarly, cannot know exactly what the Soviets will do. It therefore can only do its best to make sure that aggression is a very unattractive choice for the Soviets no matter how the Soviets make their decisions.

Some argue that if the extended deterrent is to be truly credible, the United States must be able to greatly erode the Soviet assured destruction capability, either by preemptive counterforce attacks on Soviet missiles, by incorporating significant defenses (civil, air, and ballistic missile), or both. As Colin Gray has put it,

> . . . if U.S. strategic nuclear forces are to be politically relevant in future crises, the American homeland has to be physically defended. It is unreasonable to ask an American President to wage an acute crisis, or the early stages of a central war, while he is fearful of being responsible for the loss of more than 100 million Americans. If escalation discipline is to be imposed upon the Soviet Union, even in the direst situations, potential damage to North America has to be limited . . .[13]

On the other hand, if the Soviets wish to avoid such "escalation discipline," they have a strong incentive to try to assure the penetration of their forces through such U.S. defenses—to see to it that the United States does *not* come to believe that damage can be limited.

[13]Colin Gray, "Nuclear Strategy: the Case for a Theory of Victory," *International Security*, vol. 4, No. 1, summer 1979, p. 84.

CURRENT U.S. STRATEGIC NUCLEAR POLICY

Current U.S. nuclear strategy is a balance between attempting to minimize the risk of nuclear war, on the one hand, and attempting to prevent the coercion or intimidation of the United States and its allies, on the other.[14] Secretary of Defense Caspar Weinberger listed the five highest priority national security objectives of the United States in his Report to the Congress for Fiscal Year 1984. The three that directly concern strategic nuclear weap-

[14]Henry Kissinger wrote in 1957 that " . . . the enormity of modern weapons makes the thought of war repugnant, but the refusal to run any risks would amount to giving the Soviet rulers a blank check . . . " (c.f. *Nuclear Weapons and Foreign Policy* (New York: Harper & Brothers, 1957), p. 7).

ons are quite similiar to the formulations of previous administrations:

- To deter military attack by the U.S.S.R. and its allies against the United States, its allies, and other friendly countries; and to deter, or to counter, use of Soviet military power to coerce or intimidate our friends and allies.
- In the event of an attack, to deny the enemy his objectives and bring a rapid end to the conflict on terms favorable to our interests; and to maintain the political and territorial integrity of the United States and its allies.
- To promote meaningful and verifiable mutual reductions in nuclear and conventional forces through negotiations with the Soviet Union and the Warsaw Pact, respectively; and to discourage further proliferation of nuclear weapons throughout the world.[15]

The strategy adopted to achieve these objectives is based on three major principles, stated in the same report, that are also a continuation of longstanding policies:

- First, our strategy is *defensive*. It excludes the possibility that the United States would initiate a war or launch a preemptive strike against the forces or territories of other nations.
- Second, our strategy is to deter war. The *deterrent* nature of our strategy is closely related to our defensive stance. We maintain a nuclear and conventional force posture designed to convince any potential adversary that the cost of aggression would be too high to justify an attack.
- Third, should deterrence fail, our strategy is to *restore peace on favorable terms*. In responding to an enemy attack, we must defeat the attack and achieve our national objectives while limiting—to the extent possible—the scope of the conflict. We would seek to deny the enemy his political and military goals and to counterattack with sufficient strength to terminate hostilities at the lowest possible level of damage to the United States and its allies.[16]

While catastrophic failure of this strategy would be clear, its success is hard to quantify. "We can never really measure how much aggression we have deterred, or how much peace we have preserved," wrote Secretary of Defense Weinberger. "These are intangible—until they are lost."[17]

Countervailing Strategy

In 1980, after having conducted a comprehensive review of U.S. strategic policy, President Carter issued Presidential Directive 59 which formally codified a "countervailing" strategy. As described by Secretary Brown in his Report to Congress for Fiscal Year 1982, the countervailing strategy is based on two fundamental principles:

> *The first is that, because it is a strategy of deterrence, the countervailing strategy is designed with the Soviets in mind.* Not only must we have the forces, the doctrine, and the will to retaliate if attacked, we must convince the Soviets, *in advance*, that we do. Because it is designed to deter the Soviets, our strategic doctrine must take account of what we know about Soviet perspectives on these issues, for, by definition, deterrence requires shaping Soviet assessments about the risks of war .... We may, and we do, think our models are more accurate, but theirs are the reality deterrence drives us to consider ....
>
> *The second basic point is that, because the world is constantly changing, our strategy evolves slowly, almost continually, over time to adapt to changes in U.S. technology and military capabilities, as well as Soviet technology, military capabilities, and strategic doctrine.*[18]

In particular, countervailing strategy intends to make clear to the Soviets that:

> ... no course of aggression by them that led to use of nuclear weapons, on any scale of attack and at any stage of conflict, could lead to victory, however they may define victory. Besides our power to devastate the full target system of the U.S.S.R., the United States

[15]Caspar W. Weinberger, Secretary of Defense, *Annual Report to the Congress, Fiscal Year 1984* (referred to as *DOD FY84 Annual Report*), Feb. 1, 1983, p. 16.

[16]Ibid., p. 32 (emphasis in original).

[17]Weinberger, *DOD FY85 Annual Report*, Feb. 1, 1984, p. 8.

[18]Harold, Brown, Secretary of Defense, *Annual Report to the Congress, Fiscal Year 1982*, Jan. 19, 1981, p. 38 (emphasis in original).

would have the option for more selective, lesser retaliatory attacks that would exact a prohibitively high price from the things the Soviet leadership prizes most—political and military control, nuclear and conventional forces, and the economic base needed to sustain a war.[19]

Seeking to incorporate flexibility and encompassing many options and target sets, the countervailing strategy continues to be the basis for U.S. strategic nuclear policy.[20]

Strategic Stability

American nuclear strategy has placed high priority on strategic stability. Most often, the term "stability" used alone has stood for *crisis stability*, which describes a situation in which, in times of crisis or high tension, no country would see the advantages of attacking first with nuclear weapons as outweighing the disadvantages. Crisis stability depends on the force structures and doctrines of both sides and on each side's perception of the other. The lower the degree of crisis stability, the greater the risk that a power would preempt if it perceived that it were likely to be attacked. This is not to argue that it is U.S. policy to consider a preemptive strike, but Soviet perceptions of such a possibility might increase a Soviet inclination to preempt under some circumstances. President Reagan's Commission on Strategic Forces (the Scowcroft Commission) stated that:

> . . . stability should be the primary objective both of the modernization of our strategic forces and of our arms control proposals. Our arms control proposals and our strategic arms programs . . . should work together to permit us, and encourage the Soviets, to move in directions that reduce or eliminate the advantage of aggression and also reduce the risk of war by accident or miscalculation.[21]

Another type of stability is *arms race stability*, in which there are minimal incentives for the United States and U.S.S.R. to continually update or expand their strategic arsenals in order to compensate for developments by the opposite side. The assumption underlying the concept of arms race stability is that deployments on one side may lead the other to counter-deployments which in turn stimulate new deployments by the first.

U.S. Force Requirements and Posture

According to Secretary of Defense Weinberger, present U.S. countervailing strategy places five specific requirements on strategic nuclear forces:[22]

1. **Flexibility:** " . . . A continuum of options, ranging from use of small numbers of strategic and/or theater nuclear weapons aimed at narrowly defined targets, to employment of large portions of our nuclear forces against a broad spectrum of targets."
2. **Escalation Control:** " . . . We must convince the enemy that further escalation will not result in achievement of his objectives, that it will not mean 'success,' but rather additional costs."
3. **Survivability and Endurance:** " . . . The key to escalation control is the survivability and endurance of our nuclear forces and the supporting communications, command and control, and intelligence (C^3I) capabilities."
4. **Targeting Objectives:** "We must have the ability to destroy elements of four general categories of Soviet targets." These are strategic nuclear forces, other military forces, leadership and control, and the industrial and economic base.
5. **Reserve Forces:** "Our planning must provide for the designation and employment

[19]Ibid., p. 39.

[20]In 1982, Secretary Weinberger told the Senate Committee on Foreign Relations that Reagan Administration policy "does not change substantially or materially the policy set out" in P.D. 59, and that "the essential strategic doctrine set out in P.D. 59 remains." ("U.S. Strategic Doctrine," hearing before the Committee on Foreign Relations, United States Senate, 97th Cong., 2d sess., Dec. 14, 1982, p. 99. See also an insert for the record outlining nuclear policy differences between the Carter and Reagan Administrations on p. 100.)

[21]April 1983 Report of the President's Commission on Strategic Forces (referred to hereafter as the Scowcroft Commission Report I), p. 3.

[22]Ibid., pp. 40-41.

of adequate, survivable, and enduring reserve forces and the supporting C³I systems both during and after a protracted conflict."

To attempt to satisfy these requirements, the United States maintains a *triad* of strategic offensive weapons systems consisting of long-range bombers, submarine-launched ballistic missiles (SLBMs), and land-based intercontinental ballistic missiles (ICBMs). These systems carry thousands of nuclear warheads in ballistic missile reentry vehicles, bombs, cruise missiles, and short-range air-to-ground missiles. There are thousands more nonstrategic nuclear warheads including those in artillery shells, bombs carried by tactical air forces, short- and medium-range rockets, and intermediate-range rockets and cruise missiles. However, weapons considered nonstrategic by the United States, such as the Pershing II intermediate-range ballistic missile, can reach Soviet territory and are considered to be strategic by the Soviets.

Characteristics such as survivability, basing, penetration modes, range, yield, accuracy, time of flight, independence from enemy warning systems, and ease of command and control distinguish the various strategic weapons systems. U.S. administrations have put high value on maintaining this diversity in nuclear forces. The triad, wrote the Scowcroft Commission, serves several important purposes:

> First, the existence of several strategic forces requires the Soviets to solve a number of different problems in their efforts to plan how they might try to overcome them. Our objective, after all, is to make their planning of any such attack as difficult as we can . . .
>
> Second, the different components of our strategic forces would force the Soviets, if they were to contemplate an all-out attack, to make choices which would lead them to reduce significantly their effectiveness against one component in order to attack another . . .
>
> The third purpose served by having multiple components in our strategic forces is that each component has unique properties not present in the others . . .[23]

Submarines, the Scowcroft Commission noted, can remain hidden for months at a time. Bombers can be launched upon warning without being irrevocably committed to an attack, and they have very high accuracy against a variety of targets. ICBMs "have advantages in command and control, in the ability to be retargeted readily, and in accuracy. This means ICBMs are especially effective in deterring Soviet threats of massive conventional or limited nuclear attacks, because they could most credibly respond promptly and controllably against specific military targets and promptly disrupt an attack on us or our allies."[24]

The countervailing strategy does not require that the U.S. force structure mirror that of the Soviets or vice versa, provided that the overall military capability of the United States is not allowed to become inferior to that of the Soviet Union, in either reality or appearance. "Indeed," wrote Secretary Brown, "in some sense, the political advantages of being seen as the superior strategic power are more real and more usable than the military advantages of in fact being superior in one measure or another."[25]

The Strategic Balance

Soviet strategic nuclear forces in fact do not mirror those of the United States. In particular, the Soviet allocation of warheads among types of delivery vehicles is quite different than that of the United States. The final report of the Scowcroft Commission discussed the asymmetry between U.S. and U.S.S.R. strategic forces, along with the problems of comparing the two:

> In the United States the strategic advantages of diversity, our own military tradition as an air and naval power, plus a certain amount of interservice competition, produced strong strategic bomber and submarine forces, as well as a land-based ICBM force . . .
>
> Soviet strategic forces developed along very different lines . . . Geography and history

[23]Scowcroft Commission Report I, pp. 7-8.

[24]Ibid., p. 8.

[25]Brown, *DOD FY82 Annual Report*, p. 43.

have made Russia a continental land power, with a tradition of heavy emphasis on massive artillery forces. As might have been expected under such circumstances, the development of Soviet strategic nuclear forces has been heavily oriented toward ICBM weapons . . .

The result of all these differing traditions and technical capabilities is strategic forces which are very dissimilar. In addition, each strategic force component has its own strengths and weaknesses, which tend to be different from those of the other components. This, in turn, makes force structures very difficult to compare and each side tends to stress certain aspects of the force posture of the other as more menacing.[26]

In comparing the strategic nuclear capability of the United States with that of the Soviet Union, both Carter and Reagan Administrations agreed that "the era of U.S. superiority is long past."[27] However, they differed significantly in their interpretations of what followed. The Carter Administration held that "parity—not U.S. inferiority—has replaced [U.S.] superiority, and the United States and the Soviet Union are roughly equal in strategic nuclear power."[28] Two years later, on the other hand, the Reagan Administration maintained that "the Soviets have acquired a margin of nuclear superiority in most important categories."[29] Both Administrations undertook strategic modernization programs to redress what were seen at least as adverse trends in the military balance, if not adverse situations.

Current U.S. Attitude Towards Active Defenses

BMD is currently not included in the U.S. strategic posture, while air defenses are minimal (but being upgraded). Previous U.S. Administrations have agreed that a condition in which both the United States and the Soviet Union refrained from instituting nationwide ballistic missile defenses was preferable to one in which either (and therefore most likely, both) attempted to do so. That situation was codified in 1972 by ratification of the ABM Treaty. The single BMD installation permitted the United States by the 1974 Protocol to the ABM Treaty was decommissioned in 1976 after it was determined that the limited benefit provided by such a highly constrained system did not justify the expense of maintaining it. Extensive air defenses in the absence of effective BMD were similarly held not to be worthwhile.

Like its predecessors, the Carter Administration viewed the ABM Treaty as being "to the benefit of strategic stability and deterrence."[30] The reasoning leading to this assessment found nationwide defenses to be destabilizing in that they call into question the ability of nuclear weapons to threaten destruction of assets that a potential attacker values highly. Defenses were not judged to be cost-effective in that they would merely force the Soviets to increase their offensive forces to maintain whatever level of damage expectancy had previously been thought sufficient—increases which would cost less than our defenses.

During the Carter Administration, BMD research permitted by the ABM Treaty was actively pursued as a hedge against possible Soviet developments. It focused on point defense capabilities for hardened targets, particularly MX missiles deceptively deployed in Multiple Protective Structures, and on nonnuclear destruction of ICBMs outside the Earth's atmosphere. The preferential defense possible with MPS basing of MX made BMD a logical choice for responding to Soviet warhead proliferation beyond the SALT II limits.

The Reagan Administration differs from previous ones in its conception of the role that defenses might play in future nuclear strategy and in its planning for BMD research and development. It has initiated a broad-ranging investigation into the role of and possibilities for

[26]March 1984 Report of the President's Commission on Strategic Forces (Scowcroft Commission Report II), Mar. 21, 1984, p. 4.

[27]Brown, op. cit., p. 43.

[28]Ibid., p. 43.

[29]Weinberger, *DOD FY84 Annual Report*, p. 34.

[30]Brown, op. cit., p. 51.

strategic defense. It has not, however, adopted a doctrine incorporating such defenses. Its position is that such a decision should await the completion of a BMD research and technology development program which could take at least 10 years.[31] At that time, a future President and Congress could decide whether to proceed to develop, test, and deploy one or more BMD systems. Meanwhile, the United States and the Soviet Union in 1982 "each reaffirmed its commitment to the aims and objectives of the Treaty,"[32] and President Reagan has further made clear that U.S. BMD research will be "consistent with our obligations" under that Treaty.[33]

[31]Time estimates given by Administration representatives vary. Ambassador Paul Nitze estimated that it will take "at least ten years" to determine whether sufficiently capable ballistic missile defenses can be built [quoted by Walter Pincus in "Decade of Study Seen for 'Star Wars'," *The Washington Post*, Apr. 27, 1985].

[32]Joint communique issued at the conclusion of the 5-year review of the ABM Treaty, reported in "SCC Completes Review of ABM Treaty," *Daily Bulletin (U.S. Mission, Geneva)*, Dec. 16, 1982 (quoted by George Schneiter in "The ABM Treaty Today," chapter 6 of *Ballistic Missile Defense*, Ashton B. Carter and David N. Schwartz (eds.) (Washington, DC: The Brookings Institution, 1984), p. 236.

[33]President Reagan's speech on Military Spending and Defense Technologies, Mar. 23, 1983.

Under President Reagan, the Department of Defense has initiated the "Strategic Defense Initiative" (SDI), a comprehensive program "to develop key technologies associated with concepts for defense against ballistic missiles" whose ultimate goal is "to eliminate the threat posed by nuclear ballistic missiles and increase the contribution of defensive systems to U.S. and allied security."[34] Although the SDI research and technology development program is intended to comply initially with the restrictions of the 1972 ABM Treaty, the development, testing, or deployment of BMD systems investigated under the SDI would require modification of or withdrawal from that treaty. The SDI differs substantially from previous BMD efforts in that: 1) it shifts emphasis from near-term, almost proven technologies to relatively high risk but conceivably higher payoff ones; and 2) it significantly increases both the funding and attention given to BMD research. Nuclear strategies incorporating BMD systems of the sort to be investigated under the SDI could be quite different from this nation's current strategy.

[34]Caspar W. Weinberger, Secretary of Defense, *Strategic Defense Initiative Organization (SDIO) Charter*, Apr. 24, 1984.

COMMON CRITICISMS OF U.S. NUCLEAR STRATEGY

This nation's strategic nuclear doctrine has continually evolved, but it has not been dramatically changed in the last 20 years.[35] Despite this consensus, various analysts have suggested either further modifications or major revisions to it to redress perceived weaknesses. In many cases, differing recommenda-tions stem from differen... premises...

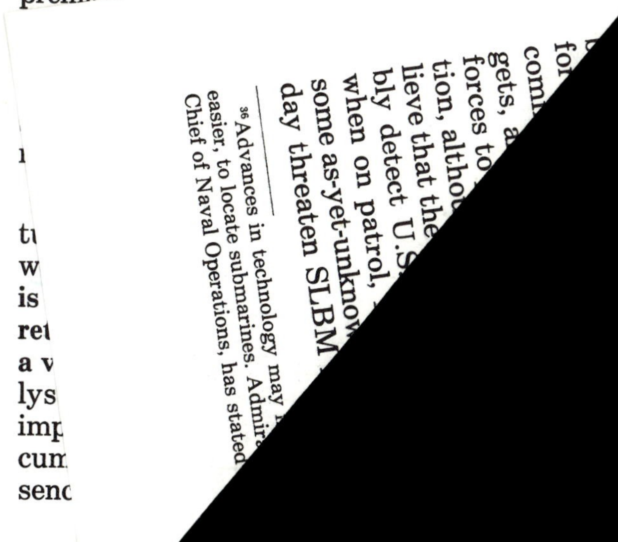

[35]Desmond Ball "Targeting for Strategic Deterrence" (London: International Institute for Strategic Studies, Adelphi Paper No. 43, 1983), discusses the evolution of U.S. strategic nuclear targeting policy over the last 40 years. He finds that although the numbers of targets and the packages of targeting options available to the President have changed dramatically over that period, the actual character of those targets has remained remarkably consistent. He argues that those changes that have occurred in targeting policies and plans are the result of many factors, including the changing nature of Soviet targets, better U.S. intelligence about those targets, and changes in U.S. force capabilities, and that changes in avowed U.S. national security policy have been "perhaps one of the least important" of those factors.

various ways or carried out (with the aid of arms control agreements) at substantially smaller force levels. Others who basically agree with the premises underlying current strategy foresee difficulty in maintaining its viability in the face of continual technological evolution, particularly on the part of the Soviet Union. Some of the latter see a potential role for ballistic missile defense in enhancing the U.S. deterrent posture.

Still others hold fundamentally different assumptions than those on which current strategy is based. Their concern is to modify existing strategy in accordance with a different set of premises.

Maintaining Current Strategy

Technological evolution influences strategy both by changing what is seen as possible ("technology push") and what is viewed to be necessary ("requirements pull"). On the "technology push" side, for example, many believe that we now have the potential to develop ballistic missile defenses which are considerably more capable than could be considered years ago. Such advances have been one of the major motivations for the requesting of this report, and they will be discussed in further detail in chapters 7 and 8.

Technology is also advancing in areas other than ballistic missile defense, and contributes to the "requirements pull" that some believe will mandate changes to our strategy. In particular, Soviet ability to harden and make mo-[illegible]ile elements of their land-based strategic [illegible]ces, and their efforts towards hardening [illegible]mand and control facilities and other tar-[illegible]ll serve to degrade the ability of U.S. [illegible] place these targets at risk. In addi-[illegible]gh there is as yet no reason to be-[illegible] Soviets will ever be able to relia-[illegible] ballistic missile submarines [illegible]it cannot be ruled out that [illegible]n technology might some-[illegible] invulnerability.[36] Space systems today are able to enhance the effectiveness of terrestrial forces, and this ability will no doubt be accentuated in the future. Combined with political factors such as the Soviet ability to proliferate military forces taken with what is perceived to be U.S. reluctance to do the same, these actual and possible technological trends lead some analysts to question whether the "countervailing strategy" can be maintained without significant change into the indefinite future.

Proposals for change vary. Some include defenses; others do not. Some would emphasize U.S. technological strengths to maximum advantage in the military competition between the United States and the U.S.S.R., including uses in areas (e.g., ballistic missile defense) now closed off by mutual agreement. Others would incorporate active defenses in our strategic posture but would not otherwise introduce major changes to U.S. strategy. Still others would eschew active defense, preferring to retain the ABM Treaty as one of a number of means to manage the overall military competition via arms control and other political and diplomatic measures. The specific means by which defense could augment our present strategy or support a transition to another are discussed below and in the following chapter (chapter 5). Discussions of how such transitions might evolve are presented in chapters 6 and 9.

Alternative U.S. Strategies

In addition to those advocating modifications to current strategy, there are those who differ with basic assumptions central to that strategy and who therefore offer alternatives. Three such alternatives are presented below.

One group believes that current strategy does not sufficiently recognize what they see as the inherent opposition between minimizing the risk of nuclear war, on the one hand, and preparing to fight one, on the other. There-

[illegible]make it harder, rather than [illegible]l James D. Watkins, the [illegible] that "... when people ask, 'Aren't the oceans getting more transparent,' we say, 'No way, they are getting more opaque,' because we're learning more about them all the time." *The Washington Post*, Mar. 22, 1985, p. A10.

fore, they see that the balance mentioned previously between war-fighting capability and crisis instability is swinging dangerously towards instability, and that weapons systems that could improve the ability to fight a nuclear war could also make such a war more likely to occur. Alternatively, they may believe that existing plans for prosecuting a nuclear war overestimate the probability that those things which the war would be defending would survive the war at all. **These analysts recommend that the United States pursue a strategy which we will label "retaliation only."**

A second group of strategists believes instead that present strategy does not sufficiently recognize the essential equivalence between deterring war and preparing to fight war. Moreover, existing strategy does not offer a coherent picture of what it would consider victory, and it cannot be expected to effectively deter an opponent who, it is argued, would have a very clear conception of his strategic objectives in war. These strategists advocate adopting what might be called a "**prevailing**" strategy.

Finally, there are strategists who think that this country should not and need not accept having its continued survival contingent on the decisions of others. They argue that no matter how strong our deterrent strategy can be made, should it fail (whether due to accident, miscalculation, or just poor design), the results would be catastrophic. **They moreover argue that we have, or will have, the means to develop defenses (possibly augmented by stringent offensive force limitations) which can remove, or substantially reduce, the ability of others to destroy this country. Discussion of such "defense dominant" strategies concludes the alternatives presented below.**

Retaliation-Only

"Retaliation-only" strategists question whether any military utility at all can be derived from nuclear weapons which justify the risks inherent in planning to use them in battle, short of retaliating against nuclear attack.[37] Although their prescriptions for change differ, they are based on a fundamental premise similar to that stated by Robert McNamara:

> I do not believe we can avoid serious and unacceptable risk of nuclear war until we recognize—and until we base all our military plans, defense budgets, weapon deployments, and arms negotiations on the recognition—that *nuclear weapons serve no military purpose whatsoever. They are totally useless—except only to deter one's opponent from using them.*[38]

Accordingly, "retaliation-only" strategists adopt the principle of "no first use" of nuclear weapons, which in some versions would be stated publicly and in others would be left silently ambiguous.[39] Starting with that premise, retaliation-only strategists can go in two different directions. In the first, a variety of nuclear weapons with flexible targeting options would be retained in order to display the capability of responding in kind to any level of nuclear attack. There would be no immediate requirement to reduce the number of warheads existing today (although should Soviet forces be reduced, U.S. forces could be reduced accordingly.) However, nuclear forces under this strategy would differ qualitatively from today's forces in that weapons would not be given **prompt hard-target kill capability**—a capability needed in order to conduct a successful preemptive attack on enemy nuclear forces. Attacks on a wide variety of military forces would still be possible under such a strategy using those weapons having **slow** hard-target kill capability. This strategy would therefore

[37] They therefore differ from the current "countervailing" strategy, which requires some measure of war-fighting capability for escalation control. On p. 40 of his Annual Report for Fiscal Year 1982, Secretary of Defense Harold Brown emphasized two points which he had made "repeatedly and publicly":

> First, I remain highly skeptical that escalation of a limited nuclear exhange can be controlled, or that it can be stopped short of an all-out, massive exchange.
>
> Second, even given that belief, I am convinced that we must do everything we can to make such escalation control possible, that opting out of this effort and consciously resigning ourselves to the inevitability of such escalation is a serious abdication of the awesome responsibilities nuclear weapons, and the unbelievable damage their uncontrolled use would create, thrust upon us.

[38] Robert S. McNamara, "The Military Role of Nuclear Weapons: Perceptions and Misperceptions," *Foreign Affairs*, fall 1983, p. 60 (emphasis in original).

[39] Of course, even a public statement leaves some ambiguity—no matter what our doctrine, it would remain physically possible to use nuclear weapons in a first strike, and the Soviets would have to worry about this possibility.

be able to maintain some degree of war-fighting potential, but would significantly lessen the degree to which that potential could be used (or would appear capable of use) in a first strike.

In the second variation, often called *minimum deterrence*, only those weapons which would be needed to threaten a number of high-value targets—cities, for example—would be retained. The number and nature of those targets would be selected to threaten enough destruction to deter a potential attacker from initiating a nuclear strike. Opinions differ as to the exact size of "minimum," but no definition of a minimum deterrent would require thousands of warheads on a multiplicity of delivery vehicles.

What would be essential in either version would be that the nuclear weapons that were retained include (in the first case) or constitute (in the second) an invulnerable, second-strike force. The size of this force would be determined in the first case by being able to retaliate for whatever form of attack had been executed initially, and in the second by being able to destroy with high confidence that set of targets judged to provide minimum deterrence. To the extent that the retaliatory weapons were vulnerable, or to the extent that a potential attacker posessed defenses, the second-strike force would either need to expand in size or increase its invulnerability and penetrativeness in order to maintain a minimum deterrent threat.

Should the Soviets acquire defenses so effective that even this minimum deterrent retaliation could not be executed with high confidence, and were the United States unable to penetrate, evade, or neutralize these defenses effectively, then the fundamental premise of promising nuclear retaliation for nuclear attack could not be assured, and strategies based primarily on the threat of retaliation would no longer be viable. On the other hand, if the United States and the Soviet Union had equal offensive and defensive capabilities (and if the survivability of offensive forces did not depend on defenses), retaliation might still be credible. However, uncertainties in each side's evaluation of the opposing side's defense might make assuring an equivalent retaliation difficult.

Since a "retaliation-only" strategy explicitly denies use of nuclear weapons in response to conventional attack, some other way of fulfilling U.S. defense commitments to its NATO allies must be found (e.g., augmentation of conventional forces in Europe). Furthermore, a "minimum deterrence" strategy, presumably using far fewer weapons than are presently in the U.S. arsenal and probably embodying a much more limited repertoire of nuclear responses, must ensure that all opponents remain firmly convinced that any use of nuclear weapons will be met with a retaliatory response. If retaliatory threats are not credible, then potential attackers may gamble that retaliation might not be carried out and they may not be deterred successfully.

One suggested implementation of a "retaliation-only" deterrent strategy[40] (similar to minimum deterrence as described above in its force employment policy but not necessarily in the size of its arsenal) would eliminate all tactical and theater-level nuclear weapons. It would retain only an invulnerable, second-strike force of central strategic weapons which would not be given the combination of yield, accuracy, and quantity needed to pose a threat to the retaliatory capability of the other side. Their survivability would be critical, and it could be enhanced by deploying them in a redundant manner similar to that of the present triad. Flexibility in responding to nuclear attack could be maintained, in that the attacked nation would have options ranging from delivering a single retaliatory weapon to launching its entire strategic arsenal.

Critics of "retaliation-only" strategists believe that there may not be effective alternatives to the threat of first use to deter attack on NATO, that such strategies (in particular

[40]Richard L. Garwin, "Reducing Dependence on Nuclear Weapons: A Second Nuclear Regime," *Nuclear Weapons and World Politics*, 1980s Project/Council on Foreign Relations (New York: McGraw-Hill Book Co., 1977), pp. 83-147.

the "minimum deterrence" approach) would not credibly deter attack since potential adversaries might not believe the United States would actually carry out its retaliatory threats, and that such strategies do not provide sufficient opportunity to terminate hostilities on favorable terms should deterrence fail.

Prevailing

A quite different proposed change to current doctrine would push in the opposite direction from the recommendations of "retaliation-only" strategists, towards the formulation of more credible plans for the use of nuclear weapons in wartime. These strategists believe that, in a world where adversaries possess nuclear weapons and may well believe in their military utility, it is not sufficient for the United States merely to seek to deny the enemy his political and military goals should war break out. Credible deterrence requires that we plan in the event of war to "secure the achievement of Western political purposes at a military, economic, and social cost commensurate with the stakes of the conflict."[41]

Where some see the uncertainties inherent in estimating outcomes of nuclear war to be so great, and the potential damage so devastating, that there is little to be gained in trying now to affect the nature of a post-war world, a "prevailing" strategy focuses specifically on the conduct of a nuclear war, and is based on consideration of how such a war might end. It would agree with the countervailing school (and the "no prompt hard-target kill" option of the "retaliation-only" school) that

> . . . the deterrent effect of our strategic forces is not something separate and apart from the ability of those forces to be used against the tools by which the Soviet leaders maintain their power. Deterrence, on the contrary, requires military effectiveness.[42]

However, to change from the current strategy towards a prevailing one, the United States

> . . . must set its planning sights considerably beyond developing a defense posture that will simply deny victory to the enemy. To prevail in stressful circumstances the United States must be able to defend itself against nuclear attack.[43]

Credibility that deterrent threats would actually be carried out would result not so much from flexibility in strategic planning or response options as it would from the "Soviet belief, or strong suspicion, that the United States could fight and win the military conflict and hold down its societal damage to a tolerable level."[44] As a result, such credibility that we would use nuclear weapons to retaliate would be greater than it is in our current, undefended posture.

Clearly, determining "tolerable" levels of damage "commensurate with the stakes of the conflict," in addition to predicting potential levels of attack, will be needed in order to specify the defensive capability required by such a strategy. Effective air defense, civil defense, and ballistic missile defenses would all be required were defending a major portion of population and economic and industrial infrastructure to be a high priority. Offensive force requirements for such a strategy would depend on the set of targets in the Soviet Union (their number, hardness, and location), and would depend critically on the level to which these targets were defended.

"Prevailing" strategists directly address the problem of extended deterrence by recommending sufficient damage-limitation capability (passive defense, active defense, or preemptive attack) to make believable the threat that the United States would use central strategic forces in circumstances other than responding to nuclear attack. If the Soviet Union were convinced that a defended United States believed it could use tactical or even strategic

[41]Colin S. Gray, *Nuclear Strategy and Nuclear Planning*, Philadelphia Policy Papers (Philadelphia, PA: Foreign Policy Research Institute, 1984), p. 2.

[42]Scowcroft Commission Report I, p. 7.

[43]Gray, op. cit., p. 2.

[44]Ibid., p. 3.

nuclear weapons in defense of NATO Europe without leading to unacceptable devastation of the United States, the Soviets might be more likely to believe that conventional attack against NATO would lead to the use of nuclear weapons against the Soviet Union.

One essential factor in establishing defense requirements for a prevailing strategy is determination of how much damage to the United States can be tolerated in pursuit of those objectives that strategic nuclear forces will be employed to attain or preserve. Another is the degree of U.S. military superiority such a strategy would require, and whether such a strategy would be viable without it. From 1945 until the early 1960s, U.S. strategic superiority was such that this country had the capability to adopt a "prevailing" strategy; adopting one today in the light of existing Soviet forces poses an entirely different set of challenges.

Critics of "prevailing" strategies argue that the United States has no guarantee of being able to attain or maintain the degree of military superiority necessarily to implement them, and that these strategies are equivalent to destroying the Soviet "deterrent," which the Soviets have the will and the technology to prevent.

Defense Dominance

The "countervailing," "retaliation-only," and "prevailing" strategies described so far are characterized by the policies they recommend for employing offensive forces. Although there are also differences between them in the roles that defenses play, it is primarily the role of the offense that distinguishes them. In contrast, defenses supplant, more than they augment, offensive forces in "defense-dominant" strategies. President Reagan's speech of March 23, 1983, and his Strategic Defense Initiative, have greatly stimulated discussion about the feasibility of attaining such a long-term goal. However, since a defense-dominated world is "too distant a technical prospect to be a very active player in the U.S. strategy debate as yet,"[45] there is not so widely developed a body of strategic thought on this alternative as there is concerning some of the others.

Proponents of "defense-dominant" strategists see defenses as lessening both the probability of nuclear war and the damage that would be done by such a war, should it occur. They also see such strategies as being moral, in that defending through active defense is preferable to defending through terrorism—the ultimate mechanism by which deterrence through threat of retaliation operates. In a "defense-dominant" world, the probability of war would be lessened since the attacker, less certain of achieving his objectives, would be less likely to attack in the first place. Two factors would lessen the attacker's confidence in success. For one, it would be much more difficult to destroy all his intended targets, directly frustrating his objectives. Probably more importantly, though, he would not be able to plan an effective attack since he would not know *in advance* which warheads will penetrate the defense. Defenses will contribute uncertainty to an attack in addition to defeating part of it. In addition, if war nevertheless were to break out in a "defense-dominant" world, its consequences might be less severe than they would be in any of the other cases described here.

In a way, "defense-dominant" and "retaliation-only" strategists share a common goal: a world in which the only plausible use for a strategic nuclear weapon is in retaliation for the use of another. However, adherents of the "reliation-only" strategy believe that we are already in such a world although our offensive strategy does not recognize it, and that BMD might destabilize the situation; supporters of the former believe that the Soviet Union, at least, finds "military utility" in ballistic missiles and that only BMD can ensure that all sides will perceive the use of nuclear weapons as truly and clearly irrational for all sides. Moreover, they argue, at the very highest

[45]Ibid., p. 3.

levels of defensive capability, even an irrational decision by the Soviets would not lead to the destruction of U.S. society. Indeed, if defense dominance became total, we could consider strategy of "assured survival" in which retaliation became unnecessary because we had confidence that no Soviet nuclear attack of any kind could succeed.

However, to the extent that defenses on both sides lessen the utility and the probability of preemptive nuclear attack, they will interfere with any other roles assigned to offensive strategic forces. This is, after all, the point. In particular, if conventional attack on Europe is deterred by the ultimate threat of escalation to central strategic exchange, then lessening the effectiveness of strategic forces may lessen their deterrent value, possibly **increasing** the likelihood of conventional war in Europe. A "defense-dominant" strategy, like a "retaliation-only" one, must solve the problem of deterring conventional attack without nuclear weapons.

Unless a defense can be deployed which is so effective that the Soviet nuclear arsenal becomes irrelevant, the Soviet response will be the key to the success of a "defense-dominant" strategy. Such a strategy will either attempt to force the U.S.S.R. to unilaterally avoid strategies which the United States believes to be particularly dangerous, or it will seek cooperation with the Soviet Union in order to be implemented in a coordinated, mutual manner.

The degree to which the Soviet Union, and other nuclear powers, would cooperate in a transition to a defense-dominated world is therefore crucial. The Soviets will choose to cooperate in such a transition either if they conclude that such a world is preferable to the present situation, or if they decide that defensive measures will prove to be so cost-effective that they recognize the futility of offensive/defensive competition.[46] In either case, they might be expected to be amenable to regulating the defensive buildup and controlling offensive arms.

Critics of "defense-dominant" strategies argue that it is by no means clear that defensive technologies capable of supporting such strategies can be developed, that such strategies raise the risk of both preemptive nuclear attack and conventional war, and that nobody knows how a coordinated transition to defense-dominance could ever be carried out.

[46]Cost-effectiveness is not the only criterion on which the Soviets will base their decision to cooperate in a defensive transition. Others include total resource base, total defensive system affordability, ability to redirect civilian resources to the military, and relative utility of offensive forces vs. defensive forces for geopolitical ambitions. Internal Soviet politics and interservice rivalries may also play a role.

POTENTIAL CONTRIBUTIONS OF BALLISTIC MISSILE DEFENSE

Current (Countervailing) Strategy

The overall contribution that defenses can make to current strategy depends on whether the benefits of implementing defenses are seen to outweigh the advantages to the United States of having the Soviet Union refrain from building defenses or adding to offenses. Should that be the case, and should defenses able to provide those benefits prove to be technically feasible, there are several roles that defenses might play in a strategy similar to the current one:

- BMD might enhance deterrence by increasing the difficulty a potential attacker would have in achieving military objectives, strengthening "deterrence by denial." Defenses would also introduce uncertainty into attack plans, lessening the attacker's confidence in achieving his goals as well as reducing his ability to do so.

- "Deterrence by retaliation," the underpinning of our current posture, might be strengthened by protecting our own retaliatory forces against preemptive attack.
- Our ability to project military power abroad, or alternatively our ability to prevent adversaries from doing so, would also be enhanced were our conventional military forces defended against preemptive nuclear attack.
- Certain deployments of BMD might raise the threshold of nuclear war by removing the military utility of small nuclear attacks, and they might also protect against small accidental or nonsuperpower attacks.
- To the extent that assets including, but not limited to, military forces could be defended, our retaliatory threats might be more credible because we might be perceived as having less to lose should our retaliation provoke further attack.

All of these benefits, of course, become liabilities when the tables are turned and we face Soviet defenses: a rational decision requires the two to be balanced off against one another. Stability issues, in particular, are discussed in greater detail in chapter 6.

Since the United States and the U.S.S.R. have different nuclear doctrines and force structures, it might be that similar defensive capabilities would confer asymmetric benefits to the two sides. For example, since our strategy "excludes the possibility that the United States would initiate a war or launch a preemptive strike,"[47] making a first strike more difficult might seem to confuse Soviet plans more than U.S. ones.

However, in addition to depending on one's conception of Soviet attack plans, any such analysis must take the problem of extended deterrence fully into account. To the extent that strategic nuclear weapons lose their military utility, they lose their power to affect the likelihood or outcome of a conventional war in Europe. If the superpowers are able to defend themselves better than Europe can be defended,[48] nuclear war in Europe might become more likely rather than less. Soviet nuclear weapons aimed towards Western Europe would retain a degree of effectiveness lost to those fired back in retaliation. On the other hand, effective homeland defense of the United States might strengthen the credibility of extended deterrence. Any net assessment requires consideration of the relative effectiveness of U.S. and Soviet defenses against their respective offensive threats.

Further complicating BMD's effect on extended deterrence are the independent French and British nuclear forces. At present consisting of SLBMs, intermediate-range nuclear missiles (IRBMs), and a few bombers, they are far less extensive than the U.S. and Soviet nuclear arsenals. More because of their small size than because of the reasons discussed in the previous footnote, the French and British nuclear forces would be more easily negated than the superpower arsenals, further stressing the extended deterrent demand on the U.S. central strategic forces.

[47]Weinberger, *DOD FY 1984 Annual Report*, p. 32

[48]There are reasons both for why this should and should not be the case. Since the short- and intermediate-range ballistic missiles threatening Europe arrive at their targets traveling more slowly than ICBM warheads, they might be more easily destroyed by terminal defenses. On the other hand, since a shorter-range missile reenters the atmosphere at a steeper angle than does an ICBM, the *vertical component* of its velocity (its rate of descent) is comparable to that of an ICBM. Therefore, the time from when atmospheric effects begin to separate warheads from decoys to time when the weapons arrive on target is about the same for a shorter range system as it is for an ICBM. As a result, screening out the decoys and intercepting the actual warheads in their terminal phase will not necessarily be easier for shorter range systems. Furthermore, the total flight time of a shorter range missile, and consequently the period during which it might be destroyed in midcourse, is much less than that of an ICBM. Those short-range systems never exiting the Earth's atmosphere will not be vulnerable to certain directed-energy weapons at all. Tactical and theater-range systems are likely to be less extensively MIRVed, lessening the advantage of destroying them in boost phase. In addition, delivery systems other than ballistic missiles (e.g., bombers, cruise missiles, artillery, or even covertly placed mines) can more easily be used against European targets than against the superpowers, so defense systems other than BMD would need to be compared as well as BMD effectiveness in order to determine whether Europe were better or worse defended than the superpowers.

Chapter 5 looks at the relationship between strategic objectives for BMD and the capabilities of BMD systems.

Retaliation-Only

In practice, most adherents to the "retaliation-only" school of strategy see only a limited role for ballistic missile defense in "retaliation-only" strategies. This point of view probably stems from a concern that the introduction of defenses into the strategic equation could lead to dangerous instabilities for crises or for the arms race in general. If each side were intent on maintaining substantial offensive capabilities, defenses would only be tolerable to the extent that the United States could be assured that its retaliatory capabilities were not undermined. Defenses consistent with this principle, for example defending retaliatory forces, would be acceptable and possibly beneficial. However, comprehensive, areawide defenses would be not be compatible with preserving retaliatory capability unless the net effectivenesss of the offenses on both sides were approximately equivalent.

Ballistic missile defense could serve a similar role in a world which went much further towards nuclear disarmament. Even if possession of nuclear weapons should be renounced, the possibility of building them cannot be eliminated. One vision of a nuclear-free world[49] would have nations retain their weapons design and production facilities as a hedge against sudden development of nuclear weapons by other states. To guard against surprise attack, these facilities would be protected by active defenses.

Prevailing

Defenses are necessary to make a "prevailing" strategy viable. However, they are not sufficient. In order to impose escalation dominance, it would very likely be necessary that overall U.S. capability, offensive and defensive, be superior to that of the Soviet Union. **The success of a "prevailing" strategy, then, depends on the ability of the United States to maintain this superiority.** Defenses in a "prevailing" strategy would protect strategic offensive forces, deny the Soviets success in their attack plans, and lessen "self-deterrence" by which U.S. leaders would be unwilling to use U.S. strategic offensive forces for fear of incurring unacceptable retaliation. However, Soviet defenses would limit the effectiveness of those offensive forces, and they would complicate the extended deterrence problem as discussed in the "Current Strategy" section above.

Defense Dominance

Defenses are not only necessary but also preeminent in these strategies. Going beyond enhancing the "denial" aspect of deterrence as we now know it, a "defense-dominant" strategy relies on defenses while the role of offenses is greatly reduced. Neither side could count on achieving any military objectives by using ballistic missiles. Attacks intended only to do general societal damage, although possible with all but extremely capable defenses, would be highly irrational. If defenses could be brought to a high enough level of performance, even the capacity to do societal damage might be greatly reduced. Then, U.S. survival would not depend on Soviet rationality, but would be assured by our ability to intercept even an irrational attack.

Imposing a "defense-dominant" strategy on an uncooperative adversary requires an extremely high level of defensive capability. Reagan Administration officials have suggested that effective U.S. defenses might offer the Soviets incentives to reduce their offensive forces. Against increaingly constrained offensive forces, any defense would be more effective.

In chapter 5, we look more closely at how various levels of BMD capability, if technically feasible, might play roles in U.S. nuclear strategy, current or prospective.

[49]Jonathan Schell, *The Abolition* (New York: Alfred A. Knopf, 1984).

Chapter 5

BMD Capabilities and the Strategic Balance

Contents

Chapter 5

BMD Capabilities and the Strategic Balance

INTRODUCTION

Since the President's March 23, 1983, speech there has been much discussion of the strategic implications of the steps along the way to his goal. In that speech he announced his "... ultimate goal of eliminating the threat posed by strategic nuclear missiles," while recalling the need to "... remain constant in preserving the nuclear deterrent and preserving a solid capability for flexible response."[1] He also warned that the pairing of offensive and defensive systems "can be viewed as fostering an aggressive policy."

Among those who see potential value in developing BMD there are some who argue that only a realistic prospect of defending the U.S. population against an all-out Soviet attack can justify both a major change in strategic direction and the massive program that developing and deploying BMD would entail. In their opinion, the United States has little or nothing to gain—and perhaps much to lose—by building less effective defenses. As they see it, in an attempt to reach the President's goal the U.S. strategic position may worsen before it gets better, since the Soviets also can be expected to build defenses. Other supporters of BMD maintain that the United States can benefit from any level of strategic defense and that U.S. security will improve as the strategic balance moves from offense-dominance toward defense-dominance. Critics of the President's Strategic Defense Initiative (SDI) Program maintain that his ultimate goal is unattainable and that little could be gained by building lesser levels of defense.

[1]Relevant sections of that speech reproduced in app H.

Almost all observers agree that reaching or approaching the President's goal would require a lengthy and complex transition period.[2] They do not all agree that we can be sure that the transition could ever be completed. **But whether partially effective defenses are viewed as transitional stages or as the most we can in practice achieve, it is important to understand the strategic implications of various levels of U.S. strategic defense and Soviet strategic defense. This chapter discusses how various levels of U.S. and Soviet defense capability might affect the strategic balance as well as the choice of strategy available to the United States. Of particular interest are the implications for a transition from a condition of offense-dominance to one of defense-dominance.**

To understand how U.S. and Soviet strategic defenses can affect the strategic balance, it is necessary to be able to specify what each defense can accomplish against the other's offense. Saying that we have a BMD system that can destroy some number of Soviet reentry vehicles (RVs), or that it has a given leakage rate, tells us little by itself. What it could accomplish would depend on how many ballistic missile weapons the Soviets had, what other nuclear delivery systems they had, how they attacked, and how we defended.

We cannot specify now what offensive and defensive weapons systems the two sides will deploy in future decades. What we can do, *for the sake of analysis*, is postulate various levels

[2]In the President's March 23, 1983, speech he predicted that it will take years, probably decades, of effort on all fronts. For a discussion of the Administration's scenario for the transition, see ch. 9.

of effectiveness for strategic defenses. These levels might or might not be achieved in the face of the other side's offensive measures. But assessment of the BMD issue requires analysis of the strategic implications of various defense capabilities *if we could have them*.

This chapter is divided into two parts. The first examines and explains factors that affect strategic defense capability: the major components of strategic defense (BMD, air defense, civil defense, etc.); basic modes for operating BMD; and the structure and possible evolution of strategic offensive forces. In the second part we specify illustrative levels of defense capability and discuss their implications.

THE COMPONENTS OF STRATEGIC DEFENSE CAPABILITY

The Role of BMD in Strategic Defense

Defense—reducing the damage an opponent can do in an attack—can be divided into three broad categories: passive defense, active defense, and preemptive destruction. Even if the Soviet Union struck first and the United States retaliated, the United States could attempt to defend against follow-on strikes by preemptive destruction of Soviet weapons not employed in the first strike.

Active defenses such as BMD and air defense seek to prevent launched weapons from reaching their intended targets, either by destroying them or by disrupting their operation. **If the goal of the defense is to prevent all weapons—or the great majority of them—from reaching their targets, both BMD and air defense would be required. If, on the other hand, the goal is only to reduce the number of weapons reaching their targets, or to reduce the number of arriving weapons that have properties unique to ballistic missiles, then BMD alone may be sufficient.**

Ballistic missiles are unique in their ability to kill targets at intercontinental ranges promptly —within 30 minutes or less of launch. Because of their accuracy, ICBMs can kill *hard targets* promptly, and it is likely that in the future SLBMS will also have that capability. Bombers and cruise missiles can kill hard targets, but it takes them much longer to reach their targets. Assuming that bombers and cruise missiles can be detected hours before they reach their targets, being able to defend against ballistic missiles would mean having hours rather than minutes to take steps such as getting command authorities to safety, activating civil defense procedures, and deciding to launch a retaliatory strike. Today, ballistic missiles are the largest part of the Soviet strategic nuclear threat against the United States.

An alternative to reducing the number of weapons reaching their targets is to reduce the effectiveness of each weapon. This might be accomplished by passive defense techniques. We currently use some passive defenses to protect elements of our strategic forces. ICBM silos are hardened to reduce the effectiveness of nuclear weapons detonating nearby. Submarines are hidden in the open ocean to preclude the Soviets successfully barraging their deployment areas with nuclear weapons. Bombers, their tankers, and airborne command posts can be sent aloft so that very large areas would have to be barraged to destroy them. Civil defense applies similar techniques—dispersal and sheltering—to protecting civilians. Civil defense cannot protect the buildings and other structures within the cities.

Passive and active defenses can be alternative means to reach the same ends, or they can be combined. In general, the more ambitious the goal the more likely it is that some combination of both would be required. For example, consider population defense. The U.S. population is not evenly distributed over the United States, but is highly concentrated in cities. These cities are soft targets. A few nuclear weapons delivered in any way against a major population center would kill millions of people. City defense would therefore require both BMD and air defense. That defense would have to be extremely capable to ensure that *no* weapons got through, especially if the Soviets launched many at each city they attacked. One reason an attack would kill so many people is that the population is concentrated, unprotected, in a small area. Even with highly effective active defenses, it would probably also be necessary to use civil defense to reduce the number of people killed by the few weapons that might reach the target. Civil defense alone would probably be of limited value against a large attack, since destruction could be spread over a large area. Active defense would have to be extremely good to protect against that attack. However, together the two types of defense might be much more effective than either alone.

Operating Modes for BMD

Depending on its design, a BMD system may operate in one of several different modes. The simplest operating mode might be called "random subtractive." In this case the defense would shoot at as many enemy reentry vehicles (RVs) as possible, with no attempt to distinguish among them.[3] Random subtractive defenses can be characterized by a kill probability (i.e., the probability that any given RV is stopped by the defense), or, alternatively, by a "leakage rate" (the probability that any given RV gets through the defense). The kill probability is the same regardless of where the RV is aimed. It would depend on the size of the attack and the time over which it occurs.

At the other extreme is the "completely preferential" defense. A completely preferential defense shoots only at selected RVs, and can select them for maximum effect. A completely preferential defense can determine where all the RVs in an attack are aimed and can allocate its weapons so that all the RVs aimed at selected targets are destroyed, thereby saving those targets. This kind of defense would maximize the ability to save targets. In reality, a completely preferential defense would be very difficult—if not impossible—to achieve. Nevertheless, the idea is a convenient analytical tool because it represents the best that any defense could do.

A more likely situation would be "semi-preferential" defense. A semi-preferential defense would also shoot only at selected RVs, but the defense capability allocated to defend any particular target would be determined before the attack. To operate semi-preferentially the defense would have to be able to determine where individual RVs are aimed, but unlike a complete preferential defense it would not have to determine where they are all aimed before it begins firing. A semi-preferential defense would be less efficient than a completely preferential defense: some targets would be over-defended while others might be under-defended. When the defender has a semi-preferential defense, the attacker and defender play a double-blind game. Each allocates its weapons according to how it thinks the other will. This introduces an uncertainty into predictions of the outcome beyond the uncertainty stemming from ignorance of the precise capabilities of the offensive and defensive weapons. However, if the attacker knows how well each target is defended, or if he can destroy some of the defensive system with his RVs or other weapons, some of the advantage of a semi-preferential defense is lost because the attacker knows exactly how many RVs to allocate in order to overcome the defense and achieve his attack goal.[4]

[3]Reentry vehicles carry the nuclear weapons. They are dispensed from ballistic missiles above the atmosphere, and are designed to shield the weapons from the effects of reentry into the atmosphere. A random subtractive defense would probably attempt to distinguish RVs from decoys.

[4]Although semi-preferential and completely preferential defenses can increase the number of assets (ICBM silos, for example) surviving a large-scale attack, neither can necessarily provide *enduring* survival. If the offense can exhaust the defense and determine which targets were not destroyed in the first strike, it can reattack those targets in a follow-on attack. A determination of targets surviving the first strike might be made by visual (or photo) reconnaissance. It has been suggested that it could be done more rapidly with space-based sensors that can accurately locate nuclear detonations, or by fitting warheads with devices that broadcast their location just prior to detonation. In planning his follow-on strike, the attacker would want to have some estimate of his opponent's remaining defense capability.

Photo credits: U.S. Air Force

Test reentry vehicles being loaded on "MX" or "Peacekeeper" nosecone indicate general scale and appearance of "RVs" referred to in this and other chapters.

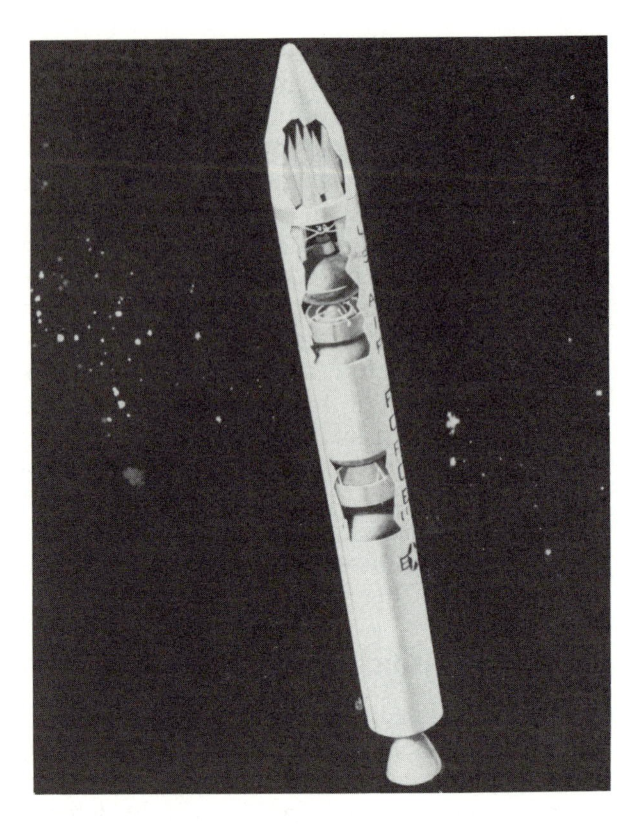

Artist's concept of the missile, right, shows in cut-away how multiple, independently targetable reentry vehicles (MIRVs) are positioned on upper stage (post boost vehicle, or "PBV") of the rocket.

The key to a preferential (or semi-preferential) defense is the ability to destroy any RV that is shot at with near 100 percent confidence. This ability might come from highly capable interceptors or from using a less capable defense to shoot several times against each of the selected attacking weapons. In this latter case, a random subtractive defense might kill more attacking weapons than a preferential defense, but would save fewer of the targets from destruction.

Distinctions between random subtractive and preferential defenses are most important for light and moderate defenses. Defenses that can destroy in excess of 90 percent of an attacker's RVs are likely to be random subtractive. If the defense were composed of highly capable interceptors each of which had close to a 100 percent probability of killing any RV that was shot at, it would be unlikely that the defender would only build enough interceptors to destroy 90 percent or 95 percent of the attacker's force. It is more likely that he would build more than enough interceptors to kill all of the attacker's RVs. If, on the other hand, the single shot kill probability were substantially less than 100 percent, but the defender had enough shots to assure a very high kill probability against 90 percent of the attacking RVs, he could achieve a kill probability almost as high against all of the attacker's RVs. Shooting at all of them would simplify his battle management problem but not concede any targets to the attacker.[5]

[5]For example, assume that the attacker has 10,000 RVs and the defender can shoot a total of 81,000 times with a .4 kill probability per shot. If he elects to shoot nine times at each of 9,000 RVs he can achieve a .99 kill probability against each RV and expect to kill 8,910 RVs. If he elects to shoot eight times against each of 10,000 RVs, he can achieve a .98 kill probability, which is not substantially different, and can expect to kill 9,800 RVs. If, on the other hand, he has only 8,000 shots rather than 81,000, by shooting eight times at each of 1,000 RVs, he can be reasonably sure of saving some targets. If he shoots randomly, he will kill more RVs, but he is unlikely to save many targets if several RVs are used against each.

BMD Operating Modes

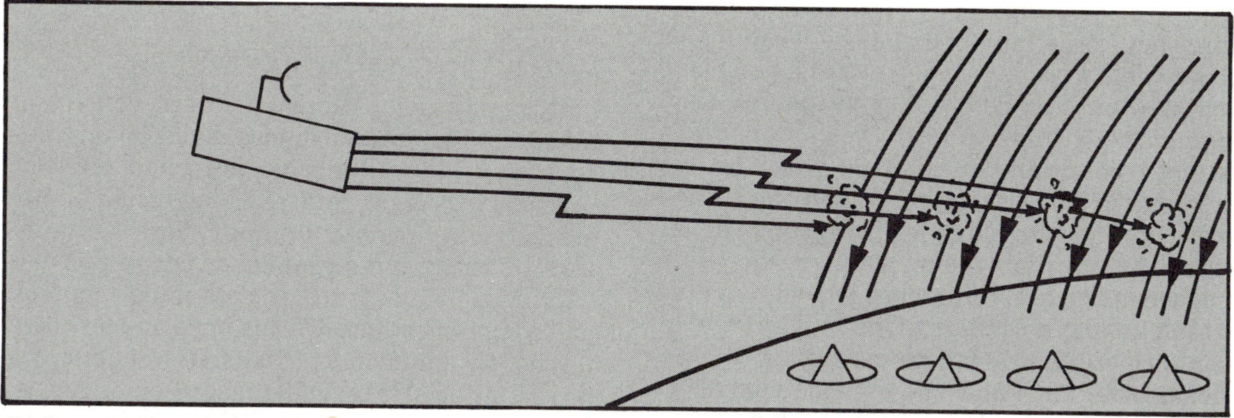

Random subtractive defense. The defense cannot distinguish among RVs, and therefore attacks them randomly. This shows one possible outcome. All the silos are destroyed. It is possible, although not very likely, that all the RVs aimed at one silo would be shot, thereby saving the silo.

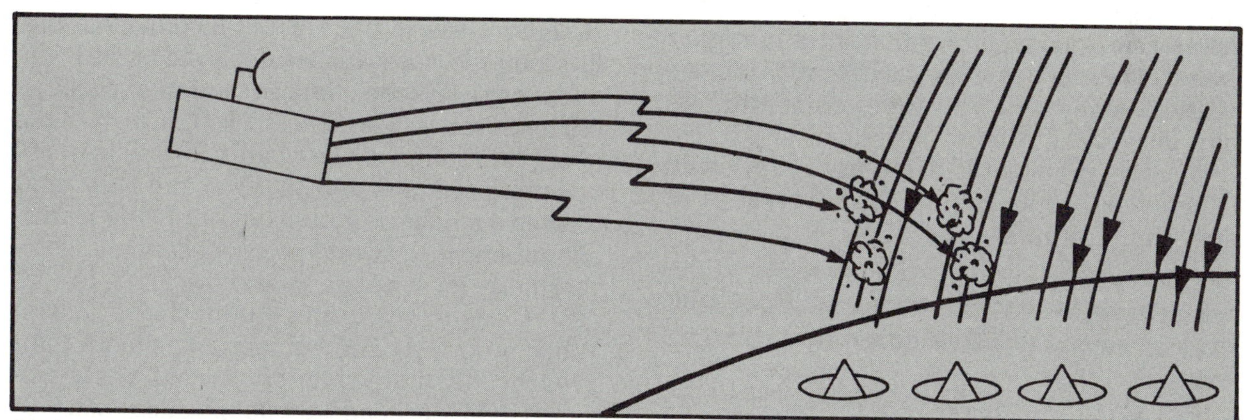

Semi-preferential defense. The defense can determine the targets of the individual RVs, but cannot determine where all the RVs are going before some of them reach their targets. The defender decides in advance of the attack how many shots to defend each silo with. In this case he has not allocated enough, and the silos are destroyed. If the attacker had decided to use fewer RVs against each defended silo, those silos would have survived.

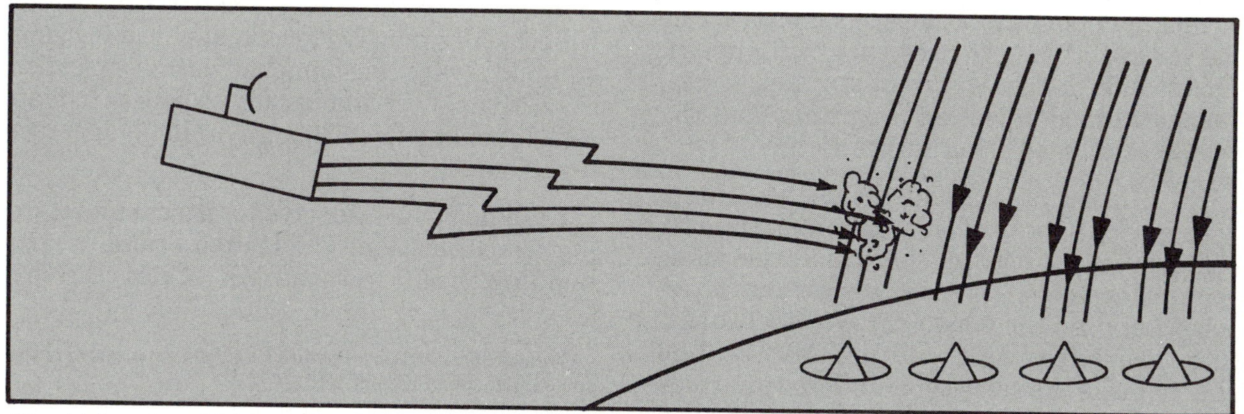

Completely preferential defense. The defense can determine where all the RVs are aimed before it has to shoot. In this case the defender elects to use all his defensive capability to ensure that the leftmost silo survives.

SOURCE: Office of Technology Assessment.

Defenses exact an "attack price." By reducing the effectiveness of the attacker's weapons, they force him to use more weapons to achieve his attack goal. The attack price can be raised both by destroying weapons and by forcing the attacker to waste weapons because he does not know in advance which of his weapons will be destroyed. When the attack price exceeds the number of weapons the attacker has available, survivability increases and the attacker's confidence in achieving his attack goals decreases. One advantage of semi-preferential defenses (assuming the attacker does not know the defense allocation and cannot destroy the defense) is that the attacker does not know which targets he will have to attack with additional weapons and how much defense each target will have. Hence, the attack price may be substantially higher than the number of RVs the defender can actually destroy. Furthermore, some targets may survive at almost any level of attack. Completely preferential defenses, if they could be achieved, could ensure the survival of at least some targets.

Current Strategic Forces and Possible Future Developments

The value of future strategic defenses to the United States would be highly dependent on the nature of future offensive forces, but it is difficult to predict with any confidence what those offenses would be. Moreover, the nature of future defenses is at least as uncertain as the nature of the offenses they will oppose. This section discusses current strategic forces, near-term modernization programs, and the problems of predicting future forces.

The United States and the Soviet Union both have a variety of strategic nuclear delivery systems. Although both use the same types of weapons, there are important differences between the ways the two sides structure their forces. In the future, both will be able to make the same types of force improvements. However, if history is any guide, we can expect the two sides to exploit their opportunities in different ways. Projecting force structures more than a few years into the future is highly speculative.

Current Forces and Near-Term Trends

U.S. strategic offensive forces consist of about 1,000 intercontinental ballistic missiles (ICBMs); 600 submarine launched ballistic missiles (SLBMs); and 325 long-range bombers carrying gravity bombs, short-range attack missiles, and air-launched cruise missiles (ALCMs).[6] The strategic command, control, and communications (C^3) system manages these forces. In addition to offensive weapons, we have limited strategic defenses—small air defense and passive defense of strategic forces.[7]

Soviet strategic forces include the same major elements, but with different emphases. Roughly three fourths (about 6,000) of our ballistic missile warheads are deployed on SLBMs, and we try to keep a large fraction of these at sea at all times. The Soviets, by contrast, have about two-thirds of their ballistic missile weapons mounted on 1,400 ICBMs, and they tend to keep a smaller fraction of their ballistic missile submarines at sea during peacetime.[8] Furthermore, most of the Soviet firepower is concentrated on fewer than half of their ICBMs. While no U.S. ICBM currently has more than 3 warheads, more than 80 percent of the Soviet ICBM warheads are on missiles with 6 to 10 warheads each.[9] Soviet ICBM silos are generally thought to be harder (i.e., more resistant to nuclear attack) than U.S. silos.[10] The U.S. strategic bomber force is a substantial leg of the triad of offensive forces, and is now being equipped with air-launched cruise missiles (ALCMs). Soviet bomber forces appear to play a less prominent—although probably increasing—role.

Both nations' offensive forces are undergoing modernization. The United States is now building the 10 warhead MX ICBM, the B-1

[6] *United States Military Posture FY1986*, Organization of the Joint Chiefs of Staff, pp. 19-33.

[7] Ibid., p. 33.

[8] *Soviet Military Power*, Department of Defense, 1985, p. 29.

[9] Ibid., p. 29.

[10] Ibid., p. 29.

bomber, the ALCM, the submarine-launched cruise missile (SLCM), and the Trident I (C-4) SLBM. A small single warhead ICBM—possibly for mobile deployment—an "advanced technology" bomber, an advanced ALCM, and the Trident II (D-5) SLBM are in various stages of development. The D-5 will have longer range and higher accuracy than the C-4.[11] The Soviets have in development a single warhead ICBM and a 10 warhead ICBM—both believed to be for mobile deployment—a new SLBM, a long-range bomber similar to the B-1, and several cruise missiles. They are building a new class of ballistic missile submarine, as well as a new variant of their existing long-range bomber, the BEAR.[12]

The Soviets have put much more emphasis on strategic defense than the United States has. In the aftermath of the ABM Treaty and its protocol, the Soviets chose to build and maintain the one ABM site permitted, a limited ballistic missile defense of the Moscow area. They have emphasized both homeland air defense and civil defense. The Organization of the Joint Chiefs of Staff estimates that Soviet strategic air defenses consist of 6,300 radars, 9,600 missile launchers, and 1,200 interceptor aircraft, including six new types of aircraft deployed since 1975.[13] The United States saw no purpose in maintaining an operational BMD as constrained by the treaty, and little purpose in building extensive defenses against bombers as long as we had no defense against Soviet missiles. The U.S. air defense system consists of about 100 radars and 300 interceptor aircraft. Both radars and aircraft are currently being upgraded with modern equipment.[14]

A coordinated nuclear strike requires a functioning command system that can communicate with the forces and exercise control. According to the Organization of the Joint Chiefs of Staff:

> The Soviets expect to be able to communicate with their forces during a strategic nuclear exchange and to direct all operations. Toward this end, the Soviets have constructed hardened, deep-underground facilities for their primary military authorities. The Soviets have developed air- and ground-mobile systems that can serve as alternate command posts if primary sites are destroyed. Soviet systems emphasize survivability, redundancy, and flexibility and provide extensive internetting of communications from the high Soviet command to lower echelons.[15]

The U.S. and Soviet C^3 systems have many features in common. However, recent unclassified publications have reported that parts of the U.S. system are soft, few in number, and easy to locate.[16] In a nuclear attack the C^3 system would be a prime target.

Projecting Forces Into the Future

According to the Administration, decisions to begin full-scale development of a BMD system might be made in the early to mid-1990s. These decisions could be expected to produce initial deployments during the first decade of the next century. More extensive and more technologically advanced systems could be expected to follow according to a time scale roughly marked in decades.

The strategic offensive forces that those defenses face could be very different from today's. By 2005, almost all currently deployed forces would have been replaced, and many of those now in early production or in development would be in the process of replacement. By 2020, most systems deployed by the turn of the century would have been replaced. While we can predict with moderate certainty the rate at which individual units will be replaced, it is much more difficult to predict how different those replacements will be. While we may have some confidence that we can predict many of the technical options for future forces, we cannot confidently predict

[11] *DOD Annual Report to the Congress Fiscal Year 1986*, p. 52.

[12] *Soviet Military Power*, Department of Defense 1985, pp. 29-36.

[13] *United States Military Posture FY1986*, Organization of the Joint Chiefs of Staff, p. 33.

[14] Ibid., pp. 31-33.

[15] *United States Military Posture FY1986*, Organization of the Joint Chiefs of Staff, p. 28.

[16] Detailed information that could allow one to distinguish major differences between U.S. and Soviet C^3 systems is classified.

which will prove workable and which will not, which will be exploited and which will not. The offenses faced by future strategic defenses *could* be very much different from today's forces (although they need not necessarily be), and the serious pursuit of strategic defenses will influence the nature of the offenses.

Offensive forces might evolve in response to actual or anticipated BMD developments in three general ways: proliferation of weapons; changes in the mix of weapons in the force; and improvements or changes in existing weapon types.

An obvious, "brute force" response to defense is to build more offense. The Soviets have said that that is what they would do. Ballistic missile warheads can be proliferated by adding more boosters, or by increasing the number of warheads carried by each booster. Between 1980 and 1984 the Soviets built 875 ICBMs, 950 SLBMs, and 2,175 theater-range ballistic missiles, an average of 800 new ballistic missiles per year.[17] Much of this production has apparently gone to replacing existing missiles as they age. However, this production rate indicates a capacity to increase their force levels and to modernize by incorporating countermeasures. The Soviet SS-18, their largest ICBM (currently restricted to 10 warheads under the terms of the SALT II Treaty), is reported to have eight times the throwweight of the U.S. Minuteman III.[18] Under the terms of SALT II the United States reserves the right to deploy Minuteman III with seven warheads, the maximum number with which it has been tested.[19] This indicates considerable room for expansion in the number of weapons carried by the SS-18.[20]

Another possible response to BMD development would be to deemphasize weapons that BMD might be effective against and to increase the role of other weapons. If these other weapons were less effective, less threatening, or less destabilizing in a crisis than ICBMs, then building BMD would have accomplished something. However, the nature of the strategic relationship would have changed. The Soviets might emphasize bombers and cruise missiles. They currently have two types of bombers in production (including the BACKFIRE, whose range is a matter of controversy), and one in flight test. They have four cruise missiles in development, including two large missiles that are probably for long-range operation.[21] Another possibility might be the deployment of shorter range ballistic missiles on submarines and other platforms close to the United States. Finally, less conventional weapons might be used, such as orbital bombing systems (now prohibited by the Outer Space Treaty), and very high-speed aerodynamic vehicles that are launched on ballistic missile boosters but stay within the atmosphere.

Technology may offer a variety of methods to improve the ability of ballistic missiles and their warheads to penetrate defenses. It may also offer counters to those countermeasures.

Photo credit: U.S. Department of Defense

Artist's concept of Soviet BEAR bomber launching cruise missile. One possible effect of BMD deployments might be to lead the Soviets to emphasize bombers and cruise missiles for delivery of strategic nuclear weapons. Higher levels of strategic defense protection for the United States would require effective air defenses in addition to BMD.

[17] *Soviet Military Power,* 1985, p. 38.

[18] *U.S.-Soviet Military Balance,* John Collins, Elizabeth Ann Severns, Congressional Research Service, 1980, Book II, p. 123.

[19] First agreed statement to paragraph 10, Article IV.

[20] The SS-19, which currently carries up to six RVs, has almost half the throwweight of the SS-18. See Collins, op. cit.

[21] *Soviet Military Power,* 1985, p. 35.

In some cases the countermeasures will win and in others they will be overwhelmed by the counter-countermeasures. However, without knowing in advance what the countermeasures and counter-countermeasures are likely to be, evaluating the effectiveness of the defense will be difficult.

While it is important to understand the range of options the Soviets would have available to them, it would probably be erroneous to assume that they could and would exploit all of them to the fullest. Soviet efforts would be limited by the resources they could allocate to strategic forces and by their rate of success in new developments. They might also be limited by arms control agreements. This range of available options, however, implies a broad range of uncertainty about future forces.

The value of BMD to the United States may also be affected by technical advances that offer ways to improve the ability of potential targets to survive a ballistic missile attack. Mobility and hardening of ICBMs and other potential targets are obvious examples, but others may emerge. While these developments will not directly affect the ability of a ballistic missile defense to destroy enemy missiles, they may reduce the payoff for doing so, and therefore affect the potential value of BMD.

HOW BMD MIGHT AFFECT THE STRATEGIC BALANCE

Assessments of the value to the United States of acquiring BMD rest on comparing what the strategic balance might be like if BMD were built to what it would be like without BMD. In order to make this assessment, we would like to know:

1. how the balance is currently assessed and how it might evolve in the continued absence of BMD;
2. how we might use various levels of BMD if we had them; and
3. what the balance might be like if the United States and the Soviets had various levels of BMD.

These topics are the subject of this section.

The Current Strategic Balance

U.S. strategic planning is based in part on a "worst case" massive Soviet strike on the United States. While other scenarios are certainly conceivable, this one is considered to be the most stressing to our forces. In overview the scenario is simple. The Soviets launch a large strike against a full range of militarily important targets in the United States, withholding some of their forces as a strategic reserve. The strategic reserve could have many purposes, but a primary one would be to retain a threat to our cities as a deterrent to the United States retaliating against Soviet cities. A major purpose of the first strike would be to limit our ability to retaliate. Therefore, they would attack our ICBMs, bomber bases, and ballistic missile submarine bases. They would also attack the C^3 system in an effort to "disconnect" the surviving forces, or "decapitate" the United States. Attention has primarily focused on an attack on U.S. ICBMs, which the Soviets would attack with their own ICBMs.

Whatever U.S. forces survived could be used in a second strike against a full range of targets in the U.S.S.R. The Soviets might then use some part of their reserve forces in a follow-on strike, to which we might respond, and so on. Any attempt to construct a detailed scenario and predict its outcome would be very uncertain. No one really knows how well systems would operate in a nuclear conflict, let alone how military and civilian leaders would act. There are wide differences of opinion on basic issues like whether Soviet leaders are likely to be bold or conservative, and what it takes to deter them from attacking. (See chapter 4.)

Today, deterrence of a Soviet strike rests on the Soviets believing that there is a high prob-

ability that if they struck, thousands of U.S. nuclear weapons would reach targets in the U.S.S.R. in a retaliatory strike. Deterrence is aided by many factors that limit Soviet confidence in their ability to achieve their first strike goals, including limiting U.S. retaliatory capability.

Although analysts disagree over the credibility of this deterrent and its continued credibility into the future, everyone would agree that thousands of U.S. nuclear weapons would survive—primarily on submarines at sea, but also on bombers that are launched successfully and on those ICBMs that survive the attack. Analysts disagree on such issues as whether enough of our C^3 system would survive to support timely employment of those weapons, how effective SLBMs can be, how many bombers would survive the first strike and the Soviet air defenses, and whether the United States would actually retaliate knowing that the Soviets had a large reserve force.

Some argue that the degree to which the Soviets could be confident that they could destroy a large part of our ICBM force has been vastly overestimated. They point out that the Soviets do not really know how capable their weapons are or how hard our silos are, and cannot be sure that we would not launch our weapons when we detect a massive ICBM launch. As they see it, this uncertainty would contribute to deterrence. In this view, the Soviets would only launch a nuclear strike as a desperate act, not as a rational instrument of policy.

Others argue that Soviet capabilities may be even greater than official estimates state. They assess a Soviet capability to accomplish militarily useful missions and to limit damage to themselves through offensive strikes and various defensive measures. As they see it, the Soviets plan for the possibility of a nuclear war as they would plan for any war, taking account of the risks and opportunities. Once in a nuclear war, they would attempt to achieve certain objectives. In this view, the Soviets would attempt to fight and win a nuclear war, if necessary, despite the risks and uncertainties.

As chapter 4 discussed, some Soviet actions and statements are consistent with a first strike posture, while others are consistent with a retaliatory posture. Whether or not they expect to attack first, it would be imprudent for them to ignore the possibility that the United States might strike first. The Soviets are likely to be highly suspicious of developments that appear to increase their vulnerability to a U.S. first strike.

As these differences of view illustrate, assessing the outcome of a nuclear exchange (or assessing the credibility of our deterrent) is uncertain. Combining many uncertain factors leads to a wide range of possible answers. Different predispositions lead different analysts to draw very different conclusions from the same range of answers. Some point out that planners, wishing to be safe and plan conservatively, will make the least favorable assumptions. A U.S. or Soviet planner would be likely to assess the adversary as more capable than he assesses himself. Others point out that wars are often won by bold, decisive, actions.

Possible Future Development of the Offensive Balance

It is extremely difficult to project forces decades into the future. Whether in the absence of defense the strategic balance would become more or less favorable to the United States than it currently is depends on the results of several competing developments. Therefore, not unexpectedly, some analysts foresee the U.S. position improving while others think it will worsen. Soviet ICBMs may become more accurate, reliable, responsive, and numerous. On the other hand, their targets may become more difficult to destroy. Mobility, proliferation, and hardening can all be applied to making U.S. ICBMs and C^3 components more survivable. Improvements in SLBMs may give them capabilities similar to those of ICBMs. Soviet air defenses may improve, but U.S. bombers and cruise missiles are becoming more capable. Submarines may become harder or easier to find and destroy. On the other hand, some kinds of targets in the U.S.S.R. may become more difficult to find and destroy, reducing the value of U.S. retaliatory forces.

What Might We Want BMD to Defend?

We could build BMD to defend either military assets—such as ICBMs and other weapons and our strategic C^3 system—or our cities (or both). We might defend our military assets in an effort to improve our ability to absorb a first strike and retaliate, or to reduce Soviet confidence that their first strike could destroy their chosen target set. We might defend our cities to shift the basis of our security away from the threat of retaliation.

The capability to protect our cities would mean a major shift in our strategy away from retaliation and toward assured survival. However, we could not abandon retaliation until defenses gave us confidence that they could assure a high degree of protection. Protecting cities requires an extremely capable defense. Opinion differs as to how many nuclear explosions in populated areas in time of war would lead to unacceptable or intolerable damage. However, that number would be at most tens of weapons out of an attack measured in thousands. **A defense that let through no more than 1 percent of the attack—and perhaps far less than that—would be required before the basis of our strategy could shift away from retaliation.**

Protecting military assets puts less stringent requirements on defense than protecting cities does, since, for example, losing a few ICBMs would mean much less than losing a few cities. If we could protect enough of our retaliatory forces that it would no longer be worthwhile for the Soviets to attack them, we might change the strategic balance in a major way. Our security might depend somewhat less on the threat of retaliation, since there would be less reason for the Soviets to attack in the first place, but we would still need a retaliatory capability. Protecting our military forces to this degree would still require very capable defenses, but not as capable as those required to protect cities. At lesser levels of defense, our security would still rely heavily on the threat of retaliation. (If offenses could be reduced by negotiation to extremely low levels, much less capable defenses would be required to produce the same results. See chapter 6.)

The implications of various levels of defense for the dependence of our security on retaliation and protection are shown schematically in figure 5-1. We could completely abandon our reliance on threatening retaliation only if our defense were nearly "perfect." To do so at lower levels of defense would risk giving the Soviets an important advantage: they could threaten considerable damage to the United States with no risk of nuclear retaliation. Even a few nuclear weapons penetrating our defense could devastate several cities. At what point reliance on retaliation and reliance on protection would be equal is, of course, debatable, but it would have to be at a very high level of defense. A defense that allowed even a few percent of a large Soviet attack to reach our cities would provide little security directly through protection, although it might contribute to deterring the attack.

Some observers argue that the Soviets would be deterred from attacking U.S. cities if it were to cost them about 100 weapons for each

Figure 5-1.—Notional Reliance of U.S. Security on Protection and Threat of Retaliation

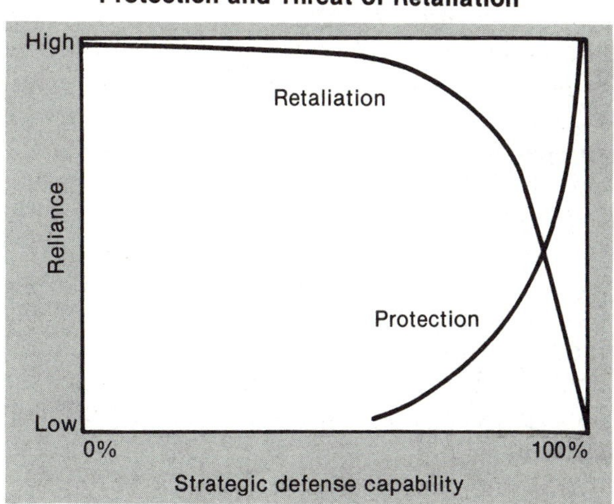

SOURCE: Office of Technology Assessment.

weapon that reached its target. If the United States were to abandon the threat of retaliation under these circumstances it would have to decide that this high cost would be sufficient to deter attack under all circumstances—that the Soviets would not attack even if we could not retaliate against them. To others, it seems likely that we would still have to rely heavily on the threat of retaliation for deterrence even if defenses were highly capable. Similarly, a defense that could preclude a meaningful attack on our military forces might aid deterrence by removing a major incentive for the Soviets to strike in time of crisis, but deterrence of an attack on our cities for whatever reasons would have to rely primarily on a threat of retaliation. Some argue that if the Soviets could not achieve some military objective—e.g., limiting damage to themselves—they would be much less likely to start a nuclear war.

Less capable defenses, although they could not prevent the Soviets from destroying large portions of our retaliatory forces, might still have a role by protecting some of those forces. Defending the forces could directly increase the number surviving the attack. Defending cities might indirectly increase the number of forces surviving the attack, since in order to maintain some minimum threat to our defended cities, the Soviets would have to reserve more weapons for that role leaving fewer available to attack our forces. The more weapons surviving the attack, the more we have available to retaliate with. **However, if the Soviets also have defenses, their defenses will reduce the number of U.S. weapons that survive and penetrate to their targets. Thus, whether or not BMD enhances the U.S. retaliatory force will depend on the capabilities of both U.S. and Soviet defenses.** Clearly, all other factors being equal, for a *given level* of Soviet defense we have a greater retaliatory capability if we defend our forces than if we do not, but it is not necessarily true that our retaliatory capability is greater if both have defenses than if neither does.

This can be illustrated as follows. Consider a Soviet first strike that includes an attack on our ICBMs. For a given U.S. defense capability used to defend ICBMs, the greatest number of U.S. ICBM RVs would survive a Soviet attack if the defense could operate completely preferentially. With completely preferential defenses the United States would be able to allocate the defense in response to the actual attack. Therefore we would be free to arrange our defense to achieve the greatest number of surviving RVs.[22] Furthermore, if our defense were completely preferential, the Soviets' best tactic would be to attack all the silos with the same number of RVs.[23] In this case, the number of RVs the defense could save from destruction would be the number of RVs residing in the silos the defense has the capacity to protect. The fraction of the silos that could be protected would be simply the fraction of the Soviet RVs aimed at the silos that the defense could destroy. If, for example, the defense could destroy 25 percent of the attacking RVs, it would preferentially destroy all the RVs aimed at 25 percent of the silos and save 25 percent of the silos. The number of U.S. RVs available to retaliate with would be the number of ICBM RVs the defense saved plus the number of SLBM RVs at sea. The number of U.S. RVs that survived the attack and penetrated to targets in the U.S.S.R. would be the number that survived minus the number the Soviet defense had killed.

---

[23]If they were to attack some silos more heavily than others, we could defend the more lightly attacked silos and save a greater number from destruction. For example, if they attacked 1,000 silos with 4,000 RVs, 4 per silo, and we could defend preferentially against 1,000 RVs, we could save 250 silos from destruction. We would destroy the 1,000 RVs aimed at those silos. If they attacked 500 silos with 2 RVs each and 500 silos with 6 RVs each, we could destroy the 1,000 RVs aimed at the first 500 silos, and save all of them. As long as the defense was completely preferential we could always defend the most lightly attacked silos first. Therefore, their best tactic would be to attack them all uniformly.

Illustration of the Effect of Defending ICBM Launchers on Retaliatory Capacity
Illustration of the Effect of Adding BMD

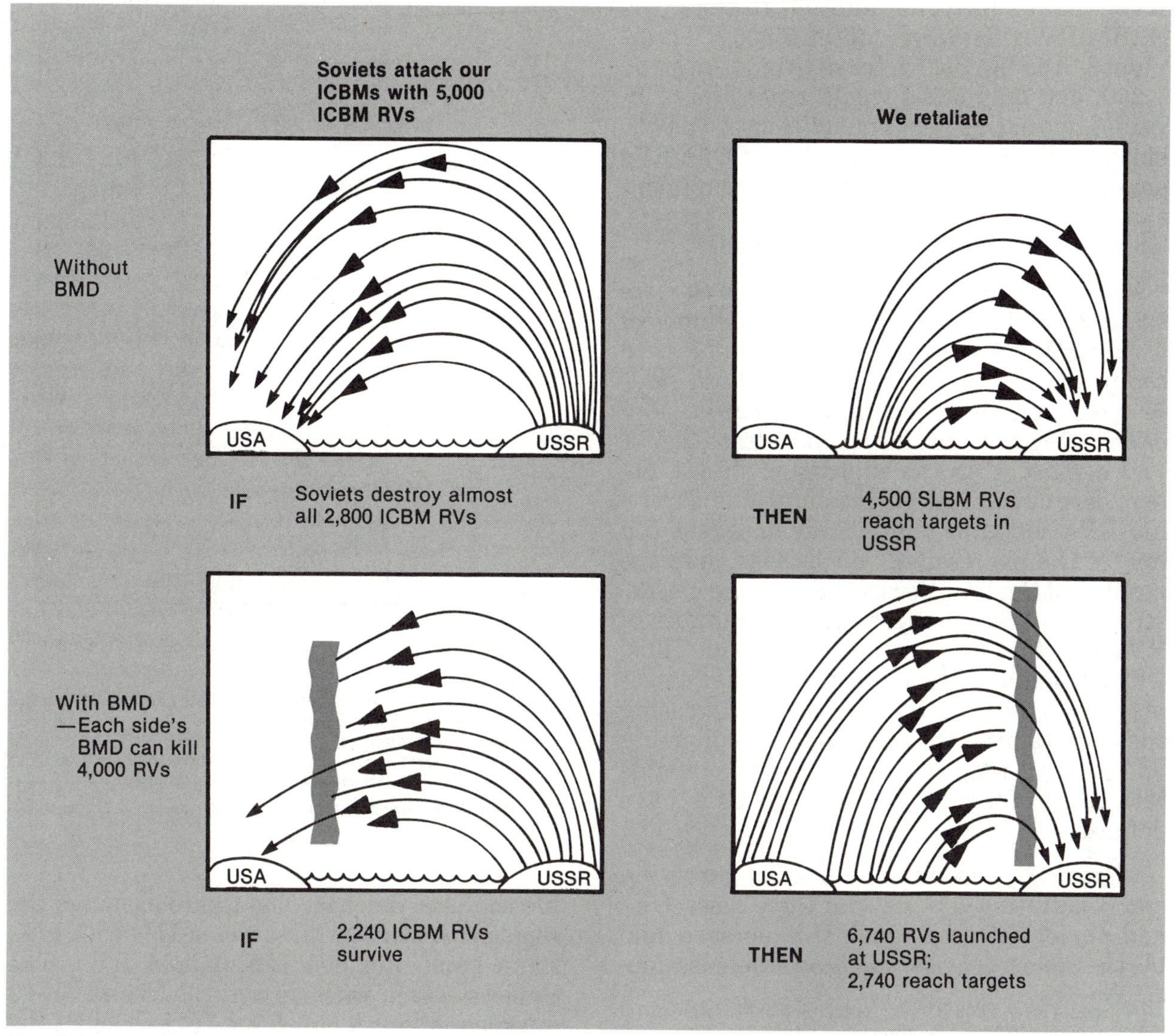

Note: Each symbol represents 500 RVs
SOURCE: Office of Technology Assessment.

Therefore, the number of U.S. RVs that could survive and penetrate is given by the expression:

$$\frac{\text{(U.S. ICBM RVs)} \times \text{(Soviet countersilo RVs intercepted)}}{\text{(Soviet RVs that attack U.S. ICBM silos)}} + \text{(U.S. SLBM RVs at sea)} - \text{(U.S. RVs destroyed by Soviet defense)}$$

The expression is independent of how the Soviet defense is operated. If the U.S. defense were not completely preferential, fewer RVs would survive and penetrate.

A specific example can illuminate the meaning of this expression. If the Soviets attacked our 1,000 undefended ICBM silos with 5,000 SS-18 and SS-19 ICBM RVs, they would probably destroy almost all of them. We could retaliate with our surviving SLBM RVs, perhaps 4,500. In the absence of Soviet BMD, almost all of these would reach their targets. If both sides had BMD capable of destroying 1,250 RVs, our preferential defense could protect 250 silos from destruction. Assuming we had a total of 2,800 ICBM RVs, 700 ICBM RVs would survive in addition to the 4,500 SLBM RVs.[24] However, of the 5,200 RVs surviving, the Soviet defense would destroy 1,250, and only 3,950 would reach their targets. Thus in this case, equal defenses on both sides would increase the number of U.S. RVs surviving the attack, but reduce the number that survive and penetrate to their targets. This is indicative of a general trend.

As long as the number of RVs the Soviets attack our silos with exceeds the number of U.S. ICBM RVs—which is quite possible with today's forces—adding defense to both sides in equal increments will decrease the number of U.S. RVs that survive and penetrate the Soviet defense.[25] As the expression shows, under these circumstances the number of surviving RVs added by the U.S. defense (the top line of the expression) will be less than the number of Soviet RVs the U.S. defense can destroy, and therefore less than the number of RVs lost to an equal Soviet defense. Thus, while defending U.S. ICBMs and other assets may aid deterrence, for example by increasing the uncertainties the Soviets face in planning an attack, it may well decrease our available retaliatory force if the Soviets also have defense.

Figure 5-2 shows the number of surviving and penetrating U.S. RVs for three cases: U.S. and Soviet defenses equal; U.S. defense double the Soviet defense; and Soviet defense double the U.S. defense. The figure assumes the replacement of 100 Minuteman IIIs with MX, for a total of 2,800 U.S. ICBM RVs. The Soviets attack with current SS-18 and SS-19 missiles, about 5,000 RVs. If the Soviet defense equals the U.S. defense in number of RVs it can destroy, larger defenses mean fewer U.S. RVs penetrating to their targets. This problem might be redressed by a defense asymmetry favoring the United States. However, in this case it would require a U.S. advantage of approximately two to one. Similarly, a major Soviet advantage in defense could result in large reductions in U.S. ballistic missile retaliatory capability. Unless the U.S. defenses could be operated completely preferentially, the number of surviving and penetrat-

Figure 5-2.—How Ballistic Missile Defense Affects U.S. Ballistic Missile Retaliatory Capability

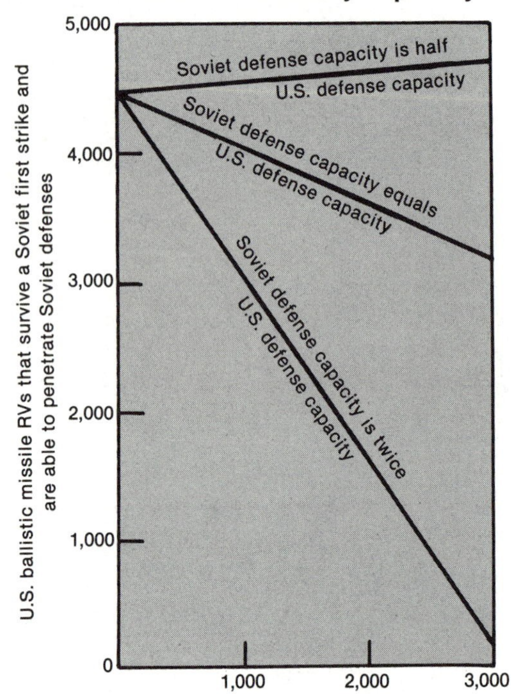

Assumptions:
- 5,000 Soviet RVs shoot at U.S. ICBMs carrying 2,800 RVs
- U.S. has 4,500 SLBM RVs at sea[1]
- U.S. defends completely preferentially
- Not shown: U.S. bomber forces—survivability will be affected by U.S. BMD and Soviet air defense

[1] *Modernizing U.S. Strategic Offensive Forces,* Congressional Budget Office, May 1983.

---

[24]We could have 2,800 RVs by replacing 100 Minuteman III missiles with 100 MX. The force would be:

Missile Type	*Number*	*RV/missile*	*Total RVs*
MX	100	10	1,000
Minuteman III	450	3	1,350
Minuteman II	450	1	450
Total	1,000		2,800

It is likely that under these circumstances the United States would defend MX silos first, defend MMIII only after all MX had been defended, and defend MMII only after all MX and MMIII had been defended. This would produce considerably more than 700 RVs surviving if the Soviets attacked all silos uniformly. However, it is also likely that the Soviets would anticipate that the United States would allocate its defenses in this manner and would allocate its attack accordingly.

[25]The OTA staff is indebted to Glenn Kent of RAND for calling this to our attention.

ing U.S. RVs would be less than the number shown in the figure.[26]

Any reduction in U.S. ballistic missile retaliatory capability would have to be evaluated within the context of the total U.S. retaliatory force, including air-breathing weapons, and what it could accomplish. Opinions vary widely about the significance of various size reductions in numbers of retaliatory RVs for the U.S. deterrent and the ability to respond to a first strike.

Whether one believes that having defenses on both sides—when the result would be a reduction in the number of surviving and penetrating U.S. RVs—would aid deterrence, detract from deterrence, or have little or no effect on deterrence will depend on certain underlying attitudes and assumptions, as discussed below (see pp. 111-112). For example, some believe that the assets that the U.S. defense might protect—including but not limited to ICBM RVs—would be much more valuable than the RVs lost to Soviet defenses. Others believe that only a few surviving and penetrating RVs are sufficient to deter and that additional retaliatory forces beyond that small number add little to our deterrent. Neither of these groups would be likely to view the reduction in U.S. retaliatory capability as being significant for deterrence.

Levels of Strategic Defense Capability

This chapter specifies strategic defense in terms of its "net defense capability." Net defense capability is what the defense can do, taking into account all its characteristics as well as those of the other side's offense. It will depend on a number of factors, including the opponent's offense, the components of the defense, and the basic mode of operation of the defense. In many cases, the same net defense capability can be arrived at in several ways. Since we are dealing with a time in the indefinite future for which we can predict neither offenses nor defenses with any certainty, we do not specify the architecture of the defenses or address the feasibility of obtaining them.

Ballistic missile defense alone could not provide a complete strategic defense of either the United States or our NATO Allies. Weapons other than ballistic missiles are part of the threat. Furthermore, passive defense techniques—e.g., civil defense—are potentially available either to augment active defenses (i.e., BMD and air defense) or to provide alternative means to the same defensive goals. This section is a general discussion of strategic defense. However, this report, like the current national debate, focuses primarily on BMD and on defense of the United States against ICBMs and SLBMs.

Drawing on the considerations discussed in the preceding section, we can identify five levels of protection against nuclear weapons to aid in understanding the implications of U.S. and Soviet defenses. These are listed in table 5-1. These are not absolute levels of defense, but rather net defense capability. The defensive system that the United States requires to achieve level 1, for example, may be larger or more capable than the defensive system the Soviets would need to achieve the same level. Given the imbalance in ICBMs, this would certainly be true today. Furthermore, the defense required to achieve a given level can change as the offenses change. The defense that the United States requires to achieve level 1 in 1995 may be very different from the defense required to achieve level 1 in 2015. Negotiated reductions of offensive forces could raise the defense level without changes in the defense systems. Increases or qualitative improvements in the offense could lower the defense level.

In the *offense-dominated region*, the strategic relationship would remain basically as it is today. Although by adding defense we might make it more difficult for the Soviets to attack our military assets, the addition of defense

[26] Under these same conditions—Soviet ICBM RVs outnumber U.S. ICBM RVs—adding equal defenses on both sides would increase the number of Soviet ICBM RVs that would survive a U.S. first strike and penetrate U.S. defenses. Put another way, it would decrease the ability of the United States to limit damage to itself by a first strike. OTA is not suggesting that the United States has a first strike posture, or that we should develop one.

Table 5-1.—Levels of Defense Capability

Region	Level	Description
Offense-dominated	0 no defense	
	1 "some ICBMs"*	A defense capable of ensuring the survival of a useful fraction of the ICBMs, but not capable of protecting cities
Transition	2 "either/or"	A defense (including BMD) that can ensure the survival of most ICBMs or a high degree of urban survival against a follow-on (or simultaneous) attack, but not both
Defense-dominated	3 "most ICBMs/some cities"	A defense that ensures a high level of survival of military targets. Massive damage can only be obtained by concentrating the entire offense against cities
	4 "extremely capable"	Ensures a high level of urban survival against a full attack. The attacker cannot have high confidence that any cities can be destroyed

*Terms in quotes are a shorthand used to identify the levels.

NOTE: For simplicity the chapter often divides targets into ICBMs and cities. There are, of course, many other types of targets that might be attacked, but discussing them all in each case would greatly expand the text. ICBMs are representative of strategic military targets (although by no means an accurate model of them all). "Cities" is typically used as a short hand for people, economic assets, and social structure. A level 1 defense, for example, might be used to defend the C^3 system rather than the ICBMs.

could preclude neither a militarily useful strike nor the destruction of our cities. Similarly, the Soviets would still know that we could absorb a first strike and be able to devastate them. Thus in this region, the offenses (including retaliatory capability) dominate the strategic balance. In an offense-dominated situation, the value of strategic defenses to the United States would have to be judged on the basis of how well they supported our ability to absorb a first strike and retaliate (or supported Soviet perceptions that they could not prevent us from doing so), weighing the effects of our defenses against the effects of any Soviet defenses. In deciding whether defenses are worthwhile in an offense-dominated posture—other than as a part of the transition to higher levels of defense—it would be necessary to weigh whatever they might contribute to our retaliatory capability against the cost of building the defense. It would also be important to compare the effort to build such a defense with alternative ways to achieve survivability of our deterrent.

In the *defense-dominated region*, defenses would severely limit the ability to use offenses. At level 3, the probability that the attacker could cause any useful level of damage to military targets would be so small that he would be limited to attacking cities. He could not hope to use his offensive forces to reduce the other's ability to retaliate. For a level 3 defense, air defense would certainly be needed in addition to BMD. At level 4, the defender would approach a condition of assured survival, but widespread civil defense would almost certainly have to play a prominent role along with BMD and air defenses.

If one side had level 3 or level 4 defenses and the other had no defense or very little defense, the side with the heavy defense could have a very significant advantage. It could attack the other and do a very good job of defending against any retaliatory attack. The level 3 defense, which could not preclude major urban damage from a full-scale strike, might be able to defend almost completely against a retaliatory strike by a force that had been significantly reduced by a preemptive strike. From the perspective of the weaker side this could be a very dangerous situation: its ability to deter an attack by the other could be seriously in doubt. The stronger side might have the forces to adopt a prevailing strategy. However, some observers believe that the probability of completely defending against the retaliatory strike would have to be very high before the stronger side could be said to have an exploitable advantage.[27]

Because of all the uncertainties in predicting the outcome of a nuclear attack, it would be difficult for the defender to know with great confidence that he had indeed achieved a level 4 defense. A small number of weapons leaking through the defense would spell the difference between assured survival and widespread destruction. Two different sublevels—4A and 4B—can be identified. At level 4A the attacker has only low confidence that his strike against the defender's cities will cause unacceptable destruction. He would arrive at this assessment by making "offense-conservative" assumptions, giving the defense the benefit of the doubt. At this level, the low prospect of success would contribute to deterrence, but the defender—making "defense conservative" assumptions—would probably want to maintain his retaliatory threat. If he did not, the attacker would have little to lose by attacking despite his low expectation of success. At level 4B, the defender would be confident that his criteria for "assured survival" were met. He could then abandon his retaliatory threat. That level of confidence could probably only be achieved with a defense system believed effective even under the most conservative assumptions about enemy offenses.

The *transition region* encompasses those situations in which neither the offenses nor the defenses clearly dominate. The attacker would be much less confident of accomplishing most of his attack goals than he would be when offenses dominate, and the defender would be much less confident in his ability to deny the attacker major attack goals than he would be when defenses dominate.

[27]Some strongly disagree with the assessment that this situation would provide the attacker with an exploitable advantage. They argue that unless the defense were perfect, the attacker could not be sure that no nuclear weapons would reach his territory. The possibility that he might suffer some retaliation would still be a powerful deterrent.

A level 2 defense might operate semi-preferentially. The defender could choose to use this defense to defend his military assets or his cities, and the attacker would not know its allocation in advance. Other types of defenses, such as short-range fixed terminal defenses, lack this flexibility and could not be used to produce a level 2 defense. Building these other types of defenses might avoid the problems of the transition region. However, building a dominant defense would be more difficult with a defense that could not react flexibly to an attack.

With a level 2 defense the defender may be able to protect a great number of his assets, but he cannot come close to protecting them all from a full strike. In particular, he might save most of his ICBMs from destruction, but he could not protect his cities at the same time. Alternatively, he might sacrifice his ICBMs while gaining a high degree of urban survival, providing the attack on his cities was sufficiently limited. (The attacker might limit his attack on cities in order to increase his chances of destroying military targets he thought to be defended.) The defender might, however, do a good job of protecting against a strike conducted with much less than the full offensive force, such as a retaliatory strike with a force that had been seriously reduced by a first strike. *Perhaps the most important characteristic of the transition region is that if one side has a level 2 defense—and especially if both do—there is a wide range of possible outcomes of a nuclear exchange. Furthermore, the outcome would be especially difficult to predict in advance because it would depend on how each side chose to allocate its weapons, and each would allocate its weapons based at least in part on how it thought the other would allocate.*[28]

[28]The outcome would depend on: the mode of operation of the defense; whether the defense were limited in capacity or engagement rate; whether the attacker elected to attack all at once, or to attack only military targets in a first strike and keep reserve forces to threaten cities with a follow-on strike; how much defense the defender allocated to cities and how much he allocated to military assets; how the defense was distributed among military targets; how good the attacker's and defender's intelligence estimates were regarding each other's capabilities and plans; how the attacker distributed his attack among and within target sets; etc.

Achieving a level 2 defense would probably require defenses against air-breathing weapons as well as BMD. We would not be able to defend our cities against a follow-on attack if we could not defend against bomber forces.

If we had a level 2 defense and we knew how the Soviets planned to attack, we might deny them success in destroying some target sets, but they would be successful against others. Even if we didn't know their plans, we might guess correctly enough to accomplish this. If they knew our defense plans, they could guarantee success against some target sets, but others would survive. If neither knew what the other intended to do, it would be very difficult to predict the result, except in the situation in which the Soviets concentrated their attack very heavily against a limited number of targets in order to ensure that they killed those targets while conceding the survival of the rest.

This situation could introduce great uncertainty into Soviet attack planning, and that uncertainty ought to enhance deterrence. If the Soviets kept a large reserve in order to be able to threaten our cities, they would have to reduce their attack on our forces. However, if we used our defenses to protect our forces, their counterforce strike might accomplish nothing. On the other hand, if they attacked our forces very heavily but we reserved our defenses to defend our cities, they might find themselves without a credible threat to our cities.

If the Soviets had a level 2 defense, it might appear to us to be part of a first strike posture, because if they could significantly reduce our forces by a first strike, they might be able to achieve high, or even total, success in defending against our "ragged" retaliation. Clearly, we would be alarmed by this prospect.[29] (Indeed, we would be alarmed by indications that the Soviets *thought* they had such a capability, even if we thought that in reality they did not.) Similarly, if the United States had a level 2 defense, the Soviets might suspect the United States of seeking a first strike posture. Whether or not we intended ever to strike first, this situation would be a very uncomfortable one for the Soviets.

If we and the Soviets both had level 2 defenses, our defenses might reduce the confidence of the Soviets that they could in fact successfully strike first, but we could not necessarily preclude it. Because of the uncertainties each side faced in planning its offense and defense, **the broad range of possible outcomes might well include a successful Soviet first strike.**

The Effect of U.S. and Soviet Defense Levels on the Strategic Balance

Figure 5-3 illustrates schematically how different levels of U.S. and Soviet defense capability might affect the strategic balance.

Figure 5-3.—How Strategic Defense Might Affect the Strategic Balance

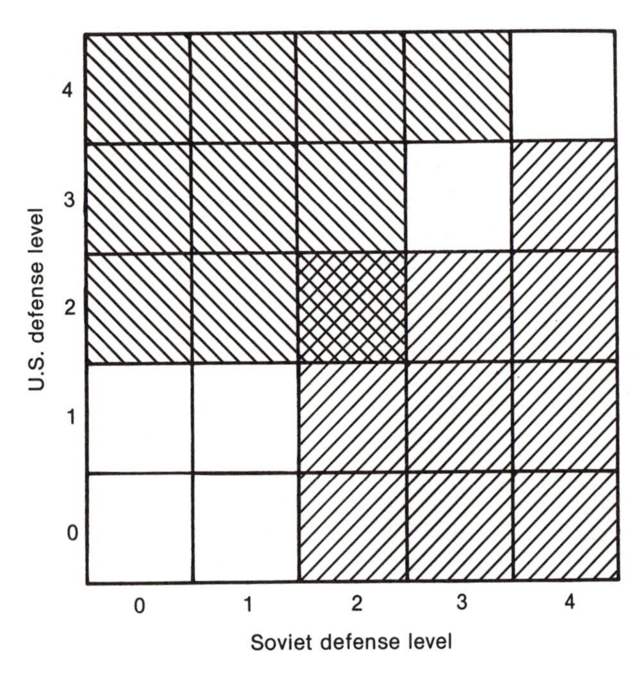

Possible U.S. exploitable advantage 

Possible Soviet exploitable advantage

SOURCE: Office of Technology Assessment.

[29] Some believe that the Soviets could not be highly confident in their ability to do this, and would be effectively deterred.

If one side had a level 4 capability and the other had a lesser defense, the side with level 4 might have a clear advantage. It could cause heavy damage to the other, which in turn could do very little in retaliation. The greater the disparity, the more the options the stronger side would have. If one side had a level 3 defense and the other had level 1 or less, the stronger side might have a strong capability to attack the other and defend successfully against the retaliatory attack. The level 3 defense, which would let enough of a full-scale counter-city attack through to cause significant damage, might be much more capable against a retaliatory strike that had been reduced by a counterforce strike. Opinions differ as to whether this would constitute an exploitable advantage. If the other had a level 2 defense, the stronger might be able to do this, but not necessarily. As discussed in the preceding section, if one side had a level 2 defense, and the other had either level 2 or less, it may be able to attack successfully. However, it would be much less sure of its ability to do so than it would be if it had a level 3 defense.

We would, of course, wish to avoid situations in which the Soviets had (or thought they had) an exploitable advantage. Both situations of approximate parity and those in which the United States would have an advantage could be acceptable, except to those who believe that the United States should strive to achieve a clear advantage over the Soviets. However, if the transition to a mutually defended world is to be a managed, cooperative one, it is unlikely that the Soviets would agree to let situations of clear U.S. advantage emerge. If the evolution were not cooperative, we should expect that the Soviets would do everything in their power to prevent a U.S. advantage. Therefore, the regions of primary interest lie along the diagonal of the square in figure 5-3, where the two sides have equal defense capability. **However, as discussed earlier, situations of equal levels of defense capability on the two sides are not necessarily situations of equal defense systems. If equivalent defense systems were added to today's offensive forces, it would result in unequal levels of defense capability.**

Offense-Dominated

The differences among the four offense-dominated situations shown in figure 5-3 (the United States having level 0 or 1 and the Soviets having level 0 or 1) are a matter of some controversy. The root of that controversy is found in differing assessments of the current situation and its evolution in the absence of BMD deployments.

Some believe that the current situation is acceptable, and likely to remain so—or improve—in the foreseeable future. Others think that the current situation is acceptable, but that the trends are adverse. In this view, sooner or later the strategic balance could become dangerously disadvantageous to the United States. Still others believe that the trends are disadvantageous and the balance has already tilted against us. These groups will differ in their assessments of the four offense-dominated situations.[30]

Those who believe that the current strategic balance is acceptable and likely to remain so believe that the Soviets know it is highly likely that were they to attack, thousands of U.S. nuclear weapons would survive and would be launched back at them. In this view **the damage that those thousands of weapons could do would be so overwhelming that hundreds or even a few thousand more or less would make little difference. This damage would far outweigh anything the Soviets might hope to gain by attacking.** Those holding this position see nothing in the future that would erode this situation, and some developments that reinforce it.

Those who see the current situation as eroding, point to Soviet developments that increase their ability to destroy our forces and our ability to use them, as well as active and passive defensive measures that decrease the effectiveness of our weapons against important targets. Some state that the Soviet leadership has a different value system than the United

[30]Yet another group finds the entire situation in which our security rests on a threat of retaliation to be unacceptable, and therefore may care little about changes that affect our ability to retaliate.

States does and therefore would not be as strongly deterred by certain threats as would the United States. Some fear that the Soviets would expect a U.S. President, faced with a threat to more than half the U.S. population from a Soviet follow-on strike, to be deterred from responding to a first strike.

From this perspective, a Soviet first strike might destroy most of our ICBMs, bombers and tanker aircraft, and all submarines in port. Soviet anti-submarine warfare before and after the strike might destroy some of the submarines at sea. The strike might also destroy most of our leadership and its strategic communication system, making a coordinated response impossible. Most important would be this decapitation and the loss of the ICBMs which are uniquely prompt and capable of killing hardened targets. In this view, Soviet defensive measures—including civil defense and similar passive defenses—would effectively protect their leadership and its communication system, as well as reduce civilian casualties and protect important war-related industrial capacity. These defenses would raise the price of attacking important targets to such a high level that our remaining forces would be incapable of covering the intended target set. Some who hold this position see current Soviet offensive and defensive developments as exacerbating the situation to an alarming extent.

If this is the case, a level 1 defense could perform a valuable function for the United States even if the strategic balance remained strongly offense-dominated. If used to defend our command, control, and communication system (C^3) it might reduce or eliminate the Soviet ability to decapitate the United States. It could contribute to the survival of ICBMs and bombers. Denying the Soviets high confidence that they could decapitate the United States and eliminate the ICBMs would more than compensate for the decrease in the ability of U.S. weapons to reach their targets due to similar Soviet defenses. (However, if the Soviets had "level 1" defenses and we did not, it would make a bad situation that much worse). Some of those holding this point of view argue that a combination of limited active defense and strong passive measures (such as mobile ICBM basing) could make it more or less impossible for the Soviets to achieve any militarily significant goal with a first strike. In effect, a level 3 defense might result from strong passive measures and the synergism between active and passive defenses. They see this as strongly enhancing deterrence by eliminating a major incentive for the Soviets to strike.

Those who believe the current balance is acceptable concede that a Soviet first strike would reduce the number of weapons the United States could retaliate with. But they think that the Soviets could only have very low confidence in their ability to decapitate the United States. In this view, neither United States nor Soviet defenses would make much difference as long as the balance remained offense-dominated. Some also point out that unless the structure of U.S. and Soviet offensive forces changes in a major way, deploying similar defenses on both sides is likely to result only in a decrease in U.S. retaliatory capability, which they see as reducing deterrence.

Whether or not those holding these two points of view would agree that building level 1 defenses would be worthwhile as a step to more capable defense, they disagree fundamentally over the value of having a level 1 defense. Those who see the current strategic balance as unsatisfactory or as eroding in dangerous ways see level 1 defenses as enhancing deterrence. Those who see the current strategic balance as satisfactory fear that level 1 defenses on both sides could harm our deterrent posture.

If one believes that having a level 1 defense would be useful, two other important questions need to be addressed. First, how much would the level 1 defense be worth? Second, are there less costly—or otherwise more attractive—ways to achieve the same benefit? For example, would passive measures suffice?

Defense-Dominated

If both sides had level 3 defenses, each would have an assured retaliatory capability.

The strategic relationship would be one often referred to as "mutual assured destruction." Each side would have the ability to inflict widespread damage on the other, but could not prohibit the other from doing so in return. Nuclear weapons would pose the ultimate threat against populations and societies, but would have little or no use as a military tool. If both had level 4 defenses, the strategic balance would approach one of "mutual assured survival."

Whether a condition of mutual assured destruction is desirable, and whether that condition would differ from the present situation (or the future in the absence of strategic defense) are both issues of contention. A related issue is whether or not passive measures that increase the survivability of important assets could bring this situation about without investing heavily in BMD and air defense. As discussed above, most observers believe that the United States now has the ability to assure destruction of the U.S.S.R. in retaliation for a strike on the United States, and many believe it is likely to retain that capability into the future. Others believe the opposite.

Few would argue that assured survival is not preferable to assured destruction. The issue is whether assured survival is attainable by the technological approaches being pursued under SDI. The basic question the SDI program is supposed to answer is: how capable a strategic defense can we produce and what would it cost to get that defense? Answers to these questions do not as yet exist, and probably will not for a number of years. Later chapters discuss the types of BMD system capabilities that might be used to support assured survival.

It is difficult to define specifically what assured survival is. Some would argue that survival is assured only if the probability that one or more weapons will reach the United States is very low (or, alternatively, that there is high confidence that no weapons can be expected to penetrate the defense). Others argue that survival is assured if society survives and the economy recovers in some number of years. In this case, the United States might survive despite the detonation of tens of weapons. (Appendix D illustrates how urban destruction might be related to the effectiveness of strategic defense.)

If our defenses could keep the probability very low that even one nuclear weapon would reach the United States, our security would be largely independent of the level of Soviet defense. If we could expect tens of weapons to reach the United States, the level of damage we could inflict on the U.S.S.R. would be relevant. If we could "survive" that level of destruction, but the U.S.S.R. could go undamaged, they might have a significant political advantage.

Assured survival would probably be impossible to achieve if the Soviets were determined to deny it to us. By improving or adding to their offense, they could increase the number of weapons penetrating to the United States, forcing us to increase our defense, and so on. Another basic problem would be the difficulty of knowing with high confidence how well our defense would actually perform against their offense, since it could never be tested and we could never know in great detail the working of their offensive weapons. Many who advocate assured survival envisage it being achieved by agreements that limit offensive levels far below defensive capabilities.

The Problem of Transition

Both the offense-dominated region and the defense-dominated region are regions of crisis stability. Neither side would have the ability to damage the other with a first strike and defend completely against the retaliatory strike. Therefore, neither has an incentive either to try it or to take action to prevent the other from doing so.[31] However, in order to reach the defense-dominated region, the strategic balance is likely to pass through the transition

[31]Some maintain that in the absence of U.S. defenses the Soviets might be able to strike the United States in such a way that the United States would be either unable or unwilling to respond, despite the fact that the Soviets could not prevent a large number of weapons from penetrating if we did retaliate.

region in which each side *may* have the capability to strike the other and defend completely (or nearly so) against the retaliatory strike. In this region there would be great uncertainty in predicting the outcome of a nuclear exchange, because it would depend strongly on how each side allocated its offense and defense. Each could be very mistaken in its assessment of how the other would make its allocations; therefore each could have a very different assessment of the outcome of an exchange.

While we would not necessarily fare worse in a nuclear exchange under these circumstances than if there were no defenses—indeed we might fare considerably better—we *might* fare worse. The Soviets *might* be able to strike first and defend completely against our retaliatory strike. In one view, the expanded uncertainty in the minds of the Soviets regarding the outcome of a nuclear exchange would aid deterrence. In another view, the possibility that the Soviets could strike first and suffer no damage, a possibility that does not exist if the offenses dominate, would undermine deterrence.[32] The knowledge by each side that the other might be able to strike and suffer no retaliation has important implications for stability.

The problem of passing through this transition region has been described as follows:

> A third potential source of instability could arise during that phase of a transition when strategic defenses would be capable of effective area defense against an offensive threat that had been degraded by a previous first strike. Assuming that a comprehensive defense cannot spring forth fully formed, like Athena from the head of Zeus, both superpowers are likely to pass through such a transitional phase unless precautions are taken well in advance. The possibility that they might pass through such a phase (roughly) simultaneously makes this situation potentially even more dangerous. The premium for striking first, and the penalty for waiting, could be powerfully destabilizing factors—particularly during an acute crisis.[33]

Precautions to avoid instability might include measures other than active defense that could reduce the ability of both sides to launch a first strike and defend successfully against the retaliation. Passive defense to reduce the effectiveness of a first strike might be one such measure. A shift in both sides' arsenals to much greater emphasis on air-breathing weapons might be another. Because of their slow speed, bombers and cruise missiles pose less of a massive first strike threat than ballistic missiles do, provided they can be detected when they are still far from their targets. Longer warning time provides more time to get bombers safely aloft and to launch a retaliatory strike.

Some implications of this transition problem are explored more fully in chapter 6.

The Effect of U.S. and Soviet Defenses on U.S. Strategy Choices

In chapter 4 we discussed at some length both our present countervailing strategy, and three suggested alternatives—a "retaliation-only" strategy, a "prevailing" strategy, and a strategy based on defense dominance. If the defense were to dominate, two strategies would be possible. If our defense were extremely capable, we may be able to adopt an "assured survival" strategy. Otherwise, defense dominance could enforce a "retaliation-only" strategy by limiting any strike to urban targets only. This section discusses which strategies are available to us for the various combinations of U.S. and Soviet defense capability that are shown in figure 5-3.

A retaliation strategy would require that some number of U.S. weapons survive a Soviet first strike and penetrate to their targets. It is beyond the scope of this report to calculate the number required, and indeed advocates of this strategy differ on the retaliatory

[32] Some believe that while the United States currently has the weapons to retaliate for a first strike, the Soviets may believe that the United States lacks the will to retaliate.

[33] Keith B. Payne, "Strategic Defenses and Stability," *Orbis*, summer 1984, p. 217.

capability required to make the threat of retaliation credible. If we can absorb the first strike and inflict great damage on the Soviets, we can have a retaliation strategy.

Countervailing would require more capability than retaliation only. In order to countervail, we must be able to execute a strike that will deny the Soviets their goals, or inflict damage beyond the value of whatever the Soviets might hope to gain. Countervailing would require more surviving and penetrating RVs than retaliating would, and it would probably require more specific capabilities to deliver those RVs against military targets. In the absence of Soviet defenses, neither countervailing nor retaliation would require U.S. strategic defense. However, U.S. defense might contribute to the extent that it helped assure that a sufficient number of weapons could survive. On the other hand, neither of these strategies would be viable if Soviet defenses could prevent the required number of RVs from surviving and penetrating to their targets.

In order to prevail, we would have to be able to defeat the Soviets while keeping our losses at a "tolerable" level. Prevailing would require even greater capability in the force that survives and penetrates than countervailing would. Perhaps more significantly, it would also require a substantial defense of the United States in order to keep our losses "tolerable." The conditions for adopting a prevailing strategy are probably the most stringent, since it would require both that we achieve a high level of protection against Soviet attack and that we have substantial capability to penetrate Soviet defense. Conditions that would support a prevailing strategy would also support a countervailing or retaliation-only strategy. Conditions that would support a countervailing strategy would also support a retaliation-only strategy.

An assured survival strategy would require an even higher level of U.S. defense capability than prevailing would. An assured survival strategy would not require a retaliatory capability, so it could tolerate Soviet defenses that kept the number of penetrating U.S. RVs very low.

Each of these strategies would generate requirements for the capabilities of U.S. offensive forces and (in the case of prevailing and assured survival) for limits on the amount of damage the Soviets could inflict on the United States. These, in turn, would be determined at least in part by U.S. and Soviet defense capabilities. Thus, which strategies the United States could adopt are dictated by the defense levels on both sides. For example, a Soviet defense that prevented attack of military targets (i.e., level 3 or 4) would generally limit the United States to a retaliation-only strategy. Similarly, with a level 1 defense we could not limit the damage to ourselves to a "tolerable" level, and therefore could not have a prevailing strategy. Figure 5-4 shows which strategy choices would be permitted by various combinations of U.S. and Soviet defenses. Appendix E explains how this figure was generated.

From this figure, we can make the following observations:

Figure 5-4.—How Strategic Defense Might Affect U.S. Strategy Options

U.S. strategy options

U.S. defense level \ Soviet defense level	0	1	2	3	4
4	A P	A P	A P	A P?[a]	A R
3	P[b]	P[b]	C	R	R?[c]
2	P?[b]	C	?	R	R?[c]
1	C	C	R	R	R?[c]
0	C	R	R	R	R?[c]

A - Possible assured survival
P - Option for prevailing, countervailing, or retaliation-only
C - Option for countervailing or retaliation-only
R - Retaliation-only
? - Unclear

[a] U.S. has a possibly large advantage, but little capability for attack of military targets.
[b] Option for prevailing only if U.S. strikes first.
[c] Ability to retaliate is in question.

SOURCE: Office of Technology Assessment.

For the United States to have the option of a prevailing strategy, we would need a level 3 or 4 defense while the Soviets had a substantially less capable defense.[34] If we had a level 4 defense we could prevail even if they struck first. However, if we had a level 4 defense it seems highly unlikely that they would strike. If we had a level 3 defense we could prevail only if we struck first; if the Soviets struck first our level 3 defense would not limit damage to a tolerable level, but if we struck first it would. Developing the option for a prevailing strategy requires that our defense deployments substantially outpace Soviet offense deployments, and that our offense deployments limit the capability of Soviet defenses.

For the United States to have a countervailing strategy, as we now do, the Soviet defense must be at level 2 or less. Enough Soviet defense precludes the existing U.S. strategy regardless of U.S. defense. This is one reason that BMD is opposed by some who see the existing strategy as the least dangerous option. Furthermore, countervailing requires that the U.S. defense level not be less than the Soviet defense level. We cannot fall substantially behind in an offense/defense arms race and maintain our current strategy.

A retaliation-only strategy is always possible unless the Soviets have level 4 defense. However, if the Soviets were to reach level 3 and we did not, we would be threatening retaliation from a position of inferiority. The combination of a Soviet first strike and the Soviet defense might limit our retaliation to a very low level (or even preclude it), but our defense would not be sufficient to keep the Soviets from inflicting great damage on the United States.

[34]Or possibly a level 2 defense if the Soviets have no defense.

CONCLUSION

Opinions will differ over whether the levels of defense capability discussed in this chapter are worth striving for. Other important factors will also influence decisions on the value and desirability of attempting to reach these defense levels. As the discussion of the transition region pointed out, crisis stability will be an important issue, as will the problem of engineering a cooperative negotiated transition. Since some defense deployments can provide incentives to compete as well as incentives to cooperate, arms race stability will also be an issue. Finally, cost and feasibility must be taken into account. These subjects are addressed in subsequent chapters.

Chapter 6

Crisis Stability, Arms Race Stability, and Arms Control Issues

Contents

Chapter 6

Crisis Stability, Arms Race Stability, and Arms Control Issues

INTRODUCTION

The preceding chapters discussed how adding ballistic missile defense to U.S. forces might affect U.S. strategy. This chapter will address the relation of BMD deployments to three other force posture issues: crisis stability, arms race stability, and arms control. *Crisis stability* is the degree to which strategic force characteristics might, in a crisis situation, reduce incentives to initiate the use of nuclear weapons. *Arms race stability* involves the effect of planned deployments on the scope and pace of the arms race. *Arms control* has been pursued in the past as a way of trying to enhance these two kinds of strategic stability. If the United States and the Soviet Union decide in the future to deploy new BMD systems, new arms control agreements may be even more important for avoiding serious instabilities, particulary during transitional stages.

CRISIS STABILITY

It is widely believed that a nuclear war would be most likely to occur as the result of escalation of a U.S.-Soviet confrontation during a severe crisis. Such a crisis could result from a deliberate act of aggression by the Soviet Union against the United States or its allies, but it could also arise from a dispute triggered by some third-country actions which involve the perceived vital interests of the superpowers. The likelihood that such a crisis would result in nuclear war not only would depend on the political and military situation at the time, but might also be influenced by the nature of the strategic forces deployed beforehand by each side. In addition, crisis instability can also motivate arms race instability by inducing remedial arms acquisitions by one side or the other. Hence, in deciding whether to develop and deploy a new weapon system, an important question is whether the new system will add incentives or disincentives for using nuclear weapons in a crisis. Before we address this question with respect to various kinds of BMD deployments, the general nature of crisis stability will be described.[1]

A decision to initiate a nuclear attack would depend on several factors, including the circumstances leading up to the crisis, the personal attributes of the leaders, their perception of each country's military capabilities and vulnerabilities, their perception of their adversary's incentives and intentions, and the doctrines of the two countries regarding nuclear strategy. Most specialists believe these doctrines differ between the two countries in important ways. For example, the U.S. contingency plans for first use of nuclear weapons contemplate a possible "flexible response" to Soviet aggression; i.e., a relatively small-scale initial use of nuclear weapons with the hope of avoiding escalation to a large-scale nuclear

[1]See app. L for a list of references on crisis stability and other aspects of strategic nuclear policy. App. M lists references to a range of views on Soviet strategic policy.

exchange.[2] As noted in chapter 4, the declared policy of the United States precludes a preemptive strike.[3] The issue here, however, is not whether American leaders would continue this policy in future crises, but whether the Soviets would believe that they would.

On the other hand, Soviet doctrines for dealing with crisis contingencies are thought by many analysts to include the option of launching a massive preemptive attack against all targetable U.S. nuclear forces (ICBM silos, bomber bases, command and control sites, etc.).[4] The Soviets also place greater reliance than we do on civil defense and air defense to help reduce the damage from a nuclear attack. The Soviets have declared that they will not be the first to use nuclear weapons. But it remains possible that, faced with the prospect of defeat in a nonnuclear conflict they consider of vital importance, the Soviets would decide to initiate a limited nuclear attack. However, if they believed that the escalation process was likely to lead to a full-scale U.S. attack, they might decide to preempt with a massive strategic attack.

Whatever the current Soviet doctrine really is, future crises could face Soviet leaders with decisions on whether to initiate a nuclear attack. In each case, the Soviet leader would have to balance his perception of the risks of striking first against his perception of the risks that the United States might strike first. The smaller he judged the chances of avoiding nuclear war altogether, and the larger he judged the advantages of striking first rather than second, the more incentive he would have to strike first. Hence crisis stability can be increased by force structures that minimize the difference in the results of striking first or second (e.g., by deployment of retaliatory forces that are invulnerable to a first strike). Minimizing this difference for *both* sides would reduce a Soviet leader's incentive to strike first in two ways. It would not only reduce his perception of the advantages of striking first, but would also reduce his fear that the United States had a strong incentive to strike first.

The analysis below is not intended to imply symmetry between the way American and Soviet leaders would make such decisions, nor is it intended to examine all of the factors that would be involved. It will focus on only one of those factors: how such decisions might be influenced by the nuclear force structures on each side.[5] Crisis stability is not absolute; it is a matter of degree. It is determined by how great a net disincentive for either side to strike first arises from the force structures of both sides.[6]

Weapon systems are considered destabilizing if in a crisis they would add significant incentives to initiate a nuclear attack, and particularly to attack quickly before there is much time to collect reliable information and carefully weigh all available options and their consequences. In the current U.S.-Soviet strategic relationship, crisis stability is enhanced to the extent that each side possesses substan-

[2]For a detailed discussion of how command and control vulnerabilities could severely limit U.S. options in a crisis, regardless of declaratory policies and doctrines, see Daniel Ford, *The Button: The Pentagon's Strategic Command and Control System*, (New York: Simon and Schuster, 1985). (Also published in *The New Yorker*, Apr. 1 and 8, 1985.)

[3]For example, the FY 1984 annual report of the Department of Defense states:

> Our strategy excludes the possibility that the United States would initiate war. The United States would use its military strength only in response to aggression, not to preempt it. Once an aggressor had initiated an attack, however, the principle of non-aggression would not impose a purely defensive strategy in fighting back.

Caspar W. Weinberger (*Annual Report of the Secretary of Defense to the Congress, Fiscal Year 1984*, Feb. 1, 1983, p. 33.)

[4]See discussion of Soviet strategic doctrine in ch. 4.

[5]It is quite possible that a leader's perception of the degree of crisis stability at a particular time could influence his willingness to risk actions that might cause a crisis to arise.

[6]Some analysts prefer to define strategic stability more broadly than as comprising crisis stability and arms race stability only. For example, Colin S. Gray has proposed a concept of stability which requires that Western governments acquire plausible "prospects of both defeating their enemy (on his own terms) and ensuring Western political-social survival and recovery." See Colin S. Gray, "Strategic Stability Reconsidered," *Daedalus*, fall 1970. Gray suggests that any NATO force structure short of that, such as the current force structure, may be insufficient to deter Soviet attack. He argues that a stable strategic balance is one that would permit the United States to:

> Initiate central strategic nuclear employment in expectation of gain . . . Seize and hold a position of 'escalation dominance,' [and] Deter Soviet escalation, or counterescalation, by a potent threat posed to the most vital assets of the Soviet state and by the ability of the United States to limit damage to itself.

Obviously this concept precludes *mutual* U.S. and Soviet strategic deterrence, which Gray refers to as "strategic stalemate."

tial retaliatory forces that are invulnerable to a first strike. Specifically, the retaliatory weapons and their associated command and control chain must be survivable, and the weapons must be able to reach their targets. On the other hand, weapon systems with a substantial capability to attack the other side's retaliatory forces, such as large numbers of highly accurate MIRVed ICBMs, detract from crisis stability.

There are different views regarding the applicability of the above analysis to future Soviet behavior in a crisis. It may be that in the future Soviet leaders would be sufficiently deterred from a preemptive strike if most Soviet cities, industrial facilities, and "soft" military targets remained as vulnerable to a retaliatory strike as they now are. Alternatively, it may be that a successful Soviet strike against U.S. ICBMs only would oblige the United States to choose between surrender and the mutual suicide of a U.S. second strike against Soviet urban-industrial targets followed by a Soviet "third strike" against U.S. cities and industry. Moreover, if the Soviet leaders thought a preemptive strike could destroy most of the U.S. ICBMs, and thus reduce the expected damage to such "hard" targets as Soviet missile silos and military and political command bunkers, their tools of control and power, they might decide to risk the loss of Soviet cities and strike first.[7]

It should be recognized that neither country's strategic nuclear forces are structured to maximize crisis stability, since both sides plan their forces to try to satisfy several other strategic policy objectives as well—objectives which may compete with the crisis stability objective. For example, both superpowers have developed "counterforce" capabilities, designed to reduce damage to themselves if deterrence should fail, and to provide war-fighting ability to try to limit hostilities and "prevail" in a nuclear war. Moreover, their ability to use nuclear forces serves to deter them from conventional attacks on each other, on their rival's allies, or in Third World areas susceptible to superpower confrontation. Views of strategic analysts differ on the relative importance of these competing policy objectives for each side, depending on their different assumptions as to, for example, the motivations and policies of the adversary and the feasibility of controlling the course of a nuclear war after it starts.

Moreover, force deployments are sometimes a response not so much to national strategic needs as to strong domestic political pressures to increase military budgets, develop and exploit new weapon technologies, or deploy weapon systems primarily because the adversary is doing so.[8]

The Current Situation and Future Prospects

The U.S. SLBM force is generally considered stabilizing to the extent that a Soviet leader would not think that a Soviet preemptive strike could destroy many of the U.S. SLBMs at sea and thereby prevent massive retaliation from them.[9] Conversely, to the extent that fixed-base U.S. ICBMs are perceived as relatively more vulnerable to attack, they tend to reduce crisis stability somewhat because of at least some uncertainty on each side as to the importance the other side attaches

[7]This option would presumably become less attractive as U.S. SLBM accuracy improved.

[8]For general discussions of such pressures, see: Gordon Adams, *The Iron Triangle: The Politics of Defense Contracting* (New Brunswick, NJ: Transaction Books, 1981); Andrew Cockburn, *The Threat: Inside the Soviet Military Machine* (New York: Random House, 1983); Miroslav Nincic, *The Arms Race: The Political Economy of Military Growth* (New York: Praeger Publishing, Inc., 1982); Marshall D. Shulman, "The Effect of ABM on U.S.-Soviet Relations," *ABM: An Evaluation of the Decision to Deploy an Antiballistic Missile System*, Abram Chayes and Jerome B. Wiesner (eds.) (New York: Harper & Row, 1969); Adam Yarmolinsky, "The Problem of Momentum," Ibid.; Ernest J.Yanarella, *The Missile Defense Controversy: Strategy, Technology, and Politics, 1955-1972* (Lexington, KY: University Press of Kentucky, 1977).

For discussions of the effects of such pressures on the Strategic Defense Initiative, see: William D. Hartung, et al., *The Strategic Defense Initiative: Costs, Contractors and Consequences* (New York: Council on Economic Priorities, 1985); and Fred Kaplan, "The 'Star Wars' Tent Holds Many Players," *Boston Globe*, Mar. 17, 1985.

[9]Views differ on the degree to which this Soviet perception would be affected by the possible vulnerability of the communication links between the submarines and the national command authority.

to such vulnerability.[10] As just noted, the views of U.S. commentators differ as to whether the Soviets would think the damage from U.S. ICBMs would be a significant addition to the overwhelming damage they would suffer from a full-scale SLBM retaliatory attack. This damage will extend to hard targets as well when the United States deploys its highly accurate Trident II SLBMs.

There are reasons to believe that current U.S. and Soviet strategic force structures are at least for now fairly stabilizing (although they include some elements that detract from crisis stability). Despite their considerable counterforce capabilities, each side has the ability to inflict devastating retaliatory damage after a full-scale first strike by the other side. Therefore, in a crisis neither leader would rationally perceive that the advantage in firing first outweighed the imperative to make every possible effort to avoid nuclear war altogether, and both leaders would have available the option of taking time to attempt to de-escalate the crisis.

The need to maintain adequately invulnerable retaliatory nuclear forces for decades to come is often cited by those who advocate BMD deployment to protect U.S. ICBM silos.[11] It is therefore relevant to review briefly the degree to which our current retaliatory forces are secure against attack, and the prospects for the future.

Presidential Science Advisor George Keyworth II has stated, "... our submarines, while as survivable today as ever, could well be threatened in coming years by the incredibly rapid advances we're seeing these days in data processing technologies."[12] According to President Reagan's Commission on Strategic Forces (the Scowcroft Commission), "... ballistic missile submarine forces will have a high degree of survivability for a long time."[13] (The Commission also recommended starting research on smaller submarines, each carrying fewer missiles than the Trident, as a hedge against possible Soviet progress in anti-submarine warfare.) Admiral James D. Watkins, the Chief of Naval Operations, has been quoted as follows: "... when people ask 'Aren't the oceans getting more transparent?' we say 'No way, they're getting more opaque ...' So the ability to track submarines—we don't see that as being a threat to our forces until the turn of the century or later, depending on what kind of breakthroughs we might find at the end of this decade or into the next decade."[14] According to press reports, Congress has asked the Central Intelligence Agency to carry out a comprehensive study of submarine detectability.[15]

The following testimony on this subject was given June 26, 1985, to two Senate subcommittees by Robert M. Gates, Deputy Director for Intelligence, Central Intelligence Agency:

> The Soviets still lack effective means to locate U.S. ballistic missile submarines [SSBNs] at sea. We expect them to continue to pursue vigorously all antisubmarine warfare (ASW) technologies as potential solutions to the problems of countering U.S. SSBNs and defending their own SSBNs against U.S. attack submarines. We are concerned about the energetic Soviet ASW research and technology efforts. However, we do not believe there is a realistic possibility that the Soviets will be able to deploy in the 1990s a system that

[10] There is considerable controversy as to how many U.S. ICBMs would actually survive a Soviet preemptive attack, given the inherent uncertainties in missile accuracy, missile reliability, and coordination of such an unprecedented, untested, and massive operation. See, for example, Matthew Bunn and Kosta Tsipis, "Ballistic Missile Guidance and Technical Uncertainties of Countersilo Attacks," *Report No. 9*, Program in Science and Technology for International Security, Massachusetts Institute of Technology, Cambridge, MA, August 1983; Matthew Bunn and Kosta Tsipis, "The Uncertainties of a Preemptive Nuclear Attack," *Scientific American*, November 1983; Les AuCoin, "Nailing Shut the Window of Vulnerability" *Arms Control Today*, September 1984; J. Edward Anderson, "First Strike: Myth or Reality," *Bulletin of the Atomic Scientists*, November 1981; John D. Steinbruner and Thomas M. Garwin, "Strategic Vulnerability: The Balance Between Prudence and Paranoia," *International Security*, summer 1976.

[11] Some of these advocates also attach importance to maintaining a prompt hard-target kill capability.

[12] Speech June 23, 1984, at the University of Virginia.

[13] *Report of the President's Commission on Strategic Forces*, chaired by Brent Scowcroft, Apr. 6, 1983. The recommendations in this report were endorsed by President Reagan on Apr. 19, 1983.

[14] *The Washington Post*, Mar. 22, 1985, p. A10.

[15] *The Washington Post*, June 6, 1985, p. A1.

could pose any significant threat to U.S. SSBNs on patrol.[16]

The Scowcroft Commission's report emphasizes that the U.S. secure retaliatory deterrent does not depend on our SLBMs alone, but on the synergistic capabilities of the triad of SLBMs, ICBMs, and long-range bombers.[17] For example, if the Soviets should decide to attack U.S. bomber bases and ICBM silos with simultaneous detonations, many of our bombers would have been alerted by detection of the first Soviet missile launch and would have escaped before their bases were struck.[18] If, on the other hand, the Soviets chose to launch their close-in SLBMs against our bomber bases at the same moment as they launched their ICBMs, hoping thereby to reach our bomber bases before the bombers had time to escape, we could launch our ICBMs after the bomber bases were hit but before the Soviet ICBMs could reach our ICBM silos. This would be launch *after* attack.[19] Of course, neither side can be sure that the other would not launch its ICBMs on warning that the other side's ICBMs were in flight.[20]

Soviet strategic forces currently possess considerable survivability, albeit with less redundancy than U.S. forces. The U.S.S.R. has missile-carrying submarines on sea patrol. It has such a large number of ICBM warheads that a substantial number could be expected to survive a U.S. attack on them.

[16]Unclassified prepared testimony before a joint session of the Subcommittee on Strategic and Theater Nuclear Forces of the Senate Armed Services Committee and the Defense Subcommittee of the Senate Committee on Appropriations, June 26, 1985.

[17]To further reinforce the survivability of the ICBM portion of the triad in future years, the Scowcroft Commission recommended development of a small, mobile, single-warhead ICBM. It also recommended continued modernization of the U.S. bomber and air-launched cruise missile force.

[18]Under crisis conditions, more bombers than usual would probably be in a state of alert.

[19]Views differ on the degree to which vulnerability of the U.S. ICBM command and control chain could affect this scenario. See Daniel Ford, op. cit.

[20]See Richard L. Garwin, "Launch Under Attack to Redress Minuteman Vulnerability?" *International Security*, winter 1979/80, pp. 117-139.

Effects of BMD Deployment on Crisis Stability

Whether various kinds of BMD deployment would tend to increase or decrease crisis stability depends on:

- the types and levels of BMD deployment on each side (e.g., whether the BMD is deployed to defend cities, strategic forces, or conventional forces);
- the types and levels of air defense and civil defense on each side;
- the types and levels of offensive strategic forces on each side (including those deployed in response to the defensive deployments);
- the survivability of each side's defensive and offensive systems;
- the perceptions (correct or not) of the top leaders of each side as to the capabilities of each side's offensive and defensive forces;
- the perceptions (probably very uncertain) of the top leaders of each side as to how the other side would allocate its offenses and defenses as between cities and strategic forces.

It is necessary to assess not only whether, on balance, a particular BMD deployment would do more to increase or to decrease crisis stability, but also whether the net effect of the BMD deployment on crisis stability would be significant in comparison to the effects of the offensive force structures. The analysis of crisis stability with BMD is far more complicated than is the case in the absence of BMD. *For reasons discussed below, we conclude that the net effects that various types and levels of BMD deployment would have on crisis stability are far too complex to analyze adequately within the scope of this study.*

Accordingly, the following discussion will not attempt to reach detailed net judgments. Rather, it will use the examples of BMD capability presented in chapter 5 to illustrate some ways in which certain types of BMD deployment could tend to increase or decrease crisis stability, and to indicate why a realistic analysis would have to be highly extensive and complex. As in chapter 5, we assume in these examples comparable levels of BMD capability on both sides unless stated otherwise, and we assume that the postulated level of BMD performance is technically attainable and sustainable in the face of the adversary's countermeasures and offensive augmentations. **For the time being we disregard questions of technical feasibility and cost.**

Level 1: Defense of Some ICBMs

Insofar as the vulnerability of ICBM silos or other hardened, redundant military targets is a destabilizing factor, the ability on both sides to defend some of these kinds of targets should be crisis-stabilizing.[21]

ICBMs have unique properties that some believe make them especially valuable. Currently, they are the only intercontinental-range weapons with enough accuracy to destroy hardened targets within 30 minutes (as opposed to several hours for bombers). Since they are based on national territory, they are potentially the easiest strategic weapons to maintain in an alert status and to communicate with reliably. (As other weapons evolve, these advantages may erode.) In typical analyses, it is usually assumed that a Soviet first strike would be carried out in large part to destroy as many of the U.S. strategic forces as possible, especially the ICBMs and their command chain. The U.S. Department of Defense estimates that currently part of the Soviet SS-18 ICBM force alone could destroy more than 80 percent of the U.S. ICBM silos.[22] Thus, judged *solely* by its effect on the ability of the Soviets to confidently destroy U.S. ICBMs in a first strike, U.S. BMD of missile silos could have a stabilizing effect.[23]

Views differ on how significant the stabilizing effect would be. On the one hand, those who believe that the threat of retaliation by U.S. SLBMs and bombers might not, for various reasons, deter the Soviets from attacking our ICBMs also believe that the survivability of our ICBMs is an important element in assessing crisis stability. If defenses for ICBMs also increased the potential survivability of the U.S. strategic command and control system, then the credibility of the U.S. ability to retaliate against a Soviet attack might also be somewhat increased. On the other hand, those who believe that the threat of retaliation by U.S. SLBMs and bombers would suffice to deter a Soviet attack on our ICBMs also believe that the uncertainty of success that BMD could add to deterrence of such an attack would be marginal or nil.

It must be remembered, however, that surviving U.S. forces would have to face Soviet defenses against a retaliatory attack. As noted in chapter 5, as long as the Soviets were willing and able to expend more nuclear warheads attacking our missiles than we have warheads on those missiles, the *net* effect of symmetrical defenses on both sides would be to *reduce* the total size of the potential U.S. retaliation.[24] Thus it is not clear that the uncertainties introduced by BMD into Soviet offensive planning would outweigh the fact that they could still use offenses and defenses to reduce the U.S. retaliatory potential.

Both the United States and the Soviet Union are currently taking measures other than BMD deployment to reduce their ICBM vulnerability, such as hardening silos and control bunkers and developing mobile ICBMs. Insofar as these measures are effective for the United States, Soviet offenses will have a reduced "first-strike" capability. Depending on what the ICBM survivability measures are, defenses may also then be a less significant potential element in the protection of ICBMs.

---

[21]Some argue that the ability to disrupt a Soviet missile attack on the U.S. nuclear command, control, and communications (C^3) system would greatly strengthen deterrence of a Soviet first strike. But unless that C^3 system is redundant and attack resistant (in the way that the system of 1000 Minuteman missile silos is), modest levels of BMD protection may not do much to improve its survivability.

[22]*Soviet Military Power*, U.S. Department of Defense, 1985, p. 30.

[23]A comparable effect might be achieved with a less vulnerable ICBM basing mode.

[24]This would be true *unless* Soviet defenses were strictly dedicated only to defending targets the United States would not be attacking in a retaliatory strike—i.e., empty Soviet missile silos.

Photo credit: U.S. Air Force

Artist's concept of U.S. "MX" or "Peacekeeper" ICBM to be deployed in silos now housing Minuteman missiles. The Minuteman silo will not be hardened above current levels, but better protection for the new missile will result from the new shock isolation system and the launch canister that holds the missile before launch. This mode of deployment would not appear to substantially reduce the estimated Soviet ability to destroy U.S. land-based ICBMs in a first strike. In the future, new techniques promising to make silos up to 20 to 25 times "harder" than current levels may offer more protection.

Level 2: Either/Or

(Defenses—including BMD—able to ensure the survival of most land-based ICBMs **or** a high degree of urban survival against a follow-on (or simultaneous attack), but not both.) As indicated in chapter 5, there would be a far more serious potential for crisis instability if both sides had a "Level 2" strategic defense capability. It ought to be a stabilizing factor that the Soviets would be less certain that an attack on U.S. ICBMs would succeed . On the other hand, at "Level 2" there would be at least the possibility—not previously available—that a first strike combined with defenses could keep damage from a retaliatory strike to a relatively low level. Worst of all, it is possible that both sides could arrive at a highly unstable situation in which each could have a chance of assuring its own survival by striking first, *and only by striking first*. This situation could occur even if the Soviets and the United States had approximately equivalent defensive capabilities. Under that circumstance, uncertainties on each side about the actual capabilities of the other could be especially high and could intensify mutual suspicions.

We would like to be able to discern the *net* effect on crisis stability of deploying BMD on both sides, and to identify potential areas of instability to be avoided as defensive and offensive forces evolve on both sides. This would depend partly on speculation as to how future leaders on both sides would weigh various factors when making decisons. It is possible, however, to throw some light on this important issue by assessing a large number of possible cases. That assessment would require a detailed specification of the defensive and offensive capabilities and the options they provide each side, as well as an exploration of the tactical choices each has in allocating its defense and offense under representative circumstances. In addition, because crisis stability depends so much on perception, it would be important to consider how each side might think the other would use its defense.

Level 3: Effective Defense of Most ICBMs, Some Cities

If both sides had ballistic missile and air defenses that could unconditionally deny the other side the ability to destroy most land-based ICBMs in their silos, but could not deny them the ability to destroy many of one's cities *if all the offenses were concentrated on cities*, crisis stability should be quite high. The advantages of attacking first should be marginal, the threat of retaliatory destruction still substantial.[25]

Level 4: Extremely Capable Defense

At a level of defense at which few or no military targets and few or no cities could be destroyed, there would be little incentive to strike first. An aggressor calculating that he might in some way deliver a few weapons on enemy territory might have to contend with a risk that the victim might be able to retaliate on a similar level. Striking first would probably not reduce such retaliatory capabilities. Hence crisis stability, strictly defined, would be high. But other kinds of strategic instability could arise from the possibility of nuclear weapons smuggled into U.S. cities with no assurance the the United States could retaliate against such an attack.

Special Cases

City Defense or ICBM Defense.—As shown in chapter 5, defenses that could be allocated to defending *either* retaliatory forces *or* cities would lead to a complex range of possible outcomes of a nuclear exchange. Defenses *able to defend only retaliatory forces* should be relatively stabilizing; they would not raise the prospect of a first strike against missiles followed by an effective defense against a "ragged retaliation." Defenses *able to defend only cities* but leaving retaliatory forces unprotected would be destabilizing, because they would place a premium on striking the unprotected forces, thus increasing the incentive to use those forces before they were destroyed.

The latter situation may not be purely speculative. The U.S. BMD debate has focused mostly on far-term deployments of BMD systems based on advanced technologies. However, if the Soviets were to deploy BMD, they might well elect to begin with extensive deployments of ground-based rocket interceptors of the types they have already deployed around Moscow.[26] Each interceptor deployment would be restricted to defending a definite area. Using such technology would oblige the Soviets, in peacetime, to choose among defending their cities (as the system now deployed near Moscow does), defending their ICBM silos, or defending both. If the Soviets chose to defend only cities, whatever inclinations they had before toward preemptive strategic attack could be strengthened: they would have the incentive described above to use rather than lose their ICBMs.

[25]This situation would be equivalent to one in which neither side had defenses and both sides had deployed most of their offensive nuclear forces in an invulnerable basing mode.

[26]*Soviet Military Power*, 1985, op. cit., pp. 46-48.

Asymmetric Defenses.—If the Soviet Union had BMD and air defenses that were substantially more effective than those of the United States, crisis stability would be reduced. In this case, the Soviets might calculate that by striking first, they could sufficiently penetrate U.S. defenses to weaken the U.S. retaliatory response, and then use their own BMD to deal with that response.

Conversely, a substantial U.S. advantage in BMD and air defense capability could cause the Soviets to fear that the United States is more likely to strike first. They might fear such an attack particularly if they believed U.S. defenses to be able to intercept nearly all the Soviet weapons that could survive a U.S. first strike, thus largely avoiding Soviet retaliation. Fearing this, the Soviets might calculate that a Soviet preemptive attack could possibly reduce the ultimate damage that the Soviet Union might suffer, or at least draw down U.S. defenses to the point where remaining Soviet forces could threaten a subsequent high damage attack on the United States. The latter threat, they might calculate, could also induce U.S. leaders to restrain their retaliation for the initial attack.

Alternatively, if the Soviets could be persuaded that U.S. policy would not permit a U.S. preemptive strike, whatever the apparent incentives, then a U.S. advantage in defense capabilities should contribute to stability.

BMD System Survivability.—One criterion for a BMD system which many Administration officials have cited is system *survivability*—the ability of the system to perform at desired levels despite direct attack on its components. We may take it for granted that neither side would deploy a BMD system which could obviously be rendered ineffective by enemy attack. Rather, the question would be about the degrees of confidence on each side about the *continuing* survivability of its own and the other side's defensive systems.

Ambassador Paul Nitze has said, "The technologies must produce defensive systems that are survivable; if not, the defenses would themselves be tempting targets for a first strike. This would decrease, rather than enhance, stability."[27] This point has also been stressed by other Administration spokesmen.

Whether an attack on a defensive system were part of an ICBM attack or not, it could leave the attacked side defenseless. The attacker, on the other hand, would be at least partially defended[28] against retaliation—even if the victim of attack launched ICBMs before they could be destroyed. Whether anti-BMD attacks could be prevented from escalating to attacks on silos or cities is difficult to predict.

If both sides had vulnerable BMD systems, the net result of simultaneous successful attacks on both systems could be to leave the two sides in an offensive stand-off similar to the one existing now. However, an extremely unstable situation would arise if each side's space-based BMD system were vulnerable to attack from the other's BMD system and only to that system. Each would then have powerful incentives to "use or lose" his system, to attack before the other side did. The one that struck first might substantially disarm the other side.

It is also important that the capabilities of a BMD system not be subject to degradation from an attack by ballistic missiles or airborne nuclear weapons. A nuclear first strike could be better planned, coordinated, and executed than a retaliatory strike. Even if both sides began with comparable BMD capabilities, the premium on preemptive attack would be high if a first strike had a much higher probability of penetrating enemy defenses than did the retaliation.

Automatic Command and Control.—A space-based BMD system, especially one targetted against missiles in their boost phase, would have to have some form of automated command and control if it is to respond in time to engage its targets. There are arguments

[27]Speech to the Philadelphia World Affairs Council on Feb. 20, 1985.

[28]Some believe that there would be virtually no strategic advantage in having a defense of cities that is only partially effective.

that this would be a source of instability, and other arguments that it might be stabilizing.

If the automated system malfunctioned, or if some unanticipated situation arose for which the system had not been programmed, the system could respond in a way that set off a fatal chain of action and reaction in the strategic forces on the two sides. Or, automation might be a stabilizing factor, because having an automated system forces planners to think in advance about what situations the system might have to respond to, and how they would want it to respond. Even if the system were not automated, leaders would still have to respond to the same situations in a very short time, and therefore they should have developed their responses beforehand. However, while many contingencies could be imagined and programmed into the system, there would be some practical limit on the number which would be feasible to include.

Transition Periods.—At present, when each side has thousands of offensive nuclear warheads and essentially no defenses ("offense dominance"), the mutual threat of retaliation provides a relatively high degree of crisis stability. Conversely, if each side were able to obtain virtually perfect defenses against all types of nuclear weapon delivery, there would be a very low probability that even a single nuclear weapon could reach its target, and the situation ("assured survival") would also be relatively stable. But, as the analysis of "Level 2" of BMD capability in chapter 5 suggests, the *transition* from the current situation to one of defense dominance could require passing through an interim stage which might be very unstable. Since that interim period might last for many years, there could be a serious risk that a crisis would arise during that period.

BMD DEPLOYMENT AND ARMS RACE STABILITY

The strategic nuclear force postures of the United States and the Soviet Union are shaped by both internal and external factors. The internal factors may be political, bureaucratic, economic, and technological. The chief external factor for each side is the other side's force posture, both current and forecast: the adversary's forces may present threats to counter, incentives to reduce disparities, or opportunities to seek strategic advantage. One issue to consider in deciding to deploy a weapon is what kind of reaction it is likely to evoke from the other side. If a deployment on one side is likely to lead to a responding deployment on the other side which is in turn likely to induce a still higher level of deployment on the first side, the first side's deployment might be seen as "destabilizing" the arms competition.

A destabilized arms competition might not necessarily be a bad thing for U.S. national security. For example, if we and the Soviets entered into a competition in *defensive* strategic systems (e.g., BMD) but the deployments on the two sides did not lead to *offensive* increases, the race in defensive systems might be self-stabilizing. That is, if each side could reach a high degree of protection against the other's offenses, the competition might wind down. Alternatively, we might see it to be in our advantage to begin an arms race if we were sure we would "win" at acceptable cost. That is, if superior technology, for example, could give us a permanent strategic advantage over the Soviets, we might want to engage them in a race which would give us long-lasting escalation dominance over them and might even force them into expenditures so heavy as to draw away from their conventional armed strength.

On the other hand, a destabilized strategic arms competition could prove both costly and indecisive. We could spend billions on new weapons but find that our strategic position relative to the Soviets was about the same as or worse than when we started. Moreover, the ongoing competition could lead to deployments on one side or both that reduced crisis stability as well. In general, past strategic arms control agreements with the Soviets have (at least on the U.S. side) been intended to add at least some stability to a continuing competition.

As noted in chapter 3, many are dissatisfied with the results of arms control thus far. In one view, the strategic arms competition is already unstable, due largely to Soviet initiatives over the past decade. In this view, the Soviet deployment of many accurate ICBM warheads threatens the survivability of the ICBM leg of the U.S. nuclear triad and of the command and control system which would direct a U.S. retaliatory attack. In addition, Soviet air defenses, civil defense activities, and sheltering of key leadership facilities would lessen the effectiveness of a U.S. retaliatory attack. A Soviet breakout from the ABM Treaty would further weaken the deterrent effect of the U.S. threat of retaliation. Responding to Soviet activities, the United States is increasing the accuracy of its own ICBMs and SLBMs, improving its bomber force, and accelerating BMD research. Thus, in one view, the possible destabilizing effects of future BMD deployment on the arms race will have to be considered in the context of the instabilities which will exist in any case.

Responses to BMD Deployment

One can imagine a variety of Soviet responses to U.S. BMD deployments (and vice-versa). Some of these responses might be stabilizing, others more destabilizing. In rough order of increasing destabilization, the range of imaginable Soviet responses follows.

Negotiation

If the Soviets could be persuaded to negotiate the transition to a world in which ballistic missile defenses played an important strategic role, the process might be a stable one. Each side would agree to reduce offensive nuclear capabilities, or at least not to increase them, while building up defenses. The stable conclusion would be that each side's offensive threat to the other would be reduced and neither felt compelled to try to negate the other side's defenses.

Tacit Stabilization

If the United States began to deploy BMD unilaterally and the Soviets followed suit, there might still be a stable competition. Each might find the reduction in its own offensive capabilities against the other acceptable because the other's was also proportionately reduced. A situation similar to the negotiated one above might be reached, but as the result of mutual unilateral calculations rather than joint decision.

Maintenance of the Offense

The Soviets might decide that it was worthwhile to try to maintain or restore their offensive capabilities by countering the U.S. BMD system. As Presidential arms control advisor Paul Nitze has said:

> New defensive systems must also be cost effective at the margin, that is, it must be cheap enough to add additional defensive capability so that the other side has no incentive to add additional offensive capability to overcome the defense. **If this criterion is not met, the defensive system could encourage a proliferation of countermeasures and additional offensive weapons to overcome deployed defense, instead of a redirection of effort from offense to defense.**[29]

There are several ways the Soviets could try to maintain offensive capabilities, and some of these ways could lead to a destabilizing arms race. In fact, the Soviets have explicitly announced that they intend to preserve their offense capabilities in the face of any U.S. defense.[30] Possible means of maintaining the of-

---

[29]Paul H. Nitze, Speech to the Philadelphia World Affairs Council on Feb. 20, 1985. Emphasis added.

[30]For example, Soviet General Nikolai Chervov told reporters that to counter U.S. efforts in space, "... we will have both an increase in offensive strategic weapons, and correspondingly we will take certain defensive measures." (*The Washington Post*, June 9, 1985, p. A-1.)

fense, which are not at all mutually exclusive, include:

Deployment of Passive Countermeasures.—Such countermeasures as decoy weapons, deception of BMD sensors, or altered ballistic missile flight characteristics could require the U.S. to respond with additional BMD system components or with technological changes in the BMD system. If corresponding increments of U.S. defense were less costly for the United States to add than the increments of offensive countermeasure were for the Soviets to add, this measure might not be destabilizing—or not for long. The Soviets ought to see that there would be no point in a continued offense-defense competition, because no gains in capability would be possible. On the other hand, if the cost-exchange ratio between defense and offense were somewhat ambiguous, the two sides might go through many expensive rounds of offensive and defensive countermeasure before the futility of further counteractions was obvious. A competition involving defensive systems and offensive countermeasures could be costly, though probably not as costly as one involving defenses and additional offensive weapons. If weapons were cheaper than countermeasures (per warhead penetrating the defense), than one would probably just add weapons.

Active Countermeasures: Attacking the Defense.—If the Soviets believed that vital components of a U.S. BMD system were vulnerable to attack, they might deploy weapons designed to weaken or disable the BMD system. We have noted in the first section of this chapter the potential for crisis instability if either side had plausible chances of a successful attack on the other's BMD system. A part of the cost-exchange ratio calculation for a decision to deploy a BMD system would be an assessment of the cost of defending the system as opposed to the cost of attacking it. Unless the Soviets were persuaded early on that the survivability of the U.S. BMD system could not be seriously threatened within the limits of Soviet resources, a costly race of deployments of anti-BMD weapons and anti-anti-BMD weapons might result.

Increasing the Numbers of Offensive Weapons.—Again, the cost-exchange ratio between increments of defense and of offensive countermeasures would have to favor the defense if the race were not to go on expensively, indecisively and indefinitely. It is also possible that the Soviets might decide to try to maintain some level of *net* offensive capability even at a cost higher than the corresponding U.S. defenses. If, on the other hand, the United States were willing to match Soviet expenditures, the Soviets in the long run would see their net offensive capability decline. In the meanwhile, however, additional Soviet offensive weapons could be destabilizing in another way: if the United States perceived the additional Soviet weapons as upsetting the balance of U.S.-Soviet offensive forces, the United States would have an incentive to respond with offensive additions of its own.

Circumventing the Defense.—If defenses clearly had the advantage over ballistic missiles, the Soviets might try to compensate for their declining strategic nuclear offensive capabilities by deploying other means of deliv-

Table 6-1.—Missile Production: U.S.S.R. and NATO[a]

Soviet missile production rates in the 1980s indicate a substantial capability to respond to U.S. BMD deployments with additional offensive missile deployments, should the Soviets choose that option.

	U.S.S.R.					NATO
Missile type	1980	1981	1982	1983	1984	1984
ICBMs	250	200	175	150	100	0
LRINF[b]	100	100	100	125	150	70
SRBMs[c]	300	300	300	350	350	0
SLCMs	750	750	800	800	850	665
SLBMs	200	175	175	200	200	80

[a]Revised to reflect current total production information. Includes United States; excludes France and Spain.
[b]LRINF—Long Range Intermediate Nuclear Forces
[c]SRBM—Short Range Ballistic Missile.

SOURCE: U.S. Department of Defense.

Photo credit: U.S. Department of Defense

The Soviet submarine-launched SS-NX-21 cruise missile, in development, has a range of 3,000 kilometers and can be fired from standard-size Soviet submarine torpedo tubes. Deployment of additional cruise missiles might be one Soviet response to declining strategic nuclear offensive capabilities imposed by a U.S. BMD system.

ery, such as bombers or long-range cruise missiles. If the United States believed that these slower moving delivery systems did not pose a disarming first-strike threat, it might decide not to add defenses against them, or it might decide to add only equivalent U.S. offensive forces. On the other hand, the United States might decide to build defenses against these other systems. The possibilities for competitive action and reaction described thus far for ballistic missiles and BMD might then manifest themselves for air-breathing weapons and air defenses. That is, the competition might prove stable or unstable.

Defense Plus Offense

Should the Soviets respond to U.S. BMD deployments with both a Soviet BMD deployment and additions to Soviet offensive forces, the United States might feel compelled to add to its own offensive forces to try to maintain a parity of offensive capability with the Soviets.

If each side were making different calculations about the cost-exchange ratio between its own offenses and the other side's defenses, the situation could become quite complex. Indeed, it is not clear that in practice either side will have an accurate knowledge of the the other side's current and future BMD capabilities or of the cost-exchange ratios. As was illustrated in chapter 5, the side with inferior defenses could see its situation as so disadvantageous as to call for substantial efforts to catch up, regardless of cost. But if the defensive and offensive capabilities of the two sides are not well understood by both, one side or both might see the other as having—or seeking—an advantage.

Additional Observations

A Problem of Timing

If the current BMD research and development program demonstrated in a few years that BMD deployments could lead to a safer world, the United States would certainly want to alter the current treaty regime banning all but very limited BMD deployments. But we would want to avoid a breakdown of that regime *before* the research and development program is concluded. This might be true for at least two reasons.

First, the Soviets appear to have maintained a technology base for a large-scale deployment of current-generation BMD systems, and in the short run of a few years they might attain a noticeable advantage in BMD deployments over the United States. This could lead to the kind of crisis instability discussed in the first section of this chapter.

Second, it is possible that the U.S. BMD research program may show that effective BMD is not feasible and should not be deployed. But, if in the meantime the ABM Treaty regime limiting BMD had been abandoned, the United States might consider it necessary nevertheless to deploy additions to its offense to counter Soviet BMD, and perhaps to deploy defenses as well, just to maintain the current strategic balance.

Limited BMD Systems

As was noted in chapter 5 and again in the first part of this chapter, BMD systems which could defend ICBMs or cities but not both would be potentially destabilizing. If the United States did not decide to pursue high levels of BMD capability, but had the limited objective of a defense of its land-based ICBMs, there would still be potential arms race instabilities. *The United States would have to be very careful to configure the BMD system so that its purpose was unambiguously the localized defense of hardened targets.* Otherwise, the Soviets might see the system as the core of a much broader defense, and take anticipatory countermeasures to maintain their own offensive threat. The United States would have to react accordingly, increasing its defensive forces, its offensive forces, or both.[31] Nor might the United States feel secure if the Soviets were to respond to a U.S. missile-site defense by expanding and spreading the system now deployed around Moscow. Designed to protect selected regions rather than just Soviet missile silos, such an expanded system would degrade the retaliatory threat residing in the ICBMs that the United States was defending. An expansion of U.S. offenses, defenses, or both might be taken in response.

Instabilities With Either/Or Defenses

We noted in chapter 5 and in the first section of this chapter that special instabilities may arise if both sides have what we call "Level 2" defense capabilities—the ability to protect most ICBM silos, or many cities, but not both. The danger is that one side may perceive the other to have the possibility of launching a very effective first strike against the other's retaliatory force and then defending very effectively against a "ragged" retaliation. Such perceptions would lead to very great pressures to remove the possibility of such a strike by increasing offenses to restore the credibility of the retaliatory deterrent.

[31] Since it takes several years to develop and deploy major weapon systems, each side tends to plan and build its systems on the basis of what it thinks is the largest deployment the other side might be able to field several years in the future.

BMD DEPLOYMENT AND ARMS CONTROL

The Importance of a Negotiated Transition

Administration officials have stressed the importance of a substantially favorable cost-exchange ratio between defense and offense as an incentive for the Soviets to agree to negotiate the reduction of offenses.[32] Some believe that without such an incentive—i.e., without clear evidence that ballistic missiles are being made economically obsolete by defenses—the Soviets may never agree to deep offensive reductions. On the other hand, it should be noted that if the United States and the Soviet Union could agree that it was desirable to reduce offenses and increase defenses, then a favorable cost-exchange ratio would not be a prerequisite to moving in that direction. Mutual offensive reductions could be the main instrument for increasing the effectiveness of defenses: the less formidable the offensive threat, the less capable the defenses would need to be.

Recently, Administration spokesmen have emphasized the importance of negotiating with the Soviet Union about the transition to a strategic relationship in which BMD plays a significant role. As Presidential national security advisor Robert McFarlane said,

> There is a relationship between reductions of offensive systems and the integration of defensive systems because of the potentially destabilizing effect of either side achieving a first-strike capability through possession of both. So our policy must be to first establish agreement between ourselves and the Russians on the value of defensive systems. Once we have reached agreement on that, then we must establish a path for the integration of these defensive systems into the force structure that will be stable.[33]

There is a degree of paradox associated with the uncertainties that BMD deployment could introduce in the calculations of the two sides. On the one hand, increased uncertainty about the likelihood of successful attacks could increase crisis stability by making the aggressor less willing to gamble on a favorable outcome from a first strike. On the other hand, in the face of growing uncertainty about the effectiveness of its military forces, each side will have an incentive to try to reduce that uncertainty by deploying additional offensive and defensive weapons and countermeasures.

In the absence of coordinated structuring of defenses and offenses on the two sides, the United States would have to anticipate and adapt in advance to a wide range of potential Soviet responses. Even if the cost-exchange ratio between defense and offense favored the defense, the transition period could bring a costly arms competition until the effects of the cost-exchange ratio asserted themselves.

Arms control has been one measure pursued by the United States to try to enhance crisis stability and arms race stability.[34] Crisis stability may be enhanced if the United States and the Soviet Union can negotiate force structures or mutual procedures (e.g., the hot line) which might reduce incentives in a crisis to strike. Slowing the arms race may be possible if the two can agree to limit weapon deployments which might accelerate the competition. Arms limitations can also add a certain amount of predictability to the force structure planning on each side, reducing the steps each

[32] As noted earlier, there are likely to be large uncertainties in calculating such ratios, and the two sides may well assess them differently.

[33] As interviewed in *U.S. News and World Report*, Mar. 18, 1985, p. 26. In a similar statement, Kenneth Adelman, Director of the U.S. Arms Control and Disarmament Agency, said in February 1985:

> We must scrupulously guard against a vicious cycle of defensive efforts—even research for defense—spurring the other side on to more offensive weapons in order to saturate prospective defenses, and so on, and so on. That snowball effect would undercut stability and weaken deterrence.
>
> That risk can be reduced and managed through the kind of overall strategic discussions Secretary Shultz launched in Geneva last month and that Ambassador Kampelman will take up further when the arms talks begin again next month. This type of exchange with the Soviet Union—an in-depth dialog about critical strategic relationships, strategic concepts, strategic stability—is indispensable to an effective SDI approach.

(Speech before the International Institute for Strategic Studies, Feb. 13, 1985.)

[34] For a discussion of the objectives of arms control, see National Academy of Sciences, *Nuclear Arms Control: Background and Issues* (Washington, DC: National Academy Press, 1985), pp. 4-6.

might feel compelled to take in *anticipation* of what the other *might* do in the future.

A negotiated transition to a U.S.-Soviet strategic relationship in which BMD plays an important role would be an arms control arrangement intended, like earlier ones, to enhance strategic stability. The two sides would first need to agree in principle that there should be such a negotiated transition. According to Secretary of State George P. Shultz:

> As our [BMD] research proceeds and both nations thus gain a better sense of the future prospects, the Soviets should see the advantages of agreed ground rules to ensure that any phasing in of defensive systems will be orderly, predictable, and stabilizing. The alternative—an unconstrained environment—would be neither in their interest nor in ours.[35]

Soviet acceptance of such ground rules may not come easily. The public position of the Soviet Union thus far is that BMD deployments (beyond what is now allowed by the ABM Treaty) would evoke an offensive response and make arms control impossible. (See appendix K for various Soviet statements on this subject.)

We do not know whether the Soviet public position is purely propaganda posturing intended solely to undercut the Strategic Defense Initiative. There is general agreement that at least the *initial* Soviet response to U.S. BMD deployments would be to try to restore their own offensive capabilities. The United States might decide to deploy BMD because it believed that it would make further offensive deployments by the Soviets futile. What is difficult to predict is when or whether the Soviets might arrive at the same conclusion. Until they did, they might engage in a substantial offensive build-up.

Once agreement in principle to a negotiated transition had been arrived at, the stages of the transition would have to be defined. The ultimate goal may be to reach a state in which greatly reduced offenses are believed to be highly unlikely to penetrate very effective defenses. Before that stage is reached, a reasonable intermediate stage would be one in which defenses could prevent offenses from effectively attacking military targets, even though cities might still be vulnerable. A stage to be avoided, however, is one in which the Soviet Union, for example, might be able to use offensive missiles to weaken the U.S. retaliatory force, then defend very well against a "ragged" U.S. retaliation. (See discussion above, p. 125 and in chapter 5). In that stage it would be very difficult for the United States to agree to further offensive reductions when it already feared the possibility of the Soviets defending successfully against a U.S. retaliation.

To avoid that and the other kinds of instabilities discussed in this chapter, the two sides would need to agree on the orderly accumulation of comparable ballistic missile defense (and, possibly, air defense) capabilities. They would need to agree on comparable, mutually acceptable, offensive capabilities. Without such agreed levels of capability, each side might see the other as having or seeking military advantages.

Working Out the Details

As with past agreements on offensive and defensive arms, agreements on acceptable levels of offensive and defensive capability would probably have to be translated into agreements on some specifications of the weapons systems each side could deploy. The current ABM Treaty, for example, specifies what kinds and numbers of components of a BMD system, are acceptable. This agreement was possible in part because the BMD systems of the two sides were roughly similar in principle and because the permitted BMD deployments kept actual capabilities almost negligibly small.

Agreements to phase in increasingly higher levels of BMD capability together would be far more challenging. One problem seen in previous arms control negotiations could be particularly severe: that of asymmetries in the forces on the two sides. The United States and the Soviet Union have in the past found it difficult to agree on what mixes of ICBMs, SLBMs, and aircraft on the two sides con-

[35]Speech in Austin, Texas, Mar. 28, 1985.

stituted equivalent nuclear offensive forces. At least in the early transition stages, force asymmetries could remain a serious problem. As long as the number of ballistic missile warheads that the Soviets could use to attack U.S. missile silos exceeds the number of U.S. warheads in those silos, equal defense capabilities would not have equal strategic significance. Equal defenses would reduce the net number of U.S. retaliatory weapons surviving a Soviet first strike and penetrating Soviet defenses (see chapter 5).

Different technical approaches to BMD and different levels of technological accomplishment would also complicate calculations of equivalence. Moreover, those differences would exacerbate the problem of assuring adequate verification that one side or the other did not have significantly more capable defensive systems. Although President Reagan has suggested the possibility of the United States sharing BMD technologies with the Soviets,[36] many are skeptical that this would or should ever happen. They point out, for example, that the more the Soviets knew about the details of a U.S. BMD system, the easier it would be for them to devise effective countermeasures to overcome it.

The two sides might also have difficulty agreeing on which approaches to BMD are acceptable. The Strategic Defense Initiative is currently emphasizing nonnuclear defenses. The kind of system the Soviets have deployed and are currently best prepared to expand uses nuclear warheads. In the longer term, however, perhaps the two sides' technological approaches to BMD problems might converge.

If the two sides were to reach the ultimate stage of deeply reduced offensive missiles and aircraft and highly effective defenses, yet another potential problem would still have to be considered. That is, some residual uncertainties would likely remain about Soviet capabilities and intentions. They might be suspected of working on or actually achieving some effective countermeasure to a key part of our defensive shield. There would always be the possibility of smuggled nuclear weapons, secretly implanted in U.S. cities. The United States, for its part, might have no comparable retaliatory threat. This situation would leave the United States open at some point to Soviet nuclear blackmail.

Potential Effects of the Absence of a Negotiated Transition

The deployment of BMD in the absence of a negotiated transition would mean, in effect, that the United States and the Soviet Union would have abandoned the ABM Treaty but not replaced it with a new arms control regime for BMD. The potential diplomatic and broader arms control consequences of such a course deserve consideration.

Offensive Arms Limitations

Negotiations on offensive arms limitations without regulation of defensive deployments could be extremely difficult. Unless each side was absolutely convinced that it could not afford to deploy offenses that would counter the other's defenses, it would have a strong incentive to increase, rather than decrease, offensive arms levels.

U.S.-Soviet Relations

To the degree that arms competition adds to tension in the U.S.-Soviet relationship (some would argue that the arms competition is solely a result, not a cause, of the tension between the two political systems), a BMD-offense competition could make U.S.-Soviet agreements in other areas more difficult.

U.S.-Allied Relations

To the extent that U.S. allies see the ABM Treaty (or would see its successor) as central

[36]In replying to a statement that the proposed U.S. SDI program is seen in Moscow as an attempt by the United States to regain strategic superiority by making the Soviet Union vulnerable to a first strike, Secretary of Defense Caspar Weinberger said:

> My response is that is not only totally wrong, but it's conclusively proved to be wrong by the President's offer to share this with the world if we can get it. If we wanted a war-winning capability through this means, we wouldn't be talking about sharing it with the world.

(ABC Network television program "The Fire Unleashed," June 6, 1985.)

to an arms control process that they wish to sustain, and to the extent that they saw the United States as responsible for its abandonment, U.S.-allied relations could be strained.

Nuclear Non-Proliferation Treaty

Some nonnuclear-weapon states that signed the Nuclear Non-Proliferation Treaty take seriously the obligations assumed in that accord by the nuclear-weapon states to try to make further progress in nuclear arms control. If they saw abandonment of the ABM Treaty (without replacement by a new arms control regime) as a major step away from that promise, then their adherence to the Non-Proliferation Treaty could be called into question. (This risk is discussed further in appendix C.)

In sum, there appears to be a broad and, cumulatively, compelling set of reasons to support recent Administration emphasis on the importance of a negotiated transition to U.S.-Soviet BMD deployments, should a deployment decision be made. For either side to proceed to deployment of BMD outside the context of an arms control arrangement effectively governing offensive and defensive arms on both sides could lead to serious strategic instabilities. Whether such a negotiated transition is possible remains to be seen. But because both sides plan strategic forces several years in advance, the negotiations would probably have to begin during the research and development stage, not in the stage of first BMD deployments. Indeed, any decision about BMD deployment should take into consideration the realistic prospects for such negotiations. The relationship between research and development and arms control is discussed in chapter 10 of this report.

CONCLUSIONS

A complete analysis of the potential impact of BMD deployments on crisis stability would have to include, *inter alia*, a large and complex exploration of the potential outcomes of nuclear exchanges between the Soviet Union and the United States given various levels and kinds of offense and defense on the two sides. Such an exploration would require highly sophisticated "exchange model" calculations to simulate the possible impact on outcomes of such factors as:

- asymmetries in the offensive and defensive force structures of the two sides;
- uncertainties on each side about the offensive and defensive capabilities of the other side;
- varying degrees of ability on each side to defend certain types of targets "preferentially" (see chapter 5); and
- the differences made by the size of attack and rate of attack defended against on the numbers of warheads each side could intercept.

Such an analysis would require extensive computing resources and many hundreds of "runs" of the model. Carrying out this analysis would not *prove* that the net consequences of deploying various levels of BMD would be positive or negative. But it would be one contribution toward such assessments and might help to avoid particularly unstable relationships between the force structures on the two sides. Potential crisis instabilities may not only be risky in themselves, but may induce arms race instabilities, as one side or the other adds new forces in an attempt to remedy what they would consider to be a dangerous strategic disadvantage. Moreover, exploration of the strategic implications of various levels of offense and defense would be an important preparation for attempting to negotiate with the Soviets on a transition to a world of increased defenses and reduced offenses.

Congress may wish to see that credible and thorough strategic analyses have been performed well before it must decide whether to authorize BMD programs beyond the research stage.

Chapter 7

Ballistic Missile Defense Technologies

Contents

Tables

Figures

Chapter 7
Ballistic Missile Defense Technologies

INTRODUCTION

This chapter and chapter 8 describe the technologies applicable to ballistic missile defense and point out some of the uncertainties that further research may hope to resolve. Ballistic missile defense **technologies** and ballistic missile defense **policies**, of course, are interdependent. BMD policy choices, the subject of the preceding chapters of this report, are constrained by the state of our technology. At the same time, however, policy decisions influence technological advances by providing (or withholding) resources and incentives to extend our knowledge and capabilities.[1]

Feasibility

The overall feasibility of ballistic missile defense technologies involves a set of related issues which become increasingly harder to answer definitively. **Scientific feasibility**—whether or not something is physically possible—is obviously necessary for any BMD concept, but it is by no means sufficient. **Technical** and **economic feasibility** questions go on to ask whether a device permitted by the laws of nature can actually be built at a reasonable cost within a reasonable amount of time. Assuming that a system can be designed and built according to specifications, **operational feasibility** issues address the questions of whether it can actually be deployed, tested, maintained, and operated with a high degree of confidence.

Overriding all of these considerations is the issue which forms the crux of the BMD technical debate: **any effective BMD system must be "robust," in that it must operate and endure against a reactive adversary intent on defeating it.** The dynamic competition between offensive and defensive technologies—among measures, countermeasures, and counter-countermeasures—forces the successful development and implementation of BMD technologies to be far more than a purely technological accomplishment, such as reaching the moon or splitting the atomic nucleus. The moon and the nucleus did not hide, run away, or shoot back.

Evaluating the robustness of a prospective defensive system requires making assumptions about the motivations and relative technical skills of the two sides. It also requires a clear conception of the system's intent. Is a successful defense one which can defeat a given threat and **deters** threat growth? Or is it one which can defeat the threat and **provokes** growth, forcing the Soviets to spend a lot of money?

Failure to take full account of the offense-defense competition can lead to what has been called the "fallacy of the last move," in which some action is evaluated as if the strategic competition were frozen immediately afterwards. However, although the concept of a "last move" in the competition between offense and defense does not make sense, the starting point of such a competition is well defined. Massive, diverse, and highly effective offensive forces dominate the strategic relationship today. From that starting point, advanced defensive technology and advanced offensive technology will evolve together, in the absence of political agreements to regulate that competition. **If both offenses and defenses evolve at comparable rates, the present dominance of the offense will clearly be maintained.** Economic questions are as important as technical ones, since the outcome of a technological competition depends in part on who is better able to pay for it. These economic questions are discussed further in chapter 8.

If it turns out that offensive technologies have developed so far along their learning curve that their rate of continued technical progress slows, evolving defensive technolo-

[1]For example, technology in the area of pollution control has primarily been driven by policy decisions.

gies might make progress in eroding the great distance currently existing between the two. The relevant question is whether it is likely that an arms development competition will close that gap. **To say that the time is ripe for offense dominance to give way to defense dominance either prejudges the outcome of this technological competition, or assumes that a political agreement will be reached which will ensure that defenses catch up and overtake offenses.**

Technological Prediction

Even aside from the all-important question of effectiveness against a reactive opponent, predicting future technical feasibility is a difficult business. Experts can have hunches and gut feelings, and they can make elaborate technical calculations. However, firm answers cannot be obtained without experimentation. No one, regardless of technical credentials or creative ability, is an expert when it comes to predicting the future. Secretary of Defense Caspar Weinberger has called attention to Albert Einstein's 1932 observation that "there is not the slightest indication that [nuclear] energy will ever be obtainable."[2] Arms Control and Disarmament Agency Director Kenneth Adelman has similarly recalled the warning that Admiral Leahy, President Truman's Chief of Staff, gave the President in 1945: "The [atomic] bomb will never go off, and I speak as an expert in explosives."[3] Adelman warned that technical critics of the Strategic Defense Initiative "may well turn out to be just as shortsighted in retrospect as many of their predecessors have been in hindsight today."

In the context of the time, however, Einstein was correct. The "indications" that nuclear energy might be obtainable had yet to be discovered. If a major effort to develop nuclear power had been undertaken before basic research had revealed the phenomenon of heavy element fission, it might have focused on the wrong end of the periodic table and floundered for years.

[2]Speech before the Foreign Press Center, Dec. 19, 1984.

[3]Quoted by Ambassador Adelman in "SDI: Setting the Record Straight," speech before the Baltimore Council on Foreign Affairs, Aug. 7, 1985.

When the "experts" do make mistakes, they err in both directions. Just as breakthroughs have been made which were previously predicted to be impossible, other foretold inevitabilities have never come to pass. For example, General David Sarnoff, Chairman of the Board of the RCA Corp., claimed in 1955 that "[I]t can be taken for granted that before 1980 ships, aircraft, locomotives and even automobiles will be atomically fuelled." John von Neumann, the father of the modern computer and a member of the Atomic Energy Commission, stated the following year that "[a] few decades hence, energy may be free—just like the unmetered air."*

One way to attack the question of predicting the feasibility of a given technical accomplishment (setting aside the question of reactive opponent) is to specify a time limit. Is a Boeing 747 airliner feasible today? Of course—it has been in service for more than 15 years.[4] Would it have been feasible in 1940? No—unless 30 years were allotted at that time for its development, including some unanticipated and rather fundamental inventions that made it possible, and provided that large expenditures were allocated for producing and operating its predecessors.

A more relevant measure is to ask whether progress can be accelerated significantly by a crash ("technology-limited") program. Perhaps a 747 could have been developed by 1955 or 1965 if doing so had been a compelling national priority. However, attempting to build one before all of the required technologies had matured to their 1970 levels would probably have produced a very different airplane at much greater expense.

Organization of This Chapter

This chapter introduces the technological components which might contribute to future

*All quotes in the preceding three paragraphs can be found in *The Experts Speak: The Definitive Compendium of Authoritative Disinformation,* by Christopher Cerf and Victor Navasky, a joint project of *The Nation* magazine and the Institute of Expertology (New York: Pantheon Books, 1984).

[4]Pan Am introduced the Boeing 747 airliner to commercial service with a flight from New York to London on Jan. 22, 1970.

ballistic missile defense systems. It reviews the characteristics of many of the relevant technologies and outlines the key uncertainties concerning those technologies' potentials. Readers who do not wish to immerse themselves in technological details are invited to concentrate on the sections labeled "Issues," with the descriptions preceding those sections used as reference material. In addition, the following chapter (chapter 8) summarizes some of the major issues of technological feasibility.

This chapter also examines how the technological building blocks need to be put together, and it introduces some of the systems issues relevant to integrating the pieces into a coherent whole. It does *not* attempt to predict exactly how each of the technologies will evolve, and it compares different contenders for some given task to each other only in a general way.

The chapter was prepared with access to classified materials. For the most part, those classified data concerned schedules, budgets, and technical matters too detailed for this discussion. A few relevant classified details and concepts are discussed in a classified annex.

BMD TECHNOLOGIES

Overview

BMD Concepts

Ballistic missile defense systems as described by the ABM Treaty and as primarily pursued prior to 1980 consisted of ground-based interceptors of various ranges supported by ground-based radars. These systems would attempt to intercept ballistic reentry vehicles (RVs) as they descended toward the United States, either prior to or just after they reentered the Earth's atmosphere. Current BMD concepts posit systems that can intercept ballistic missiles and their RVs at all stages of their flight, from shortly after launch to just prior to detonation.

Layered Defenses

The basic concept is to use layered defenses, which provide the defense with several opportunities to attack the incoming warheads. Early layers would reduce the number of warheads that later layers would have to handle; later layers would "mop up" those that get through the early layers.

It has become convenient to discuss defensive layers which are associated with each phase of a ballistic missile's flight. Since the missile has different properties in each phase, different defensive components are associated with the different phases.

The first opportunity to engage the missile would be in its *boost phase*, when the ICBM's booster motor is burning. A second layer might operate in the *post-boost phase* after the booster has dropped away, leaving a *post-boost vehicle* (PBV or *bus*) which aims the individual warheads at their targets and lets them go. Decoys and other defense penetration aids can also be dispensed by the PBV during this phase. The post-boost is followed by a *midcourse phase* of up to 20 minutes in length during which the RVs and decoys coast towards their targets; the last phase is the *terminal or reentry phase*, lasting less than a minute, which starts when the RVs reenter the Earth's atmosphere and the lighter decoys burn up.

Operation of each layer is controlled by a *battle management* system, which also coordinates between the layers and provides overall supervision and control.

Properties of the Phases

Boost Phase.—During boost phase, the hot gases in the booster's exhaust produce a large, easily detected infrared signal, or *signature*, especially as the rocket rises above the clouds and the denser layers of the atmosphere. For current missiles, the boost phase lasts 3 to 5 minutes.[5] However, not all this period is avail-

[5]See J. C. Fletcher, "The Technologies for Ballistic Missile Defense," *Issues in Science and Technology*, fall 1984, p. 15.

Figure 7-1.—Multilayered Space Defense

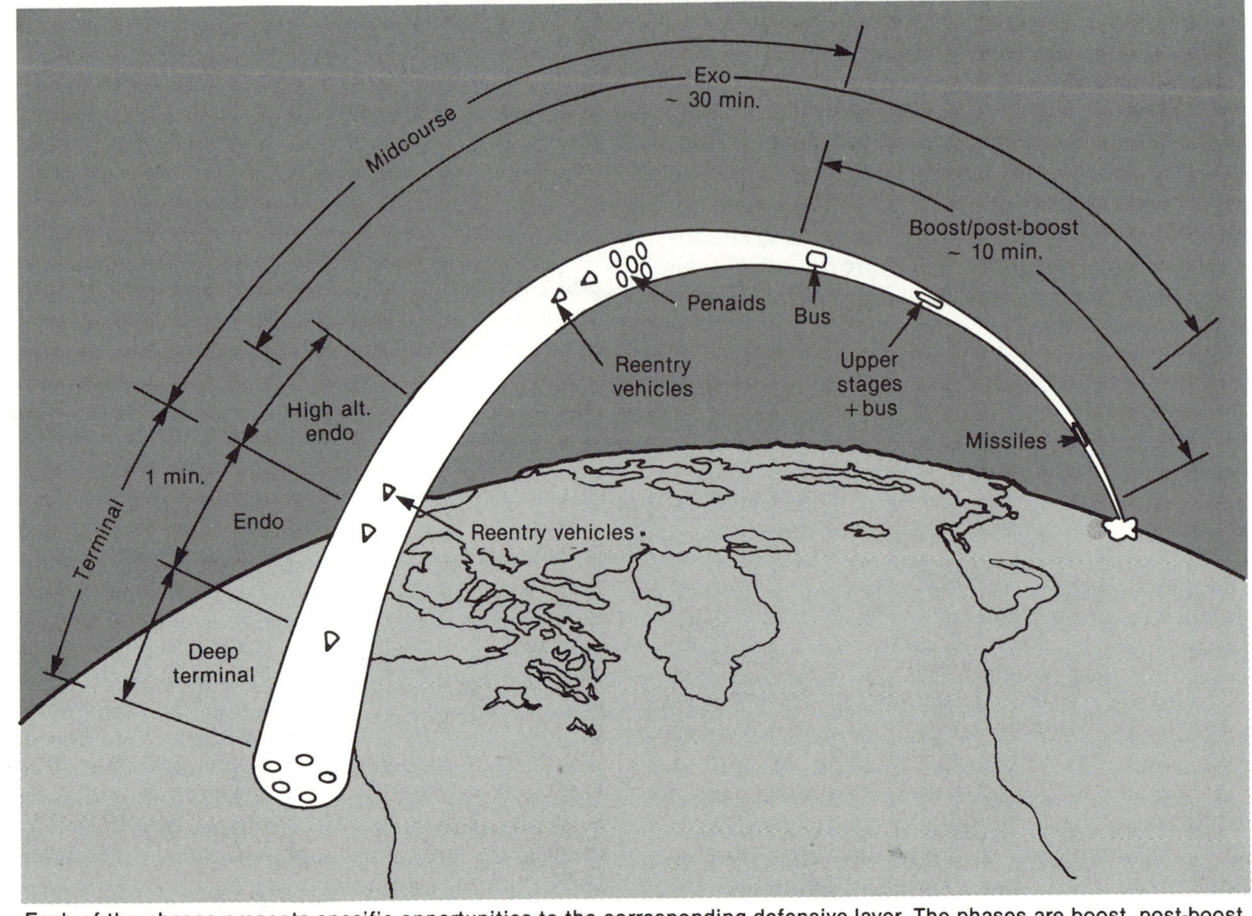

Each of the phases presents specific opportunities to the corresponding defensive layer. The phases are boost, post-boost, midcourse and terminal.

SOURCE: U.S. Department of Defense.

able for the defense to attack the boosters. The defensive system must first detect the launch, determine that there is actually an attack in progress, decide to engage the boosters, and allocate defensive weapons platforms to the boosters. How much time this would take depends on how automated the system would be and how quickly decisions could be made. **In particular, requiring that human intervention be necessary before the defense can commence firing imposes extreme time constraints on command and control procedures.**[6] Possibly, a succession of alert conditions could be established which, under day-to-day conditions, would require human intervention before boost-phase defenses could engage an attack (leaving surprise attacks to be handled by later layers), but which in times of crisis would permit the defense to engage boosters autonomously if an attack were detected. Such a procedure, of course, would increase the incentive for surprise attack.

Successful engagement in the boost phase can provide a high degree of *leverage*—i.e., the destruction of one booster results in the destruction of all its RVs and decoys. There is also another sort of leverage involved—the boosters are much more vulnerable than the RVs, providing another advantage to attacking the boost phase.

[6]Requiring human decisionmaking could pose even more severe problems in terms of platform self-defense, when only a very short time may be available to characterize and engage attacking weapons before they come within lethal range.

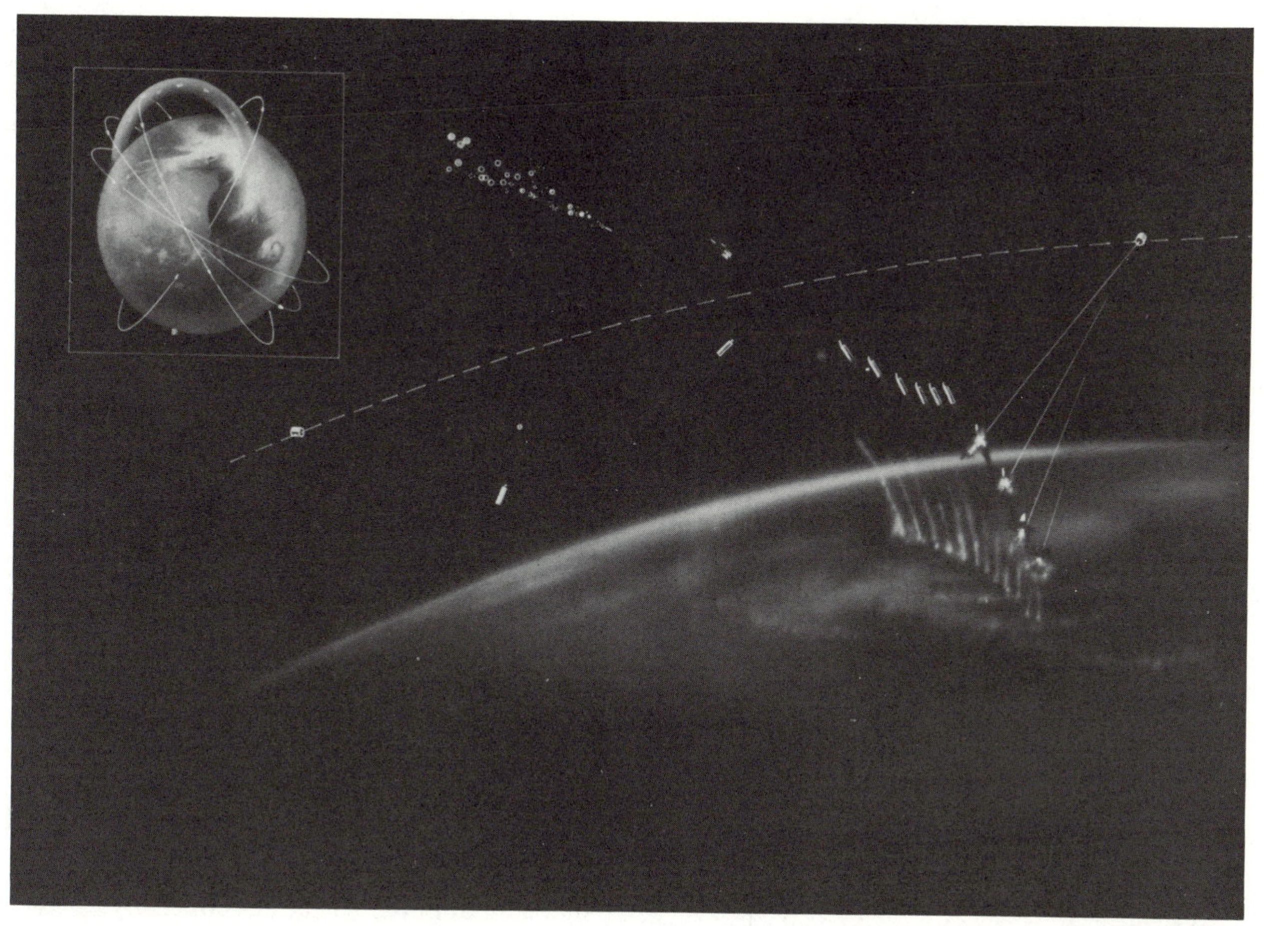

Photo credit: Lawrence Livermore National Laboratory

Boost-phase intercept would provide great leverage, since each booster may contain many targets and decoys. In the illustration, some boosters, penetrating this layer of the defense, have deployed RVs and decoys. Inset shows "threat tube" of Soviet missiles and orbits of some of the satellites in the defense constellation. The number of battle stations required depends on weapon range (related to brightness for directed-energy weapons), retarget time, number of targets, and time available for attacking the targets.

One consequence of the high leverage of boost-phase defenses is that small errors in boost-phase BMD performance are magnified to become larger errors in the later phases. Each missile that survives the boost phase may ultimately produce hundreds of objects (RVs and decoys) that must be tracked, discriminated, and attacked by later layers. Because effective discrimination is vital to the success of the midcourse layer, successful midcourse defense may be tightly linked to the success of the boost phase.

By their nature, boost-phase defenses have little ability to defend *selectively*. While a booster is burning, it may be possible to determine where it came from, where in general it is headed, and what kind of missile it is. However, until the individual RVs are released by the PBV, their specific targets cannot be determined. Therefore, boost-phase defenses cannot effectively conduct *preferential defense*, in which limited defensive resources are concentrated on defending only some sites at the expense of permitting attacks on others to continue unimpeded. In this manner, a limited defensive capability can be used to save a greater number of sites than would be possible with a random allocation of defenses (see discussion of preferential defense in chapter 5).[7]

Even without the ability to conduct preferential defense, a boost-phase defense which eliminates some fraction of an attack can deny the offense the ability to conduct a highly structured attack which requires warheads to arrive at specified targets in a precise order. Such an attack would be much more difficult to carry out if some fraction of the offensive boosters were intercepted by the boost-phase layer. A structured attack, intended to blind or destroy components of the defensive system, might make it easier to penetrate later defensive layers. Therefore, the inability to conduct such an attack might make those later layers more effective.

Post-Boost Phase.—The post-boost phase may last as long as 6 minutes, but it could be much shorter. As this phase progresses, the PBV dispenses RVs and decoys and therefore loses value as a target. Therefore, leverage is high at the beginning of the post-boost phase and low at the end. Although the individual RVs on each PBV are relatively "hard" (difficult to destroy with certain types of defensive weapons), the PBVs themselves might be "softer," adding to the leverage of the post-boost phase. However, if a PBV is disabled without disabling the RVs still attached to it, those RVs may still have to be handled by later defensive layers.[8]

The post-boost defense has more time to get ready than the boost-phase defense (since it can get ready during the boost phase), and it may have more total time to engage each individual target. However, targets must be engaged early in the post-boost phase to achieve best results. As in the case of boost-phase defenses, small errors in post-boost performance can have larger consequences in the later phases. Unlike boost phase, however, the performance of the post-boost phase depends on *when* targets are killed as well as on *how many* are killed. Since targets become less valuable as time goes on, uncertainties in timing will affect overall post-boost-phase performance. Selectivity in the post-boost-phase defense—the ability to conduct preferential defense—is similar to that of the boost phase.

Midcourse Phase.—Most of an RV's flight time is spent in the midcourse, the period between release from the bus and reentry into the Earth's atmosphere. This period lasts about 20 minutes for ICBM RVs; it may be much shorter for SLBM RVs. Although there is much more time to find and engage targets in the midcourse than in the earlier phases, there is also much more to do. Before a target can be engaged, it must be discriminated from decoys and possibly from debris; imperfect discrimination capability will result in shooting at objects that are not really targets and in withholding fire on objects that should actually be attacked. To kill the 10 RVs carried on one SS-18 in boost-phase, the defense must find and destroy one target in the few minutes that the boost phase lasts. To kill the same number of RVs in midcourse, the defense must sort through possibly hundreds of objects in order to find and destroy the 10 RVs in 20 minutes. The rate of activity required in midcourse could, therefore, be the same or higher than in the boost phase. Of course, in a massive launch, the number of targets and decoys would be thousands of times larger.

Leverage is low in the midcourse, but midcourse defense does have the potential for being selective. Destinations of individual RVs can be determined once the RVs have separated from the bus.

Terminal Phase.—The terminal (or reentry) phase is very short. If hardened targets are defended, defensive intercepts can occur at

[7]Only redundant targets, where the loss of many or most can be tolerated, are logical candidates for preferential defense. ICBM silos are perhaps the best example, since only a small fraction of them still carry enormous destructive potential.

[8]RVs which have not been properly dispensed by the PBV will not be accurately aimed, although they may nevertheless be armed and left on a trajectory which will take them somewhere within the borders of the country being attacked. If their mission requires high accuracy, such RVs will have been rendered ineffective; if their mission merely requires that they reach some region of the country and detonate, they may still be able to accomplish their task.

fairly low altitudes, since hardened targets by definition are designed to survive nearby nuclear explosions. However, if soft targets are to be protected, intercepts must take place at a higher altitude.[9] As few as tens of seconds would be available between the time reentry began (or more accurately, the time that atmospheric effects begin to sort out decoys from RVs) and the time that terminal interceptors would have to be launched in order to destroy the RV at a sufficiently high altitude. However, the terminal defense would have almost 30 minutes to get ready to engage the RVs. Time pressures would be minimized if earlier phases had identified and tracked those RVs which they failed to destroy, and were able to "hand off" this trajectory information to the terminal defense. This tactic assumes good discrimination and kill effectiveness in the earlier layers.

Advantages and Disadvantages of Layered Defenses

Multi-layered defenses have the potential for performing much better than single layer defenses. First, several layers of moderate effectiveness which combine to produce a total defense of high effectiveness will, in general, be easier to design and build than a single layer having the same resultant effectiveness. Second, multi-layered systems, in theory, are more robust than single-layered systems, especially if each layer employs different technologies and different designs. In that case, offensive developments which degrade one layer might not severely affect later layers. Third, the presence of early layers—the boost and post-boost layers—reduces the burden on the later layers. The number of objects the midcourse defense has to handle is cut in half if the early layers kill half of the missiles. Finally, building several layers allows the designer to take advantage of whatever unique advantages each layer provides. For example, a multi-layered defense could have both the leverage of the boost-phase defense and the selectivity of the midcourse. A single layer defense could have only one or the other.

However, there are drawbacks as well as advantages to layered defenses. The most obvious problem is that four layers are likely to cost more than one layer—especially if the layers are completely independent—although perhaps less than a smaller number of layers which were as effective as the total of the four. Second, the degree to which the layers can combine to produce high effectiveness will depend on how independent the layers are. To take an extreme example, if all layers depend on the same sensor system and that sensor system fails, all the layers will fail. The same holds true for battle management algorithms or other shared resources. **The leakage rates of the individual layers of a layered defense can be multiplied together to give the total leakage rate only if the individual layers are totally independent and share no common elements.** Otherwise, leakage through early layers may not be fully compensated for by later layers.

The robustness of the system against the loss (or severe degradation) of one layer will depend on how much capacity is built into the system to compensate for that loss. The layers must be able to take advantage of the other layers without being overly dependent on them. For example, if boost and post-boost defenses permit twice the expected number of objects to reach midcourse, and if that in turn substantially degrades the midcourse defense's ability to sort objects, the midcourse may let through not only the additional RVs but also many of the ones it would otherwise have intercepted.

In practice, it will be impossible to know in advance exactly how effective any layer will be; there will probably be large uncertainties in predicting how well it will work against an actual attack. Those uncertainties, however, will be viewed differently by the two sides. From the defensive point of view, extra capacity will be required in each layer in order to hedge against the possibility that it (or the other layers) will not perform as well as anticipated. From the offensive view, however, uncertainties will make it more difficult to de-

[9]Even if the defensive interceptor is nonnuclear, the attacking warhead may be *salvage-fused* to detonate if intercepted. Therefore, intercepts must take place higher above soft targets than they need to above hardened ones.

stroy (or penetrate) the defense with high confidence.

The significance of a degradation in capability will depend on the goal of the defense. It makes little difference whether an ICBM silo defense is 40 or 50 percent effective. If one were only interested in providing a survivable deterrent, rather than defending populations, these concerns regarding the vulnerability of one of several layers would be relatively unimportant. However, the difference between a 90 percent effective city defense and a 99.9 percent effective city defense is a hundredfold increase in the number of weapons reaching U.S. cities; this could make the difference between the survival and the destruction of our civilization.

Individual Tasks of Each Layer

Each layer must perform the following tasks:

- *Surveillance and Acquisition:* Attacks must be detected, and the number, location, and probable destination of all threatening objects must be determined.
- *Discrimination:* Actual missiles, busses, and warheads must be distinguished from nonthreatening decoys and other debris.
- *Pointing and Tracking:* Targets must be tracked with whatever precision is required by the weapon designated to destroy that target, and that tracking information must be communicated to the defensive weapon.
- *Target Destruction*: A defensive weapon must deliver sufficient energy to a target rapidly enough to destroy it.
- *Kill Assessment*: Those targets that have been successfully destroyed must be identified and distinguished from survivors. In addition, if it can be determined why a targeted warhead was not destroyed (incorrect pointing, for example), this information can be used for a subsequent attack.

The above tasks all involve processing either information or energy. **Sensors** collect signals or radiation emitted by or reflected from targets. These are processed to yield information about the individual targets. Sensor and data processing technologies are therefore crucial to an advanced ballistic missile defense. When targets have been identified and assigned to weapons systems, energy stored in the weapons must be converted to a form which can be delivered to the target in sufficient quantity rapidly enough to destroy it. Various types of **directed-energy (beam) weapons** and **kinetic-energy (projectile) weapons** have been proposed for this role.

Technological candidates for sensors, processors, and weapons are described in this chapter. The battle management issues involved in coordinating and integrating these "building blocks" into a complete, functioning system are also discussed, along with possible offensive responses or countermeasures. Some of the logistical issues involved in constructing and operating such a system are noted as well. Further discussion of the feasibility and operational issues is presented in chapter 8.

Weapon Kill Mechanisms

Introduction and Types of Kill

The Strategic Defense Initiative Organization is investigating the feasibility of many types of weapons. The type that has been publicized the most, possibly because it appears to be the most exotic, is the directed-energy weapon. Although this class of weapon is only one potential facet of the SDI, it could possibly become the centerpiece of some of the defensive layers. The advantage of such weapons is clear: killing energy is delivered at or near the speed of light, and, for typical BMD distances, arrives at the target in less than a tenth of a second.

Concepts under investigation in this area include several types of laser and particle beam weapons. For weapons purposes, the relevant criteria used to determine the usefulness of the different technologies mostly concern their ability to neutralize targets in a small amount of time (seconds, at the most). Another consideration is the capability for kill assessment after the target has been engaged. This latter question depends in part on the target; it may be a booster rocket stage, a post-boost vehicle, or an RV. Enemy satellites could also be targets.

There are three types of kill mechanisms by which directed-energy systems can act: 1) *functional kill*, 2) *thermal kill*, and 3) *impulse kill*.

The *functional kill* mechanism, pertinent to particle beam or microwave weapons, prevents an offensive weapon from operating correctly without necessarily destroying it. Subatomic particles with kinetic energies of a few hundred million electron-volts[10] (MeV) can penetrate at least several centimeters of dense materials, or tens of centimeters of typical aerospace materials. Therefore, sensitive electronic components deep inside the target can be altered or destroyed. However, it may not be immediately apparent to an outside observer that a kill has occurred. A kill of this sort may be referred to as a "soft," i.e., initially unobservable, kill.

To disable boosters by *thermal* means, a nominal range of 1 to 100 kilojoules of energy deposited per square centimeter (kJ/cm^2) of target has been taken as an estimate in the literature. This energy must be delivered quickly—if the time needed to deliver a lethal amount of energy is very long (hundreds of seconds or more), the heated area of the booster may have time to conduct away much of the energy being directed at it and may then not fail. The actual value of a lethal energy dose for a given target depends on many factors, including material, surface properties, and mechanical stress. This energy will raise the surface temperature of the target sufficiently to weaken or deform it, allowing internal forces to cause a catastrophic failure. The ability of a given technology to effect a thermal kill depends on the power levels attainable, the focusing ability of the weapon, and the distance from the target.

[10]An electron-volt is the amount of energy an electron can pick up from a 1-volt battery.

Impulse kill does not achieve its goal by heating the target, but by depositing energy in a powerful pulse on its surface. A mechanical shock wave is driven through the target, collapsing it.

Lasers

A laser is a device which produces a coherent beam of electromagnetic radiation at a well-defined wavelength. Coherence means that all the waves of radiation are in step, crest-to-crest and trough-to-trough, and maintain this alignment over time. When they strike a surface, the effects are greater than would be the case for incoherent radiation. The intensity of incoherent radiation is limited by the temperature of the object producing that radiation; there is no such limit to laser radiation. The radiation may be in the infrared, visible, ultraviolet, or X-ray regions of the electromagnetic spectrum.

Lasing occurs when more of the lasing material's molecules (or atoms) are in an "excited" higher energy state, and fewer in a lower energy state, than is normally the case. When an excited molecule drops back to a lower energy state, it emits radiation at a precisely defined wavelength. This radiation stimulates other molecules to do exactly the same thing. They drop back to the same lower state, emitting radiation in step with the original radiation and having the same wavelength. This effect quickly spreads throughout the lasing material (the *lasant*), and a laser beam is produced. Mirrors are usually placed at each end of a resonant cavity which contains the lasant. They reflect the radiation back and forth in order to stimulate further emission along a very narrow range of angles.

A major task is to arrange the molecules of the lasant so that there is a "population inversion"—i.e., so that there are more molecules in an excited state than in a state of lower energy. A suitable lasant must be found, and the energy needed to "pump" it—i.e., to raise its molecules to the upper laser state—must be provided. There are several ways to provide the energy for this purpose. Some lasers use chemical energy (in the form of a chemical reaction which produces molecules in an excited state); others use electrical energy. The characteristic wavelength produced depends on the material used as a lasant, and it is determined by the difference in energies between the upper and lower states.

The effectiveness of a laser as a beam weapon depends on the rate and amount of energy which can be delivered per unit area on a target. This quantity is determined by the laser power, the distance to the target, and the degree to which the beam can be focused on the target. Effectiveness also depends on the retarget time.

All electromagnetic radiation, even focused radiation, eventually spreads out with distance. This spreading, known as **diffraction**, results in a beam which becomes less intense as it travels out from its source; the maximum possible intensity of a laser beam (assuming the greatest possible degree of focusing) falls off as the square of the distance from the laser. The amount of diffraction depends on both the wavelength of the radiation and the diameter of the mirror, with the minimum possible spreading angle in radians[11] being equal to about 1.2 times the ratio of the wavelength of the radiation to the diameter of the laser aperture.[12] This angle of spreading is an ideal limit, assuming perfect optics and perfect focusing. An important consequence is that the smaller the wavelength, or the larger the laser mirror diameter, the less spreading occurs. To reduce diffraction, and therefore to reduce the beam size on target and deliver more energy per unit area, wavelength should be minimized and mirror size should be maximized.[13] Of course, in choosing these parameters, one is limited by physical and engineering constraints.

Smaller wavelengths, while allowing smaller mirrors for the same amount of spreading, also impose more stringent tolerances on the quality of the optics used. The size of the irregularities in the optics must be much less than one wavelength of the radiation used.

Aiming radiation at a moving target thousands of kilometers away requires highly accurate tracking and pointing. Typically, a beam spot of roughly a meter in diameter is envisioned for attacking today's missiles in the boost phase. To hit a target with an error of tenths of a meter at a distance of thousands of kilometers (km) requires aiming accuracy of about a tenth of a microradian. This is equivalent to hitting a television set in Los Angeles with a beam fired from directly over New York City.

In order to complete a thermal kill, the beam has to dwell on the target long enough to deposit a lethal amount of energy. Tracking accuracy must therefore be maintained over that interval. During one second, an ICBM may travel from 1 to 7 km (depending on when it is engaged) and can sweep through an angle (as observed by the laser weapon) of up to about 3,000 microradians.

Chemical Lasers:

Description.—Chemical lasers use the energy from a chemical reaction between two fuels to produce laser radiation. The most mature chemical laser technology for high-powered lasers is the hydrogen fluoride (HF) or the deuterium fluoride (DF) laser, in which hydrogen (deuterium) and fluorine combine to form hydrogen (deuterium) fluoride. Relatively high levels of power have already been produced in this type of laser, although a major scale-up from these levels is still needed

[11] One radian is an angle of 360/2π (about 57.3) degrees. It is defined as the angle subtended by that portion of the circumference of a circle having a length equal to the circle's radius.

[12] Ashton Carter, *Directed Energy Missile Defense in Space*, background paper prepared for the Office of Technology Assessment (Washington, DC: U.S. Government Printing Office, April 1984), p. 17, takes the spreading angle to be 1.2 times the wavelength divided by the mirror diameter. It is noted that the full angle subtended by the null ring of the Airy disk diffraction pattern requires a multiplier of 2.4. However, most of the energy is contained within a diameter only half as big; therefore, 1.2 is taken as the multiplier. This assumption is favorable to the laser technology, since not all of the beam's energy is contained within this angle; the lethality of the actual beam will thus be slightly less than the estimates in this section.

[13] For example, the minimum diffraction angle for a wavelength (in the infrared) of 3 microns (millionths of a meter), using a perfect mirror 10 meters in diameter, would be 3.6×10^{-7} radians, or 0.36 microradians. Even this small angle, however, would result in a beam spot of about 1 meter diameter at a distance of 3,000 km.

before power levels necessary for BMD can be obtained. The HF (DF) wavelength is 2.7 (3.8) microns (millionths of a meter). Other chemical lasers at different wavelengths are under consideration, such as oxygen-iodine (1.3 microns wavelength), iodine fluoride (0.65 and 0.72 microns), and nitrogen oxide (0.24 microns).

For illustrative purposes, we can look at potential requirements for an HF laser. As mentioned above, the diffraction phenomenon sets a lower limit on angular spreading. Approaching this limit requires a significant technical effort; for illustrative purposes, the following example assumes that this theoretical limit can be attained in practice.

The beam from an HF laser with a 10-meter diameter mirror would have a minimum angular spread of $1.2 \times 2.7 \times 10^{-6}/10 = 0.32$ microradians. At a distance of 1,000 km, therefore, a spot size of about 0.3 meters diameter would be produced. A laser of 20 megawatts (MW) output power would have an intensity of 25 kW/cm^2 at this distance. A watt is a joule per second. Therefore, exposures of 0.04 to 4 seconds would be required to reach the level of 1 to 100 kJ/cm^2. At a distance of 2,000 kilometers, exposures four times as long would be required. At 100 kilometers, the required times would be 100 times shorter.[14]

The length of time required to deliver a lethal amount of energy is inversely proportional to the power of the laser, if the other parameters are held constant. Thus, if a 20 MW laser were to be replaced by a 40 MW laser of the same wavelength and mirror diameter, the required dwell time would be cut in half.

We could also increase the diameter of the mirror to extend lethal range. By doubling the mirror diameter to 20 meters, the spreading angle is reduced by a factor of 2, and the delivered intensity increases by a factor of 4. Therefore, the lethal range of a laser weapon grows as the laser power and as the square of the mirror diameter.

Issues.—Several technical questions must be resolved in order to demonstrate the feasibility of the chemical laser approach. The required laser power levels must be approached closely enough to assure that no significant engineering problems will prevent scaling up to the full required power. Mirrors of the required dimensions and quality must be constructed and tested at high power levels. The total system of the required power, optical quality, and physical size must be robust enough for transport to on-orbit position. The atmospheric absorption of HF infrared radiation does not permit ground-basing (although this is not necessarily true for other wavelengths under consideration). Finally, the physical characteristics of the system should permit the installation of many units in orbit, given the transport shuttle capabilities likely to be available within two decades or so. To deal with Soviet countermeasures making their boosters more resistant to attack, it would be necessary to increase greatly brightness levels over those needed to counter existing Soviet ICBMs. Such devices would be several orders of magnitude beyond present capabilities, and would require reducing the laser wavelength, increasing laser power, increasing the size of the optics, or some combination of the three.

Excimer Lasers:

Description.—Another promising area of laser research is the excimer laser. An "excimer" is an excited dimer, or two-atom molecule, typically consisting of a noble gas (e.g., argon, krypton, xenon) atom and a halogen (e.g., chlorine, fluorine) atom. In an excited state, these two atoms can form a bound molecular system. When the molecule drops to a ground state, it rapidly disassociates into two separate

[14]The device parameters used here are only intended as examples. Although relatively powerful HF lasers have been constructed and operated, none has yet come close to a 20 MW rating. The largest telescope mirror in the United States, at the Mt. Palomar observatory, is 200 inches (5.08 m) in diameter; a telescope using a 10-meter diameter mirror is currently being designed by astronomers at the University of California. The diameter of the Space Telescope mirror is 2.5 meters. No mirrors of this size have yet handled megawatts of electromagnetic radiation. There are, however, no obvious technical bars to prevent either the laser or the mirrors from being developed.

atoms: noble gases do not form stable molecules in the ground state. The excited population of excimer molecules is produced by a pulsed electrical discharge process, rather than by a continuous chemical reaction. The light produced, therefore, occurs in pulses. After the pulse of laser radiation is produced, the process repeats with a new electrical discharge, leading to another "pumping" of excited dimer molecules. Relative to HF lasers, excimer lasers have the advantage of a shorter wavelength (typically 0.3 to 0.5 microns in the near ultraviolet to visible region of the spectrum), which greatly reduces the size requirements on mirrors. As a result, however, the optical requirements for mirror uniformity are that much more stringent.

The reduced requirement on mirror size, if such mirrors can be made, is a significant advantage over longer wavelength options. The fact that the wavelength is only about one-tenth that of infrared lasers means that, for a given range, the mirror's diameter need only be one tenth that required for chemical lasers. The area, then, would only be one-hundredth as large. Since the thickness of a mirror, including its support structure, can be kept fairly constant over a substantial range of diameters, its weight will be approximately proportional to its area. Excimer laser mirrors, then, with one-tenth the diameter, may weigh only on the order of one-hundredth as much as HF laser mirrors with the same capability. To see the advantage of the shorter wavelength laser, consider the hypothetical example of placing a laser in geosynchronous orbit where it could always see all of the Soviet missile fields. The distance of effectiveness would have to be about 40,000 kilometers. At that range, an HF laser would require a perfect mirror of about 130 meters in diameter to keep the beam size down to 1 meter in diameter at the target. This is infeasible for the foreseeable future. However, an excimer laser would require a mirror of the order of "only" 15 meters in diameter for the same size beam spot.

Placing any sort of a BMD-capable laser in geosynchronous orbit may, however, be impractical. Instead, a ground-based laser might be aimed at one or more mirrors in geosynchronous orbit. This scheme would have the obvious advantage of utilizing a ground-based power source, allowing continuous operation for long periods of time and reducing the weight placed into geosynchronous orbit. There would have to be several lasers, since cloud cover could render some of them useless at a given moment. The geostationary **relay** mirrors would reflect the beams from the ground lasers onto smaller **battle** mirrors in low-earth orbit, which, in turn, would track individual targets and redirect the laser beam to them.[15] The geostationary relay mirror would have to be much larger in diameter and much more complicated if it were to attack boosters directly without the help of lower orbit battle mirrors. Using the battle mirrors, the relay mirror need not track individual targets at all. A constellation of battle mirrors would be needed so that enough of them could always be on station to deal with missiles launched from all possible launch sites. Note that this scheme is impractical for long wavelength lasers since the required mirror sizes are so large.

Even at short wavelengths, a very powerful laser would be needed in order to travel through the atmosphere and bounce off several mirrors while retaining its lethality. (The additional spreading introduced by making the beam travel out to geosynchronous orbit and back can be compensated for by making the relay mirrors sufficiently large in diameter.) A very large quantity of power (hundreds of megawatts) would have to be available on short notice to each of the ground-based lasers.

In order to compensate for atmospheric distortions, a technique known as "adaptive optics" is being developed. A pilot laser beam

[15]Relay mirrors could be in lower orbit than geosynchronous, and could therefore be somewhat smaller than they would have to be if they were in geosynchronous orbit. However, since mirrors in lower orbit would not remain over the same spot on the ground, enough would be required so that one or, preferably, more would always be in a position to relay the laser beam to an appropriate battle mirror. In addition, if the relay mirror orbits were too low, more than one bounce would be required to direct the beam from a ground laser to a battle mirror on the other side of the Earth.

Figure 7-2.—Compensating for Atmospheric Distortion With Adaptive Optics

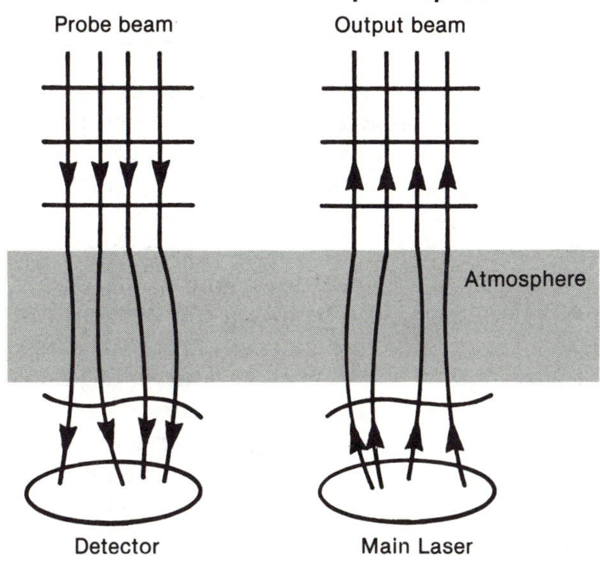

Phase information from an incoming probe beam reveals atmospheric distortion. Large ground-based laser beam is distorted in the opposite sense to cancel out the atmospheric effects.

SOURCE: Office of Technology Assessment.

sent from the space mirror would be detected at the ground-based laser. Information from the beam would reveal the pattern of phase shifts caused by atmospheric distortions. As a result, corrections could be applied to the beam generated by the ground laser, possibly by distorting a mirror at many points over its surface, in such a way as to compensate for the atmospheric effects. Atmospheric disturbances typically occur over times which are on the order of tenths of a second, during which time the pilot beam can travel through the relevant part of the atmosphere (the 20 km nearest the surface), the main laser beam can be corrected for the distortion, and the beam can propagate back out through the atmosphere, all before the disturbance changes significantly. As atmospheric distortions change, the optics would automatically compensate.

Issues.—To determine the feasibility of this ground-laser/space-mirror approach, the adaptive optics method of compensating for atmospheric distortion must be examined for high power levels over long distances. The ability to compensate for atmospheric effects in the presence of such an intense beam has not yet been demonstrated. Mirrors of the required size and robustness, and satisfying the exacting tolerances, must be constructed. They must withstand intense laser beams without distorting significantly or failing. Unlike the HF case, only a few very large ones need be made. However, many more smaller battle mirrors, each with a diameter of about 5 meters, would be needed. With countermeasures by the offense, this number could increase further.

A possibly significant problem for this system, as well as any other directed-energy system, is the need to retarget from one object to another in 1 second or less. This requirement may be quite difficult to meet; greater retargeting times would not rule out given weapons systems, but could imply the need for far larger constellation sizes (see section on System Architecture, p. 179 ff.). If only small retargeting angles are needed for a single satellite, fast retargeting may be easier to attain; this possibility is under investigation.

Another issue to be resolved is the ability to produce excimer lasers of the power levels required. Current excimers are several orders of magnitude smaller than requirements for SDI applications. Large amounts of power will have to be delivered in short pulses. The required power levels will depend on the results of research in the various fields. Excimer lasers tend to have a high weight-to-output power ratio, which would make them more problematic for space-basing. This would not affect a ground-based mode, where weight is less of a consideration.

Free-Electron Lasers:

Description.—When the paths of charged particles are bent by a magnetic field, they emit radiation. The recently developed free-electron laser uses this principle in an innovative way to produce laser radiation. A beam of electrons is passed through a periodically-varying magnetic field. The radiation produced can provide an intense coherent beam. In the free-electron laser, the interaction of the electrons and the magnetic field replaces the

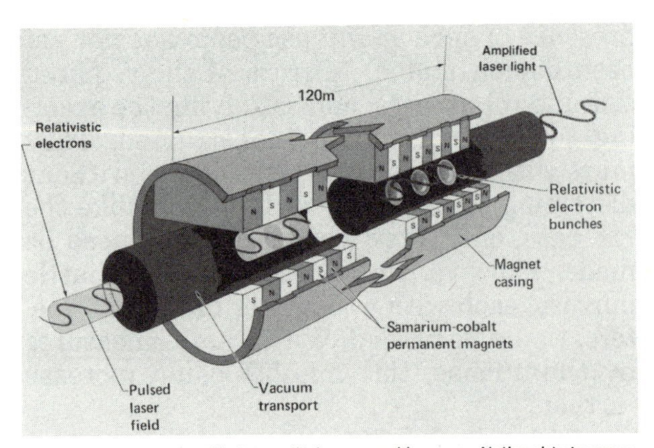

Photo credit: Lawrence Livermore National Laboratory

A schematic drawing of a free-electron laser with field-alternating "wiggler" magnets. The magnetic field forces the electron beam to follow a sinusoidal path, causing them to emit laser radiation. In this drawing, a low-power laser beam is amplified by the electrons.

excited energy levels of a lasant as the source of coherent radiation. The wavelength, which depends on the periodicity of the magnetic field and on the electron energy, can be changed as desired by varying either one. Such a laser operating in the visible could be ground-based, using space-based reflectors to reach targets beyond the horizon (see previous section). High energy efficiencies could also permit the possibility of space-based lasers.

In addition to the advantages of good beam quality and high energy efficiencies which are obtainable, the free-electron laser also has the advantage of being able to use a relatively mature technology: that of the particle accelerator.

Issues.—The process is potentially more energy efficient than other schemes, and it has the significant advantage of frequency tunability. The technology is in its infancy, although progressing rapidly, and much research effort is needed to determine its potential for application as a directed-energy weapon. The SDI program is investigating whether power levels can be scaled up by many orders of magnitude at useful wavelengths. It is also studying, as in the case of excimer lasers, whether window and mirror materials can be developed which are capable of withstanding the intense laser beam while maintaining the required optical quality.

X-Ray Lasers:

Description.—Like the free-electron laser, X-ray lasers are relatively new. The "pumping" of the lasant material to an excited state would be accomplished by intense sources of radiation, such as a nearby nuclear explosion, an optical laser, or some other source. Should a nuclear explosion be used to pump an X-ray laser, that laser could be lethal to a target even if the energy conversion process were very inefficient since the energy produced in just a small nuclear explosion is still very large. The U.S. Department of Energy is investigating the feasibility of developing nuclear-pumped X-ray laser weapons; however, it has classified virtually all details of this research other than its existence.

An advantage of a nuclear-pumped X-ray laser weapon would be that it would have the potential for killing many targets using multiple beams, providing high leverage and countering attempts to saturate the defense. A disadvantage of such a weapon would be that it could be fired only once—the explosion that powered such a weapon would very shortly afterwards destroy it. Such a weapon would not be able to assess damage and fire again, although a second weapon could certainly do so.

There are natural limits on the distance to which X-rays can propagate within the atmosphere, where they are rapidly absorbed. Since an X-ray laser used for a boost-phase defense must therefore wait for a booster to climb higher than the minimum altitude to which the X-rays can reach (which depends on parameters such as X-ray intensity, wavelength, and incident angle), the time available for the defense to act is reduced.

A conceivable mode for use of X-ray lasers, assuming that they would be developed as weapons, is the pop-up technique. The relatively low weight of such a weapon system could contribute to the desirability of such an architecture. The lasers could be deployed on specially developed submarine-launched missiles.

When the defense receives notice of an attack, it could launch its pop-up weapons to an altitude sufficient to attack one or several ICBM boosters or post-boost vehicles after they have risen above the minimum engagement altitude. In such a system, the weapons would not have to be deployed in space, avoiding a serious vulnerability problem faced by space-laser or ground-laser/space-mirror schemes. Also, deploying nuclear-pumped weapons on submarines, rather than in space, would avoid violating the Outer Space Treaty (see appendix C). Other military applications of this technology may be possible, but are beyond the scope of this report.

Issues.—The first question to be resolved is whether the X-ray laser can be developed to the point of use as a weapon. The efficiency of the conversion process and the possibility of achieving adequate levels of brightness are major issues. If a pop-up mode were to be investigated, secondary systems-related questions would also arise. Since the X-ray lasers would have to be popped up after a Soviet launch, their boosters would have to be substantially faster and more rapidly accelerating than the Soviets', which would have a head start. This means that the pop-up would have to burn much more fuel per unit payload weight than its target. A system would have to be developed with an almost instantaneous response time, including high-quality communications links between the orbiting satellite sensors and the submarines. Further, a submarine, which might only be able to fire one rocket at a time with a delay between successive launches, could become a "sitting duck" once it had revealed its location by firing. The practicality of a global scheme involving pop-up X-ray lasers of this type is doubtful.

Particle Beams:

Description.—Unlike electromagnetic radiation, which consists of pure energy, beams of subatomic particles consist of bits of matter. They can be protons, electrons, neutral atoms, heavy ions, or more exotic types. They are accelerated to velocities approaching the speed of light by electric fields in particle accelerators. Accelerators for diverse types of particles exist in various sizes all over the world. They are used for fundamental research in the areas of solid state, nuclear, and, above all, high energy physics. It is known and understood how to produce and accelerate all manner of atomic and subatomic particles. The challenge for weapons purposes is to produce beams of very high intensity which are also extremely collimated—i.e., which are narrow and have a small spreading angle, so that the particles move in very nearly parallel paths. Accelerators producing such beams must be light enough to be placed in space economically, and they must be very reliable.

In order to accelerate particles with electric fields, one must use charged particles. However, over long distances, a charged beam will bend in the Earth's magnetic field, presenting formidable difficulties in targeting. Inhomogeneities in the Earth's field can render it virtually impossible to direct a beam at a target spot of a meter size, or so, at the distance of thousands of kilometers.

Preliminary experiments in laboratories have led to some hope that for low-Earth orbit altitudes, it may be possible to use a charged beam by means of the following mechanism. A laser first ionizes a straight path through the rarefied near-space environment. Then, an electron beam is fired along this channel, with the positively charged gas ions providing an electrostatic restoring force which compensates for the bending forces of the Earth's magnetic field. Demonstrating the ultimate practicality of such a scheme requires further resolution of a number of issues.

A less exotic solution is to produce a neutral beam, unaffected by magnetic fields, which will travel in a straight line and be more easily directable to a target. To make a neutral hydrogen beam, for example, a large number of hydrogen ions is created by attaching an extra electron to neutral hydrogen atoms. The charged ions (H^-) are then accelerated by electric fields, and, after exiting the accelerator, are neutralized by one of a variety of techniques. The extra electron can be knocked off by passing the ion beam through a small amount of matter, for example, or it can be stripped off by means of appropriately tuned laser radiation. In either case, a set of neutral hydrogen atoms is again produced. Now, however, they are traveling together in very large numbers at nearly the speed of light.

This beam of neutral particles contains a significant amount of energy, can penetrate several centimeters into virtually any material, and can penetrate typical aerospace materials to a depth of tens of centimeters. Because of this penetrating power, neutral particle beams may be difficult to countermeasure. The hydrogen atom's electron is quickly stripped off as the atom enters the target. The bare proton which remains deposits its energy more or less uniformly along its path through the material, with a slight enhancement in the small region where it finally comes to rest. If the beam strikes electronic circuits, such as those in guidance systems, they can be fatally altered or destroyed, rendering the target "stupid" and unable to function in its programmed way. An energy deposition of some tens of joules per square centimeter could be sufficient to destroy unprotected electronics in a target. The deliverable energy requirements for this type of weapon, therefore, are considerably less stringent than in the case of lasers, where 1 to 100 kJ/cm^2 (10 to 1,000 MJ/m^2) may be required. At higher levels of neutral particle beam energy deposition, objects can be melted and explosives detonated.

Such a weapon, however, can work only outside the atmosphere: even a small amount of air will strip off the electrons, resulting in a beam of charged protons. These will be bent by the Earth's magnetic field and will also be scattered by collisions with atmospheric molecules. As a result, the beam will not be effective against targets below about 100 km.

Issues.—The energy of each particle in a particle beam is not a serious problem. Energies of several hundred million electron-volts are usually discussed in this context, which are well within the limits of current capabilities for the particles in question. In fact, the largest accelerators today are able to reach energies a thousand times higher, although at lower intensities.

Individual particle energy, however, is only one of the parameters that determines a particle beam's effectiveness as a weapon. Another is the beam *current*, which is proportional to the number of particles per second in the beam. The beam current (in amperes[16]) multiplied by the energy of each particle (in electron-volts) gives the total beam power (in watts). The **brightness** of a particle beam, which determines the power that can be delivered per unit area at a given distance, depends on the beam power and additionally on how tightly the beam can be focused. At present, particle beams are closest to the required level of brightness of all the directed energy options generally discussed for SDI.

One problem with a particle beam weapon will be kill assessment. A beam intensity sufficient to disable its target's electronics may not be sufficient to produce external effects which are immediately observable. This drawback might impose severe problems regarding targeting decisions if neutral beams are to be used in such a "soft," functional kill mode. To provide externally observable effects, brightness levels perhaps a thousand times greater might be required, unless the guidance system of a booster or post-boost vehicle were struck in such a way as to cause obvious trajectory modifications.

A related problem is tracking the beam. If the beam misses the target, it will be very difficult to know where it went; even if it strikes the target, it may not be visible at "soft" kill intensities. "Open-loop" pointing, in which one measures the direction of the beam as it exits the accelerator with great precision, is a possible solution, but it remains to be demonstrated.

A further problem would be presented if the electronic components of beam weapon targets were hardened against radiation. Circuits using gallium-arsenide (GaAs) technology could be as much as 1,000 times more resistant to radiation than commonly existing cir-

[16]Neutral beams, which are not charged, technically carry no electrical current. The intensity of a neutral particle beam in amperes is the electrical current that the beam would carry if each particle in it had the charge of one electron.

cuits based on silicon technology.[17] Such hardening could increase resistance to ionizing radiation by a factor of up to 1,000, stressing further the energy delivery requirement of a particle beam weapon. The system hardness, however, may not be increased by the same large factor as the component hardness.

Another serious issue is whether an accelerator can be constructed which will be light enough to permit many to be placed in orbit, and which will retain the ability to function reliably after long periods of dormancy. While current Earth-bound accelerators approach the necessary beam intensities, they are large and heavy and also require maintenance and repair at irregular intervals. Space-qualified equivalents would need to be much lighter, much more reliable (since they would be harder to fix), and like other space-based assets, protected against attack.

Photo credit: Los Alamos National Laboratory

The radiofrequency quadrupole (RFQ) at the White Horse facility at Los Alamos National Laboratory. White Horse is a project to develop intense neutral particle beams. The RFQ is a device which can accelerate a beam of particles in a much shorter distance than previous acceleration techniques could. This helps make it possible to build accelerators which are much smaller in size, a necessary requirement for space-basing.

Kinetic Nonnuclear Kill

Boost, Post-Boost, and Midcourse Phase:

Description.—A classic method of destroying a target is simply to hit it with another object having a large velocity relative to the target. This method utilizes **kinetic energy**, or energy of motion, and has been used for a good many millenia in forms such as rocks, catapults, arrows, and bullets. Missiles and RVs travel at high speed, typically several kilometers per second; they can be killed quite effectively by colliding with something else moving at a significantly different velocity. The problem lies in arranging the collision—in reaching the missile or warhead and hitting it.

This technique could be implemented by a constellation of space-based battle stations, each containing a large number of small rockets or electromagnetically launched projectiles. Satellite sensors would detect a launch and would hand tracking information over to the battle station. The rockets would be assigned to, aimed at, and launched towards their targets. When close enough, homing detectors on the projectiles would be used to direct them to their targets. The kill could be by means of striking the target directly or by detonating an explosive near it, sending fragments into it. (Outside the atmosphere, of course, an explosion does not produce a shock wave, so the fragments would be necessary for a kill.)

For attacking boosters, the bright infrared signal from the rocket plume serves as a target for a short-wave infrared homing device;

[17]Gallium arsenide (GaAs) technology shows promise for making circuits which are faster, as well as much more radiation-resistant, than circuits based on silicon technology. It is therefore a topic of intense research interest by the Department of Defense as well as by private industry. GaAs technology will likely be considerably more expensive than equivalent silicon circuits, primarily because silicon technology has had the benefit of decades of intensive research. However, it will probably become the technology of choice for many applications requiring speed or radiation-resistance.

however, the projectile will probably need to correct for the distance between the booster body and that portion of the exhaust plume flame or engine nozzle which emits most brightly in the infrared. For post-boost vehicles or RVs, short-wave infrared sensors could not be used since the PBVs (except during short bursts of their rocket motors) and RVs would be at much lower temperatures than the boosters and would radiate much weaker signals at longer wavelengths. Cooled, long-wave infrared detectors could be used during the post-boost and midcourse phases, but other sensing devices might be required because the relatively cool targets may not be readily detectable by infrared means against the background of the Earth. Short- and long-wave infrared detectors, along with other sensors, are discussed further in the following section on sensors (p. 159 ff.).

Issues.—A major question is whether the space-based rocket could reach the booster before burnout. Basing at altitudes of about 400 kilometers has been discussed. Soviet SS-18 rockets burn out in about 300 seconds at an altitude of about 400 km. If the satellite platforms carrying the interceptor rockets were based at an orbital altitude of about 400 km, an interceptor could travel horizontally for up to 300 seconds to reach a booster if it could be launched at the same time that the booster was. If the platform is based at a higher altitude, the interceptor will have to shoot down to reach the booster and will not have as far a horizontal range.

If the interceptors had a burnout velocity of 10 km/sec,[18] each could travel about 3,000 km in 300 seconds, giving them a useful range. A terminal velocity this high for a chemical rocket would imply a very small ratio of payload weight to fuel weight; this would be compensated for, in part, by multi-staging, but the need for a low-cost lift capability to place the interceptors and their fuel in orbit would nevertheless be manifested.

[18]For comparison, a satellite in low-Earth orbit travels at about 8 km/sec.

If the length of the boost phase were reduced, interceptor ranges would be correspondingly reduced. The MX missile burns out in about 180 seconds. The Soviets are currently testing an MX-like ICBM (the SS-X-24) which would therefore effectively shorten the maximum interceptor range from that attainable against an SS-18. A fast-burn booster would reduce the effective range still further.

Another issue for infrared homing devices involves the ability of such detectors to function in the upper atmosphere. Since friction with the atmosphere will heat the skin of the interceptor rocket, any infrared detector will have to look out through a very hot window. Windows which do not emit much infrared radiation even when heated to high temperatures will be required. If homing interceptors cannot be made to operate in the atmosphere, it will become necessary to wait until boosters have left the atmosphere before intercepting them. Alternatively, interceptors might dispense with homing sensors, being guided by commands sent from other satellites better able to track the boosters ("command guidance").

Homing outside the atmosphere has already been demonstrated in a test configuration. The Homing Overlay Experiment conducted by the U.S. Army in June 1984 demonstrated the ability to find a cool target outside the atmosphere against the cold background of space, and to home in on it accurately enough to collide with it. Similar technology is utilized by the U.S. Air Force's air-launched ASAT weapon.

Interceptors could attain higher velocities using a developing technology: the electromagnetic railgun. An intense magnetic field is used to impart large velocities to electrically conducting projectiles; the conductor can be formed by ionizing a substance which might be an insulator in its normal state. Speeds of greater than 20 km/sec or more are envisioned. Such techniques would appear promising because they could greatly extend the range of interceptors based in space. However, attaining such a high velocity by the time the projectile has left the gun requires accelerations hundreds of thousands of times that of gravity.

Problems to solve, besides the actual proof of principle at these high velocities, are the development of large power sources, power delivery in short pulses, good recoil momentum compensation so as not to degrade pointing capabilities, and the development of materials and guidance systems of very low mass which can survive the rapid accelerations needed. The ability to refire rapidly and accurately would also have to be developed.

Terminal Phase:

Description.—In the terminal intercept phase, kinetic-energy interceptors could be very high acceleration rockets located near the sites to be defended. If nonnuclear, they would kill by striking their targets or by detonation and fragmentation near the target. Phased-array radars (electronically steered and able to shift rapidly from one target to another) would track RVs and decoys and give pointing information to the interceptors, which would then home in on their targets with radar or infrared sensors.

Another possible technique is the "swarmjet" proposal. A large number of small rockets is fired in the direction of an incoming RV towards a region 50 m in diameter at a range of 1 km from the defended site. If properly timed, the swarm would have a high probability of destroying the RV. Since intercepts take place close to the ground, decoys will have already burned up during reentry and will not be a problem. However, the attacking warhead may be salvage-fused to detonate when intercepted, and since intercepts will take place relatively close to the defended object, a "swarmjet" defense would only be suitable for defending hardened targets able to survive a nearby nuclear explosion.

Issues.—The ability of a nonnuclear homing device to kill an RV outside the atmosphere has been shown. However, in addition, either the interceptor or (more likely) the overall battle management system must be able to discriminate between decoys and RVs (see discussion on discrimination, p. 162 ff.). For intercepts deep within the atmosphere, the atmosphere itself will screen out decoys.

Cost-exchange issues will be very important. For a given hypothetical defense system, the cost of large numbers of interceptor vehicles will have to be compared to the cost of additional incoming RVs. Crucial to the cost calculations is the defensive system's footprint, i.e., the size of its defended area. The larger the footprint, the fewer systems are needed.

Overall, the state of the art of terminal interceptors, with the associated sensors and battle management systems, is closer to practicality than many other BMD technologies. However, these technologies at present are best applied to hardened targets. Discrimination is easier because intercepts can be delayed, and a far smaller volume of space would have to be covered. Near-term technology may be capable of defending hardened targets against a significant fraction of incoming RVs. However, the detonation resulting from the first intercept (in case the target were salvage-fused, giving a nuclear explosion upon impact) could make subsequent intercepts difficult. These problems could be mitigated by hardening sensors and by providing high levels of redundancy.

Interceptors would themselves have to be placed in hardened sites in order to remain operational in the case of a nearby nuclear explosion. The survival of some fraction of the targets could thus probably be assured, unless a very large number of RVs per target were attacking.

Soft targets, however, would be more difficult to defend. As has been stated, the higher intercept altitudes needed to protect soft targets make discrimination harder and also require defending a much larger volume.

Nuclear Kill:

Description.—The discontinued U.S. Safeguard ABM system used a nuclear warhead to kill incoming RVs. Such a system was desirable when homing systems could not approach closely enough to kill by impact or by explosion and fragmentation. The Low Altitude Defense System (LoADS), which used nuclear-armed interceptors for protecting hard

Photo credit: U.S. Department of Defense

Launched in November, 1984 from White Sands Missile Range, New Mexico, as part of the Small Radar Homing Technology (SRHIT, pronounced "S-R-hit") Program. A novel steering steering system is being used.

targets, had been under development until it was recently deemphasized under SDI. In principle, nuclear-armed missiles, representing a mature technology, could soon be operational as elements of a terminal defense. Although major uncertainties still remain concerning the operation of such a system in the presence of many nuclear detonations, improved sensors, radar tracking, and communications would result in a more effective system now than could have been built in the early 1970s. These improvements would make nuclear interceptors a possible fall-back position for terminal defenses, in case serious impediments develop in adapting nonnuclear kill technologies for that purpose. For example, maneuverable reentry vehicles (MaRVs) might be able to evade interceptors to the extent that a nonnuclear kill vehicle could not approach within lethal range. The greater kill radius of a nuclear warhead might compensate for inability to achieve a close approach.

There could also be uses for nuclear kill against space-based defenses. Space mines, or weapons placed in orbit with the purpose of detonating on command to destroy enemy battle stations or to neutralize satellite sensors, could be nuclear-armed. They could also be salvage-fused so that, once within lethal range of a potential target, they could destroy that target even if attacked themselves.

In terms of killing attacking missiles or RVs, nuclear kills may be less desirable, since they might not destroy more than one target at a time but could complicate other defensive actions by damaging or blinding elements of the defensive system.

Issues.—When using nuclear interceptors in the terminal phase, difficulties could arise from collateral damage or blinding of the defense's own radar tracking system and communications. Such use implies the need for hardened electronics, robust radar tracking, and effective battle management to minimize collateral damage. Homing systems that would permit use of very small nuclear weapons could mitigate some of these effects, particularly for exoatmospheric interception. If incoming missiles are salvage-fused, however, the environment would be stressful to defensive battle management whether or not the defenses use nuclear-armed missiles. It should be remembered that an advantage of nuclear kill is that the technology is essentially currently available.

Sensors and Data Processing

Advances in sensors and in data processing technology—in the ability to acquire and manipulate information—have had at least as much to do with the resurgent interest in ballistic missile defense as have advances in weaponry. In addition to their key roles in BMD technologies, sensors and data processors probably have greater general application than advanced weapons concepts in other military (and of course in civilian) applications.

All of the functions of a BMD system, save target destruction, involve primarily sensors and processors. As sensors acquire more and more data and incorporate greater amounts of processing directly within the sensing components, it becomes increasingly difficult to separate these two functions. Perhaps a better breakdown would be sensors (including processing), and other data processing activity, such as battle management or command and control. Battle management will be discussed in a later section on "System Architecture"; the following discussion will concentrate on the data acquisition and manipulation performed by the sensors of a ballistic missile defense system.

Sensors can further be broken down into surveillance and acquisition sensors, whose primary function is to notice threatening objects and determine their approximate location, and higher resolution sensors, which investigate these objects in much greater detail.

Surveillance and Acquisition

There are a number of technological candidates for performing surveillance and acquisition functions. They are distinguished in this section by the phase in which they would most appropriately be used. (Sensors used for discrimination, as opposed to surveillance, are discussed in the next section. See p. 162 ff.).

Boost Phase.—The hot gases exhausting from an ICBM booster motor emit hundreds of kilowatts at short- and medium-wave infrared (SWIR and MWIR) wavelengths of a few microns. This radiation can be detected by sensors at great distances. Both the United States and the Soviet Union now obtain early warning of ballistic missile launch by sensing the infrared radiation from these exhaust plumes; U.S. early warning satellites are at geosynchronous orbit 36,000 km above the Equator, while their Soviet counterparts travel in highly elliptical orbits which are at even higher altitudes when over the United States.

The launch detection sensors characterize the approximate size and trajectories of the ICBM attack in order to "hand off" the suspected targets to systems having higher resolution, which can examine the objects and aim at the threatening ones. In addition, some discrimination can be done at the earliest stages of detection, depending on the spatial and spectral resolution of these early warning sensors and the image processing software used with them. If the infrared sources are not moving, or are not moving towards defended areas, then they do not pose a threat.

If a boost-phase layer is present in the defense system, it will only have a few minutes after launch detection in which to destroy the climbing boosters. Once infrared sources are detected and are identified to be ICBMs, the detection sensors will "hand off" their tracks to the pointing and tracking sensors associated with each weapon system.

Post-Boost and Midcourse Phases.—Surveillance requirements become considerably more difficult in the post-boost and midcourse phases because the objects to be detected are no longer necessarily associated with conspicuous infrared sources. By the end of the boost phase, all the ICBM booster stages have burnt out and dropped off, leaving the post-boost vehicle. The PBV then dispenses reentry vehicles and decoys, changing course slightly to aim each weapon individually before letting it go.[19]

The PBVs, decoys, and deployed RVs can be detected by their own radiation, rather than that emitted by their hot exhaust gases. Objects more or less at room temperature emit **long-wavelength infrared (LWIR) radiation** having wavelengths mostly near 10 microns. (By comparison, the Sun is hot enough to shine in the visible portion of the spectrum, with wavelengths primarily near 0.5 microns. Rocket exhaust, which is cooler than the surface of the Sun but is still much hotter than the PBVs, RVs, and decoys, emits radiation primarily in the short- and medium-wave infrared wavelengths of a few microns.)

LWIR radiation can be detected by sensors which are cooled to near absolute zero in order to prevent their own radiation from swamping the signal.[20] Such sensors can search for objects already in space (deployed warheads, satellites, decoys, or debris) without having to observe a launch, and they can provide independent backup for the launch surveillance systems.

In addition to detecting deployed warheads which may have escaped launch detection, a BMD system must do **space surveillance** in order to keep track of threats to itself. Before launching nuclear weapons against terrestrial targets, the offense may choose to attack the defensive system directly in order to damage or disable it. Therefore, the BMD system must keep track of satellites which could attack from a long range (those suspected of holding nuclear warheads or directed-energy weapons), objects which need to approach closely in order to attack (ground-based interceptors, or space-based nonnuclear ones), and satellites called **space mines** which, if allowed by the defense, could constantly trail system components and detonate on command, destroying both themselves and the BMD component.

LWIR sensors are useful for such space surveillance systems. Any object in space for a long enough period of time will reach a steady temperature when the rate at which it absorbs energy (either from the Sun or radiated up from the Earth) equals the rate at which it emits LWIR radiation. The amount of absorbed power, which tends to heat the object up, depends on the surface properties and the surface area of the object; the amount of radiated power, which tends to cool the object down, depends on its surface properties, its size, and its temperature. For example, the average temperature of the Earth itself is set by the balance between absorbed sunlight and emitted LWIR; a similar process goes on for all satellites.

In addition, any equipment on a satellite that uses electrical power will dissipate heat, further raising the satellite's temperature. Although modifying the satellite's surface properties can lessen the amount of LWIR power radiated at a given wavelength, doing so would also increase the object's reflectivity at those wavelengths.[21] Emission and reflection cannot be minimized simultaneously, and lessening one will increase the other.

As stated above, the Earth itself is a powerful LWIR emitter. It will be hard for LWIR surveillance sensors to pick out satellites when seen against this background. Therefore, space-based LWIR surveillance systems will look away from the Earth, spotting objects against the cold background of space. Low-orbiting satellites can only be detected from space by looking just over the Earth's horizon.

[19]The PBV itself intermittently fires small rocket motors to maneuver. These rockets emit detectable short- and medium-wavelength infrared radiation, but only when firing. Their radiation is not nearly as bright as that produced by boosters.

[20]Putting such sensors in space systems requires long-lived (at least several years), lightweight, and low-power **cryogenic** refrigerators capable of keeping them at their operating temperature.

[21]A highly reflective surface only absorbs a small fraction of the power striking it. However, since the ability of an object to **emit** power at a given wavelength is directly proportional to how well it can **absorb** that wavelength, a reflective object does not emit well, either.

A piece of metal left out in the sun will heat up, even though it is highly reflective, because it radiates even less power in the infrared than it absorbs in sunlight. The infrared power emitted by an object increases rapidly as the object gets hotter (doubling the temperature above absolute zero increases emitted power by a factor of 16), so the metal heats up until it can radiate away as much power as it absorbs.

Figure 7-3.—Long- and Short-Wave Infrared Detection

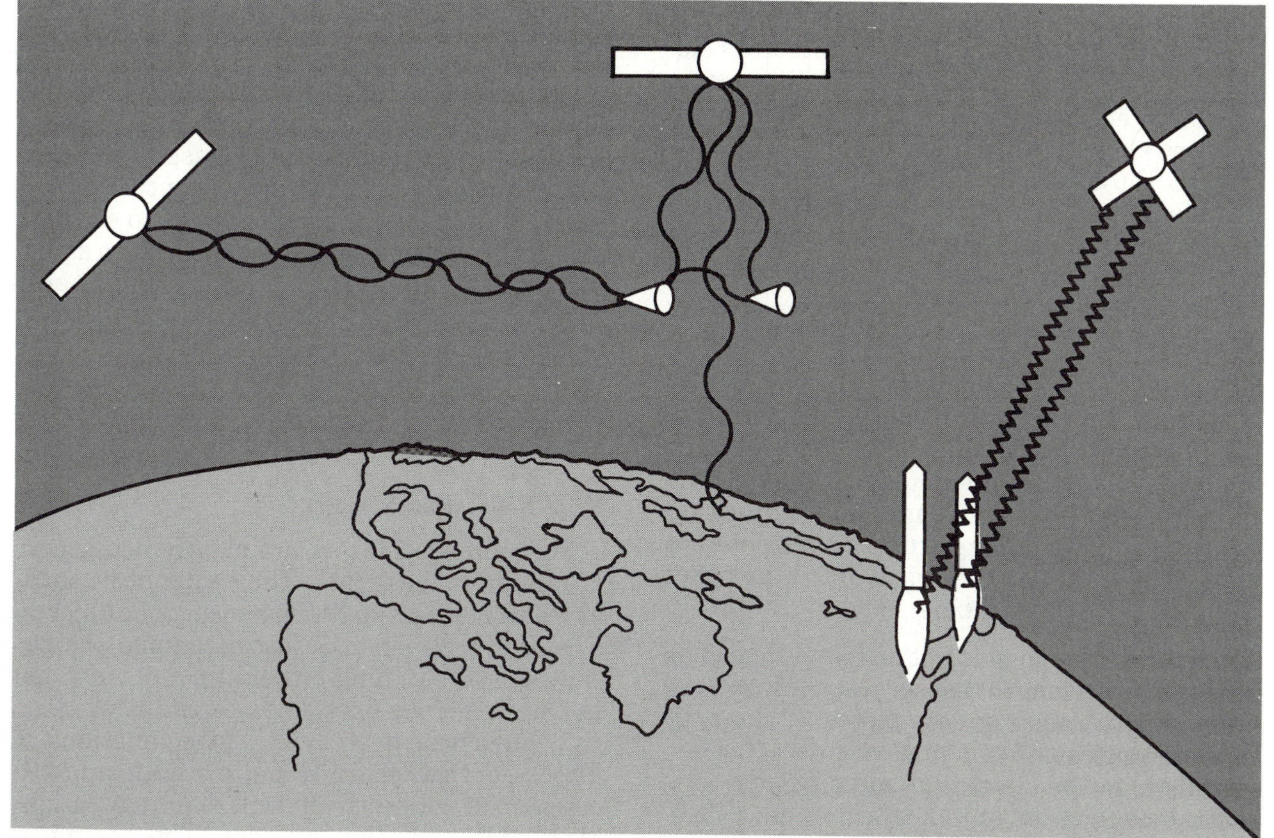

The hotter the object, the shorter the average wavelength of infrared radiation which it characteristically emits. Rocket plumes, having temperatures in the range of thousands of degrees Celsius, emit primarily in the short-wave infrared region, and are readily detectable against the Earth's background. Objects which are roughly at room temperature, such as reentry vehicles, emit mostly longer wavelengths, whch may be difficult to detect against similar wavelength radiation from the Earth. The surveillance satellite at left can more easily detect them against a space background.

SOURCE: Office of Technology Assessment.

Another problem may arise from the infrared backgrounds generated by nuclear explosions in the upper atmosphere. Such effects are only partially understood and may remain mysterious in the absence of experimental nuclear tests in the atmosphere. These would, of course, violate the Limited Test Ban Treaty. However, tests involving other sources of ionization are being carried out, in conjunction with computer simulations, to provide more extensive knowledge on what might happen in the upper atmosphere under such conditions.

Other techniques are available to do space surveillance. The U.S. Air Force at present uses both radar and optical observations to monitor objects in space from the ground. Due to interference from the atmosphere, LWIR sensors cannot be used efficiently on the ground. Airborne observations are, however, feasible.

Terminal Phase.—Reentry vehicles and decoys which survive the defenses long enough to reenter the atmosphere enter the terminal phase of a BMD system. Since reentry vehicles can be salvage-fused to detonate if they are attacked, interception must take place at a high enough altitude if "soft" targets below are not to be destroyed.[22] Surveillance systems

[22]This "keep-out distance" depends on the yield of the weapon and the hardness of the target. At sea level, a 1 megaton weapon will produce overpressure of 2 pounds per square inch, which structures might survive with repairable damage, at a distance of 13 km (8 miles) from the blast.

that could operate in the terminal phase include **ground-based radars** and **airborne optical and infrared detectors.** LWIR detectors, located on airplanes to provide mobility and to minimize atmospheric interference, can detect reentry vehicles which have not yet started to reenter the atmosphere, helping **exoatmospheric** interceptors to destroy them. Once the RVs have begun reentry, they heat up and start to glow, permitting shorter wavelength infrared and visible detection for **endoatmospheric** interception.

Issues.—The technology of SWIR and MWIR sensors is fairly mature. Additional software and on-board processing capability will have to be developed to do image processing. The requirements for surveillance and acquisition sensors and processing are not anticipated to stress the state of the art as much as other required BMD technologies will.

LWIR technology is not as far advanced as shorter wavelength sensor technology. As wavelength requirements increase, the task becomes more difficult since new detector materials must be developed and since the systems must operate at temperatures near absolute zero. However, LWIR space surveillance systems have been designed, and the technologies involved have been under investigation for a number of years. The data processing requirements of post-boost and midcourse phase surveillance sensors are also stressing but may not present major technical problems if computer science continues to progress over the next two decades at the same rapid rate which has been evident so far.

Radar technology of the sort applicable for terminal defense is well advanced; radars have been investigated for decades. Of particular interest is making such radars small and cheap, so that they can be proliferated (deployed in large number) to deny the offense the ability to blind the terminal defense by destroying a single, high-value radar. The wavelength at which radars can operate has decreased steadily as technology has progressed. More recently, advances in infrared technology have steadily **increased** the accessible infrared wavelengths. At present, the wavelength bands for which the two technologies can be utilized are starting to overlap, at wavelengths on the order of a millimeter.

Surveillance and acquisition sensors also have wide application beyond BMD. Space surveillance systems would be useful either to verify an anti-satellite arms control agreement, should one be concluded, or to support an ASAT weapon system, should such an agreement not be entered into.[23] Such systems also may have potential for permitting surveillance of terrestrial targets such as airplanes, but they would need to contend with the highly significant additional problem of distinguishing the target from its surroundings.

High-Resolution Sensors

Surveillance sensors are clearly necessary. However, in most cases they will not be sufficient. In addition to finding suspicious objects, a defensive system must also determine whether they are threatening or benign, aim weapons at the dangerous ones, and determine whether they have been destroyed. These functions of **discrimination, pointing and tracking**, and **kill assessment**, respectively, will require additional, higher-resolution sensors. The computational capability which can be built into these high-resolution systems could make it possible to extract useful information from the weak and/or noisy signals which they will be detecting.

Discrimination.—Each layer of a defensive system must be able to differentiate between objects which are missiles or warheads and objects which are **decoys** designed to fool the defense into treating them as if they were missiles or warheads. If the defense is unable to distinguish between the two, its job is orders of magnitude more difficult.

If the defense is to be able to discriminate effectively, it must utilize **multiple phenomenology**—repeated observations of the same ob-

[23]See U.S. Congress, Office of Technology Assessment, *Anti-Satellite Weapons, Countermeasures, and Arms Control,* OTA-ISC-281 (Washington, DC: U.S. Government Printing Office, September 1985), the companion report to the present volume, for a detailed discussion.

jects using different sensor systems and different physical principles—and it very likely will need high resolution sensors. Although decoys which duplicate one particular observable (radar cross section, temperature, size, etc.) of an actual warhead can be made relatively easily, it becomes progressively harder and harder to mimic more and more characteristics simultaneously. If enough parameters are to be duplicated, in principle it will cost as much to build a highly accurate decoy as it would just to add another RV.[24] Note, however, that the process of making decoys look more and more like warheads, or **simulation**, may not be as effective as making warheads look more and more like decoys, or **anti-simulation.** These techniques will be discussed further in "Countermeasures," below (p. 170 ff.).

It is possible, in principle, to decoy ICBM boosters. Discriminating true ICBMs from decoys could be done if accurate data on the origin, trajectory, and characteristics of each launch could be obtained by the boost-phase surveillance sensors. These operations require primarily data processing capability and would not necessarily require high resolution. The effort needed by the offense to defeat such a discrimination scheme would depend on how much data the defense were able to collect on each launch (and on how well the offense knew what the defense was looking at). Note that if a booster decoy were launched from a pad which did not have some of the characteristics of an ICBM launchpad, real missiles might then be placed on similar pads to guarantee them a free ride through at least the first defensive layer. This is an example of anti-simulation.

If ICBMs are able to penetrate the boost-phase system, each can begin to deploy tens of warheads and/or hundreds of decoys. The remaining layers, then, may have to contend with thousands of warheads and hundreds of thousands of decoys and other pieces of debris.

The defensive task is lessened if it is able to maintain "birth-to-death" tracking of all objects. On the one hand, the independence of the different layers will be compromised if later layers rely completely on the earlier ones to detect and discriminate warheads. However, if earlier observations are used to enhance later ones, instead of to replace them, independent observations of the same object can be compared.

One technique which might make discrimination easier would be direct observation of objects as they are deployed off of the PBVs. In principle, it might be possible to see balloon decoys being inflated or to notice some characteristic PBV behavior which indicates that a weapon, rather than a decoy, has just been deployed. Objects correctly determined at deployment to be decoys could therefore safely be neglected by later layers.

Imaging the RVs and decoys requires high resolution. However, the same diffraction phenomenon that limits how tightly a laser beam can be focused also limits the the angular resolution with which images can be resolved. Examining an object with 30 cm resolution (about 1 foot) from 3,000 km away requires an angular resolution of 0.1 microradian. To attain such resolution in the long-wave infrared wavelengths (about 10 microns) which are emitted by such objects, a telescope 120 meters in diameter would be required!

One way to mitigate the diffraction problem is to utilize prior information about the target. If the target's true appearance is already known, and only its precise **location** is required, the additional knowledge about its appearance makes it possible to calculate diffraction effects and remove them from the sensor image. This process could yield a more precise location than would be otherwise obtainable. On the other hand, if it is not known what the target looks like, as would be the case before it had been identified, this technique would not be applicable.

The only other way to minimize diffraction is go to shorter wavelengths. Reducing the wavelength in the above example by a factor of 50, changing the 10 micron wavelength long-wavelength infrared radiation to 0.2 mi-

[24]It is assumed in this example that adding RVs to defeat the defensive system is prohibitively expensive. If not, the offense presumably would have done just that and would not have worried about decoys in the first place.

cron ultraviolet radiation, permits the same resolution to be obtainable from a mirror 50 times smaller in diameter. However, since the objects to be observed do not emit brightly at these shorter wavelengths, an **active** system—one which illuminates the target—must be used. A **laser radar**, or **ladar**, lights up the target with a low power visible or ultraviolet laser beam while a telescope observes the reflected light. If the laser beam scans sequentially over the telescope's field of view, the laser need not illuminate that entire field at once, minimizing the required power. The wider the ladar's field of view, the less precisely it needs to know where to start looking for a target.

Under certain conditions, antennas which are physically small can have the effect of very large ones, providing high resolution at long wavelengths. Microwave wavelengths on the order of a centimeter, a thousand times longer than LWIR, would require an antenna equivalent to one 120 km long to achieve the 0.1 microradian resolution discussed above! However, very long antennas can be synthesized, in effect, if the antenna is moving. Processing together the echoes of signals emitted at different positions along the antenna's path can yield a resolution equivalent to that of a stationary antenna which is as long as the path of the moving one. Such **synthetic aperture radars** (SARs), when based on satellites typically moving at velocities of about 8 km/sec, might be applicable in high resolution imaging systems. A similar technique for achieving high resolution takes advantage of motion of the target, rather than of the antenna. Such **inverse synthetic aperture radars** (ISARs) can examine objects which are rotating or tumbling, although they cannot obtain optimal resolution on objects which are vibrating or otherwise arranged to shake.

The price paid for the higher resolution of active sensors is that they cannot operate without revealing themselves, thus warning the offense and giving it an opportunity to spoof, blind, or otherwise interfere with the sensors. To help prevent this, again at the cost of increasing complexity, the defense can separate the transmitter from the receiver(s). In a **multistatic** system, the receivers would be passive and might be able to operate without revealing their location. The transmitter, of course, would be highly visible, and the targets might still be able to know when they are under observation. However, their ability to interfere with the observations might be complicated if they did not know where the individual receivers were.

Pointing and Tracking.—Once the targets have been detected and identified, weapons must be trained upon them and fired. Pointing and tracking requirements, of course, will differ for each type of weapon. Kinetic-kill vehicles having the ability to home in on their targets need only be pointed closely enough for their on-board sensors to acquire the target. On the other hand, laser beams (except those which kill in one pulse) must be held on a single spot on the target until damage is achieved. Depending on the laser, this can require localizing a beam to the order of tens of centimeters at distances of up to thousands of kilometers, or angular resolutions of less than tenths of microradians. To obtain this resolution in the presence of diffraction, either shorter wavelength active sensors or detailed knowledge of the target itself (or both) would be required.

Part of the pointing problem is determining how far off the beam is if it misses the target. Although by far the majority of a high-quality laser beam's energy will fall within a well-defined central area, there will be radiation outside that main part. Even if the main beam does not strike the target, there will still very likely be enough radiation reflecting off the target for the pointing and tracking sensor to see and use to direct the main beam to the target.[25]

[25]Even though a laser beam travels at the speed of light (it is light), that speed is not infinite and the laser must be aimed ahead of where the object actually is at the time the laser fires. For a target 3000 km away, the target will have moved between 50 and 100 meters (depending on its velocity) in the 0.01 second that it will take the beam to reach it. Since the laser sees the target by observing light which took another 0.01 second to arrive at the laser, the target's **actual** position at the time the laser is fired is another 50 to 100 meters ahead of where it appears to be at that time.

It is harder to determine the position of a neutral particle beam in the vicinity of a target. The angle at which the beam leaves the weapon can be measured by probing the beam with a weak laser tuned near a frequency which will be easily absorbed by some of the hydrogen atoms in the neutral particle beam. How well the laser will be absorbed depends on its exact wavelength as seen by the beam atoms, which in turn depends on the particle beam's velocity and its angle with respect to the laser beam.

However, since the effects of a neutral particle beam on a target are for the most part less visible than the effects of a laser beam (the target will not reflect beam atoms back to a sensor which can see them directly, for example), putting the beam precisely on target may be more difficult than it would be for a laser. (Kill assessment for a neutral particle beam in the functional kill mode is correspondingly more difficult; see below.) The problem is lessened, however, since a neutral particle beam will likely be wider than a laser beam and therefore will not need to be so accurately pointed. A possible method of detecting whether the beam has struck the target would be to look for secondary radiation emitted from the target object. This possibility is being investigated.

Homing kinetic-kill vehicles are the most straightforward; they keep the target continuously in sight, correcting their course until impact. The sensors aboard these vehicles can be **passive**, detecting radiation from the target; **active**, illuminating their targets and detecting the reflected light; or **semi-active**, in which the vehicles would home in on reflected radiation which was originally beamed at the target by another source. These sensors would have difficulty in distinguishing between close-spaced objects, as for example in the case of several balloons tethered to an RV at a distance of a few meters.

Kill Assessment.—Determining whether or not a target has been destroyed depends on the type of weapon and on the defensive phase. An ICBM killed in boost phase will either explode or veer visibly off course, being easily detectable in either case. Kinetic kills in midcourse, whereby a projectile hits a target with a closing speed of several kilometers per second, will also be easily seen. However, since many pieces of what had been the target will continue along in more or less the original target trajectory, the battle management system must keep track of all fragments large enough to confuse subsequent sensors and weapons.

The visibility of laser kills in midcourse depends greatly on how badly the target has been damaged. If the target flies apart, its destruction will be easily discernible. However, damage which might not be easily visible may nevertheless disrupt the RVs heat shield so badly that it will not survive reentry. Such an RV, not recorded as killed, may draw additional fire from later layers even though it no longer poses a threat. Further, RVs which appear to fly apart could be merely programmed to jettison parts under attack, even though they may not be killed. This is analogous to submarines releasing oil to make attackers think they have succeeded.

Neutral particle beams used in the functional ("soft") kill mode may present the most difficult problems for kill assessment. Since neutral particle beams (NPBs) penetrate into their targets rather than depositing all their energy on the surface, damage can be done to the interior which may not be visible from the outside at all. Successful NPB attacks in the boost and post-boost phases might cause boosters or PBVs to act erratically and possibly to destroy themselves. However, the case of RVs is different. There is now no guidance on RVs, so the accuracy of an RV would be unaffected by a "soft" kill. Although the detonation mechanism could be damaged, RVs which have been successfully disabled in midcourse might not be distinguishable from live ones. An RV incorrectly assessed as live might waste resources as later layers kill it over again, and an RV incorrectly assessed as dead will do a great deal of damage if it is allowed to pass through later stages to detonate on target. Therefore, to attack RVs with NPBs, the hard kill mode, which would provide visible evidence of destruction, would be required. The use of NPBs in a high current, hard kill mode is being investigated.

Issues.—The discrimination problem is one of the most challenging technical tasks required of a ballistic missile defense. Even if some successful techniques are developed, they will remain successful only so long as the offense does not counter them by developing decoys which are not susceptible to them.

The techniques for high-resolution sensing described in this section are not so far developed as the surveillance and acquisition sensors described earlier. They are extremely computation-intensive and will depend on substantial advances in real-time processing capability.

Pointing and tracking systems capable of operating in a BMD system, particularly in the presence of a hostile enemy, have never been built. Systems having some of the required characteristics, however, do exist today. NASA's Space Telescope, utilizing a technology level which represented the state-of-the-art characteristic of the time its design was finalized, will be able to lock onto a point target with an accuracy of less than 0.05 microradians—on the order of hitting the "S" in a San Francisco stop sign from Washington.

Developing the required kill assessment techniques may be even more challenging. Not much effort has been devoted to this area until recently. Before much progress can be made in assessing whether an object has been destroyed by a given weapon, a better understanding of that particular kill mechanism may be required.

The pointing ability of candidate weapons systems, and the ability of sensors to assess their effects, will likely influence a decision on the ultimate feasibility of those weapons as much as the technical progress made on the weapons themselves.

High-Speed Processing

Many of the systems described above require extensive computational capability. Some of these computations, such as those required for synthetic-aperture radars, will be ones we already know how to do, except they will need to be done faster. Others, such as those required for interpreting images and making decisions based on those interpretations (e.g., "the first twelve objects in this field of view are decoys") will require development of new mathematical techniques and new processing concepts, in addition to high-speed processors. Advances in both hardware and software will be required; they are discussed both immediately below and in that portion of the "System Architecture" section concerning Battle Management (p. 188 ff.).

Hardware:

Description.—Data processing technology has steadily evolved at a rapid rate (figure 7-4). Although we have not reached the end of this technological evolution, we are now approaching some physical (rather than technological) limits. Processing speed is limited both by the rate at which individual computations can be done, and by the time it takes the intermediate results to move throughout the processor. The former can be improved somewhat by utilizing higher speed materials and circuit elements, but the latter is limited by the speed of light. Shrinking the overall size of circuits by moving their elements closer together mitigates that problem to some extent, but we are also approaching physical limits on miniaturization of components. Both these approaches are under investigation in DOD's Very High Speed Integrated Circuit (VHSIC) program.

When individual processors approach fundamental limits to their speed, further improvements in processing capability can be made by tying many processors together and doing many calculations at once. Such **parallel processing** is most effective for problems which lend themselves readily to being broken down into many independent pieces. There is considerable interest in developing parallel processors, and perhaps even more in inventing techniques to utilize these processors efficiently for a wide range of applications.

Another technique for very high-speed signal processing is the use of **analog** devices. In

Figure 7-4.—Onboard Signal Processing

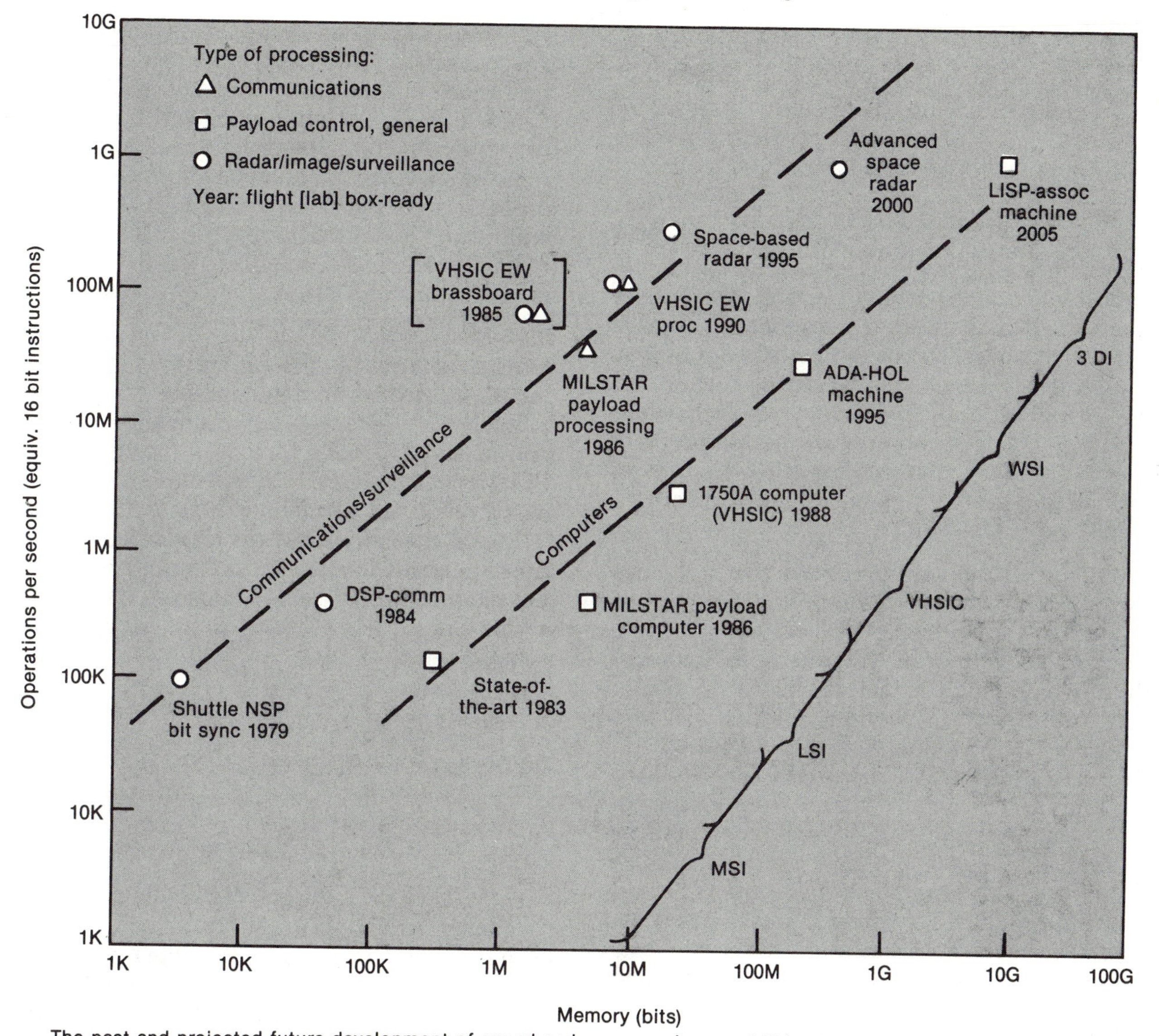

The past and projected future development of speed and memory size capabilities in signal processing, including onboard satellite processors.

SOURCE: TRW.

such a device, the data to be processed are not represented as a stream of numbers, as they would be in a **digital** processor, but rather are represented directly by some physical quantity such as the intensity of part of a laser beam. Certain manipulations of that physical system (e.g., shining that laser beam through a pinhole) are equivalent to performing calculations on the data which that physical system represents.

To give an example, a digital processor would determine the time required for a ball to fall a certain distance by solving the equations of motion for an object in a gravitational field and calculating the answer. A very simple analog approach to that problem would be to drop a ball and time it.

In this example, the computer would calculate the ball's trajectory much more rapidly than the ball could fall. However, for some specific applications, an analog calculation can be much faster than the corresponding digital one, with the greater speed usually coming at the expense of accuracy. When the calculation is amenable to analog techniques, and when great precision is not required, analog processing (called **optical processing** when the physical system is a light beam) offers tremendous speed advantages.

Issues.—Although hardware requirements for BMD processing will require technical advances beyond the present state of the art, no technological barrier yet identified appears likely to preclude development of sufficiently capable processors to do those tasks that a BMD system would need to do. In addition to operating rapidly enough, BMD processors will have to be able to operate in an environment where many nuclear weapons could be detonating in space. These bursts produce high levels of charged particles and other radiation which will severely disrupt the operation of circuits which are not **radiation-hardened**. Use of gallium arsenide (GaAs) instead of silicon holds out promise for making circuits which are both fast and radiation-hard, although these circuits would not be as small as more radiation-sensitive ones.

Reliability is also a key criterion for space system hardware. There is considerable interest in developing **fault-tolerant** processors which are able to detect and compensate for failures without significantly degrading system performance. The Department of Defense is actively investigating both radiation-hardened and fault-tolerant devices.

Software:

Description.—The task of programming a BMD system will be extremely challenging. Part of this task is developing and implementing specific algorithms which will be needed by individual components of a BMD system. Some of these tasks, such as those involving image processing, will require significant development. In several cases, full utilization of hardware advances (such as parallel processors) will be contingent upon equivalent advances in software techniques.

Other software development tasks involve not so much the implementation of specific tasks but rather the coordination and integration of the different tasks done by various components. These battle management issues are complicated by the sheer size of the job, the number of different contingencies which must be anticipated, and the inability to debug the programs under realistic conditions so that they can be relied upon to function adequately the first time.

Issues.—The issues involved in developing and testing BMD software are discussed primarily in the section on "Battle Management," p. 188 ff.

Power and Logistics

The details of the problems associated with the placement, supply, and upkeep of a space-based missile defense system depend largely on the details of the system to be employed. Here, we shall only outline the problems and the requirements for various of the possible technological options mentioned above. In no way should this outline be considered a complete treatment of the problems which must be dealt with, although the requirements listed should be considered a bare minimum for the successful deployment of a usable system.

Space Power

Description.—Large amounts of power will be required for each battle station, particularly if particle beams, electromagnetic railguns, or free-electron lasers are used. The demand for power may be on the order of tens of megawatts or more. For comparison, this power demand is roughly equivalent to that of a town with a population of at least a few thousands. For some projected applications, large quantities of power must be delivered in short surges.

Past space-based power supplies have ranged from a few watts to several kilowatts. The SP-100 project, representing an intermediate stage of development for high power space-based systems, is intended to develop a nuclear reactor of 100 kilowatts or more. In general, solar power may not be practical for demands in excess of tens of kilowatts, or for large surge requirements. Possibly, one power technology, most likely nuclear, would be used for the continuous source, and another method, perhaps stored chemical energy, could be used for the surges.

Issues.—The requirements for multimegawatt power systems in space pose engineering problems which are difficult, but within the limits of foreseeable technology. Requiring large surges would provide additional problems for power conditioning.

Minimizing the frequency of maintenance problems is also a serious issue, and one which could become dominant in developing the appropriate power supplies and conditioning. Extremely high reliability would have to be attained, considering the need for many battle stations. The Fletcher Panel wrote of requirements for a 10-year maintenance-free reliability standard for space-based computer and software systems. Placing similar demands on power sources would be a difficult problem. Since such high reliability is not cost-effective for Earth-based applications, where maintenance can readily be performed, it has not yet been developed and there is little experience to draw on. There are no obvious reasons why such reliability would be impossible, although new testing procedures may have to be developed.

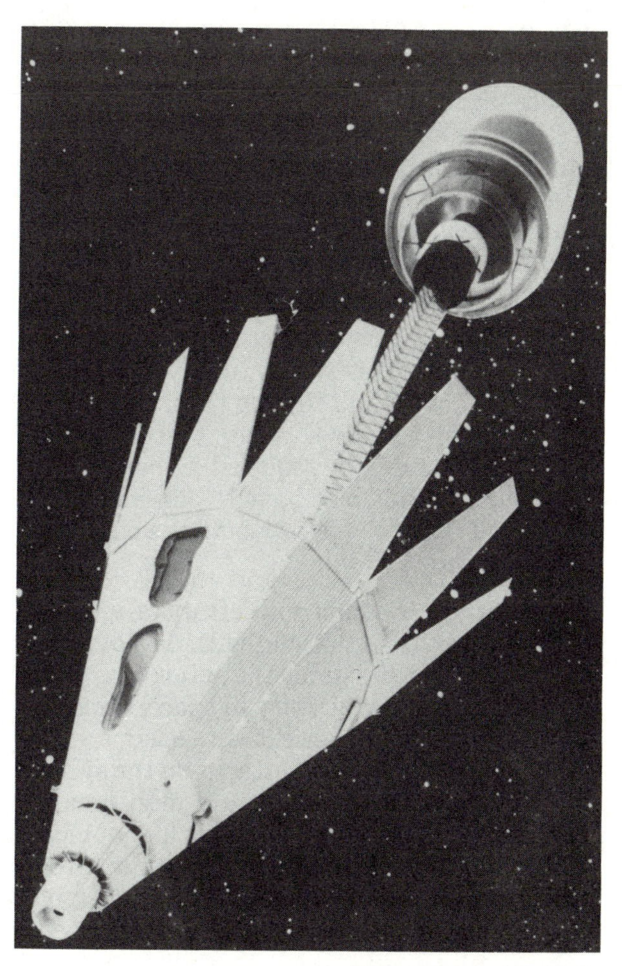

Photo credit: General Electric Co.

SP-100 Space Power Nuclear Reactor: Artist's concept of a deployed nuclear-powered electric generator in space. Power levels are designed to be in the neighborhood of 100 kW. The reactor is at the lower left of the drawing. Fingers pointing backwards from the reactor are heat radiators.

Space Logistics

Description.—Whichever weapons options may eventually be chosen, an enormous amount of mass will have to be placed in Earth orbit. Placing objects in geosynchronous orbit, of course, is more expensive than putting them in lower orbits. The Fletcher Panel declared the necessity for a new heavy-lift launch vehicle for space-based platforms of up to 100 metric tons. The space shuttle has a capacity of up to 30 metric tons for orbits of 200 km altitude or so, and less than this at higher altitudes. Additionally, there will be a need for a space transport which can travel between orbits. This would provide means of moving personnel and objects from a space station base to individual system components for the purposes of maintenance and testing. Some deployments might also require such a vehicle.

The mirrors in a laser-based system, as an example, would have to be checked periodically for operability. This would involve removing protective covers and testing the mirrors' performance with lasers. After testing, some maintenance might be required. Other weapon components would also have to be periodically tested and maintained, as would the computer hardware and software.

Altogether, the cost and effort of a space-based system does not end with deployment. Even in the absence of hostile action, there will have to be constant activity in space, occasionally with human presence, to maintain a working system. The threat of attacks on the system would require the erection and maintenance of defenses. It is also possible that vehicles used for deployment may have to have the capability of defending themselves. Alternatively, they would have to be defended by specifically designed protective satellites already in space. A significant fraction of the total payload to be launched from Earth in the early stages could be shielding. Further into the future, it is possible that near-Earth asteroids could be mined for shielding, reducing the requirements for lifting payloads from the surface of the Earth.

Issues.—The feasibility of developing some high-reliability multi-megawatt power system in time for the deployment of space-based BMD assets needs to be demonstrated. Power conditioning for burst mode operation must also be shown to be feasible if such surges are required by the chosen weapons option. The total cost of placing various possible systems in orbit will have to be estimated. For this, it will be necessary to estimate the cost per space platform and the needed constellation size. In addition, the feasibility and estimated cost for component testing, maintenance, and repair must be determined for each candidate system. Finally, estimates will have to be made of the level, cost, and feasibility of self-defense needed for a space-based system. For further discussion of testing and reliability issues, see the section on "Testing, Reliability, and Security," p. 190 ff.

COUNTERMEASURES

Countermeasures to Sensors and Discrimination

Blinding

Sensors used in ballistic missile defense rely primarily on electromagnetic radiation of diverse frequencies. Short-wave infrared radiation emanating from the booster exhaust plumes is used in the boost phase. Post-boost-phase interception will rely on more sensitive infrared detection at longer wavelengths, since the target will not be as hot and its emissions will be less intense. There is also the possibility of using radar or ladar (a technique which uses laser light in a way analogous to radar) to locate targets in the boost, post-boost, and midcourse phases. In the terminal phase, infrared, visible, and microwave wavelengths would be used to locate targets and to discriminate between decoys and real RVs. In addition, communication links could function at various radio, microwave, and possibly optical frequencies.

A generic problem with sensors is the fact that they must be very sensitive in order to perform their tasks of locating and tracking objects, often small ones, at distances of thousands of kilometers. At the same time, they must be able to resist attempts by the offense to disable or confuse them—an easier offensive task than destroying them outright.

Defensive capabilities can be compromised by neutralizing the abilities of the sensors to perform their tasks. If a sensor can be overloaded with energy, particularly at frequencies to which it is sensitive, it may be disabled. The condition may be permanent—here referred to as "blinding"—or temporary. If temporary, say for a period of seconds or minutes, the phenomenon may be referred to as "dazzling." Blinding or dazzling will be effective if the sensor is thereby unable to give correct position and/or velocity information for its targets to the needed accuracy.

There are a number of ways in which blinding or dazzling may be induced. However, sensors could be hardened against some of these effects by a variety of means.

One possible blinding technique could be the occasional nuclear detonation of an RV by the offense by salvage-fusing when attacked, by active battle management, or by preprogrammed plan. One characteristic of a nuclear explosion is the very intense electromagnetic radiation it produces at all frequencies, from gamma rays down to long wavelength radio waves. Additionally, a nuclear explosion in the upper atmosphere causes ionization glows over a range of infrared wavelengths. These glows may extend for substantial distances and persist for many seconds, possibly masking signals from potential targets.

The intense radiation from a fireball could cause problems for sensors. The first problem is one of overloading or even blinding the detector. This possibility could apply to all types of sensors, from visible and near-visible light detectors to radio and radar devices. Secondly, when the nuclear explosion is not as close, the background signal from the explosion might divert the "attention" of the sensor (depending on how "smart" it was).

However, there can be costs to the offense of employing these tactics. In addition to the chance that defensive sensors may be hardened to resist them, the offense must contend with the risk that detonating nuclear explosions during the midcourse phase could "unmask" its own decoys for a substantial distance around the explosion. There is also the possibility that some of the offense's own assets could be damaged.

Large nuclear explosives would be useful to the offensive forces during the terminal phase. When exploded high in the atmosphere, they would disturb the ionospheric layer, thus causing communication difficulties and making tracking and intercept more difficult for minutes. Exploded above the atmosphere, they would create large electromagnetic pulses, which could destroy electronics which are not adequately hardened and would threaten large power grids and power interconnections with destruction or disablement. Defense battle management and C³I might be threatened.

Nuclear weapons, judiciously used, could be a simple, brute force way of fooling or disabling some sensors. The offense decides when to use them and how many to use. Its only limitation is to avoid collateral damage to its own hardware.

Spoofing and Hiding

Another method of defeating sensors is the use of misleading signals, or the use of decoys, by the offense. This is commonly called "spoofing." For example, the characteristics of the rocket plume could be changed so that a homing sensor which compensates for the distance from the plume to the vulnerable parts of the booster would do so incorrectly, sending the weapon into space, rather than into the target. The defense's response to this strategy could be to use the infrared emission from the plume only for the initial target acquisition, and to use ladar to illuminate directly that part of the booster to be attacked. Shielding of the plume has also been discussed, although this would present engineering difficulties if all directions were to be covered.

In the midcourse phase, the key problem defined by the Defensive Technologies Study Team is the difficulty of discriminating between RVs and decoys. It is by no means clear that the possible future methods of discrimination that have been proposed and analyzed by the Fletcher Panel will be successful. Further work, both theoretical and experimental, is necessary before an informed judgment on this point can be made.

An often discussed problem for discrimination would be the possible tactic of using aluminized mylar balloons to surround both RVs and decoys. Balloon-type decoys could be very light, and could be included in payloads in quantities far in excess of 10 per RV. When a balloon is placed around the RV, the warhead is made to resemble a decoy—an example of the concealment technique called **anti-simulation.** Also, shrouds such as balloons or other configurations may be placed around, but not centered on, the RVs. This would make a kill more difficult for some kinetic-energy weapons, since the position of the RV target inside may not be known with sufficient precision.

A decoy could be given signatures which would closely match real RVs for several sensing methods, a technique known as **simulation.** There may be from 10 to 100 decoys per RV, causing an immense bookkeeping problem. Up to hundreds of thousands of objects could be involved.

A variety of measures has been suggested to overcome this problem.[26] One possibility would be to observe the deployment of the RVs and decoys with ladar during the post-boost phase, if the offense permitted such observations. The changes in post-boost vehicle velocity upon each deployment could be observed, providing a clue to the mass of the object deployed: RVs will be much more massive than decoys. Another tactic might be to attempt to discriminate between RVs and decoys by observing emission from an object at several electromagnetic wavelengths, and by inferring temperature from the radiated electromagnetic energy spectra. Prolonged observations may be needed to perform discrimination using this technique, since the *rate* of temperature change would be the discriminant.

Many other possibilities for discriminating between RVs and decoys also exist, as well as countermeasures which would make the discrimination task more difficult. Some discussion of these items is presented in the classified annex to this chapter.

[26]See discussion on "Discrimination," p. 162 ff.

Countermeasures to Weapons

Hardening

Some simple passive countermeasures to laser weapons have been suggested. One involves rotating the booster, so that the laser spot must illuminate a larger area. This would work if the period of rotation of the booster were not much longer than the necessary laser dwell time for a kill, and it would force the offensive laser to be increased in power in order to compensate. Such a countermeasure would not work in defending against a pulsed laser, which would deposit its energy in a time much less than the period of rotation of the booster.

Resistance to continuous-wave (non-pulsed) lasers could be increased by coating boosters with ablative shields which evaporate when heated, protecting the booster underneath. The booster would suffer some loss in throw-weight, but could gain some laser protection.

Post-boost vehicles could be hardened against attack, although the weight penalty could prove serious. The possible degree of attainable hardening is an issue to be investigated. RVs are hardened by design, since they must survive the high temperatures and decelerations of reentry. This does not mean that they are immune to attack, even by lasers, but that the energy required for a kill would be substantially greater than in the case of a booster. Kinetic-energy weapons are the weapons of choice for midcourse and terminal phases. Hardening against such weapons does not seem a feasible option. Particle beams would also be difficult to protect against without a great cost in weight. However, in this case, kill assessment could be a serious problem unless a hard kill mode were used.

Evading—Fast Burn

Another much-discussed countermeasure is the use of fast-burn ICBMs by the Soviet

Union. The current SS-18s burn out in about 300 seconds at an altitude of about 400 km. The boost-phase defense then has nearly 300 seconds to reach each target. If the length of the boost phase were to be reduced by one-half (to approximately that of the U.S. MXs), the defense's job would be severely complicated. The Soviets are currently testing their equivalent to the MX, the SS-X-24. If boosters are developed which burn out in the atmosphere (say, at 60 to 80 km) in 50 seconds or so, as the Fletcher Panel asserted is possible, some boost-phase defensive techniques would be seriously compromised, if not rendered unworkable. These are the particle beam, and, probably, the X-ray laser, both of which might not penetrate to the required altitude without losing the ability to kill.

The effectiveness of homing vehicles could also be impaired by fast-burn boosters, which would require the vehicles to enter the atmosphere. Their infrared homing sensors might be blinded by atmospheric heating of the windows through which the sensors must look. Since fast-burn boosters burn out both at a lower altitude and in a shorter time, homing vehicles would also have the problem of traversing a greater distance in less time. These problems could be circumvented to a degree if sufficiently accurate targeting were possible after the burn-out of the booster, and before deployment of the RVs during the post-boost phase. One would have to rely on far more sensitive infrared sensors which were capable of finding their targets against the background of the Earth. The homing technology could be strained by this requirement. Alternatively, command guidance of the homing vehicle could be used, whereby the homing vehicles are steered by commands transmitted from other satellites.

All possible systems would face the enormous problem of dealing with a boost phase lasting only one-sixth to one-half as long as the current base case time of 300 seconds. If a 1 second dwell time were necessary on average for a kill, each satellite involved in the battle would be able to handle 50 to 150 boosters, instead of the 300 that each could have handled in the full 300 seconds. The first effect would be to multiply the requirements for the size of the defense constellation by a factor of between 2 and 6.[27] Additionally, the problems of battle management would be severely exacerbated, with more permutations of tracking, target assignment, and kill assessment to accomplish in a shorter time. Again, cost-exchange arguments will have to resolve the issue of whether the offense finds it cheaper to double its fleet, or the defense finds it cheaper to compensate by increasing its constellation size.

The fast-burn booster would also severely strain the capabilities of pop-up weapons. Since the weapons would be placed on missile submarines several thousand kilometers from the ICBM fields, they would have to travel farther in less time than their quarries. This is because the weapons would have to rise high enough to clear the Earth's curvature before they would have straight line paths to their targets. Even if both hunter and hunted were launched simultaneously (which is clearly not possible), the defensive weapon would have to travel farther in the same period of time. Moreover, weapons unable to penetrate below a certain point in the atmosphere must rise high enough so that not only the target, but also the entire line of sight between the weapon and the target, is above the minimum altitude. These difficulties are mitigated if the pop-up weapons are able to detect and destroy the targets after boost phase is over, without the clear infrared plume signals.

The defense could therefore counter a fast-burn booster by improving post-boost-phase detection and kill. The offense could counter again by deploying the post-boost vehicle in much shorter times than is now the case.

In the terminal phase, there is less than 1 minute available for intercept. Countermeasures in this phase, besides the use of nuclear

[27]For some possible types of constellations, constellation size would grow by a smaller factor than that by which the boost phase were decreased. In those cases, constellation size would grow by less than the factor of 2 to 6. See section on "Constellation Size," pp. 179-186.

weapons mentioned earlier, would include the introduction of maneuverable reentry vehicles (MaRVs). Such warheads would engage in a preset series of zig-zag motions to avoid interceptors which would be unable to match the evasive maneuvers in the time required. Small movable fins or other aerodynamic techniques might be used. Counters to this tactic could involve the use of kill mechanisms (probably nuclear) with larger radii of lethality.

Countermeasures to Overall System Performance

Saturation

There are a number of ways to saturate defense systems. For example, in the terminal phase, a preferential offense could overcome defenses if the defender's object is to protect cities. The aggressor could concentrate many RVs on a few cities. If the defense wishes to protect all its cities, not knowing which of them will be attacked, it is forced to deploy defenses at all of them, resulting in the requirement for significantly more defensive forces. If it were considered acceptable to permit the destruction of some cities, the defensive requirements could be relaxed.

For defensive systems which are substantially less than 100 percent successful, one possible countermeasure is simply to increase the number of warheads. It should be remembered that the Soviet SS-18s probably have a capacity of at least 18 and perhaps 30 warheads each, well beyond the currently tested 10.[28] Therefore, a doubling of Soviet RVs could be relatively inexpensive.

For a 50 percent effective defensive system, a simple response of expanding the SS-18 capacity could restore the previous strategic balance, as far as soft target defenses (which were not "preferential defenses") are concerned. This would be the case unless a significant portion of the defensive capability were in the boost phase, in which case more boosters might have to be added by the offense to overcome the defense. If hardened targets were defended preferentially, some of them could be protected against attacks of many RVs;[29] if the defenses operated randomly, some seven or eight RVs per target would still provide high kill assurance for nearly all of the targets. One possible offensive counter strategy would therefore be to aim RVs at soft targets instead of hardened ones. Should the offense counter the defense by adding warheads, the cost of adding a significant number (although perhaps not enough to regain high confidence of killing preferentially defended targets) is likely to be far less than the cost of deploying a 50 percent effective defense. Moreover, the offensive response would require much less time to implement than the defensive system would.

If a defensive system were 80 percent effective, the needed response by the offense to accomplish the same expected damage on soft targets (again, not defended preferentially) would be to multiply the number of warheads by 5. For preferentially defended hardened targets, it would be more difficult for the offense to assure the same expected level of damage.[30] There will clearly be some point of defensive efficiency where the offense will not find it economical to respond merely by increasing warhead and missile production. An estimate of the cross-over point requires knowledge of the cost of other countermeasures which would be possible, as well as of the defensive system.

Saturation could occur particularly during the midcourse phase, where penalties to the offense are small for producing a large number of decoys. These can be cheap and light. It cannot be emphasized too strongly that the ability to discriminate in this phase is essential to the feasibility of the whole space-based BMD concept. The quality of midcourse discrimination determines the difficulty of constructing credible decoys.

---

[28]See, e.g. B. W. Bennett, *How to Assess the Survivability of U.S. ICBMs* (Santa Monica, CA: Rand Corp., 1980), p. 70; W. J. Perry, *Nuclear Arms—Ethics, Strategy, Politics, J. Woolsey (ed.)* (San Francisco: ICS Press, 1984), p. 90; Strobe Talbott, *Endgame* (New York: Harper & Row, 1980), p. 265.

[29]Even so, preferential defenses can be defeated if the offense is able to determine which warheads did not detonate on target, and is then able to retarget additional warheads on those targets.

[30]See, however, the preceding footnote.

Evading—Circumvention

Circumvention of space-based BMD could take several forms. A heavy reliance on cruise missiles or other air-breathing delivery systems, for example, would force the construction of a parallel air defense system in addition to a space-based missile defense. It is conceivable that some elements of a space-based system would be useful in such a defense, but they would be unlikely to be sufficient. An air defense could possibly be technically feasible, but would not be perfect in defending soft targets, and would be expensive. Full analysis of air defense is beyond the scope of this study.

Depressed trajectory missiles launched from submarines could pose difficulties for BMD systems. If the missiles were launched near U.S. territory, the shortened flight times could significantly strain defensive timelines.

Vehicles could be developed which never leave the atmosphere, but glide immediately after booster cut-off. They would bypass the post-boost and midcourse phases (unless those phases employed atmosphere-penetrating weapons). They might, however, be vulnerable to certain types of boost-phase defense and could be vulnerable to terminal defenses.

The boost phase might be avoided entirely by pre-positioning nuclear weapons in orbit.[31] Such weapons, if permitted to be launched and to remain in space, would bypass all but the terminal layer and perhaps the later part of the midcourse layer of a BMD system. The warning time for nuclear attack would be reduced from the 25 minutes or so of an ICBM flight (or 7 to 10 minutes of an SLBM) down to only a few minutes for reentry. In order to be viable, such weapons would have to be survivable against attack, especially since their emplacement in orbit could be considered an extremely provocative act. Although the missions and technologies of orbiting nuclear weapons for ground attack would be quite different from the mission and technologies of a space-based BMD system, some of the survivability techniques necessary for the latter might be applicable as well to the former.

The introduction of bombs into the United States by suitcases, commercial routes, or diplomatic pouches could be accomplished and, for all we know, may already have been done. Techniques for screening such devices by neutron interrogation and radiation detectors are mature technologies and would be easy to use at designated ports of entry. To cover all possible entry routes, however, including deserted coastlines, forests, and deserts along our borders, would be expensive and impractical.

Suppression

With the exception of pop-up weapon concepts, directed- and kinetic-energy weapon scenarios all postulate a large number of space-based stations, which must function continuously in order to be effective. The assets contained in these satellites may be high-powered lasers, delicate optical mirrors, a fleet of homing rockets, electromagnetic launchers, or particle accelerators. **These assets have varying degrees of sensitivity to disruption when subjected to external attack.** They would probably have to be shielded as a defensive measure. Required shielding weights could reach up to many tons for each defensive satellite station. Further in the future, as noted earlier, it may be possible to use material mined from near-Earth asteroids for shielding purposes. This would eliminate the need for putting enormous weights into orbit from the Earth's surface.

Even the best shield, however, would probably be useless against a nearby nuclear detonation (within a few kilometers or so). A serious threat to any set of satellites is therefore the concept of space mines. A salvage-fused space mine could be emplaced, if unopposed, within kill range of any ballistic missile defense satellite. Presumably, this would occur during the deployment period. The mines could already be in orbit when the defensive battle stations were deployed, and could then be

[31]Any basing of nuclear weapons in space would explicitly violate the Outer Space Treaty of 1967. However, there is no reason to suppose that the OST would necessarily be any sturdier than the 1972 ABM Treaty, which would already have been a casualty of any space-based BMD regime.

moved into position trailing those stations. The cost of a small nuclear (or conventional) device would likely be much less than that of a defensive satellite station for any system being discussed, so the cost-exchange would appear to favor the offense.

A defense against this tactic might need to rely on previously stationed defender satellites, which would be able to destroy the space mines before they approached within lethal range of their targets. The difficulty is, that these defender satellites could also be space mined: the technology for the mines could be developed in the near term, and there is little reason to suppose that, once the United States began positioning a layer of defender satellites, the Soviet Union would be unable to launch (or redeploy from higher orbit) its mines. The defenders would then have to be able to defend themselves against the mines.

The issue revolves then around the ability of mines (an easier and more accessible technology) to disable the defender satellites (a new technology) as they are being deployed for the first time, at a favorable cost-exchange ratio. They must be able to do this at a stage when there may be far fewer defenders than mines. Another important issue is the willingness of either side to initiate hostilities by attacking a suspected mine in peacetime.

Defense satellites or battle stations could also defend themselves by means of kinetic-kill vehicles (KKVs), attacking whatever objects approached. A reply could then be to exhaust the kill vehicles by means of cheap decoys, and then to send in a real mine for the kill. A counter-reply could be to use cheap KKV decoys.

The number of sensing and battle station satellites would be far less than the number of warheads and decoys. **Therefore, a directed-energy technology, even if not very effective against an offensive assault, could be deadly when used by the offense against inadequately hardened defensive space assets, provided they could be found.** There is always the possibility that these weapons could be more effective for offenses than for defenses. This is because satellites travel in known and predictable paths, and because beam weapons act nearly instantaneously. An attacker can choose the moment to strike, and can take a very long time to plan the logistics and battle management. Battle management problems for an offensive attack on sensing satellites would be minor compared to those of a defender against a ballistic missile attack, when decisions must be made in only a few seconds. Since the place and time of the attack on satellites would be up to the offense, to a large degree, the offensive forces could possibly even use land-based lasers to kill some satellites. To accomplish this, the offense would have to act when the sensors were exposed, and might only be able to deaden a few satellites in a constellation. However, for an attack to succeed, it may be sufficient to punch a "hole" in the constellation and to attack through the breach. A robust defensive system architecture would have to be resilient against such an attack.

The defense would have to develop means to hide and disguise its satellites, if possible, and decoys would have to be deployed. The sensing satellites and decoy satellites would have to be proliferated to complicate attacks on them. The extra satellites could be deployed in a dormant mode in different orbits from the active sensing satellites, ready to change orbit and come on-line when needed. Careful study would be needed to determine whether the cost-exchange arguments would favor the offensive or defensive forces in such a scenario.

As OTA's companion report on *Anti-Satellite Weapons, Countermeasures, and Arms Control* indicates, a number of advanced technologies have the potential to be used in future anti-satellite weapons which could be highly effective against current generations of satellites. Several countermeasures which could make satellites more difficult to attack are also under investigation; presumably, space-based components of a BMD system would employ such countermeasures (cf. p. 186 ff.).

Once an entire defensive constellation has been deployed, attacking parts of it could be rather difficult. A system intended to handle tens of thousands of targets, or more, might be more easily able to handle a few in self-defense. In principle, in the mid- to far-term, it might be possible that ground-based directed energy stations could damage the sensors of space assets. However, if a complete constellation were in place, defensive countermeasures could be taken. These include redundancy, the use of battle assessments by bystander members of the constellation, and counterstrikes by the defense to avoid further damage, as well as maneuvers, decoys, and anti-simulation techniques.

While the system is being deployed, components may be vulnerable. It is quite conceivable that the adversary would try to destroy the first few satellites as they were being placed in orbit. A complete system could require many scores of stations, and deploying it would take a substantial amount of time. Therefore, the opportunity will probably exist to attack when few stations are deployed. This could be accomplished with space mines which could already be in orbit, or by ground-launched missiles, possibly nuclear-tipped.

A defense against this countermeasure would be to have a smaller deployed system already in orbit, which could defend the battle stations, as noted above. Another counter might be to threaten retaliation for any hostile act against the newly deployed stations. A full analysis of such deployment battle scenarios would have to be based on more detailed deployment plans and weapons choices which have not yet been made.

Relationships Between Countermeasures

Offensive countermeasures usually provide some penalty which must be considered in evaluating the interaction of the defense and the offense. Possible countermeasures to one part of a defense system may **increase** vulnerability to other parts. A few examples might be of some interest:

- The fast-burn rocket avoids several types of defensive weapons, and it puts a severe strain on the defense by reducing the time available to attack and kill boosters and post-boost vehicles before the deployment of RVs and decoys. The throwweight penalty may be relatively small. However, the post-boost vehicle and decoys cannot usefully deploy within the atmosphere, so some period of vulnerability in the post-boost phase cannot be eliminated.
- Offensive responses which modify the timing of launches (for example, which launch all at once to put maximum stress on the defense) can interfere with structured attack plans, which then make the terminal layer more effective.
- Nuclear weapons as suppression or blinding agents could disable one's own space assets during a nuclear engagement, and thus prove harmful to the offense.
- Decoys can imitate RVs better if they contain small thrusters, for example. These would behave more like real RVs upon reentry. However, the thrusters are heavy, and thus a throwweight penalty would be incurred. Simple decoys, such as balloons which mimic the optical properties of an RV, might not also mimic other signatures such as radar cross section. More sophisticated decoys would have to be used which duplicated as many signatures as the defense measured.
- Likewise, hardening of the boosters or any other component by heat-countering ablative coatings may increase survivability against some weapons, but would reduce available throwweight for real warheads.

COUNTER-COUNTERMEASURES

The discussion immediately above, pointing out the costs to the offense of implementing countermeasures, is closely connected with the problem of **counter-countermeasures.** Possible offensive responses to a defensive system may themselves be countered by modifications to the defense.

Counters to all possible countermeasures do not now exist. Ideas have been suggested for some, but it is far too early to determine whether many have any validity at all. They cannot, however, be ruled out. The "fallacy of the last move," described in the introduction to this chapter, is just as invalid when used to show that countermeasures will always be found as it is when used to neglect the existence of countermeasures at all.

It is misleading to treat the countermeasure/counter-countermeasure competition as a game in which each side moves in turn. In actuality, a proposed defensive deployment must try to **anticipate** possible countermeasures **before** they are made. **Defensive counters to obvious offensive responses, such as increasing the number of warheads per booster, proliferating decoys, and attacking space-based assets of the defense, must clearly be available before a decision to deploy the defense is justified in the first place.** Similarly, the most effective **offensive** countermeasures will be the ones which anticipate and frustrate possible **defensive** reactions.

Some counter-countermeasures can be implemented after deployment has been made. Since neither side can anticipate nor prepare for **all** possible counters by the other, each side can hope to at least confuse the other by attempting to keep its own moves secret while at the same time trying to discover what its opponent is doing. If one side can successfully keep the other from knowing which of a number of possible approaches it might take, it might be able to force the other side to **prepare** a number of possible countermeasures while preventing it from **implementing** any of them.

The eventual outcome of this competition will depend on whose **intelligence cycle time** is shorter. The defense will win if the interval between the time it discovers that the offense is preparing a particular countermeasure and the time when it can neutralize that countermeasure is less than the time the offense requires to discover that its counter has been defeated, discard it, and prepare another. On the other hand, if the offense can constantly keep the defense one step behind, the offense will win. Note that if the defense is required to be 99 percent effective, the offense need only manage to penetrate the defense with a few per cent of its warheads in order to "defeat" the defense (e.g., to cause the defense to fail in achieving its defensive goals). In general, no clear outcome of the offense-defense competition can be predicted.

Examples of counter-countermeasures have been given already in this chapter. If the defense can develop a method to measure the mass of objects in space, the offense will not be able to use light decoys. If the defense is able to develop extremely effective post-boost and midcourse phase defenses (which would require effective discrimination or else extremely rapid weapons), it would not need to use a boost-phase layer and fast-burn boosters would be less useful. (However, the post-boost phase can also be speeded up, and the duration of midcourse phase can be adjusted somewhat by changing trajectory.) If the offense hopes to overwhelm a defense by executing a massive, simultaneous launch, defensive weapons which operate best when many boosters are available at once will be more effective.

SYSTEM ARCHITECTURE

The building blocks of a strategic defense have to be integrated into a coherent, organized system if they are to constitute a useful defense. The **system architecture** specifies the design of such a system. It denotes what sorts of components are to be included, how they are to be based, and how they will interact. The system architecture is driven by the objectives of the system and by the effectiveness of each of its parts. Cost and schedule factors also influence the system architecture, as do operational constraints imposed by those who will eventually be asked to manage and maintain any deployed BMD system.

Many of the elements required to specify a BMD system architecture are not available at present, such as a clear specification of system objectives (which must include an estimate of the threat such a system will face) and estimates of the effectiveness of various components. Further off still are estimates of the costs and times at which various levels of capability might be deployed. Extensive research not yet conducted must be undertaken to provide this information. In its absence, this study will review some aspects of a BMD system which any candidate architecture must specify. These are:

- **Size:** the defensive system must be big enough, taking the expected threat into account, to satisfy its objectives.
- **Survivability:** the defensive system must be able to survive attacks upon itself.
- **Battle Management:** various components of the system must accomplish their individual missions and must also interact with the rest of the system. Due to the overall complexity of such a system, the way in which its pieces are to act and interact must be considered at the time the system is designed. Moreover, the defensive system must be able to operate with a minimum of human intervention.

Constellation Size

One factor influencing the total cost of systems utilizing space-based weapons is the number of weapons platforms, or **constellation size.** This number by itself is no more important than other features of the defense, including the as-yet unknown unit cost of the satellites, their vulnerability to attack, and their resistance to potential countermeasures. Furthermore, weapons platforms are only one of the types of space- and ground-based components that a BMD system would require, and the number of weapons platforms needed might or might not accurately reflect the total system complexity.

Nevertheless, calculations of the number of space-based weapons platforms needed to perform boost-phase intercepts have attracted considerable attention because they provide one way to investigate how variations in the **quality** of system components, or in the demands put upon them, affect the required **quantity** of those components.

There is no "correct" constellation size. These calculations can only be done assuming hypothetical defensive capabilities and offensive threats, and different sets of assumptions will lead to different numbers of satellites. However, the way in which constellation size depends on various parameters can be determined. If values for these other parameters are assumed, the corresponding number required of defensive satellites can then be found.

Constellation size depends most directly on the number of missile boosters the defense must handle in a given amount of time. Either increasing the number of missiles or decreasing the available time will serve to increase the rate at which missiles must be destroyed, forcing the defense to grow. Other important factors influencing the size of a defensive constellation are weapon brightness (for directed-energy weapons), retarget or "slew" time, constellation altitude, threat size, and threat distribution. No simple formula relating number of defensive satellites to the offensive launch rate will be valid over the entire ranges of these other factors.

Weapon and Target Characteristics

The effectiveness of a defensive weapon, together with the vulnerability of its target, determines how long (and with what likelihood) it will take the weapon to destroy the target at a given distance. These individual kill times, divided by the total length of time available, determine the number of targets that each defensive weapon can destroy.

Directed-energy weapons are characterized by *brightness*, or how much power they can concentrate into a specified angular range. Since the maximum possible intensity of such a weapon on a target falls off as the square of the distance between the two, the time required to kill a target goes up as the square of that distance. The kill time also depends on the target's *hardness*—how much energy per unit area is necessary to destroy it. Although targets may be very sensitive to attack in certain critical spots, target hardness represents the intensity necessary, on average, to destroy the target without taking advantage of these "Achilles heels."

Kill time, then, is proportional to the target hardness J and the square of its distance R,

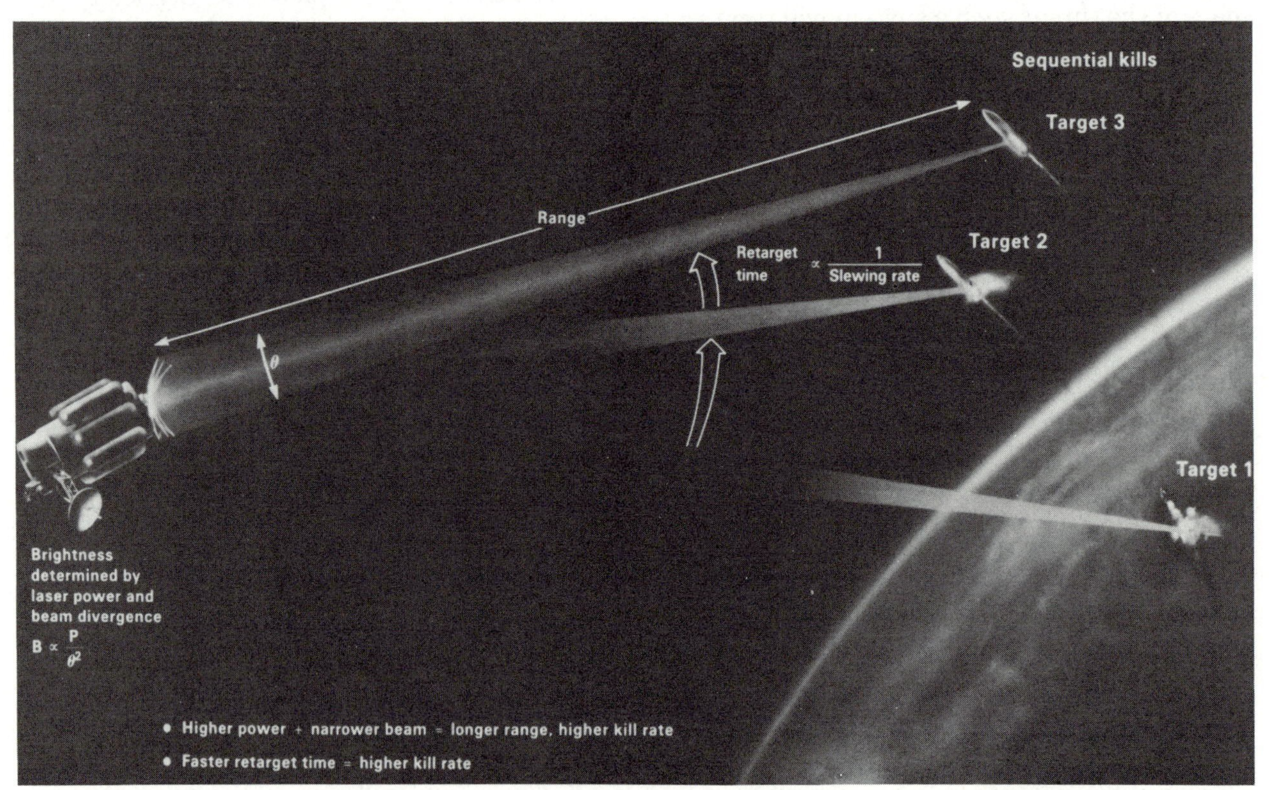

Photo credit: U.S. Department of Defense

The effectiveness of a directed energy weapon depends on its brightness and its retarget time. Greater brightness and lower retarget time mean greater effectiveness. Brightness, or the amount of power per solid angle, is determined by power and beam divergence angle. The greater the power, the greater the brightness; the smaller the divergence, the greater the brightness. The divergence, in turn, is proportional to the wavelength and inversely proportional to the diameter of the aperture. Thus, smaller wavelength means greater brightness and larger mirror diameter means greater brightness.

and inversely proportional to the weapon brightness B. To the kill time must be added the *slew* or *retarget time*, which is the interval required for the weapon to move to the target, stop, and settle down enough to fire:

$$T_{kill} = \frac{J R^2}{B} + T_{slew}.$$

To increase the number of targets that can be killed in the available time, a directed-energy weapon must either increase its brightness or decrease its slew time; the more targets each weapon can kill, the fewer weapons are needed. **Note that reducing the brightness of a directed-energy weapon by a factor of 2 has exactly the same effect on kill time as doubling the hardness. Both are equivalent to doubling the number of targets (to the extent that slew time is negligible—i.e., if a second target were put next to each existing one and the weapon could switch instantaneously from one to the other).**

Kinetic-energy weapons have a different set of characteristics from directed-energy weapons. They can kill only those targets close enough to be reached by projectiles in the available time. Increasing either the projectile velocity or the available time of engagement increases the range of each weapon and lessens the total number required. Hardness is less relevant for kinetic kill; a 1 kg projectile colliding with a booster at a relative speed of 10 km/sec carries the energy equivalent of a heavy tractor-trailer rig traveling at 140 miles per hour.

Altitude

Raising their orbits takes the defensive satellites farther away from the boosters. For directed-energy weapons where the total kill time is not dominated by the retarget time, increasing the altitude will significantly decrease each satellite's total kill rate. At the same time, however, satellites in higher orbits can see farther, putting more boosters in their field of view at any given instant. Depending on which effect is more important, increasing the altitude can either increase or decrease the total number of defensive satellites required. (One of the two example constellations presented at the end of this section gets bigger at higher altitudes; the other gets smaller.)

Depending on the target distribution and orientation, an optimum altitude can be calculated to maximize the constellation's kill rate. However, other considerations (e.g., orbital lifetime or satellite survivability) are often more important, so nonoptimal altitudes will in all likelihood have to be used.[32]

Orbit

In addition to altitude, the angle of a satellite's orbit with respect to the Equator (its *inclination angle*) affects how efficiently a satellite can cover a launch site. The satellite orbit most effectively covering a site at a given latitude has an orbital inclination equal to that latitude. For example, a satellite in polar orbit (inclination 90°) will pass over the poles (latitude 90°) on every orbit and can cover high-latitude sites efficiently. However, it will pass over a different portion of the Equator on each orbit as the Earth rotates underneath, and will therefore not often be in a position to cover a particular site at low latitudes. Conversely, a satellite in equatorial orbit passes over every point on the Equator on each orbit, but has no coverage of higher latitudes at all.

Orbital inclination is not very important for long-range weapons at high altitudes, which are able to attack boosters far from the point on the Earth's surface which is directly beneath the defensive satellite.

Mission

Obviously, a boost-phase system expected to destroy all enemy missiles at launch must be more capable than less ambitious systems which accept some leakage. However, there are more subtle effects of system mission upon system capability. A mission requirement specifying certain orbital inclinations can impose a penalty if those inclinations are not optimal for other mission requirements.

---

[32]The chosen altitude must be high enough so that residual atmospheric drag will not cause the orbit to decay too quickly (above about 300 km); survivability considerations might mandate an altitude significantly higher than that (1,000 km or more). The greater altitude would provide increased warning time in event of direct-ascent attack and might lessen the threat posed by other types of ground-based weapons.

Figure 7-5.—Orbital Inclination

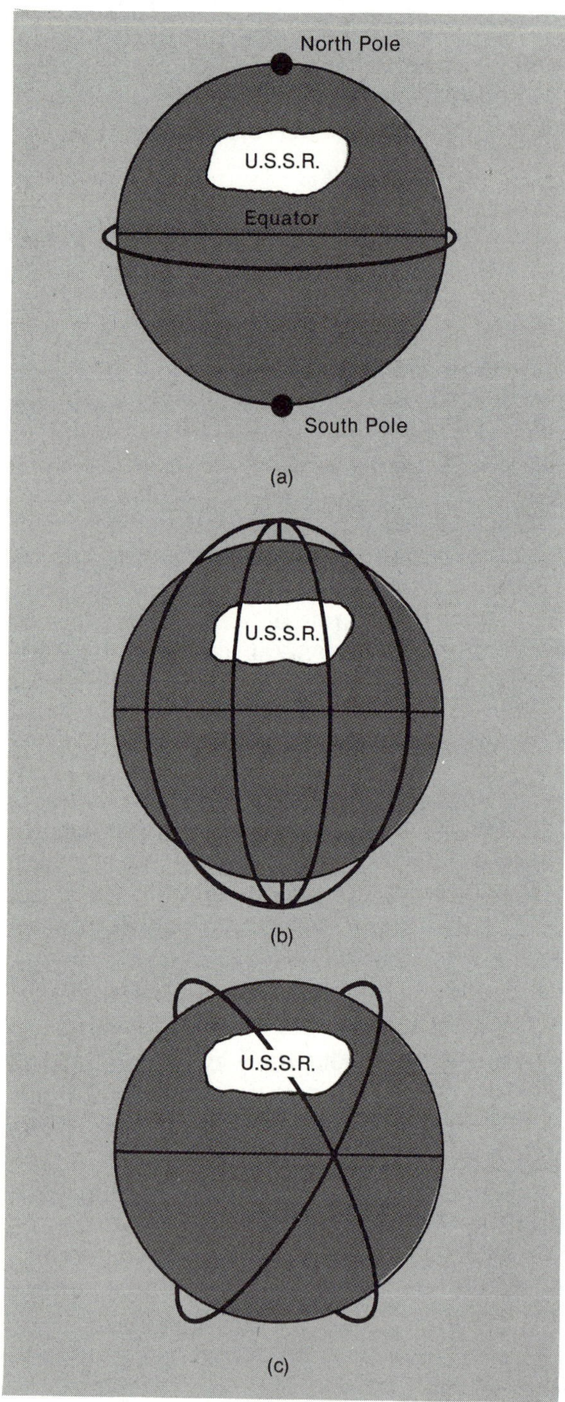

Equatorial orbits (a) give no coverage of northern latitudes. Polar orbits (b) concentrate coverage at the north pole. Inclined orbits (c) are more economical.

SOURCE: Ashton B. Carter, *Directed Energy Missile Defense in Space*, background paper prepared for the Office of Technology Assessment, OTA-BP-ISC-26, April 1984.

One example would be requiring a boost-phase defense to counter submarine-launched missiles as well as land-based ICBMs. The number of extra satellites to counter SLBMs need not be much more than the number needed only for ICBMs because most of the additional capability (in terms of weapons platforms) needed to counter SLBMs comes "for free." In a system sized to handle the existing Soviet land-based ICBMs, only a small percentage of the defensive satellites will be within range of those missile fields at any one time. The rest will be somewhere else. Those others which are over the oceans can counter SLBMs *if they are in a position to see them.* In order to cover possible Arctic Ocean deployment of Soviet SLBMs (which would have to be able to break through the polar ice cap), at least some of the defensive satellites must be in polar orbit. These satellites will be less effective against land-based ICBMs than they would be if they were in less inclined orbits which did not waste time going over the poles.[33]

Target Distribution

In the example immediately above, the capability to handle SLBMs came at almost no cost because SLBM launch areas are far from ICBM silos. Those satellites which would handle the SLBMs in an attack would probably be different from the ones handling the ICBMs, so both jobs could be done simultaneously. Similarly, should the Soviets add additional ICBMs in areas so far away from their exist-

[33]Since not all the missiles on a sub can be fired at once, and since the subs are more widely dispersed than missile silos, the rate of submarine-based missile launches per unit area of the ocean will be smaller than the corresponding rate of ICBM launches per unit area in a missile field. Therefore, SLBM launches should be easier for a boost-phase defense to handle. This becomes less true for higher altitude constellations, where each satellite defends against launches from a wider area and more satellites are in a position to shoot simultaneously at ICBMs and SLBMs. Moreover, these statements apply for simultaneous SLBM and ICBM launch. Should the Soviets be able to time SLBM launches so that they occur under defensive satellites which have already been depleted in countering ICBMs, the SLBMs would not be intercepted. However, the orbital arrangement of defensive satellites can mitigate this problem to some extent by ensuring that satellite coverage areas overlap.

Midcourse systems will have a harder job defending against SLBMs than against ICBMs, since normal SLBM flight times are shorter than those of ICBMs. SLBMs flying depressed trajectories can arrive on target even faster than those on more usual flight paths.

ing missiles (and from SLBM launch areas) that the defensive satellites needed to counter the increase were not already being used to attack existing boosters, no additional defensive capability would be needed to handle the increase.

However, targets which are close together are more difficult for the defense to handle than targets which are dispersed. Should additional missiles be added near existing ones (or near each other, if there are enough of them), new defensive satellites would have to be added to counter the increase. **Boosters located in the same general direction from a defensive weapon can be considered "near each other" if that weapon requires about the same amount of time to target and kill each of them.** For directed-energy weapons that operate by "thermal kill," this will be the case if the kill time for each booster in a group is dominated by retarget time (e.g., the time required for the beam to switch between targets is large compared to the amount of time the beam must dwell on each target) or if the dimensions of the missile field are smaller than the orbital altitude of the weapon. Note especially that missiles within *existing* ICBM missile fields are already "close together" by these criteria. Deployment in a closely-spaced basing mode such as that proposed for the MX missile ("dense pack") would be *much* closer together than required to be considered near each other.[34]

Examples

In this section, we examine two different hypothetical laser weapon systems. In the first case, the lasers generate **20 MW of power at a wavelength of 2.7 microns; this power is directed with mirrors 10 meters in diameter.** The brightness of these lasers is 2.3×10^{20} watts per steradian (w/sr).[35] Systems having these parameters, or very similar ones, have recently been discussed fairly extensively in the literature.[36]

The second case utilizes extremely bright lasers generating **25 MW of power at a wavelength of 0.25 microns. These lasers have 25-meter diameter mirrors.** The greatly reduced wavelength, in particular, yields a very large increase in brightness, since it permits the radiation to be focused to a much smaller spot.[37] An increase in brightness by a factor of 911 over the first case is thus obtained ($B = 2.1 \times 10^{23}$ w/sr).

A booster 4,000 km away could be irradiated with an intensity of 1.5 kw/cm^2 in the first case and 1,300 kw/cm^2 in the second one. If that booster had a **hardness of 10 kJ/cm^2**, the value taken for these examples, it would be destroyed in 7 seconds (ignoring retarget time) in the first case and in 8 milliseconds (similarly ignoring retarget time) in the second.

These hypothetical cases have been selected only for the purpose of demonstrating how constellation sizes vary as system parameters are changed. The actual parameters chosen do not represent an optimized system design, nor does their use imply that either system could or would be constructed. By way of reference, the first case uses space-based lasers which are much more capable than any existing ground-based ones; the second case requires great advances in optical capability beyond those needed for the first case.

Perfect optics is assumed, so that the beam spreads at the minimum diffraction-limited angle. Absorption by the atmosphere, in particular absorption by the ozone layer which would severely affect the second case, is neglected.

[34] For satellites at sufficiently high altitudes or having retarget times much longer than required dwell times, missiles anywhere in the Soviet Union would be considered close to each other.

[35] One *steradian*, a measure of *solid angle*, covers that portion of the surface of a sphere having an area equal to the square of the radius of that sphere. (This angular measure is independent of the size of the sphere; all spheres have a total solid angle of 4π steradians.) Considering a directed-energy weapon to be at the center of a sphere and aimed at some small portion of that sphere's inside surface, the weapon's brightness is given by the amount of power (in watts) the weapon can beam into a given angular range (in steradians). Brightness increases either if power increases or if the width of the beam (the beam's solid angle) decreases.

[36] For a summary of references, see Richard L. Garwin, "How Many Orbiting Lasers for Boost-Phase Intercept?" *Nature*, vol. 315, May 23, 1985, p. 286.

[37] Brightness is proportional to the square of the ratio of mirror diameter to wavelength.

For the purposes of this example (unless otherwise noted), Soviet boosters are assumed to have been replaced with hypothetical MX-like boosters having a burn time of **180 seconds** and the 10 kJ/cm^2 hardness figure given above. Such boosters would probably be more difficult for the defense to destroy than most existing Soviet boosters, but they also would be much easier to destroy than boosters that the Soviets could develop in the time it took the United States to deploy such a defensive system. In addition, it was assumed that **a spot at least 15 cm in diameter** would be required to destroy a booster. If the laser was capable of focusing to a smaller spot than that, its beam, in effect, was blurred to be 15 cm wide on target. **Only boost-phase engagements are presented here.** The effective engagement time is **150 seconds** since we assume that **the defense requires 30 seconds** to identify and assess the attack and to prepare to fire. No limitation was imposed on the resources available (e.g., power and fuel) on each station; the number of kills each satellite could make was limited only by the number of targets in view and the available time.

For most cases, **Soviet missiles were assumed to be located at 12 sites having the approximate locations of Soviet ICBM fields. Each hypothetical launch site was given 117 boosters for a total of 1,404**, approximating the size of the present Soviet ICBM force. Many cases were also run for a doubled threat where each site had 234 missiles. Defensive satellites were placed in **60°** inclined orbits, maximizing their coverage of Soviet missile fields. The lasers were credited with being equally effective against surfaces at any orientation.

Examples were also run for satellites in polar orbits and for a Soviet force concentrated at a single site—the most stressing case for a boost-phase defense. As was mentioned above, a "single site" does not necessarily indicate a high density of boosters. **In these examples, distributing all Soviet boosters over an area the size of the State of Ohio effectively puts them in a single site.** One run took a more realistic angular dependence for laser lethality which, in effect, made it easier for a laser to kill a missile firing broadside at it than firing straight down on its nose cone.

Some examples, run for both laser brightnesses, assumed that the Soviets would use "fast-burn" boosters which burned out in 80 seconds, rather than 180. A 30 second delay for identification and assessment was taken for these cases, as for the others.

The computer model used was provided by Christopher Cunningham at the Lawrence Livermore National Laboratory. A constellation of defensive satellites is specified by laser brightness, beam divergence (which depends on the ratio of mirror diameter to laser wavelength), altitude, retarget time, number of satellites, and orbital placement. For each set of conditions, the offense was assumed to launch at the moment when the defensive satellites were in the worst position to handle the attack, and the minimum defensive constellation size capable of destroying all the missiles under those circumstances was found. Constellations were not augmented to provide spares to account for satellites which would be out of service due to attack or maintenance. The following two tables present the results of the computer simulations for the two cases.

Observations

The most useful information derivable from the above tables is the relationship showing how minimum constellation size varies with orbital altitude and with retarget time. This variation is less sensitive to individual assumptions than is the actual size of the constellations, which could be increased or decreased by taking different values for other parameters. We draw the following conclusions for the first example:

1. The number of satellites needed in the constellation varies ***linearly*** with the threat size. The only exception is for very low altitude constellations (300 km) having retarget times substantially less than 1 second and attacking widely distributed boosters. In this case, the number of satellites increases less rapidly than the number of boosters. Even in this

Table 7-1.—Constellation Size Given Assumptions in Text
20 Megawatt Laser/10 Meter Mirror/2.7 micron
Booster Hardness 10 kj/cm²

Altitude	300 km			
Retarget time	0 sec	0.1 sec	1 sec	3 sec
Threat size:				
1,404	80	88	192	396
2,808	120	143	368	777
Altitude	1,000 km			
Retarget time	0 sec	0.1 sec	1 sec	3 sec
Threat size:				
1,404	*108**	117	192	336
2,808	204	224	384	667
*—Base case				
Altitude	3,000 km			
Retarget time	0 sec		1 sec	3 sec
Threat Size:				
1,404	312		357	440
2,808	620			

Excursions about base case (noted with asterisk above):

Base case (Distributed launch, 1,000 km altitude):	108
Fast burn (80 sec), 1,000 km altitude, 0 sec slew	306
Fast burn (80 sec), 1,000 km altitude, 1 sec slew	588
Fast burn (80 sec), 300 km altitude, 0 sec slew	176
"Single site" launch, 1,000 km altitude	180
"Single site" launch, 1,000 km altitude, double threat	352
"Single site" launch, 300 km altitude	132
"Single site" launch, 300 km altitude, doubled threat	234
"Single site" launch plus fast burn, 300 km altitude, 0 sec slew	352

Table 7-2.—Constellation Size Given Assumptions in Text
25 Megawatt Laser/25 Meter Mirror/0.25 micron
Booster Hardness 10 kj/cm²; Atmospheric Absorption Ignored

Altitude	1,000 km			
Retarget time	0 sec	0.1 sec	0.25 sec	1 sec
Threat size:				
1,404	10	15	25	69
2,808	14	27	42	128
Altitude	3,000 km			
Retarget time	0 sec	0.1 sec	0.25 sec	1 sec
Threat size:				
1,404	6	9	16	40
2,808	9	16	27	75
Altitude	10,000 km			
Retarget time	0 sec	0.1 sec	0.25 sec	1 sec
Threat size:				
1,404	5	8	10	26
2,808	9	13	20	49
Altitude	15,000 km			
Retarget time	0 sec	0.1 sec	0.25 sec	1 sec
Threat size:				
1,404	6	8	11	26
2,808	10	14	22	49

"Single site" launch, 3,000 km altitude, 0 sec slew	7
"Single site" launch, 3,000 km altitude, 0 sec slew, double threat	12
"Single site" launch, 15,000 km altitude, 0 sec slew	7
Fast burn, 3,000 km altitude, 0 sec slew	12
Fast burn, 3,000 km altitude, 0 sec slew, double threat	24
"Single site," fast burn, 3,000 km alt., 0 sec slew	16
"Single site," fast burn, 3,000 km alt., 0 sec slew, double threat	25

case, the constellation size varies nearly linearly with the threat size for a "single site" launch.

2. For the particular parameters chosen for hardness, boost time, etc., the constellation size must be at least 100 satellites for altitudes of 1,000 km and above.
3. Grouping boosters at a "single site" increases constellation size by about 60 percent. If the threat is doubled by placing the additional boosters at a "single site," the defensive constellation should ***more than double*** (except, possibly, for the low altitude, low retarget time case). By grouping the added boosters together, they become even harder to kill than if boosters were doubled at their existing locations.

For the second (superbright) example, we find the following:

4. The altitude is much less important than the slew time in determining constellation size. The limit on the kill rate appears to be determined by the slew time, which, when longer than 0.25 second, is much longer than the time needed to kill an individual target. The number of targets each satellite can kill is then limited, not by laser brightness, but by the time needed to retarget. In this regime, defensive constellation size scales ***linearly*** with threat size.
5. For some of the particular parameters chosen here, constellation sizes can be very small.
6. For retarget times at or below about 0.25 second with distributed launch, the defensive constellation size scales ***less than linearly*** with the threat size. In this regime, however, the difference in absolute number between the actual scaling and linear scaling is not very big.
7. For very high altitude constellations, the entire Soviet Union is effectively a "single site," and constellation size varies essentially ***linearly*** with the threat even for zero retarget time.

Further Notes.—For cases where constellation size increases linearly as threat size increases—most of the ones examined here—the use of fast burn increases the constellation size in inverse proportion to the time of engagement. **For the dimmer laser system,** putting all the Soviet boosters in one place increases the required constellation size by two-thirds, since fewer defensive satellites are in a position to attack boosters and more are therefore needed. However, depending on their attack plans, the Soviets may not want to "group" their boosters. Although such a grouping would still be large enough that it would not necessarily be any more vulnerable to attack than their existing booster distribution is, the Soviets would lose the ability to conduct certain types of structured (precision-timed) attacks. Because of flight time variations in reaching targets when the launch occurs simultaneously from one limited region, one could not simultaneously strike widely separated targets with a simultaneous launch. **For the brighter system,** there is less advantage to grouping boosters. The satellites have longer range, and exact booster placement does not matter as much.

Although not shown in the table, placing the defensive satellites in polar orbit increases the constellation size by less than 20 percent over the inclined orbit case for the lesser brightness system. Modeling the laser effectiveness by including the effects of the angle between the laser beam and the booster surface also makes less than 20 percent difference.

The results given in these tables do not apply to kinetic-energy weapons, where the important parameters are the velocity of the driven projectiles, and the rapidity of fire. A separate analysis would be needed to determine the behavior of constellation size in those cases.

Survivability

If the defensive system is itself vulnerable to attack by the offensive force, the offense can penetrate it by diverting part of its resources to attacking the defense directly, permitting the remainder to continue unimpeded. Therefore, such defenses themselves have to be effectively invulnerable to attack. Paul Nitze, chief arms control adviser to President Reagan, stated criteria for BMD which included the requirement that

> The technologies must produce defensive systems that are survivable; if not, the defenses would themselves be tempting targets for a first strike. This would decrease rather than enhance stability.[38]

General Abrahamson, director of the SDIO, similarly has recognized that

> . . . the key functional components of a defensive system must be made survivable against attack. This problem is particularly keen for defensive space assets.[39]

Some scientists have stated that a defense should not rely on space-based weapons platforms since they would be very difficult to defend. Discussing ballistic missile defense systems with a House subcommittee, Dr. Edward Teller emphasized that

> I am not talking about orbiting space laser battle stations. I am talking about third generation weapons and other instruments that pop up into space when the time to use them has come.[40]

"We need eyes in space," continued Dr. Teller, but once they are there, "our eyes are sensitive and our eyes are in danger."

However, other opinions have been quite the opposite. In an interview, Presidential Science Adviser George Keyworth remarked without elaborating that as a result of recent advances, "We possess the technology today to deal very effectively with survivability of space assets."[41] General Abrahamson, with a slightly different emphasis, stressed *functional* surviv-

[38]Speech before the Philadelphia World Affairs Council, Philadelphia, PA, Feb. 20, 1985.

[39]J. A. Abrahamson, "The Strategic Defense Initiative," *Defense/84*, August 1984, p. 8.

[40]Defense Department Authorization and Oversight Hearings on H.R. 2287, Research and Development Subcommittee, House Committee on Armed Services, Apr. 28, 1983, H.A.S.C. 98-6, Part 5 of 8, p. 1357.

[41]Interviewed in *Science and Government Report*, June 1, 1984, p. 5.

ability of space systems rather than *individual* survivability:

> An analogy can be drawn by comparing [satellites] to the evolutionary use of military aircraft during World War I . . . The fact that extremely delicate and vulnerable airplanes became legitimate military targets did not end their utility in World War I, nor in any conflict since. Both sides quickly learned how to make their airplanes survivable . . . Although these [survivability measures] did not eliminate an enemy's ability to concentrate forces to destroy a given airplane, *squadrons were so constructed that missions could be accomplished in the face of losses of large numbers of individual airplanes. All these tactics and technologies are applicable to spacecraft survivability . . .*[42]

Nevertheless, ensuring the survivability of space systems or space functions—in the presumed presence of a highly capable enemy BMD system, and in the highly stressing environment of nuclear war—is as challenging a task as it is crucial. **After all, many of the technological advances required to destroy ballistic missiles could also be effective against satellites, and in many cases satellites prove much easier to destroy.** Satellites in orbit today are more fragile than ballistic missile boosters, which in turn are easier to destroy than reentry vehicles. Satellites of the future need not be so fragile, but hardening them will impose costs and may interfere with their function. Sensor satellites in high orbits may be concealable, at least for a while; concealing battle station satellites in lower orbits would be quite difficult and such satellites would be even more difficult to conceal if they carried extensive shielding.

Furthermore, the presence of orbiting BMD components greatly increases the incentive for the other side to develop highly capable ASATs to negate those components. The defense must constantly maintain full capability to defeat an attack; depending on defensive system design, the offense may need only to "punch a hole" through the defense in order to challenge its effectiveness seriously. The offense can choose both the time and the place of the attack, and might have the advantage of surprise.

Passive and active measures can both be used to improve space system survivability. Some relevant technologies are given below; none are applicable in all cases, and all impose costs and/or are themselves vulnerable to countermeasures. Much more information on these techniques and technologies can be found in chapter 4 of OTA's companion study of *Anti-Satellite Weapons, Countermeasures, and Arms Control*. See also the section earlier (p. 175) on "Suppression" as a countermeasure to BMD.

Passive Measures

- **Hiding:** Satellites can be made more difficult to detect. For example, they could be miniaturized or stored in a tightly folded configuration and deployed only just before use, or they could be hidden either in very distant orbits or very close ones, where they might be hard to detect against the Earth's background.
- **Deception:** Satellites can be simulated with decoys or hidden in clouds of aerosols or chaff.
- **Maneuvering:** Satellites can evade attackers.
- **Hardening:** Satellites can be made resistant to attack.
- **Proliferation:** Satellites can be replicated, and damaged satellites can either be replenished with on-orbit spares or reconstituted from the ground.

Active Measures

- **Jamming:** Satellites can interfere with the sensors of attackers by overwhelming or saturating them.
- **Spoofing:** They can fool attackers by emitting or broadcasting deceptive signals.
- **Counterattack:** Highly capable BMD weapons can be trained on attackers; alternatively, armed defensive satellites (DSATs) can be provided to escort and defend other satellites.

[42]Abrahamson, op. cit., p. 9 (emphasis added).

Secrecy

No matter what combination of active and passive measures is utilized, protecting satellites from a hostile and responsive opponent will be an interactive process. (See section on "Counter-Countermeasures," p. 178.) In a competition where different techniques and measures may suddenly be introduced, it becomes very important for each side to keep its opponent from learning in advance what it is doing. It is equally important to learn as much as possible about what the opponent is doing.

Security therefore becomes especially important in ensuring survivability. However, it will be more important to protect those items which the opponent cannot easily find out for himself, such as battle tactics or the locations of "hidden" satellites, than it will be to protect those things that will become obvious when a system is deployed. That one or another of the techniques under "active measures" or "passive measures" above is intended to be used in a defensive constellation should not be particularly sensitive; exactly how that technique is to be implemented and under what circumstances it is to be used would be.

However, any system that relies **solely** on keeping some particular piece of information secret has a catastrophic failure mode should that information be revealed.

Survivability—Summary

Overall, the Fletcher Committee concluded that

> Survivability of the system components is a critical issue whose resolution requires a combination of technologies and tactics that remain to be worked out.[43]

From what OTA has been able to determine, examining data on both an unclassified and a classified basis, the "technologies and tactics" required to resolve system survivability issues still "remain to be worked out." Either the work done so far has been so highly classified that OTA has not been granted access to it, or else it has not materially affected this conclusion. It is likely that Congress would require assurance that those survivability issues have indeed been satisfactorily addressed before agreeing to fund a full-scale development system.

Battle Management

Definition

Battle management is concerned with the allocation of resources. A ballistic missile defense system consists of a number of sensors and weapons, each having a finite amount of available power or fuel. Engagements against attacking ICBMs take place in a region of space and an interval of time determined by geometry, weapon capability, and the attacker's strategy. The defensive components (sensors and weapons) which provide coverage of that region have to do their jobs within the available time. The battle management system—the set of rules specifying the operation of and the relationships between system components, and the computers which process those rules—must ensure that the overall defensive mission is accomplished successfully.

This job is a demanding one. The first major conclusion of the Defensive Technologies Study Team subpanel on Battle Management, Communications, and Data Processing was that:

> Specifying, generating, testing, and maintaining the software for a battle management system will be a task that far exceeds in complexity and difficulty any that has yet been accomplished in the production of civil or military software systems.[44]

[43] *The Strategic Defense Initiative: Defensive Technologies Study* (Washington, DC: Department of Defense, Office of the Under Secretary for Research and Engineering, March 1984), p. 5.

[44] *Battle Management, Communications, and Data Processing*, B. McMillan, Panel Chairman, vol. V of *Report of the Study on Eliminating the Threat Posed by Nuclear Ballistic Missiles*, J. C. Fletcher, Study Chairman (Washington, DC: Department of Defense, Defensive Technologies Study Team, February, 1984), p. 4. This volume will be referred to as the DTST Report, vol. V. Unlike the other six volumes of the DTST report, vol. V is entirely unclassified. Distribution is limited to U.S. Government agencies or as directed by the Assistant for Directed Energy Weapons, OUSDRE/ADEW, The Pentagon, Washington, DC, 20301.

Hardware

Although the hardware required for BMD battle management also exceeds the present state-of-the-art, the panel recognized that "the basic technology is evolving rapidly and is likely to be available when needed."[45] In addition, the panel found that technology exists today to transmit data between system components at the rates which a BMD system would require.

Battle management functions are done both within a given defensive layer and across different layers. Each layer must perform acquisition and tracking, target discrimination and classification, and resource allocation. Across layers, the defensive system must provide overall surveillance, specify rules of engagement, delegate control, coordinate between layers, trade off defending Earth targets against defending itself, and furnish current assessments of the state of the defense system and the battle. This last function includes selecting relevant portions of a much larger set of data and presenting it to the command authority.

Software

The subpanel of the DTST estimated that for the system to monitor 30,000 objects (not a highly conservative number; values of 100,000 and 300,000 were also considered), it would need to maintain a **track file** of about 6 million bits of information, or about 200 bits per track. This amount of data is about the same as would be contained on 350 double-spaced typed pages, and could be transmitted within a second at the data rates considered by the panel.

The software required, however, was estimated to be three to five times more complex than what was the largest similar existing military software system—that controlling the Safeguard ABM system developed in the late 1960s and early 1970s. That project, constituting just the terminal and late midcourse layers of a BMD system, required over 2 million lines of computer programming, leading the DTST to estimate that over 10 million lines of code might be required for the BMD systems which they investigated. More recent estimates have gone considerably higher.

[45]Ibid., conclusion 4, p. 6.

Such a software system "will be larger, more complex, and have to meet more stringent controls than any software system previously built,"[46] reported the subpanel. No one person will be able to comprehend or oversee the entire system, and developing such a system will itself likely require the development of automated programming techniques. Computers will not only be needed to create the final BMD program but will also have to subject it to exhaustive reliability testing. One analyst[47] has expressed the view that, in the absence of "extensive operational testing in realistic environments," it would be essentially impossible to produce error-free software of the size and complexity required. This argument claims that it would be impossible otherwise to be sure that all catastrophic design flaws had been eliminated, even if automated programming techniques were applied.

Decentralization and Survivability

The data processing requirements for a system of this complexity must be distributed among the system elements. This decentralization serves both to minimize the amount of data which must be passed from component to component as well as to enhance survivability by eliminating indispensable elements. Having a surveillance sensor, for example, process each raw image locally and transmit only the position of a target to a weapon, rather than transmitting the entire sensor field of view, cuts down greatly on the transmitted data rate and lessens the risk that the system could be paralyzed by failure of a central processor. With such a decentralized architecture, the Fletcher subpanel concluded that:

> . . . it appears possible to design a battle management system having a structure that can survive battle damage as well as other parts of the BMD system do.[48]

[46]Ibid., p. 45.

[47]Herbert Lin, "Software for Ballistic Missile Defense," (Cambridge, MA: MIT Center for International Studies, July 1985).

[48]DTST Report, vol. V, op cit., conclusion 9, p. 9.

Command and Control

Given the short times and the large number of individual tasks which need to be done in a ballistic missile defense engagement, the system must be designed to run as much as possible without human intervention. Command and control of such a system must also be highly reliable. The Fletcher subpanel declared that:

> No BMD system will be acceptable to the leaders and the voters of the United States unless it is widely believed that the system will be safe in peacetime and will operate effectively when needed.[49]

Even if the weapons utilized are incapable of causing mass destruction should they be fired in error, the activitation of a defense system (like the placing on alert of strategic nuclear forces today) would almost certainly be noticed by the other side, and could instigate or escalate a dangerous crisis situation. This danger might be mitigated by the adoption of mutual confidence-building measures by the United States and the Soviet Union.

Testing, Reliability, and Security

These last issues of reliable control exist today with respect to strategic offensive nuclear forces. However, the Fletcher subpanel highlighted them because of the "unprecedented complexity" of the BMD mission and because these issues bear directly on the design of the battle management system itself. Another problem existing today, but aggravated in a BMD system, is the inability to test it under realistic circumstances. "There will be no way, short of conducting a war, to test fully a deployed BMD system," wrote the subpanel. They concluded that:

> The problem of realistically testing an entire system, end-to-end, has no complete technical solution. The credibility of a deployed system must be established by credible testing of subsystems and partial functions and by continuous monitoring of its operations and health during peacetime.[50]

In addition to reliability, a battle management system must obviously resist attempts at penetration or subversion. This requirement mandates that extreme attention be given to overall system security. However, the subpanel realized that:

> There is no technical way to design absolute safety, security, or survivability into the functions of weapons release and ordnance safety. Standards of adequacy must, in the end, be established by fiat, based upon an informed consensus and judgment of risks.[51]

OTA concurs with these conclusions of the Fletcher subpanel.

[49]Ibid., p. 6-7.

[50]Ibid., conclusion 6, p. 7.

[51]Ibid., conclusion 7, p. 8.

NON-BMD APPLICATIONS

The same characteristics of BMD technologies which enable them to intercept and destroy ballistic missile attacks will also provide the capability to accomplish other military missions, including offensive ones. If not designed into the defensive system from the beginning, these other military missions may not be effectively performed; however, the technologies used to construct BMD systems might nevertheless also find use in different systems better suited to these other missions.

To understand fully the possible implications of deploying a BMD system, it is important to recognize the additional, non-BMD, contributions that BMD technologies could make to our strategy. Perhaps more importantly, we must understand what capabilities we might, de jure or de facto, have to concede to the Soviet Union, were we to decide that a mutually defended world was preferable to a mutually vulnerable one.

What Is "Offense?"

No military system easily lends itself to being characterized as strictly "offensive" or "defensive." A system's **capabilities**, such as overwhelming destructive power in the case of a nuclear weapon, can provide some clues. But in the final analysis, it is a weapon's **use** which becomes offensive or defensive, and even that is not unambiguous. A nuclear weapon used in an unprovoked attack on another country is an offensive instrument; one used to **deter** such an attack plays a defensive role. When a retaliatory weapon is actually used, the distinction becomes very difficult to make.

Ballistic missile defense technologies, which would not only be incapable of causing mass destruction but which would be able to prevent it, could be characterized as being primarily defensive. But there are also offensive roles in which they could be used—some inherent in any defensive system and others for which technologies developed for BMD might be well suited for, even if a BMD system itself did not seek to fulfill them. Some offensive roles, such as use to support a first-strike attack by blunting what would be a ragged retaliation, have been mentioned earlier in this report (c.f. chapter 6) and will not be discussed further. Other aspects, however, will be presented in the following discussion.

Inherent Capabilities

Ballistic missile defense systems involve the precise application of power at long range. Depending on the characteristics of the sensors and weapons systems, the targets of that power might be many things other than ballistic missiles.

ASAT

Any BMD system will need to protect its space-based components against potential ASAT attack and will almost certainly require ASAT capability to defend itself. Since the same technologies applicable to boost-phase and midcourse defense can be adapted for ASAT, and since ASAT attack is a potent BMD countermeasure, the BMD mission and the ASAT mission are closely coupled. The connection is discussed elsewhere in this report (e.g., "Suppression," p. 175 ff.) and in the companion volume to this report, *Anti-Satellite Weapons, Countermeasures, and Arms Control.*

Anti-Aircraft

Those BMD weapons able to penetrate the atmosphere would be able to attack targets in the atmosphere if those targets could be located. Neutral particle beams and X-ray lasers, which cannot penetrate the atmosphere to low altitudes, could not be used in this role. Neither could kinetic-kill vehicles unless they could be made to function through reentry. Visible lasers, however, could attack aircraft targets in the absence of clouds.

Perhaps the more difficult part of the anti-aircraft mission will be finding the targets. Detecting an aircraft against a warm and cluttered Earth background is harder than spotting a satellite against the cold and relatively empty background of space. Cruise missiles, being smaller, will be even harder to find; application of "stealth" technology complicates the task still further.

Nevertheless, technology that is potentially capable of detecting aircraft from space is now under investigation. One of the objectives of the **Teal Ruby** sensor, a focal-plane mosaic array containing on the order of 100,000 infrared detectors which is scheduled to be deployed by the Space Shuttle, is to "provide proof of concept of stepstare mosaic for aircraft detection and tracking"[52] from space. Devices based on that technology could be powerful surveillance tools, and in conjunction with atmosphere-penetrating weapons could be effective against airplanes.

Attacking airplanes which are over the territory of another country is at present very difficult. If that task were made easier, it could

[52]Rockwell International Satellite Systems Division diagram.

Photo credit: U.S. Air Force

TEAL RUBY will develop technology for a space-based surveillance system capable of detecting and tracking aircraft near the Earth's surface.

have profound strategic implications. Crucial U.S. command and control functions are now conducted aboard airborne command posts, which are mobile and difficult to find. These planes supplement ground facilities, which at present are vulnerable to nuclear attack. Should these aircraft also become vulnerable to Soviet attack, the command and control structure of our nuclear forces would be seriously weakened and would have to be redesigned.

A BMD system does not have to be able to attack aircraft. However, should one be developed, the advantages of also providing it with anti-aircraft capability may be compelling.

Precision Ground Attack

Weapons which penetrate into the atmosphere can also attack targets on the ground. There are at present lots of other ways to destroy terrestrial targets, so ground attack missions might not be an attractive option for a ballistic missile defense. Furthermore, ground targets would probably be easier to protect than ICBMs or satellites from the types of damage that a ballistic missile defense-capable system could inflict. It may therefore be the case that space-based weapons systems would be grossly inefficient for attacking targets on Earth. However, such a system might provide the ability to do so with essentially no warning, which no existing weapon can do; moreover, even if the system is deployed for other reasons, some amount of ground-attack potential might nevertheless be present. Although the optimal weapons for space-to-earth attack may not be the best for ballistic missile defense, they would probably be difficult to ban under an arms control regime which allowed space-based BMD weapons.

CONCLUSION

There is a wide variety of technologies which could, in principle, be assembled to form a space-based BMD system. Candidate technologies for kill mechanisms include various types of lasers, kinetic-energy weapons, and particle beams. No physical law would prevent the construction of a workable system, consisting of boost phase, post-boost phase, midcourse, and reentry phase layers. Each technology, however, is limited by physical laws. These limitations complicate, but do not eliminate, the possibility of a working system based on that technology. Such problems relate, for example, to: the limitation on the distance traveled in the time available, due to finite velocities (kinetic-energy weapons); inability of the energy-delivery device to penetrate the atmosphere effectively (particle beams, X-rays, possibly kinetic energy); the curvature of the Earth (pop-up systems).

For all of the methods envisioned, much research is necessary to determine scientific, engineering, and economic feasibility. All methods except the X-ray laser and very bright optical or ultraviolet lasers with very low slew times require a large number of space-based satellites with high performance reliability and with access for maintenance. To the extent that sensor satellite requirements exceed those for space-based weapons platforms, all systems will require large numbers of satellites. In general, the higher the attainable power, and the faster the retargeting time (for directed-energy weapons), the fewer battle stations are needed. Great improvements in computer speed, reliability, and durability are needed to achieve a workable system. Current research in computer hardware development gives cause for some optimism in that area. However, even greater advances are required in software capabilities.

A new space shuttle with about three times the capacity of the present one may have to be developed for most options. An alternative would be to reduce greatly the cost of placing material in orbit using the shuttle or something with roughly the same payload capability. If Soviet attack during the deployment phase is considered likely, this shuttle should be able to defend itself during and after insertion into orbit, or it must be defended by satellites already deployed.

The defensive systems discussed are yet to be proven, and are very far from being developed and deployed. In a number of essential particulars, improvements in performance of several orders of magnitude (factors of 10) will be needed.

Operational issues, rather than technical ones, may come to determine questions of technical feasibility. These operational issues are of two sorts—the ability of a defensive system to anticipate and cope with offensive countermeasures, and the confidence which defensive planners can have in a strategic defense which cannot be tested under fully realistic conditions.

An issue to be resolved is the susceptibility of sensors to defeat by various countermeasures. Their sensitive nature, required for long-distance detection, also renders them vulnerable to various levels of blinding. Another general counter-tactic is the emplacement of space mines, which can be used against sensor satellites or battle stations. For each technology there are many possible countermeasures, both active and passive, which can be taken by the offensive forces. Some are simple and straightforward, even with today's technology. Others are more complicated and would require great effort, perhaps comparable in magnitude with the technology they would counter.

Defensive systems, if deployed 10 to 20 years from now, will have to deal not only with today's countermeasures, but also with those which will exist 10 to 20 years hence. For these countermeasures, there may well be counter-countermeasures which are feasible. The eventual outcome of the contest between measures, countermeasures, and counter-countermeasures cannot be predicted now.

Chapter 8
Feasibility

Contents

Chapter 8
Feasibility

HYPOTHETICAL BMD SYSTEM

Introduction

As a way of illustrating the scope and the nature of the technical and operational feasibility issues, this chapter hypothesizes an imaginary system architecture. Since an official proposed architecture does not yet exist, the following system is presented as a structure which is at least plausible enough for illustrative purposes. We do not suggest or predict that all or even any of its parts can or will actually be proposed or built.

The example described is not intended to be definitive or exhaustive. We suggest it to convey a feeling for the nature of the problems to be resolved in planning a workable BMD.[1] Several levels of effectiveness are hypothesized. Consonant with conservative strategic planning, we assume, in outlining the system, that it must deal with Soviet force modernization and Soviet countermeasures (a "responsive threat"). It is conceivable that future levels of the Soviet offensive threat, rather than increasing, could decrease as a result of negotiation, in which case the hypothesized architecture would be more effective than otherwise.

The hypothetical BMD architecture is treated as a nested set. That is, the first system, consisting of one layer of terminal defense, is the simplest and most readily achievable; the second incorporates and extends the first by adding another layer; the third incorporates and extends the second, and so on, through the fifth system. The reader is also referred to the discussion of a layered defense in chapter 7. It is imaginable that an entire architecture could be deployed in this order. The first system could be realized soonest; the others might be added in succession, if and when the required technology is developed. There is a rough correspondence between the elements of this set of systems and the four levels of defense capability described in chapter 5. The first system might have the capability of chapter 5's Level 1. The second or, more likely, the third system might have roughly the capability of Level 2; the fourth system is meant to have the capability of Level 3, and the fifth system is meant to have the capability of Level 4.

The first layer of defense hypothesized is a terminal defense for hardened sites. The defense is not structured to defend large areas or soft targets, but rather has as its purpose the defense of a significant fraction of U.S. missile silos and hardened command and control sites. The purpose would be to provide the United States with the assured survival of a significant fraction of its land-based retaliatory force in the face of a Soviet ICBM attack, and thus bolster the other legs of our "triad" in deterring a Soviet first strike. This layer might not be very effective against a responsive threat without the presence of other layers, and, by itself, would not follow the path of current Administration policy, which is to develop methods of defending populations, not weapons.

The second level adds a layer with some midcourse capability to the terminal defense. This begins to provide some area defense and is also intended to assure the survival of a larger fraction of the U.S. retaliatory force. Any reentry vehicles (RVs) destroyed or decoys discriminated during the midcourse phase will correspondingly reduce the stress on the terminal defenses. A structured attack may be disrupted by this capability, and the overall number of targets presented to the terminal layer might be significantly reduced.

[1] cf. J.C. Fletcher, "Technologies for Strategic Defense," *Issues in Science and Technology*, vol. 1, No. 1, fall 1984, for a similar exercise.

The third level adds a significant midcourse capability. The fourth level incorporates boost-phase and post-boost-phase layers, intended to give an effective layered defense with low overall leakage.

The fifth level illustrates the magnitude of the requirements of a near-perfect ballistic missile defense. It improves capabilities for all layers, and augments terminal defenses to try to make the total leakage extremely small, having as its goal the neutralization of all incoming warheads. This level of defense would logically require that all other practical means of nuclear weapons delivery could be similarly neutralized. Otherwise, the aggressor would use those alternate delivery strategies and the advantage of this level over the fourth one would vanish.

Each section of this chapter lists a series of technical requirements to be met in order for the given system to be effective. Some of these requirements could be met today or in the near future. For example, endoatmospheric nuclear interceptors have been developed for years and would likely be able to reach significant performance levels (defending hardened targets) within a short period of time. Appropriate communication systems with survivable links are likewise highly developed and should be available very soon.

The technologies for satisfying most requirements, however, are not nearly as mature as in the above examples. One class of requirements consists of those which appear feasible, but need to be scaled up in magnitude, capability, or both. These are generally considered to be midterm prospects. A hardened system of passive sensors, adequate for some target discrimination and able to survive in a nuclear environment, could probably be developed with known technology. However, such a system would probably require many years of development and testing. Similarly, homing kinetic-energy weapons, which are relatively inexpensive and fast (5 to 8 km/sec), can almost certainly be developed, but would also require a number of years of development before effective deployment became possible.

Another class of requirements includes those technologies which still need substantial research effort in order to demonstrate feasibility. Among these would be space-based particle beams of sufficient brightness, pointing capability and kill assessment capability; lasers powerful enough for boost-phase kill; and space-based mirrors of many meters in diameter, which could aim laser beams with great accuracy in less than a second. In general, as the layers in the hypothetical system become more numerous and more complex, the corresponding requirements tend to need longer term development. For some requirements, there is general agreement on whether they could be available in the near term, in the midterm, or are still to be demonstrated, but for others, experts may disagree on the prospects for success and on the time needed for development.

There is another type of requirement which is more difficult to assess, namely, the capability of a subsystem to respond effectively to an adversary's countermeasures. The survivability of a system in a nuclear environment or under direct attack is especially difficult to gauge at this stage, particularly in the absence of a well-defined architecture and of a well-defined threat.

In addition to the development of the appropriate technology, other requirements to be met include questions of reliability and maintenance of the system's components. Discussion of these matters can be found in chapter 7, pp. 169-170 and p. 190.

Terminal Defense

This layer of defense would have to intercept incoming RVs in the last 30 to 60 seconds of flight as they reenter the atmosphere. Detection and tracking of targets in the earlier phases of their trajectories is required, but little discrimination would be possible before atmospheric reentry.

The elements of the hypothetical terminal defense system would consist of:

Table 8-1.—Hypothetical Multi-Layered BMD System

System level	System elements	Description	Comments
Level 1 **Terminal Defense** (defense of hardened sites using endoatmospheric rockets to intercept reentry vehicles (RVs) as they approach their targets)	Early warning satellites; ground-based radar; airborne optical sensors; ground-based battle management computers; fast endoatmospheric interceptors.	Warning of launch provided by high-orbit satellites: RVs detected and tracked in region of ground targets by ground radar and airborne sensors; ground computers assign interceptors to RVs; kill assessment[a] permits reassignment of defense interceptors; atmospheric interception used; air effects used to discriminate between RVs and decoys.	Homing either infrared (IR) or radar; interceptors should be relatively inexpensive, since many needed; may be nuclear or nonnuclear.
Level 2 **Light Midcourse and Terminal Defense** (additional layer added with some interception capability in midcourse and some ability to discriminate RVs from decoys in space to reduce burden on terminal layer; some area defense)	Level 1 plus: exoatmospheric homing interceptors, range hundreds of km; pop-up[b] IR sensors (possibly satellite-based instead); self-defense capability for space assets.	As in level 1 for terminal defenses; longer range interceptors added which can intercept some RVs above atmosphere, providing some area defense; this requires some discrimination capability, furnished by passive IR pop-up sensors, launched towards cloud of decoys and attacking RVs; the new layer reduces the burden on the terminal layer.	Passive IR sensors used for crude discrimination and possibly kill assessment; data base of Soviet RV and decoy signatures needed; sensors must be able to function in a hostile nuclear environment.
Level 3 **Heavier Midcourse Layer** (effective midcourse layer added, giving realistic two-layer system, with each layer highly effective)	Level 2 plus: ultraviolet laser radar (ladar) imaging on satellites; highly capable space-based battle management system; space-based kinetic energy weapons; effective self-defense in space; significant space-based power.	Satellite-based ultraviolet laser radar (ladar) used to image objects; discrimination provided by comparing images with data base of Soviet RV and decoy characteristics; RVs attacked by in-orbit kinetic-energy weapons, which also defend all space-based components of system; this level has fully developed terminal and midcourse layers, but no boost or post-boost phase defense.	Ladar imaging rapid with resolution good to 1 meter or less for adequate discrimination and birth-to-death tracking of RVs; kinetic weapon homing capability good to less than a meter.
Level 4 **Boost-Phase Plus Previous Layers** (boost-phase intercept added to kill boosters or post-boost vehicles before RVs and decoys dispersed)	Level 3 plus: ground-based high-intensity lasers (either excimer or free electron); space-based mirrors for relay and aim; high resolution tracking and imaging in boost phase; self-defense for all phases.	This level adds a boost- and post-boost-phase layer, consisting of very bright ground-based laser beams directed to their targets by orbiting mirrors; sensing by infrared sensors, imaging by ultra-violet ladar; battle management to handle all layers doing discrimination, kill assessment, and target assignments and reassignments. Boost- and post-boost-phase layers may be combined, since post-boost phase could be shortened to 10 seconds or so.	Extremely capable battle management system needed; kill assessment required for boost phase as well as midcourse.
Level 5 **Extremely Effective Layer** (Level 4 with better capability; meant to permit only minimal penetration to targets by enemy RVs)	Level 4 plus: more terminal and exoatmospheric interceptors; electromagnetic launchers for midcourse and boost-phase intercepts; large capacity space-based power; all systems extremely reliable.	More interceptors are added in terminal and midcourse layers; electromagnetic launchers used for boost, post-boost and midcourse intercepts; high capacity space power needed; all systems, including battle management must be extremely reliable.	Essentially same as Level 4, but more of it and higher reliability; newer technologies used as they become available.

[a]Kill assessment refers to the process of determining whether a struck target has been effectively disabled.
[b]Pop-up components are ground-based assets which are launched into space for action upon warning of an enemy attack.

SOURCE: Office of Technology Assessment.

- Ground-based radars, for sensing RVs and decoys as they approach.
- Several thousand fast acceleration interceptor rockets with infrared (IR) or radar homing capability; nonnuclear kill capability would be preferable; in case nonnuclear kill could be defeated by offensive countermeasures, or were too expensive, small nuclear warheads would be substituted.
- Early warning satellites to give notice of attack launch.
- Use of air-based infrared sensors to track incoming RVs and decoys at large distances.
- A battle management system, consisting of computers, sensors, and communication links, which would take data from tens or hundreds of sensors aboard satellites and on the ground, and register the reentry of the attacking objects in the upper atmosphere; it would calculate track files for thousands of such objects, and use atmospheric effects (e.g., deceleration) to discriminate between RVs and decoys in the upper atmosphere. The system would assign particular interceptors to targets identified as RVs, would determine whether or not the RVs were killed, and would revise target assignments accordingly. It would also present real-time information to command authorities on the progress of the battle.

Several of these elements are now available or could be shortly. Geosynchronous early warning satellites have been in dependable use for many years. Ground-based radar technology, capable of multi-object discrimination using atmospheric effects, now exists, for intercepts taking place at sufficiently low altitudes. In the face of an attack using nuclear precursor explosions, however, such radar could be blacked out or otherwise put out of operation in the early stages of the assault.

The development of aircraft-based infrared sensors could provide a more survivable and flexible backup for the ground-based radar. Another possibility, one within current capabilities, involves the use of hardened, disposable radars. Normally buried for protection, a few would expose themselves to attack in order to perform their tracking tasks. Those destroyed by early nuclear explosions would be replaced by others, which would rise from bunkers following the destruction of their siblings. These radars would have to be rather inexpensive, since many of them would be needed. Survivability would be provided by their numbers and distribution as well as their protective shelters.

Fast interceptors with nuclear warheads have already been developed. In order to minimize control and command problems when nuclear weapons are used, to reduce collateral damage to one's own hardware and to reduce the chances of blinding or dazzling one's own sensors, it would be preferable to use homing interceptors with nonnuclear kill devices.

To defend an area 100 km in radius, rockets would have to attain the speed of several km/sec in a matter of 10 seconds or so. This should be achievable with current technology. One could imagine, for the sake of argument, a defense of 10 such areas, in order to assure some level of retaliation by U.S. ICBMs in response to a Soviet first strike. To defend against an attack of 5,000 RVs (about half of today's Soviet strategic inventory), with the aim of assuring the temporary survival of a significant fraction of U.S. silos, a preferential defense could be used. If one were to suppose that Soviet RVs were aimed, in a random distribution, at 1,000 U.S. silos, one would anticipate 5 RVs per silo. The defense then could pick a fraction of silos to defend and assign, say, three interceptors to each RV aimed at those silos, while allowing other RVs to penetrate. The number of interceptors to be used would then depend on how many silos would be preferentially defended. The interceptors could be mobile, making it more difficult for the offense to target them. Radar units could also be mobile.

A terminal defense could be used in conjunction with multiple protective shelter basing (MPS), as was once proposed for MX missile siting.[2] Since an extensive national debate at that time resulted in the rejection of such a plan, MPS is not considered as an option in this hypothetical architecture. However, its application, together with a terminal defense, would provide great leverage if one were to defend missiles preferentially. As described above, in the case of preferential defense, some sites are defended and others are not, while the information on which sites are defended is concealed from the adversary. In this manner, a small number of interceptors could protect a smaller number of missiles from a much larger attack.

The following technical requirements need to be met for such a terminal defense system to operate successfully:

- Effective homing devices; if infrared (IR), they must avoid being swamped by the strong infrared signal emanating from the nose of the interceptor, which is heated by its rapid passage through the atmosphere; if IR or if radar they must be able to function in an environment where many nuclear explosions may be occurring.
- A communication system with survivable links between its component units, able to operate in an extremely hostile nuclear environment.
- A battle management system able to survive and function while under nuclear attack.
- Battle management sensors and computer which can discriminate accurately between decoys and RVs at an altitude high enough so that interceptors can be launched in time to reach the RVs.
- Battle management systems able to assign interceptors to targets within fractions of a second per target.
- If ground-based radars are not sufficiently effective, air-based infrared sensors able to operate successfully in a hostile nuclear environment; important in this context is the problem of "redout": "scintillation," or bright electromagnetic radiation, caused by a nuclear explosion in the upper atmosphere, which masks infrared signals from targets and could dazzle or neutralize sensors.
- The development of a relatively inexpensive homing interceptor with fast acceleration; a nuclear-tipped warhead could be necessary as a backup if a reasonably inexpensive nonnuclear kill device could not be developed.

In reacting to a defensive system which uses only terminal defenses, the Soviets could apply countermeasures which are well within the realm of today's technology. They could simply proliferate RVs with relative ease. The marginal cost-exchange ratio between offensive RV and defensive interceptor might or might not favor the interceptor. It is not obvious which side would win the economic battle on this level, or whether a cost exchange analysis alone would be the determining factor in this competition.

Another countermeasure would be the development and deployment of more sophisticated penetration aids, which could fool the defensive battle management system into thinking that many more RVs are attacking than actually is the case. A variant approach would be to try to make the RV appear to be a decoy. The objective would be to saturate the defenses, and to reduce the time available for the defense to commit and intercept. The lower the intercept altitude, the harder it would be to simulate an RV's behavior without making a decoy as heavy as an actual RV.

Yet another Soviet option would be a structured attack, where the incoming RVs, possibly fused to detonate when attacked (salvage-fused), would come in waves. The first wave would detonate at high altitudes, blinding the defenses long enough to permit subsequent waves to penetrate closer to the target. Following waves would repeat this process and, eventually, in this "laddering down," the tar-

[2]For an extensive review of MPS in the MX context, cf. U.S. Congress, Office of Technology Assessment, "MX Missile Basing," OTA-ISC-140 (Washington, DC: U.S. Government Printing Office, September 1981).

gets would be reached and destroyed. The penalty of this technique to the offense is that several RVs would need to be expended per target. Its resources are correspondingly drained. The defense can extract a high price for each defended target, thus perhaps saving nondefended targets through attrition of the offense's RVs. If the Soviets were to pursue this option, they could be expected, therefore, to make a serious effort to increase greatly the number of warheads.

A further countermeasure would be for the offense to use maneuverable reentry vehicles. This would greatly stress homing capabilities for nonnuclear kill vehicles. However, the defense could then counter with nuclear warheads, which would reduce the need for high precision homing.

In general, the technology needed for the terminal defense system is either available or could be available within the short term. However, the overall operation of such a system in an environment of multiple nuclear detonations is not well understood. The system described above would be far more robust in the face of possible short-term threat responses if supplemented by other layers.

Light Midcourse and Terminal Defense

While the requirements of the previous system could probably be met in the near term, this system and the following ones require technology which is somewhat further off. This additional layer could probably be added relatively quickly after the deployment of the previous one. Most of the technological requirements in this section should be achievable in the near to midterm.

In addition to the terminal phase described above, this level of the hypothetical system would add a set of hundreds of ground-based infrared homing interceptors, based near the borders of the United States, which are capable of exoatmospheric interception. These interceptors would have a range of many hundreds of kilometers. Their long range would make possible some level of area defense in

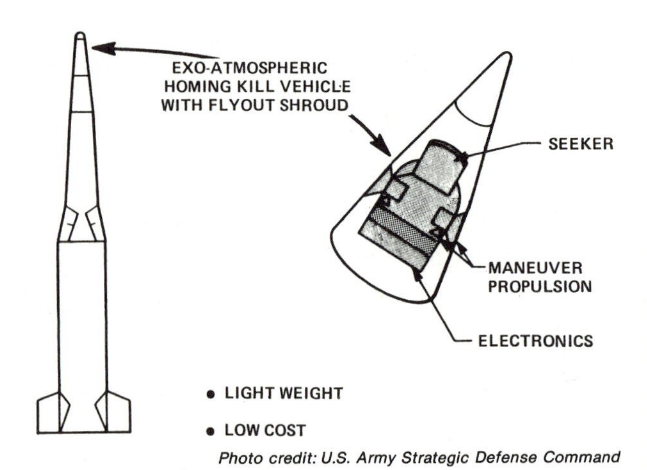

Photo credit: U.S. Army Strategic Defense Command

Exoatmospheric Reentry Intercept Subsystem (ERIS): Sketch of proposed concept for exoatmospheric interceptor.

addition to the defense of a few hard sites. This layer of the system would be intended to break up structured attacks and could relieve some of the stress from the large number of RVs and decoys which could otherwise confront the terminal defense system. The hard-target defense would therefore be more solid, and, by use of preferential defensive tactics, some soft targets could also be afforded some protection.

One possibility for a sensor system would be the deployment of perhaps 100 satellites, each equipped with sensors, which would have some ability to observe the deployment of decoys and RVs from the post-boost vehicle, possibly aiding in discriminating between the two. Perhaps a more survivable and cheaper alternative could be a set of pop-up sensors, to be launched on notice of a massive attack, which would serve the same purpose.

The sensors might be based on a passive infrared system which could be used to measure the infrared emissions of targets. Measurements at several different wavelengths might make decoying or deceptive simulations more difficult. This level of capability might be effective against certain types of simple decoys. Information on track files for targets identified as real would have to be transmitted to ground stations by links robust enough to be secure in a stressful nuclear environment.

The ground stations would relay the information to battle management computers, which would then assign targets to interceptors.

In order to build such a defense, the following requirements must be met:

- long-range interceptors with very rapid acceleration and exoatmospheric capability at relatively low cost per unit;
- passive infrared sensors which can observe characteristics of objects in midcourse with some ability to distinguish simple decoys from RVs;
- a data base of Soviet RV and decoy signatures at various wavelengths which can permit one to distinguish between the two;
- algorithms (rules incorporated in battle management decisionmaking) capable of accurate and rapid discrimination between RVs and decoys, using the data available from the sensors used in the system;
- communication links between sensors in space and stations on the ground which can function in a hostile nuclear environment, through redundancy or other means;
- the development and deployment of a constellation of satellites or pop-up rockets carrying the passive IR sensors;
- a sensor system capable of rapid return to effective operation, following nuclear detonations within the fields of view of individual elements;
- means of defending the satellite-based sensors (if used) from attack; and
- some kill assessment capability, with the ability to relay the information to ground.

The effectiveness of this system could be severely impaired by countermeasures employed to reduce the ability of the sensors to discriminate between decoys and RVs. Such countermeasures could include the use of chaff, aerosols, or other concealment strategies. It is also important to emphasize that the sensor system would have to be robust enough to return to operation rapidly if dazzled by nuclear detonations. This is because targets may be salvage-fused, or may be programmed to detonate at appropriate times in order to confuse defenses. The homing devices on the interceptors may not need to be as robust as the battle management sensors, in this respect, since only those explosions within the narrow field of view of a given interceptor's homing system would be of concern.

Heavier Midcourse Layer

To the terminal layer and light midcourse layer, one might add a space-based midcourse defense layer. The weapons of such a layer could supplement the ground-based exoatmospheric interceptors described in the previous section. More sophisticated space-based sensors might substitute for the infrared sensors of the previous system.

Such a layer would greatly relieve the stress on the terminal layer for three reasons: first, the total number of objects to be tracked and attacked in the terminal phase would be reduced; second, structured attacks intended to defeat the defense could be disrupted; and third, the more capable midcourse system would be better able to help discriminate between decoys and real RVs than the system described in the previous section. This information would be used by the midcourse layer and would also be passed on to the terminal layer. For this level of midcourse defense, the weapons could be space-based kinetic-energy nonnuclear kill vehicles, which are more mature than directed-energy weapons.

To function effectively, a midcourse system would have to be able to discriminate decoys from RVs. An ultraviolet (UV) imaging laser system might be used, with units based on a constellation of about 100 satellites. The exact number of satellites would depend on the angular resolution achievable and the altitudes of deployment. These would then replace the less capable sensors in the previous midcourse system. The laser imaging could be substituted for or augmented by a radar imaging system, located on the same satellites. The acquisition and tracking of the enemy targets, from post-boost vehicle (PBV) stage until atmospheric reentry, might be accomplished by a long wave infrared detection system. The sensor system would aim for birth-to-death tracking of RVs and decoys. The decoys would be identified by shape or other cues which might be detected during deployment from the PBV. Battle management computers must be able to calculate and store a track file for each object, frequently updating this file, and to hand off data on RVs and decoys, that are not intercepted in this phase, to the terminal defenses for interception there.

During the midcourse phase, the defenses would try to kill as many identified RVs as possible, and to unmask or negate decoys as well. We might postulate a kinetic-energy kill system as a moderately near-term option. Reentry vehicles would be quite difficult to kill with optical lasers since they are already hardened to survive the stresses of reentry. Neutral particle beams might be possible candidates for kill systems, but kill assessment would be a serious problem (see chapter 7). The technology for practical space-based accelerators will likely not be available in the *near-term*, particularly in view of the fact that beam intensities would have to be greatly increased from the present state of the art to assure hard (i.e., visible) kills. However, *long-term* development of such a capability is possible.

More plausible for the near term are kinetic-energy carrier satellites, with large numbers of chemically powered two-stage rockets mounted on each one. These would orbit the Earth in a constellation whose size would depend on the acceleration and terminal velocity of the interceptors. The rockets would accelerate rapidly to 5 to 8 km/sec. They would have long wave infrared homing devices capable of detecting emissions from reentry vehicles. The homing devices would need to have cryogenically cooled detectors so that the infrared radiation given off by the sensor itself would not overwhelm the signal from the RV. The interceptors would destroy the target by colliding with it or by approaching closely enough so that a fragmentation charge could disable it. The kill vehicles would receive initial guidance information from the more capable infrared tracking system located on the sensor satellites; their own homing devices would take over when they approach their targets.

Technical requirements for this kind of system include the following:

- Kinetic-energy weapons with a homing capability of within 10 to 20 cm, and which are relatively inexpensive, since tens of thousands may be needed (depending on the threat size and the acceleration and velocity capability of the rocket interceptor).
- The launching satellites must be able to defend themselves against attack.
- High-speed imaging resolution (less than 0.1 sec per image) of less than a meter at ranges of 3,000 km, in order to discriminate RVs from decoys as they are deployed from the PBV.
- A data base of decoy signatures and RV signatures which would aid in discrimination.
- Tested algorithms for accurate discrimination based on target signatures at various infrared (and possibly other) wavelengths and based on other cues (balloon inflation, PBV accelerations during deployment, etc.).
- Computing capability to calculate track files for tens of thousands of objects or more.
- Accurate kill assessment based on UV or other imaging information after apparent hits are achieved.
- Battle management capability to reassign vehicles to new targets within seconds or less, based on constantly updated kill assessments and PBV observations. For this and the following systems it may be desirable to deploy redundant battle management computers both in low-Earth orbit and in orbits beyond geosynchronous, in order to aid in survivability.
- The ability to defend the weapons and satellite-based sensors from a precursor attack.

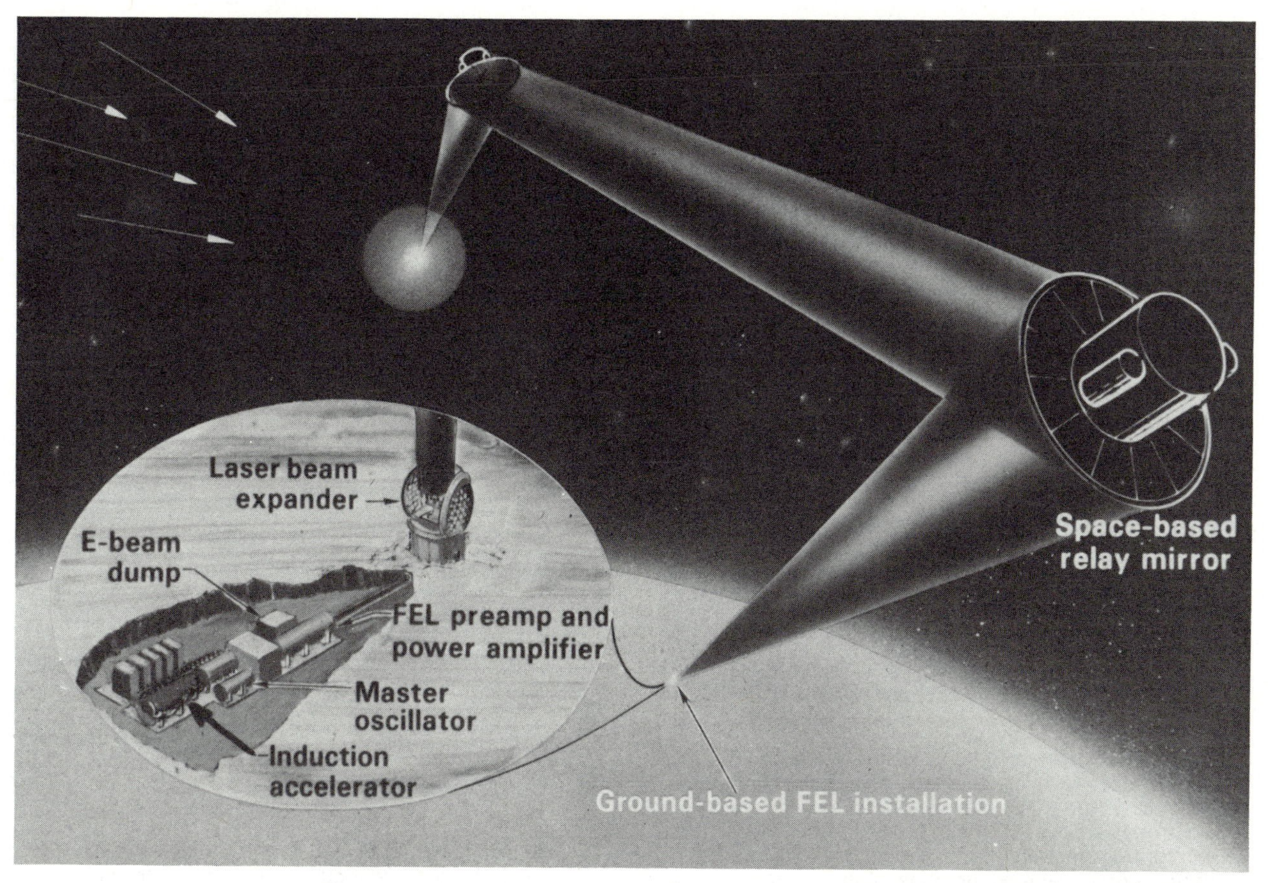

Photo credit: Lawrence Livermore National Laboratory

Artist's conception of possible missile defense system using space-based mirrors to redirect ground-based free-electron laser beam to targets in space. Space-based mirrors might also be used with other types of ground-based lasers, such as excimer lasers.

- Sufficient and reliable space-based power sources to supply energy for the sensing satellites.

An important issue is whether it is possible to image effectively the deployment of RVs and decoys from the post-boost vehicles, in the face of countermeasures achievable with current or near-term technology. More discussion of these questions may be found in the classified annex to chapters 7 and 8.

Boost-Phase Plus Previous Layers

A boost-phase defense might be added to the system described in the previous section. Effective boost-phase interception would have enormous leverage: for every kill, at least one and perhaps tens of RVs in addition to hundreds of decoys would be eliminated from the attacking force, thereby greatly reducing the stress on the succeeding layers of the BMD system.

For a boost-phase system, we hypothesize a set of ground-based excimer or free-electron lasers, with a constellation consisting of a small number of large geosynchronous orbit relay mirrors and a large number of low-orbit "battle" mirrors. Excimer or free-electron lasers were chosen over particle beams, X-ray devices, and chemical infrared lasers because of their ability to penetrate the atmosphere all the way to the ground. A ground-based sys-

Photo credit: Lawrence Livermore National Laboratory

The final accelerator module of the Advanced Test Accelerator (ATA) at Lawrence Livermore National Laboratory. The ATA produces an intense, high-energy electron beam which is used both for beam research and for free-electron laser research. It will be used in future experiments to produce high-powered free-electron laser radiation at long-wave infrared frequencies.

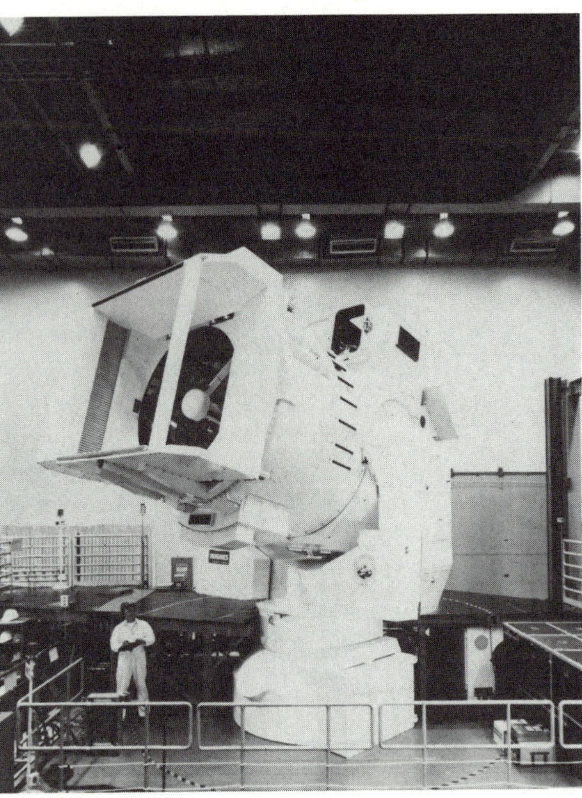

Photo credit: Department of Defense

Designed for use with high-powered lasers, this device aims, stabilizes, and focuses a laser beam to selected aimpoints. It will be used to gain experience in integrating a high-power laser with a precise beam director.

tem is easier to supply with power: it obviates the need for space-based power for the weapons of this layer.

The laser beams would be generated on the ground at a number of stations and be sent to the geosynchronous mirrors. From there, they would be directed to those low-orbit battle mirrors which are nearest the targeted boosters. These mirrors, in turn, would direct the beams onto the targets. If optically perfect, the geosynchronous mirrors would probably need an effective diameter of about 30 meters, given a laser wavelength of 0.5 microns (the requirement of a large diameter could be lowered by reducing the wavelength somewhat). The low-orbit mirrors, if optically perfect, would need to be about 5 meters in diameter. Hundreds of megawatts of electrical energy would be required to power the ground-based lasers.

In addition, adaptive optics would be needed to compensate for beam distortions introduced during passage of the radiation through the atmosphere. In one such technique, a pilot laser beam near the geosynchronous satellite would give information on atmospheric distortions along the path to the ground laser. The ground laser mirror would then be mechanically distorted in such a way as to compensate for the atmospheric effects on the laser beam.

Initial acquisition of the attacking boosters would be provided by a geosynchronous short-wave infrared satellite system, using technology similar to current U.S. capabilities. More precise tracking needed for attack by the large ground-based laser could be provided by ladar (laser-based radar, referred to in chapter 7) systems mounted on the low-altitude sensor satellites.

In order to keep the number of mirrors from reaching well into the hundreds, slew times (time required to change pointing from one target to another) will probably have to be on the order of 1 second or less.

Technical requirements for this system include:

- the development of laser beams of sufficient brightness to destroy rocket boosters after traveling from the ground to geosynchronous orbit, back to a low-orbit mirror, and then to the booster;
- the development of many high optical quality, 5-meter diameter mirrors capable of being deployed in orbit while maintaining their geometry to a small fraction of a wavelength (visible or near UV), robust enough to maintain high optical quality in a hostile nuclear environment, and able to switch from target to target in a second or less;
- the development of a few 30-meter mirrors, with the same optical and physical capabilities as the smaller mirrors (except for the retarget rate, which could be slower);
- battle mirrors inexpensive enough so that the offense cannot overwhelm the boost-phase system by merely adding more boosters: if doubling the number of boosters (or decoy boosters) requires a near doubling of the number of mirrors and associated subsystems, the cost of the mirrors and their subsystems cannot be much more than the cost of the boosters;
- defensive capability to protect mirrors and space sensors against attack, including more subtle attacks designed to deteriorate the quality of the mirrors;
- the ability to track a booster by means of ladar to an accuracy of 10 to 20 cm at a range of thousands of kilometers;
- adaptive optics for high-intensity laser beams to compensate for atmospheric turbulence—some atmospheric distortions will be caused by the passage through the atmosphere of the powerful laser radiation itself, and thus proof of ability to compensate must be accomplished at power levels approaching those used in the actual weapon;
- power supplies able to provide extremely large amounts of power on short notice to the ground-based lasers;

Figure 8-1.—Boost-Phase Intercept With Ground-Based Laser

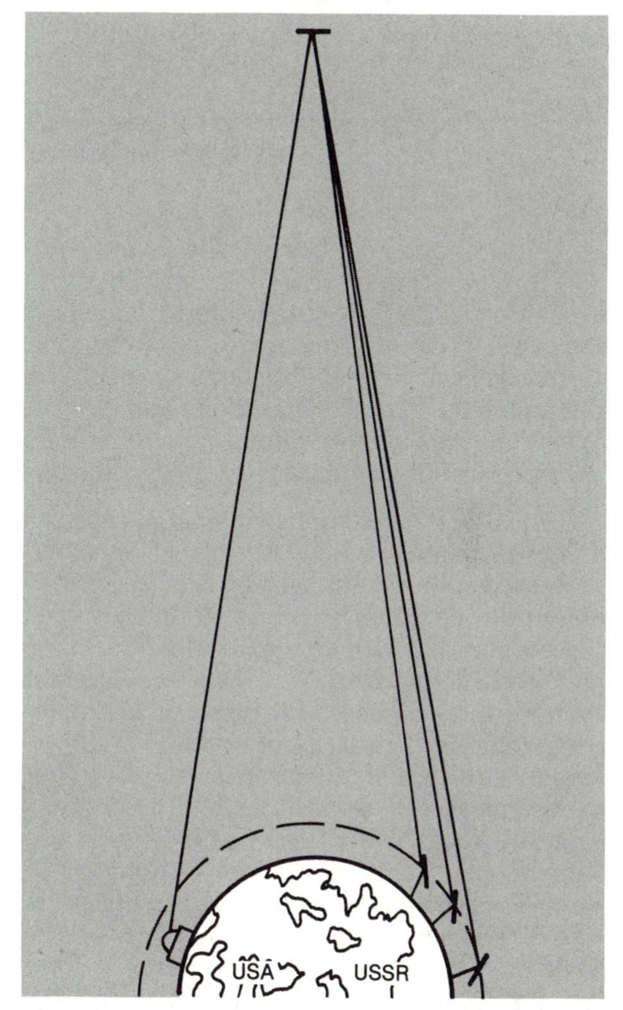

Beam from ground-based laser in the United States reflects off relay mirror in geosynchronous orbit to battle mirrors in low-Earth orbit. Battle mirrors redirect beam to ICBMs. (Geosynchronous orbit is shown to scale relative to the size of the Earth.)

SOURCE: Office of Technology Assessment.

- communication systems to link the mirrors and lasers involved with the battle management system; the links must be able to function in a hostile nuclear environment;
- overall battle management software to coordinate functions of each defensive layer; and
- some kill assessment is necessary for both booster and post-boost vehicle. In the latter case, damage may be more difficult to ascertain, yet it is important to do so. If a PBV is unable to deploy its RVs to their targets, it may still be able to reach the defender's territory with one or more live RVs. This cluster would have to be handled by subsequent layers of the defense.

A future offensive countermeasure could be a fast-burn booster with a fast post-boost deployment phase. In the system described above, one could, in principle, not only attack the booster, but also attack the post-boost vehicle as it dispenses RVs. Therefore, against current systems having long boost and post-boost phases, hundreds of seconds could be available to the defense after booster burn out to destroy at least some of the RVs before they separate from the PBV. The offense could deny most of this advantage by using fast burn boosters and a rapid dispensing technique. The period for neutralization of the booster and the PBV could be reduced from today's 700 seconds or so to only some 50 seconds (assuming that the defense needs some 20 seconds to prepare to act and to begin engaging the offensive missiles). The offense would be penalized in terms of throwweight (on the order of 10 to 20 percent) and possibly in terms of accuracy as well.

Attacking during the PBV phase might be more difficult than attacking during the boost phase, since the PBV is much more difficult to find than the booster. Fifty seconds total engagement time would greatly stress the boost-phase/post-boost-phase intercepts, and would, at the least, greatly increase the quantity of defensive space assets needed.

Extremely Effective Defense

This system is intended to provide a nearly perfect defense against ballistic missiles. It would be designed with the object of preventing all incoming warheads of a massive attack from reaching their targets. In practice, since no system is likely to function perfectly on its first use, some leakers might be expected, although it would be hoped that they would be very few in number. Rationale for including this ambitious case is given in chapter 5.

For an attack of 20,000 warheads, which assumes more than a doubling of the Soviet strategic force in response to deployment of a U.S. BMD, the leakage rate would have to be no more than one- or two-tenths of a percent, equivalent to an overall system efficiency of above 99.8 percent. To accomplish this, all of the above systems would have to work at a high level of effectiveness. If one assumed that all four layers (including the ground- and space-based midcourse layers) were totally independent of each other **(including independence of the sensors and battle management computers on which each of the layers rely)**, this could be accomplished via an 80 percent effectiveness for each layer. If one layer failed significantly, however, the others would have to be considerably more efficient, in order to maintain the extremely low leakage rate.

In addition to the elements contained in the previous system, this one would add a larger fleet of terminal and midcourse interceptors. These would have long-range capability and be relatively inexpensive. In principle, the same interceptors which were briefly described above would be appropriate, if highly proliferated. Other options for improving the effectiveness of the various layers could include improved sensors for midcourse discrimination and electromagnetic launchers for midcourse phase and possibly boost-phase intercept. These may be longer term options, depending on when the needed technologies are developed. The "conventional" kinetic-energy weapons driven by chemical rockets are effective if they can reach their targets, but they generally do not travel more quickly than 5 to 8 km/sec. Further, time constraints on the period of acceleration (at most, a few hundred seconds) could make chemical rockets less desirable than electromagnetically launched projectiles. These could accelerate far more rapidly to faster speeds, thus possessing a greater range. Systems based on them may be more survivable than laser systems because of the vulnerability of space-based optical components; hence their possible advantage for boost-phase intercept.

Since the offense could preferentially attack certain targets, and since the defense would not necessarily know ahead of time which targets would be more heavily targeted, the requirement for very low offensive penetration is quite onerous. If one assumes that the first two layers (boost-phase and early midcourse, including PBV) are 80 percent effective, the last two layers must deal with some 800 leakers. It would be desirable for each interceptor to cover large areas of the United States so that one can defend against the eventuality that one area might be more heavily attacked than its neighbors. Long range would mean that interceptors assigned to neighboring areas could come to the help of those areas whose own defenses were in danger of depletion. As an alternative to long range, many more interceptors could be deployed. The exo-atmospheric interceptors could be designed to have ranges of many hundreds of kilometers. If each rocket could defend the whole continental United States (CONUS), perhaps only 1,000 to 2,000 would be necessary.

The terminal defense interceptors would have a range of only 100 kilometers. Some 400 basing sites might be needed to protect the entire CONUS. Perhaps about 10 interceptors (amounting to a total of 4,000) might be placed at each site. Each site of comparable value should be defended to roughly equal levels, to avoid *inviting* attack by providing an "Achilles heel" of less well-defended sites. If one assumes that the long-range interceptor layer is 80 percent effective, only 160 leakers penetrate through to the terminal defense. Although most sites would be confronted with only 0 or 1 RVs, some would have to deal with 2, 3, or even 4, just because of statistical fluctuations in actual defense effectiveness. Conservatively, one would want to assign at least 2 interceptors per RV, so 10 interceptors per defended area is a safe minimum, assuming that any structure in the offensive assault is completely broken up by the earlier defensive layers.

In relying on 4,000 interceptors, the assumption is made that the previous defensive layers are each **independently** effective to a level of at least 80 percent.

In sum, for a nearly leak-proof defense, several vital needs must be satisfied:

1. A high level of boost-phase intercept effectiveness must be attained, in the face of all countermeasures, including the relatively straight-forward ones of fast-burn boosters, rapid PBV deployment, and warhead proliferation.
2. An excellent discrimination capability between light and precision decoys on the one hand, and RVs on the other, during the midcourse phase; this must be accomplished in the face of various concealment techniques, some of which are relatively well understood at the present.
3. The sensors must function nearly continuously in the face of a massive attack and in a nuclear environment.
4. The space-based assets must defend themselves or must be defended by other assets against concerted attack.
5. The communication links must function effectively in the face of attempts to interfere, concerted attack, and in a nuclear environment.

In addition to the above conditions, since a near-perfect system is envisioned, other means of nuclear delivery must be countered to a high degree of assurance. This means that an air defense system would have to be added to handle air-breathing threats (bombers, cruise missiles), and measures would have to be taken to protect against the introduction and emplacement of nuclear weapons within U.S. borders by surreptitious means. It may also be necessary to consider the application of significant civil defense measures.

Since a nearly flawless system is postulated for this level of defense, any deviation from perfection among the above conditions above would mean that the system would fail. For defenses less ambitious than this one, for example, for the previously delineated systems above, some small failures among the above five conditions could be tolerated. However, even for such nonperfect systems, a significant failure in any one category could seriously degrade the entire system effectiveness.

SURVIVABILITY

For any BMD system, both space- and Earth-based assets should be able to survive attacks on them. Survivability is a function of the mission of each asset and of the mission of the system as a whole, the threat faced by the system, and the effective redundancy of each asset in the system architecture. The *system's* survivability depends on the details of the architecture and of the threat. Some discussion of the problems involved is given in chapter 7, p. 170ff and p. 186ff in sections on survivability and countermeasures. It should be noted that the definition of the needed level of survivability depends on the policy decision regarding the system's mission; i.e., is it to be 90, 95, or 99 + percent effective?

Earth-Based Assets

Ground-based assets consist of communication links, command and control posts, ground-based interceptors, terminal radars, ground-based sensors, ground-based laser sites, and power supplies. For the purposes of this discussion, airborne sensors are considered Earth-based.

These assets must survive long enough to do their job: to provide defense at their assigned levels. Ground-based interceptors might be made survivable by proliferation (many can be deployed) and mobility. Smaller terminal radars can be proliferated, made mobile (deployed on trucks which are assigned to rove defined areas, possibly at random, to avoid being targeted) and shielded. A disposable system has already been mentioned. Larger radars can be hardened to a degree, and higher operating frequencies can be chosen in order to provide greater resistance to blackout effects that are caused by nuclear explosions. Narrow beams can be used to make radars more jam-resistant.

Communication links can be made highly redundant, and can often be direct laser links, or other narrowly focused line-of-sight links. These are nearly impossible to jam and would be resistant to stress in a nuclear environment. Electronics can be hardened to survive electromagnetic pulses, which may be induced by nuclear explosions above the atmosphere. Command and control posts might also be made highly redundant, and different basing modes (mobile and stationary) could be utilized.

Airborne sensors might achieve survivability by maintaining uncertainty as to their exact locations, and by taking appropriate hardening measures. Ground-based sensors must be made highly redundant and must be defended against intense radiation from nuclear explosions.

Ground-based laser sites pose a particular problem, since they would be large, expensive, and therefore difficult to proliferate. They would, in principle, have to be provided with heavy terminal defenses, since they would be the object of strikes early in the engagement. Some would certainly survive the beginning of the engagement. If, as would be likely, they would be attacked simultaneously with many other military assets, they could participate in early battles: they could shoot at the first wave of attacking RVs for up to about 15 minutes before the first ones reach them. However, they would certainly be high priority targets and would very likely be attacked in the first wave. To guard against easy attack by cruise missiles, they should be located far inland, so that the cruise missiles could be detected and destroyed before they were able to reach the laser. Note that a cruise missile attack might, if detected, give the defense system warning even before the launch of the offense's ICBMs takes place. This might tend to discourage such an attack.

Sufficient survivability could probably be provided to most Earth-based assets by means of redundancy, hardening, and mobility. This does not mean that these system elements would be indefinitely survivable; it does mean that they could survive long enough so that they could perform their tasks when called upon.

Space-Based Assets

There are two broad categories of space-based assets which must be protected in order to assure the survivability of the space-based components of the BMD system: sensors and weapons.

Defense of space-based assets, particularly of sensors, is more problematic than in the Earth-based case. Since satellites follow predictable trajectories (unless they are constantly maneuvering), the offense may target them relatively easily. Knowing in advance where they will be at a given time, the offense has a long time to prepare an attack against them and might even be able to do some damage before a large-scale outbreak of hostilities.

Ground-based directed-energy weapons could damage sensors which are looking at or near them (the sensors might be observing such weapons for intelligence purposes, or might be observing nearby missile fields). The sensors would be particularly susceptible to damage if the lasers operated at a frequency in the band used by the sensors. Possible countermeasures are being investigated, including redundancy of sensing satellites to mitigate against potential losses.

During engagements, space-based sensors are vulnerable to blinding or temporary blinding (dazzling) by nuclear detonations. Sensor hardening and proliferation of sensors are again possible hedges against degradation of the system as a whole.

Sensors and weapons could also be vulnerable to direct attacks by the adversary. These attacks could include nuclear attack, kinetic energy or laser attack, or attacks on the integrity of mirrors by radiation, physical or chemical means. These attacks could be delivered by direct-ascent rockets or by space-based assets. Direct-ascent interceptors, armed with nuclear weapons, could be hardened against the BMD system to levels that would not be economical for ICBMs and might survive a counter-attack by the defense for long enough to get within lethal range. Attacks against space-based BMD assets could be made after the whole space-based BMD system is deployed, or at the very beginning of BMD deployment. In the latter case, the BMD system is at its most vulnerable stage, since it might not yet be able to defend itself adequately. It should be noted that such an attack could be considered an act of war and would be risky to the attacker for that reason. However, the attacker might view the prospect of his adversary having a ballistic missile defense of even moderate capability to be a serious enough threat to its national security that the risk would be justified. In any case, the possibility of an attack at this stage must be reckoned with.

In the case of BMD weapons stations, some defense could be provided by massive shielding. These stations would probably be too large, and orbits too low to be effectively concealed. Regarding sensor stations, although shielding is a possibility, concealment is a more likely option. One vulnerability of sensors is that during an engagement they have to function, and in order to function they must be exposed. Moreover, sensor satellites carrying large optical components will have to be large and will therefore be difficult to hide or decoy.

The case of space-based mirrors, which are used to relay radiation from space- or ground-based lasers to their targets, is somewhat different from other weapons from the point of view of survivability. During peacetime, the mirrors can be covered by protective shields against both enemy attack and small meteorites. It is reasonable to suppose that these shields will have to be removed from time to time for testing and maintenance. The mirrors would probably be extremely delicate, if only because of the very high reflective quality and the high optical precision needed to function properly and remain effective when high-power laser beams are reflected from them. Protection strategies against enemy activities could include hiding by various means. Also, shields could be in place most of the time. Since the surface coatings could be vulnerable to attack by certain chemicals (possibly including rocket fuel), the covers should be well sealed.

In almost all cases of attack by approaching rockets or mines, a "shoot back" tactic would be preferable to purely passive defenses. Either the battle station or satellite which is being attacked, or previously positioned defender satellites (emplaced during the early stages of deployment of a BMD system) could act in this self-defense role. Note that such a shoot-back policy would require at least the implicit declaration of a "keep-out zone" surrounding each asset to be defended.[3]

Such an assertion of sovereignty would require the institution of a regime unlike any now existing, including that now in effect on the high seas. Under that regime, maritime powers are free to station naval forces within lethal range of each other during peacetime.

The use of kinetic weapons for shoot back is one likely defense strategy. To be effective, they would have to have a lethal range greater than that of the attacking force. Other BMD kill technologies which have been discussed could also have a lethal capability against battle stations. In determining whether it is feasible to provide sufficient defenses of one's own space assets, particularly during early phases of deployment, a complex analysis of several factors is required. These factors include system cost of defense versus offense; attack and defense tactics; decoy, hiding, and deception tactics; hardness of defense systems; and offensive capabilities.

In the absence of a BMD system architecture, it is difficult to assess accurately the ability of a BMD system to defend its space assets. It would appear, however, that when only a few space assets have been deployed, a certain advantage would lie with the offense. Then, the offense can concentrate its efforts on a small number of defense targets. Assaults can be made repeatedly until one attacker leaks through the defenses to kill its target.

The offense might be deterred from attacking the BMD space assets for the same reason that neither side will launch a first-strike nuclear attack on the other in the absence of defenses: the threat of retaliation and all-out nuclear war. However, the attacker might calculate that his adversary would not risk mutual annihilation in response to the destruction of a few (possibly only one) satellites, and may conclude that the risk of not attacking a BMD system in its early stages of deployment are greater.

[3]Asserting sovereignty over a region of space would appear to violate Article II of the Outer Space Treaty which declares that "Outer Space . . . is not subject to national appropriation by claim of sovereignty . . ." See "Arms Control and Disarmament Agreements" (Washington, DC: Arms Control and Disarmament Agency, 1982), p. 52.

FEASIBILITY QUESTIONS

Technological Feasibility

Virtually all observers have acknowledged that the technical questions bearing on the eventual feasibility of a successful BMD are complex and cannot be answered until further research has been accomplished. The Strategic Defense Initiative Organization argues that the purpose of its research is precisely to answer questions of technical feasibility: before the research is done, there will not be enough information to make a determination.

However, there are various technical issues which appear to present the greatest challenges. These have been listed in the preceding discussion of a hypothetical defense system.

The principal outstanding technical problems in the development of a multi-layer ballistic missile defense system, with a large fraction of its assets based in space, are as follows:

- the feasibility of developing a boost-phase intercept system robust enough to be effective when confronted with plausible countermeasures;
- the ability of the system to discriminate between decoys and RVs in midcourse, when confronted with plausible countermeasures;
- the development of inexpensive ground-based interceptors, meeting required specifications for midcourse exoatmospheric intercept;
- the development of affordable terminal interceptors for high endoatmospheric intercept;
- the resistance of sensors to blinding, dazzling, or spoofing in a hostile nuclear environment;
- the development of very large and complex software packages which can be trusted as sufficiently reliable to enable the United States to make major changes in defense strategy without having been tested under battle conditions;
- the ability to retarget both sensors and (in the appropriate case) directed-energy weapons in times of the order of 1 second or less;
- the ability to deliver the required amount of energy for a kill within the time allotted by the parameters set by a responsive offense. These parameters include number of boosters and RVs, the length of time they are vulnerable to attack during their flight, and their hardness, which enables them to resist attack;
- the ability of the system to defend itself against a concerted attack when fully deployed;
- the ability of the system to be deployed without being destroyed during the early stages of deployment, when the full system is not available for defense;
- the development of computer hardware with 10-year maintenance-free reliability;
- the development of robust power systems which can deliver many megawatts, which are equally maintenance-free and which can deliver large power pulses; and
- the ability of the BMD system to operate in a hostile environment which may include many nuclear explosions—an environment which is currently poorly understood and which would be difficult to duplicate experimentally.

Within each of these categories lies a myriad of precise technical issues to resolve: for example, the degree to which an ablative shield can protect a booster from attack; the resolution achievable using UV laser imaging; the amount of infrared radiation produced by a nuclear explosion in the upper atmosphere; the difficulty of building a 5 (or more) meter diameter space-deployable mirror which is optically good to a small fraction of a wavelength.

In addition to the above technical conditions, the system should be able to be developed and deployed at an affordable cost.

Several of the above technical issues involve the battle of countermeasures versus counter-countermeasures and so on. No meaningful analysis can stop at a predetermined level; otherwise the predicted outcome would be prejudged. To determine which side, offense or defense, is likely to prevail in any particular facet of this contest requires a careful and detailed analysis.

At this stage, it is too early to predict the likelihood of success in the above areas. However, failure to satisfy even one of the above list of requirements could render many versions of the space-based BMD concept impractical. It is clear that although substantial progress has been made to date, most BMD technologies require major advances in the state of the art before their feasibility can be assessed.

After another year of research, at the end of fiscal 1986, 2 years of the SDI research program will have elapsed. At this point, which is a good fraction of the time along to the much discussed decision point on further development in the early 1990s, it may be possible to have some idea of the rate at which important technical milestones are being met. An interim progress report might then contain significant indications of the viability of many important facets of the SDI project. Such a report would provide a vital input to decisions and directions regarding research funding beyond this point.

A major question is the degree to which defensive measures can outstrip offensive countermeasures in general. If one argues that the United States can maintain a 5- to 10-year technological advantage over the Soviet Union, the question reduces to: will current defensive technologies suffice to defeat countermeasures which are 5 to 10 years behind the state of the art? If the answer is not in the affirmative, or if the United States cannot achieve and maintain this level of technical advantage, the prospect of reaching a regime wherein U.S. defenses can reach and keep superiority over Soviet offenses will be dim. In such a case, U.S.-Soviet cooperation towards a mutual deployment of BMD defenses would be essential to their successful deployment, their effectiveness, or both. The Soviets would have to be persuaded that a stepwise transition to a BMD regime would be in their interest and preferable to engaging in an arms race with the United States.

Operational Feasibility

In addition to the matter of the technical feasibility of each of the components of a proposed BMD system, there is the problem of the operational feasibility of the system as a whole.

Assume that a BMD system that meets desired technical specifications can be constructed and deployed. The system would be have to be in a state of readiness for many years; that is, it would have to spring into action from a state of dormancy on very short notice. Perhaps a few days would be available, but any system requiring days of warmup would be useless against surprise attack—indeed, it could increase the incentive to conduct one. When called upon to act, some components may not be operating, since a 100 percent reliability is probably unattainable. This difficulty is countered by providing sufficient component and system redundancy. The degree of redundancy is set to counter the measured unreliability of the system.

Suppose, for example, it is calculated that 10 boost-phase intercept battle stations (one does not know in advance which 10) out of a constellation of 100 would be called on to participate in a battle, and suppose further that the reliability of each of these stations is 90 percent over a 10-year period. On the average, about one station would have to be serviced each year. If servicing is planned on a once-a-year basis, there will be times when one or perhaps two satellites are out of operation. This would imply a need for a 10 percent (or possibly 20 to 30 percent) increase in the number of satellites, to provide spares. These would need to be available to guard against the case where the nonoperating satellite may be one of the 10 that are needed to participate.

It is important to note that the number of required spares depends on the subsystem size, the number of elements of the subsystem which

would have to participate in the battle, and, above all, on the reliability of the subsystem.

Similar arguments can be adduced for each subsystem and each layer of the BMD: the satellite sensors, ground-based sensors, and all weapons systems, power supplies, computing elements, etc., must have high levels of reliability in order to avoid large increases in systems sizes. For many parts of the whole, the 90 percent maintenance-free reliability for 10 years would be desirable.

To maintain reliability, the system would have to be monitored and tested constantly, a job which would require much ground- and space-based effort. Some human intervention might be needed in both cases. Testing and subsequent repairs would be a permanent and constant feature of a large space-based BMD system.

A totally different question, and a more serious one, arises from the fact that the whole BMD system would never have been tested in a realistic battle environment before it would have to operate at a high level of effectiveness. **Without launching a massive rocket attack to test the system, replete with nuclear explosions in space and in the atmosphere, the synergistic effects of the hostile environment will not be well understood. Failures of more than one layer in this "common mode" could drastically reduce the effectiveness of the system as a whole.**

As one example, the immense battle management system, including 10 million or more lines of software code, would have to function reliably the first time it is tested under full battlefield conditions. Large computer programs generally require much time and testing to debug. The question is whether simulated testing would be adequate and trustworthy enough for the reliance which would be placed in it.

As another example, it is not certain how much scintillation will occur in the upper atmosphere as a result of nuclear explosions there. The uncertainty includes the wavelengths, the intensities, and the duration of the scintillation. The resistance of radars, sensors, and sensitive optical surfaces to nuclear explosions in space and in the upper atmosphere is uncertain. Since these effects may be difficult to study in the laboratory or in underground nuclear tests, this uncertainty may not be resolvable in the absence of extended upper atmospheric testing.

The electromagnetic pulse induction in ground- and space-based systems, which is caused by nuclear explosions in space, is similarly not fully understood. The resistance of space-based power supplies and power conditioning systems to the various types of radiation from nearby nuclear explosions may be difficult to determine. Nuclear effects may provide significant problems to the defense in this context. The ability of a system to withstand a concerted attack may not be known in advance and the outcome of such an attack could be highly dependent on the tactics used by each side.

Costs

General

Questions of cost are even more elusive at this point than questions of technical feasibility. Before the architecture of a system is defined, it is impossible to give a reasonable and credible estimate of total system costs. Estimates have ranged from tens of billions of dollars to $1 trillion and more,[4] not including operational and maintenance costs. Not surprisingly, BMD advocates tend to estimate lower numbers than opponents. Nearly all credible observers concede, however, that the system would require a very large investment.

The SDI Organization, recognizing the large uncertainties in current cost estimating, has formed a cost estimating working group. The

[4]E.g., see D. O. Graham, *High Frontier: A New National Strategy* (Washington, DC: High Frontier, 1982), p. 9; Z. Brzezinski, R. Jastrow, and M. M. Kampelman, "Defense in Space is not 'Star Wars'," *New York Times Magazine*, Jan. 27, 1985; J. Schlesinger, National Security Issues Symposium, 1984, "Space, National Security and C³I," Mitre Document M85-3, October 1984, p. 56; Department of Defense document presented to Senate Foreign Relations Committee by Senator Pressler, Apr. 24, 1984.

group is investigating possible new and better cost estimating techniques since the magnitude and novelty of the SDI may demand new estimating tools.[5] In addition to cost estimates, cost-exchange analyses will also be performed.

Cost-exchange analyses will be essential to demonstrating the attractiveness of the system. If it costs the offense less to counter a defense than it costs the defense to deploy one, it is generally not advantageous to proceed with the defense. On the other hand, suppose the offensive countermeasures cost less than the defenses, but suppose, additionally, that the United States has far greater economic resources which it can devote to the effort than the U.S.S.R. In that case, it could pay the United States to continue with the defense, even though it would be more costly than the cost to the Soviets of offsetting our move.

On the other hand, even a defense which the United States considered to be cost-effective might not be sufficient in itself to bring about a transition to defense dominance. First, Soviet calculations of cost-exchange ratios may not coincide with our own. Second, they might redirect their offensive forces along different delivery modes, concentrating on ones which could most cheaply penetrate defenses. This could change the assumptions inherent in some of the U.S. cost-exchange calculations.

More importantly, cost may not be a major determining factor for the Soviets in policy planning. They may be capable of spending more on offense than the United States spends on defense because of inflexibilities in their economic structure, because they are more easily able to direct their economy towards military expenditures, or because they may consider the benefits of maintaining their offensive capability to be of paramount national importance, thus justifying to themselves levels of expenditure that the United States would consider to be inordinate.

Indeed, the United States already considers Soviet expenditures on their offensive forces to be inordinately large. Secretary of Defense Weinberger has stated, "Whatever the reasons, the Soviets believe that their colossal military effort is worthwhile, notwithstanding the price it imposes on the Soviet society and its troubled economy."[6]

It is difficult to do cost-effectiveness analyses on systems which are on the cutting edge of future technologies for another reason: technological obsolescence. The unpredictable nature of technological progress leads to unpredictable shifts, probably major ones, in both costs and effectiveness. It is possible that no analysis which assesses systems more than a very few years in advance will have much validity.

One additional point worthy of mention is, that in working out cost exchanges between offense and defense, one must account for the fact that both the United States and the U.S.S.R. are starting from a situation where they have little or no effective defense, but very effective offenses. The funds for this level of offense have already been spent. The funds to deal with this level of offense by defensive means have not. Therefore, a massive defense expenditure would be needed initially just to counter existing offenses. Cost-exchange ratios *at the margin* are meaningful in themselves only beyond this point, when offensive additions and countermeasures are met by defensive additions and countermeasures.

Total System Cost

Rather than present an independent estimate for system cost, this report will point out the requirements which must be met in order to devise a credible estimate.

It is possible to make some simple cost-exchange arguments regarding limits on what a system or a part of a system **should** cost. The following brief discussion is only illustrative, and subject to the reservations noted above

[5]Information contained in a speech by Lt. Gen. James A. Abrahamson at the annual meeting of the American Institute of Aeronautics and Astronautics, April 1985, reported in *Military Space*, Apr. 15, 1985.

[6]C. Weinberger, *DOD FY85 Annual Report* (Washington, DC: Department of Defense, 1984), p. 26.

concerning cost-exchange concepts in the context of BMD.

Suppose one plans a boost-phase intercept system consisting of 100 battle stations. In chapter 7, it was found that, for satellite altitudes of 1,000 km or more and slew times of 0.25 seconds or more, the number of battle stations required increases nearly linearly with the number of boosters. If one were to aim for a 30:1 kill ratio of booster to deployed battle station, this implies that a battle station should *not* cost much more than 30 times the cost of a booster. If it did, the offense could, in principle, force the defense to spend much more on staying ahead of the offense than the offense would have to spend in keeping up with the defense by the simple expedient of building more boosters. A cost of $50 million per booster would mean that the defense could spend $1.5 billion per battle station, and still keep up with the offense in the cost-exchange race. A $200 million cost per booster would imply that the defense would have to keep its battle station cost below $6 billion per station to stay in the running. This would mean that a total system cost of $600 billion would still be cost-effective.

This crude argument includes many simplifying assumptions, but it may be useful as an aid to understanding the nature of cost-exchange studies.

There are ways of making rough direct cost estimates, rather than defining allowed upper limits, as above. One could estimate the fuel costs needed to place a given payload in orbit, or one could use other crude "order of magnitude" assumptions to make simple calculations. **However, none of these techniques satisfactorily accounts for technological improvements, or,** a fortiori, **possible technical breakthroughs.** The conclusion stands that it is too early to make useful estimates.

A cost analysis for a BMD system would use several tools and techniques.

First, a work breakdown structure would be made. This is a list of items needed to design and construct the system. Research, test and evaluation, maintenance, procurement, and deployment are all elements of the breakdown. As time goes on, broader categories of items are further broken down into more specific elements.

Second, cost estimating relationships are used. These are equations which use past history for estimating costs of elements in the project being investigated. One difficulty in making such estimates for highly innovative projects, such as BMD, resides in the fact that many of the estimates may depend on totally unknown or unanticipated future results of basic research. Another problem arises from the uncertainties in extrapolating costs from items in a historical data base to similar future items. For qualitatively different technologies, the accuracy of estimates derived from historical analogs may be poor.

A third tool is the use of learning curves. These are used in predicting cost reductions per unit resulting from gains in experience and volume of production.

Also, planning factors are used to predict costs. They are arithmetic factors used in making cost estimates based on general past experiences. One example would be to use the ratio of development costs to investment costs for similar programs in predicting the development cost of a new project, when the investment cost is known. This is similar to the use of cost estimating relationships noted above, but even more general and subject to error.

It is apparent that in order to use any of these tools for estimating the cost of a BMD system, the architecture and the technologies to be used will have to be defined.

The burden for providing cost estimates should be on those who maintain that an effective BMD will be affordable, including those who define potential system architectures. If one argues for the commitment of large sums of money to research in one particular area at the expense of others, with the intent of making deployment options available, one should provide a cost estimate for the eventual deployments envisioned. Clearly, if the end product appears to be prohibitively expensive, this indication would discourage decisions to fund the expensive large-scale research which might lead to this undesirable result.

In conclusion, attempts to provide a realistic and defensible cost estimate for an effective BMD system must await the presentation of a realistic and defensible suggestion for one or more alternative system architectures. At present, it is safe to say that, if, indeed, a space-based BMD system is defined which appears to be feasible, it will likely be considerably more expensive than any other weapons program yet developed.

Chapter 9
Alternative Future Scenarios

Contents

Chapter 9
Alternative Future Scenarios

INTRODUCTION

Thus far we have examined the possible applications of ballistic missile defense to various strategic purposes and the potential effects of BMD deployments on crisis stability, arms race stability, and arms control. We then examined the technologies that might be applied to ballistic missile defense. In this chapter we attempt to give the flavor of the current debate over BMD by presenting the positions of some major policy advocates. We pay particular attention to the idea of *transition*—of how and with what consequences we might move toward a world where BMD plays an important strategic role. We also look at ways that the world might evolve if the United States does not take the initiative in deploying BMD. For each picture of the future, we identify what appear to be the major assumptions upon which that picture rests. What are the key outcomes of U.S. action (or inaction) that each picture posits? What events are assumed to occur along the way to the predicted outcomes? We leave it to the reader to choose which assumptions seem most plausible.

The policy approaches reviewed here are the following:

1. The Strategic Defense Initiative approach, as defined by various Administration spokesmen;
2. An approach advocating the earliest possible deployment of space-based and ground-based BMD, as described in the writings of representatives of the "High Frontier" organization and others;
3. An "intermediate deployment" approach advocating deployment of BMD by the mid-1990s, using technology not yet available;
4. A "missile-silo-mainly" approach, advocating defenses with the limited objective of defending the U.S. land-based retaliatory force, with other targets defended only collaterally;
5. An approach aimed at strengthening the current regime banning most ballistic missile defense through current, and possibly additional, arms control measures.

After describing in this chapter the differing views of various policy advocates on longer-term objectives for BMD deployment, we will turn in chapter 10 to the immediate problem facing Congress: how to orient the U.S. BMD research program this year and in the years to come. These current decisions will be influenced by views on longer-term objectives.

ALTERNATIVE FUTURE SCENARIOS

The SDI Policy Approach

The goals of the Strategic Defense Initiative have been explained extensively in a pamphlet, *The President's Strategic Defense Initiative*, issued January 3, 1985, by the White House; in an April 1985 *Report to the Congress on the Strategic Defense Initiative*; and in articles and speeches by George A. Keyworth, II, Science Adviser to the President, Paul H. Nitze, Special Adviser to the President, Lt. General James A. Abrahamson, Director of the Strategic Defense Initiative Organization, and other Administration officials.[1]

[1]Relevant excerpts from these sources are presented in app. H. A list of statements and articles on BMD by Administration spokesmen appears in app. I.

According to these Administration spokesmen, progressive BMD deployment would be accompanied by mutual U.S.-Soviet reductions in offensive weapons. This would lead first to enhanced deterrence, and ultimately to basing national security primarily on defense. The effectiveness of the defenses would be enhanced by the reductions in offenses. Administration statements postulate the United States moving to a future defensive strategy in four phases, along the following lines:

1. Research Phase

As described by Reagan Administration officials, the SDI approach begins by launching "a broad-based, centrally managed research effort to identify and develop the key technologies necessary for an effective strategic defense." This phase, which has already begun, is expected to include a progressive series of BMD subsystem demonstrations of evolving technical capabilities. Each of these demonstrations would display a technological advance which would be militarily meaningful but which would not violate any arms control treaty provisions.

In view of the undiminished U.S. commitment to the security of its allies, the SDI research program will not confine itself to exploring BMD technologies with potential against ICBMs and SLBMs; it will also carefully examine technologies with potential against shorter-range ballistic missiles, such as those currently targeted against Western Europe. U.S. allies have been invited to participate in the SDI research program.

Principal emphasis will be placed on technologies involving nonnuclear kill concepts. Research on nuclear directed-energy weapons will also be undertaken in order to develop an understanding of the potential of this technology and as a hedge against Soviet work in this area.[2]

This research phase might last until some time in the 1990s, when the President and Congress could assess the results of the BMD research program, and then decide whether to begin full-scale engineering development of a complete BMD system. The criteria for such a decision would, of course, be determined by the President and Congress at the time the decision is made. As presently envisaged by Administration spokesmen,[3] however, this decision would not be made unless there were high confidence that the proposed system would be:

- effective in substantially reducing the counterforce capability of current and projected Soviet intercontinental, sea-launched, and theater nuclear forces;
- sufficiently survivable itself against preemptive attack;
- cost-effective on the margin; i.e. able to counter an increment in offensive countermeasures at a cost substantially less than the cost of the offensive increment; and
- able to contribute to improving the stability of the overall strategic balance at each stage of deployment.

Meanwhile, the mutual understanding that both sides were seriously pursuing strategic defense systems would force Soviet planners to rule out an effective first strike as a realistic future option, and would provide U.S. and Soviet arms control negotiators with a common limited strategic objective, retaliation, by which to discuss possible build-down of offensive nuclear arsenals.

2. Systems Development Phase

If a decision were made to go ahead, prototypes of all the required BMD components would be designed, built, and tested during the systems development phase. Meanwhile, the United States would seek Soviet agreement to phased deployment of defensive capabilities by both sides. Arms control proposals might include mutually agreed schedules for intro-

[2] *Report to the Congress on the Strategic Defense Initiative*, Department of Defense, April 1985, p. 3.

[3] See *The President's Strategic Defense Initiative* published in January 1985 by the White House, Ambassador Paul H. Nitze's speech of Feb. 20, 1985, and Ambassador Edward L. Rowny, "America's Objective in Geneva," *New York Times*, Apr. 29, 1985.

ducing the defensive systems on both sides, associated schedules for reductions in offensive ballistic missiles and other nuclear forces, confidence-building measures, and agreed constraints on devices designed specifically to attack or degrade the other side's defensive systems. The Soviet leaders should respond cooperatively to these proposals. If they did not, the United States would have to decide whether to proceed to the next phase anyway.

On this issue, Fred C. Ikle, Under Secretary of Defense for Policy, has written:

> The more the offensive armaments can be reduced by agreement, the easier and cheaper the job of providing effective defenses. Yet, to be realistic about Soviet motivations, we must seek to develop and deploy systems that can provide effective defenses even without such reductions. The United States is now pursuing new technologies that hold promise for success on the "hard road" as well. Thus, we make it all the more probable that the Soviet leaders will join us some day on the easy road of cooperation.[4]

3. Transition Phase

During this period, operational BMD systems would be deployed by the United States and the Soviet Union on an incremental, sequential basis, going up the scale of increasing capability levels suggested in chapter 5. Each added BMD system increment, in conjunction with effective and survivable offensive systems, would enhance deterrence by making each side's land-based nuclear forces more survivable, thus reducing the incentives for a preemptive first strike. The United States could also deploy BMD and air defenses to defend preferentially a limited set of either conventional military systems or populations, in the United States or overseas. At the same time, as the United States and the Soviet Union deployed BMD systems that progressively reduced the value of ballistic missiles, it is hoped that deep reductions in the numbers of ballistic missiles on each side could be negotiated and implemented.[5]

While hardened military assets could be successfully defended by these transitional BMD systems, cities would still be hostage to mutual deterrence. This fact would be crucial to stability during the transition years. But if ICBM silos were defended, the retaliatory arsenals needed for attacking cities would not have to be nearly as large as those needed to launch or survive a preemptive strike. Moreover, during this period BMD deployments might save lives and limit damage in the unlikely event —planned or accidental—that a small number of nuclear missiles were launched despite effective defenses.

During the transition period, conventional military forces might have to be improved and expanded, especially in Europe. Our defense posture would move toward much heavier reliance upon conventional, nonnuclear forces, and correspondingly less reliance on using our nuclear forces to deter conventional attacks on ourselves and our allies. This strengthened role for conventional forces would need to be supported by restoration of technological leverage. At the same time, second- and third-generation BMD technologies would begin to become available, which could in time reduce the effectiveness of strategic nuclear weapons to the point that cities could become viable candidates for defense if offensive nuclear forces were limited to low enough levels.

Most explanations of the SDI policy approach by Administration officials tend to emphasize a scenario in which the Soviet Union agrees to deep reductions in all kinds of offensive nuclear forces. In this view, a fully effective defense would be more readily achieved with deep reductions in Soviet offensive forces.[6]

---

[4] "Nuclear Strategy: Can There Be a Happy Ending?" *Foreign Affairs*, spring 1985, p. 825.

[5] According to Ambassador Paul Nitze, "We would see the transition period as a cooperative endeavor with the Soviets. Arms control would play a critical role. We would, for example, envisage continued reductions in offensive nuclear arms." (Speech in Philadelphia, Feb. 20, 1985.)

[6] See statement by SDI Director, Lt. General James A. Abrahamson, *Science*, Aug. 10, 1984, p. 601.

Soviet unwillingness to accept such reductions might therefore complicate a successful transition to the final phase described below.

Administration officials argue, however, that in that case the United States should still deploy cost-effective BMD systems (if they can be developed) to enhance the deterrent value of U.S. ICBMs and bombers. If the Soviets keep striving to overcome these U.S. BMD systems with offensive countermeasures and deployment of larger offensive forces, the United States should try to develop and deploy more capable BMD systems. If the Soviets also build up full-scale BMD systems of their own to intercept U.S. missiles, the United States should build offensive forces capable of overcoming the Soviet BMD systems. In both cases, the United States would hope that its superior technological talents and industrial resources would permit it to stay ahead.

4. Final Phase

During this period, both countries would complete deployments of highly effective, layered BMD systems to protect their own and their allies' populations, as well as their military assets. Ballistic missile force levels would "reach their negotiated nadir." If similarly effective defenses had been developed by this time against cruise missiles, bombers, and other means of nuclear attack, such defenses could also be incorporated.

Ballistic missile and air defenses that might look less than 100 percent effective in the context of an offensive exchange involving tens of thousands of warheads could be expected to perform better against an attack by only tens or hundreds of warheads. Strategic defense could therefore make possible a world effectively disarmed of nuclear weapons, yet still retaining national sovereignty and security. Thus, by the end of the final phase the United States would achieve President Reagan's ultimate goal of "eliminating the threat posed by strategic nuclear missiles." Our present reliance on offensive retaliatory forces to deter a nuclear attack would be replaced by reliance on a combination of defensive weapons and of deep mutual reductions in offensive nuclear forces.

Critical Assumptions for the SDI Policy Approach

The SDI policy approach appears to be based on the following assumptions:

Assumption 1.—There is a reasonable prospect of BMD technological developments meeting the Administration criteria of effectiveness, cost-effectiveness at the margin, and survivability.

Effectiveness.—It is assumed that the BMD systems will be effective and that possible Soviet responses to prospective U.S. BMD systems will not negate the effectiveness of those systems. The technical requirements for such effectiveness are discussed in detail in chapters 7 and 8.

Cost-effectiveness at the margin.—Increments of Soviet offense are assumed to be clearly more costly than corresponding increments of U.S. defense. Increments of U.S. offense will presumably also be more costly than corresponding increments of Soviet defense. (But if the latter were not the case, then Administration policy would be all the more likely to succeed, because continuing an offensive competition would be even more disadvantageous to the Soviets.)

If this assumption held, the Soviets would have a strong incentive to negotiate the mutual reduction of offensive forces. (As noted in chapter 6, though, the incentive may still not be sufficient.) Otherwise, in the transition phase the Soviets might well continue to attempt to counter U.S. defenses with offenses instead of seeing futility in further offensive additions.[7] However, this criterion may be very difficult to apply in practice. First, costs may not be understood that well at the time that a decision is to be made. Secondly, the answer may vary greatly depending on the re-

[7]Indeed, the full Administration scenario would seem to imply that the Soviets, too, must find a defensive system with a favorable cost-exchange ratio vis-a-vis U.S. offenses; otherwise, the United States would find itself tempted to pursue strategic superiority by adding offenses as well as defenses.

quirements placed upon the system, and the level of confidence with which those requirements must be satisfied.

On the other hand, if both sides decided that a transition to defense dominance was desirable, they could agree to reduce offenses despite unfavorable cost-exchange ratios between offense and defense.

Survivability.—See chapter 6 for a discussion of the importance of BMD system survivability to maintaining crisis stability. Administration officials have emphasized that meeting this criterion would be particularly critical to a decision to deploy BMD.

Assumption 2.—The strategic program as a whole will be affordable for the United States.

It should be noted that there is also an assumption that defenses will be affordable *up to* the margin of trade-off between the offense and the defense; i.e. that the initial investment in defense necessary to achieve the desired effectiveness against the predicted responsive Soviet offensive threat will be acceptable to the United States. Since the Soviets already have an offensive force, their initial investment in offense has been made. The total U.S. investment in defenses to counter that threat has to be considered, as well as subsequent increments of defensive improvement to counter increases in the threat.

Moreover, the costs of the transition stage—in which defensive systems are being purchased, offensive systems are being maintained and improved, and conventional forces may be needing augmentation—must also be considered. It should be noted that particularly in the early stages, even before any U.S. BMD deployments have taken place, Soviet anticipatory offensive responses are highly likely. These could, in turn, appear to require counter-balancing U.S. offensive deployments.

Another issue is whether we and our allies could afford the additional conventional forces that would probably be needed to preserve the military balance in Europe when Soviet BMD deployment diminished the credibility of nuclear deterrence of conventional aggression.

Assumption 3.—The current ABM Treaty regime can be sustained until the United States is prepared to make a BMD deployment decision.

The SDI approach appears to assume that the necessary research and testing for a U.S. deployment decision can be conducted within what both sides agree are the confines of the ABM Treaty (or that the Soviets will agree to necessary amendments). Otherwise, if the Soviets came to believe that the United States was violating the Treaty, they might not postpone their own BMD deployment decision until the United States was ready to make one. The SDI approach also appears to assume that even if no treaty violations (or amendments) are necessary to the U.S. decision, the Soviets will be willing to wait for a U.S. decision and then negotiate the transition to defenses on both sides, rather than move ahead unilaterally with their own deployment because they believe the United States will do so soon.

Judging from published U.S. Department of Defense descriptions of the current Soviet BMD program, the Soviets are now in a position to field a large-scale, ground-based system of BMD interceptors sooner than the United States could (although such a system could almost certainly be overcome by existing U.S. offensive forces). Should the Soviets begin such a deployment, however, the United States might nevertheless feel compelled to respond with hasty offensive "fixes" and investment in U.S. BMD systems with only short-term value.

Such a Soviet move could be particularly undesirable if U.S. research should show that technology will not in fact permit defenses as effective as those now hoped for. We could find ourselves in a costly offensive-defensive arms race with little hope for decisive dominance of the defense—the situation the ABM Treaty was intended to preclude.

This assumption about the short-term viability of the ABM Treaty is not necessarily essential for the long-term SDI scenario, but if it were to prove incorrect, the transition to defenses could be more difficult and pose greater risks of instability.

Assumption 4.—Arms control agreements can be formulated and negotiated that will permit graduated, mutual, deep offensive reductions and defensive deployments that are crisis-stabilizing and arms-race-stabilizing.

This assumption has three components: a U.S. desire for such agreements, a corresponding Soviet desire, and the ability of both sides to overcome the technical difficulties in reaching such agreements.

Many analysts believe that if the Soviets should conclude that the United States is likely to abandon the ABM Treaty and deploy a nationwide BMD system, they would be highly unlikely to agree to offensive force reductions.[8] Offensive arms control may therefore be difficult during this stage, because the Soviets would almost surely start to increase their offensive nuclear forces in order to counter the U.S. BMD system and maintain their offensive force capability. This appears to be a major problem confronting the U.S. negotiators in Geneva.

The SDI scenario assumes that technological developments will eventually persuade the Soviets of the futility of trying to maintain the military effectiveness of ICBMs. If this assumption proves incorrect, the demands placed on the BMD technology for effectiveness, cost-effectiveness, and survivability could be much higher. The ultimate goal of negotiated offensive reductions might come later rather than sooner, if at all.

Even with a mutual desire to negotiate a stabilizing transition to defensive deployments, the difficulties in working out such an agreement should not be underestimated. Experience with past U.S.-Soviet arms control negotiations has demonstrated that asymmetries in offensive force structures and in strategic doctrines, Soviet secrecy, and other factors have made the process of reaching agreement highly arduous. Agreement on a transition involving a fundamental change in strategic goals and drastic changes in force structures, and adequate verification thereof, would be much more complex.[9] For these reasons it might prove to be far more difficult to reach U.S.-Soviet agreement on increasing defensive and decreasing offensive deployment levels than it would be to agree to reduce offensive levels with BMD essentially banned.

Assumption 5.—The loss of the "extended deterrent" threat of U.S. offensive forces could be compensated for by either conventional force improvements or diplomatic measures to reduce the Soviet threat to U.S. allies.

For those early stages of BMD deployment intended to enhance deterrence by increasing the survivability of U.S. nuclear retaliatory capabilities, this assumption would not come into play. But as long-range offensive nuclear weapons became less effective on both sides, the possibility of escalation of theater conflicts to nuclear war would serve as less of a deterrent to Soviet aggression. The United States would presumably seek alternate means of reducing the Soviet conventional and theater-nuclear threats to U.S. allies and interests.

Assumption 6.—Political fallout from any U.S.-Soviet disputes over the SDI will be manageable.

Already, the Soviets have stepped up their allegations of America's aggressive intentions to develop a first-strike capability, to undermine strategic stability, and to increase the danger of nuclear war. They have accused the United States of planning to abrogate the ABM Treaty and thus destroy any hopes for progress in strategic arms control.

Whether Soviet efforts to lay the blame on the United States for derailing arms control efforts will gain widespread credence, particularly in Western Europe, remains to be seen.

---

[8]For example, the Hoffman Panel of experts, appointed by the Department of Defense to assist in planning the SDI program, concluded that the Soviets would be likely to respond to U.S. BMD deployment "with a continuing build-up in their long-range offensive forces." (Fred S. Hoffman, et al., *Ballistic Missile Defenses and U.S. National Security*, unclassified summary report, October 1983, p. 11.) Participants in the OTA Workshop on Soviet military doctrine and policy also judged that the first Soviet responses to U.S. BMD deployments would be to try to maintain the effectiveness of their offenses. For Soviet statements on this point, see app. K. References to studies of Soviet strategic policy are listed in app. M.

[9]The requisite elements of such a negotiated transition are discussed in ch. 6.

If they succeed, neutralism might become a stronger political movement in many NATO countries, and might well become government policy in several.

The SDI scenario assumes that there will not be unacceptable damage to the North Atlantic Alliance from neutralist tendencies or from European beliefs that a defensive system capable of really protecting Western European cities from nuclear attack is less attainable technically, financially, and politically than one for the United States. It also assumes that the British and the French would not be excessively alarmed by the prospect that loss of the ABM Treaty and subsequent Soviet BMD deployment would undermine their own nuclear deterrents.

Opposition from U.S. allies would not necessarily preclude deployment, but the Administration has stated that consultation with our allies would play an important part in a deployment decision.

Early BMD Deployment Policy Approach

Proponents of this approach propose that the United States begin immediately to deploy ballistic missile defenses as rapidly as possible, using presently available U.S. technology.[10] Their goals are to enhance the current basis of deterrence by using defense to complicate Soviet targeting and to provide some measure of protection to U.S. society should deterrence fail. Complicating Soviet targeting would enhance deterrence by increasing the uncertainties in the minds of Soviet planners regarding the outcome of planned strikes against the U.S., especially of strikes designed to achieve decisive military advantage. Their program highlights space-based BMD as a "technological end-run" around Soviet military capabilities, largely by utilizing superior U.S. computer miniaturization technology.

In order to create such a space-based BMD system quickly, the "High Frontier" organization recommends use of "essentially off-the-shelf" technology, and describes an illustrative system which would be designed to intercept Soviet ICBMs, SLBMs, MRBMs, and IRBMs in their boost and post-boost stages. Its 1982 report estimated such a BMD system to be deployable within 5 or 6 years at a cost of about $13 billion.[11]

That first-generation BMD system would be followed about five years later by a second-generation system, perhaps using laser or particle beams to attack missile warheads in their midcourse stage as well as earlier.

The advocates of the High Frontier approach emphasize, however, that these space-based BMD systems should be reinforced with a series of "collateral actions" as follows:

- point defense of U.S. ICBM silos, using a ground-based system such as the Low Altitude Defense System (LoADS) interceptors with nuclear warheads or the SWARMJET nonnuclear, high-velocity interceptor rockets, either of which they believe could be deployed within 2 years.
- a greatly enhanced civil defense program, which they believe could save a great number of lives and "protect enough essentials of our agricultural and industrial assets to give reasonable hope for the recovery of national power and our modern standard of living."
- mobile, high-performance, manned military "spaceplanes" to inspect and maintain U.S. satellites and, eventually, conduct "an active defense of U.S. installations in space."

[10]Examples of this approach are described in Daniel O. Graham, *High Frontier: A New National Strategy* (Washington, DC: High Frontier, Inc., 1982); Daniel O. Graham, *High Frontier: A Strategy for National Survival* (New York: Tom Doherty Associates, 1983); Daniel O. Graham and Gregory A. Fossedal, *A Defense That Defends* (Old Greenwich, CT: Devin-Adair, 1983); and Angelo Codevilla, "Understanding Ballistic Missile Defense," *Journal of Contemporary Studies*, winter 1984, pp. 19-35. This last article differs from the others in recommending early deployment of space-based chemical lasers, as opposed to the space-based kinetic kill vehicle constellation of "High Frontier." Most of the following discussion refers to the "High Frontier" proposals.

[11]Department of Defense officials disagree with these estimates; see app. G.

- a manned space station for testing BMD system elements, first in low orbit and later in geosynchronous orbit.
- comprehensive anti-bomber defenses,
- increased anti-submarine warfare deployments,
- substantial strengthening of U.S. offensive strategic forces, including bombers, ballistic missiles, and cruise missiles, because the need for offensive retaliatory forces would remain.

While this policy approach has some similarities to the Reagan Administration's SDI approach, it differs in several important respects. First, it is based on the belief that a BMD system using current technology would be sufficiently effective to justify its deployment at this time. Second, it does not hold out an ultimate goal of near-perfect defense of cities, but explicitly advocates deployment of only partially effective defenses on the ground that these are better than none.[12] Third, the writings of its proponents do not suggest that it could lead to arms control agreements for deep reductions in offensive nuclear forces. On the contrary, they envisage a hostile Soviet response to U.S. adoption of this policy, resulting in an intensified arms race between the two superpowers, a race they believe the United States would win by making full use of its industrial and technological superiority:

> If . . . we move strategic systems onto a fast-track, high-priority model the Soviets will have two or three years, not decades, to respond to our latest defense. And, as they begin to devise countermeasures and build the hardware to perform them, we will already be deploying the next round of strategic defenses—high energy lasers, particle beam weapons, and so on.[13]

In commenting on this prospect, Daniel O. Graham argues, "The tasks the U.S.S.R. will face if High Frontier becomes a reality require high technology on a prodigious scale. The Soviet economy, already severely strained, may well be unable to meet these requirements for high technology without disintegrating."[14] He cites reports that indicate that Soviet "military expenditures are already approaching, if not exceeding, 'the objective limits' beyond which the U.S.S.R. cannot go without serious damage to the economy as a whole, including the reproductive [sic] base crucial to the very existence of Soviet military might."[15]

[12]In *A Defense That Defends,* op. cit., Graham and Fossedal state: "In fact, there will never be a perfect defense, not against the bullet, against the tank, against nuclear weapons. What can be done is to complicate an attacker's calculations, blunt his forces, and save millions of lives" (p. 121).

[13]Graham and Fossedal, op. cit., p. 115.

Critical Assumptions for the Early Deployment Policy Approach

This policy approach appears to be based on the following assumptions:

Assumption 1.—U.S. technology is now adequate to support prompt deployment of a functional, survivable, space-based BMD system.

In the view of some Defense Department officials, the High Frontier estimates for costs and construction times for such a system are unrealistic.[16] Lt. General James A. Abrahamson, Director of the SDI Organization, has testified before the House Armed Services Committee that while many of the components that the High Frontier proposal might utilize were available, their integration into an effective and survivable weapons system would require much more study.[17] Administration officials say, in short, that several years of research are needed to assess the validity of the assumptions critical to the early deployment approach.

[14]*High Frontier* (1982), op. cit., p. 86.

[15]Ibid., p. 162.

[16]The 1982 High Frontier report (op. cit., p. 71) estimates a total cost of about $20 billion over 5 years and about $35 billion through 1990. Department of Defense studies estimated that a comparable deployment would cost from $50 to $75 billion or more, according to testimony on the DOD authorization before the Senate Armed Services Committee on Mar. 23, 1983 (p. 2668). The DOD witness, John L. Gardner, indicated reservations about the survivability of such a system, and stated:

> Before we would recommend a significant undertaking on a system like the High Frontier we believe that significantly more work would have to be done in the examination of that system from the viewpoint of its survivability and considering the kinds of responsive threats that might come at that system from the Soviet Union, were they to conclude that it represented a military threat.

See also app. G.

[17]Testimony of Feb. 21, 1985. See app. G.

A Background Paper done for OTA contains an analysis of the "High Frontier" proposal for 432 satellites carrying 1-km/sec interceptors. It demonstrates that the concept would have meager coverage of Soviet ICBM fields.[18]

Assumption 2.—The cost of such a deployment could be relatively low.

This assumption is based partly on the idea that Congress and the Administration would be willing to allow the program to proceed on a "fast track" without existing procedures for competitive bidding and administrative review, and that these procedures waste large sums of money. While such an approach might speed progress in some areas, views differ as to whether the costs would be reduced significantly.

Assumption 3.—Crisis instability will be avoided because technically superior U.S. defenses will reduce Soviet incentives to execute a preemptive attack under any plausible circumstances.

See chapter 6 for a discussion of crisis stability issues. Full success of the early deployment scenario seems to require U.S. strategic superiority from the beginning; otherwise, the United States could be at a dangerous strategic disadvantage for a considerable period. Partial success, however, might depend only on the increase in uncertainty posed by BMD to Soviet military planners.

Assumption 4.—Arms race instability will be manageable because the Soviets will not be able to afford to match U.S. technical superiority.

The long-run affordability of this policy approach seems to depend on its unaffordability for the Soviets. But if the Soviet economy cannot be forced into collapse, this approach requires that the United States be able to maintain an indefinite lead in the defense-offense competition.

Assumption 5.—The loss of the "extended deterrent" threat of U.S. offensive forces in the face of Soviet BMD could be compensated for by the increased credibility of U.S. willingness to use conventional force.

According to the High Frontier literature,

> There would be a realization that the U.S. was beginning to break out of the paralytic bonds imposed by the concept of Mutual Assured Destruction . . . there would be a restoration of the badly shaken European confidence in U.S. ability and resolve to actually use its power to preserve the Free World.[19]

Assumption 6.—Political fallout from any U.S.-Soviet disputes over the BMD plans and deployments will be manageable.

See the discussion of SDI assumptions above.

Intermediate BMD Deployment Policy Approach

The near-term strategic objectives of the Early BMD Deployment approach—enhancing the deterrence of a Soviet attack upon the United States by increasing Soviet uncertainty in their ability to accomplish military objectives in such an attack, along with providing some measure of protection to U.S. society should deterrence fail—are also sought by some who, unlike Early Deployment advocates, do not believe that existing BMD technology is adequate. Supporters of an "Intermediate BMD Deployment" approach believe that U. S. BMD deployment should not wait for the feasibility of long-term, highly effective BMD concepts to be demonstrated. However, they would not advocate that deployments start immediately. Instead, they would support U.S. deployment of BMD in the "intermediate-term"—say by the mid-1990s—of the best system that could be deployed at that time.

Advocates of "early deployment" approaches hope to deploy BMD so rapidly that the Soviets will be unable to counter it before it becomes effective. "Intermediate deployment" supporters do not expect to avoid So-

[18]Ashton B. Carter, *Directed Energy Missile Defense in Space*, OTA Background Paper, OTA-BP-ISC-26, April 1984, pp. 34-35.

[19]Graham, *High Frontier: A New National Strategy*, op. cit., p. 88.

viet countermeasures, but believe that the added uncertainty that Soviet planners and weapons designers will face against even a partially effective U.S. defense will enhance deterrence. Except for their difference in timing, the rationales and underlying assumptions of the "early deployment" and "intermediate deployment" approaches are similar.

Critical Assumptions for the Intermediate BMD Deployment Policy Approach

The critical assumptions of the Intermediate Deployment Approach closely resemble those already discussed under the Early Deployment Approach. However, the first assumption in that approach must be modified slightly, and one more added.

Assumption 1.—In the "intermediate-term" (mid 1990s), U.S. technology will be adequate to support deployment of a functional, survivable BMD system.

Such a system might eventually include space-based components, but would probably start out with ground-based terminal and perhaps midcourse interceptors. Nonnuclear interceptors would probably be desirable, but if adequate performance and confidence could not be attained with nonnuclear technologies, nuclear interceptors would be required.

Assumption 2.—Deterrence of Soviet nuclear attack does not depend critically on the number of U.S. warheads that would penetrate Soviet defenses in retaliation.

Although an intermediate deployment of ballistic missile defenses by the United States would confuse Soviet attack plans and increase the uncertainties they would face in conducting a nuclear attack upon the United States, deployment of a Soviet defense in the same time frame, using systems evolved from the current upgrade of the Moscow system, would lessen the ability of the United States to conduct a retaliatory strike. In chapter 5, it was shown that even if the United States and Soviets deploy equivalent defenses, it is quite possible that in the event of a Soviet first strike, more U.S. retaliatory warheads would be intercepted by Soviet defenses than would have been saved by U.S. defenses. In other words, fewer U.S. retaliatory warheads would reach the Soviet Union if both had such defenses than if neither did. Advocates of "intermediate-term" BMD deployment acknowledge this possibility, but believe that the uncertainties introduced into Soviet attack plans by U.S. defenses would more than compensate for whatever success Soviet defenses might have in intercepting our retaliatory strike.

Silo Defense Policy Approach

A third variant of a policy approach favoring U.S. BMD deployment differs from those described above in that it calls for defense primarily of ICBM silos. The major purpose of such a BMD system would be to enhance the survivability of the land-based leg of the U.S. deterrent triad. While in some configurations it might offer a low level of partial protection to some soft targets like cities, it would not be intended as just the first stage of a more ambitious defense. A defense primarily of missile silos would enhance the current basis of deterrence by increasing the U.S. forces expected to survive a Soviet first strike, raising the cost to the Soviets of attempting to destroy those forces, and possibly complicate Soviet targeting.

Proponents of this approach believe it would serve several strategic objectives:

- It would constitute a hedge against possible future vulnerabilities that might arise for the SLBM and strategic bomber forces.
- It would reduce the incentive to rely on a launch-on-warning strategy for U.S. ICBMs, thereby improving crisis stability.
- By adding to the uncertainties the Soviets would face when contemplating a possible preemptive strike, it might discourage such a strike.
- It would introduce a "firebreak" against limited nuclear attacks by the Soviets, requiring any effective Soviet attack to use thousands of weapons, thereby running a higher risk of heavy U.S. retaliation than would an attack of only a few.

These objectives are shared by SDI proponents for the early stages of BMD deployment. But proponents of limited BMD primarily for missile-site defense argue that their approach would lessen pressure on the Soviet Union to increase its offensive forces to maintain its assured retaliatory deterrent capability.

Proponents of the Silo Defense policy approach believe that advances in BMD technology have made the technical feasibility of effective hard-point silo defense more promising than it was in the early 1970s. They believe that significant progress has been made toward coping with problems which were particularly troublesome then, such as decoy discrimination and defense of acquisition and guidance radars.

This policy approach does not envision a transition to a fundamentally different U.S. nuclear strategy. Rather, it proposes to deal with a problem in current strategy—the vulnerability of part of the retaliatory force to preemptive attack by Soviet ICBMs. Nevertheless, it would probably require a carefully managed arrangement with the Soviets to achieve its goals without introducing undesired instabilities into the strategic relationship.

Some advocates of silo defense see it as leading to a defense-oriented world, and argue for a "defense protected build-down"(DPB), saying that:

> An orderly transition to a defense-oriented world . . . can be achieved by combining deployment of defensive weapons with a concomitant and compensating reduction of offensive weapons.[20]

They argue that a world free of the threat of nuclear destruction is an illusion, but that the level of destruction of which retaliatory weapons are capable might be reduced.

On the other hand, another advocate of limited BMD deployments has argued that the benefits of limited defense are sufficient in themselves and that BMD would not make deep reductions in offensive arms more likely:

> Since there are no foreseeable circumstances in which either side will feel secure without maintaining an assured destruction capability, the ABM [deployment] would make it unlikely that either side would be interested in negotiating reductions to low levels.[21]

This analyst does believe, however, that the Soviets might agree to modify the ABM treaty to permit limited defenses.

Critical Assumptions for the Silo Defense Policy Approach

This policy approach appears to be based on the following assumptions:

Assumption 1.—BMD for missile silo defense can be effective in the face of increased offensive forces and countermeasures.

Effective BMD for silo defense would have the effect of raising the cost (in terms of attacking reentry vehicles) of destroying each missile silo, but would not make the task impossible if there were no constraints on offensive forces.[22]

Assumption 2.—BMD is the most cost-effective means available for protecting U.S. retaliatory capabilities.

If a system intended primarily to defend missile silos is not considered to form the core of a more ambitious defense, then the cost-effectiveness of the alternatives becomes a larger consideration. For example, mobile or other deceptive basing modes for ICBMs should be considered. Depending on the predicted size of the Soviet threat, such basing modes might be the first step to take, rather

[20] Alvin M. Weinberg and Jack N. Barkenbus, "Stabilizing Star Wars," *Foreign Policy* (No. 54), spring 1984, p. 165. For an earlier argument that limited defenses could permit reduced offensive forces, see G.E., Barasch et al., "Ballistic Missile Defense: A Potential Arms-Control Initiative," Los Alamos National Laboratory Paper LA-8632, UC-2, issued January 1981.

[21] Jan Lodal, "Deterrence and Nuclear Strategy," *Daedalus*, fall 1980, p. 170.

[22] Testimony of Richard D. DeLauer, Under Secretary of Defense for Research and Engineering, Hearing before the Research and Development Subcommittee of the House Committee on Armed Services, Nov. 10, 1983, H.A.S.C. 98-21, p. 20.

than BMD.[23] Since defended U.S. ICBMs would probably have to attack defended Soviet targets, the trade-offs in U.S. retaliatory effectiveness need to be considered carefully. Arms control negotiations to reduce the Soviet counterforce threat should also receive full consideration.

Assumption 3.—The Soviets would not respond to limited U.S. BMD deployments with a large-scale BMD system intended to defend urban-industrial targets as well as ICBM silos.

Such a Soviet response would place pressure on the United States to expand our defenses, offenses, or both, just to stay in the same relative strategic position. Some advocates of limited BMD argue that the Soviets would see that, given the likelihood of effective U.S. countermeasures, trying to build very effective defenses would be futile. But the defensive system the Soviets are best prepared to deploy might not be best suited to ICBM silo defense. Should the Soviets deploy a system which looks to the United States like the base for a larger defense of Soviet territory,[24] the United States may decide to respond by building a larger offensive force. This in turn could stimulate larger Soviet offenses and defenses.

Others argue that the Soviets could be engaged in negotiations to define the kinds of defenses both sides could live with. The best way for the two sides to assure one another of the limited nature of their BMD deployments may be for them to agree to a modification of the ABM Treaty specifically to permit carefully defined silo defenses. This assumes that the Soviets, who have deployed a defense of Moscow but never utilized the ABM Treaty provision allowing them to defend silos, could be persuaded to seek a silo defense.

Assumption 4.—Neither side will perceive the limited BMD system deployed by the other as the core of a more extensive damage-limiting defense.

A tacit or a negotiated agreement to build larger, but still limited, BMD systems may be difficult to formulate. A highly-localized (site) defense by ground-based interceptors would appear to be the least ambiguous type of deployment, but might not be nearly as effective as a system with more than one layer of defense. One side or the other might not be willing to settle for a BMD with only a single terminal layer. But systems with more than one layer may appear to give one a "breakout" potential for a much more ambitious defense.

Assumption 5.—Neither side will respond to the other's limited BMD system with greatly augmented offenses.

Lt. General James A. Abrahamson, Director of the Strategic Defense Initiative Organization, has argued that a defect in setting only limited goals for a BMD system is that the Soviets are likely to devote considerable effort to countering it, whereas the promise of increasingly effective defenses would cause them to see the futility of trying to maintain offensive capabilities.[25]

Augmented Soviet offenses could cause the United States to deploy additional defenses and offenses, which in turn could stimulate further Soviet deployments.

A Non-BMD Policy Approach

Most opponents of BMD believe that, at least for the foreseeable future, U.S. policy should be to strive to continue the current situation in which neither the United States nor the U.S.S.R. deploys BMD, and offensive arms development and deployment are limited by agreement.[26]

---

[23]On various possible basing modes and the potential of preferential BMD for missile bases, see ch. 3 in U.S. Congress, Office of Technology Assessment, *MX Missile Basing*, OTA-ISC-140 (Washington, DC: U.S. Government Printing Office, September 1981).

[24]In a typical "prudent worst-case" analysis, almost any large-scale Soviet BMD deployment would probably look like that.

[25]See, for example, his testimony before the Subcommittee on Strategic and Theater Nuclear Forces, Committee on Armed Services, U.S. Senate, Feb. 21, 1985.

[26]Comprehensive descriptions of this viewpoint appear in Sidney D. Drell, Philip J. Farley, and David Holloway, *The Reagan Strategic Defense Initiative: A Technical, Political, and Arms Control Assessment* (Stanford, CA: Center for Interna-

In view of the factors which threaten the continued viability of the ABM Treaty,[27] most people who advocate preservation of the treaty believe that steps should be taken to strengthen it.[28] Their views are described in the following sections. Since there are many ideas on how this can best be done, the descriptions below include a range of proposals drawn from many sources. While this approach is generally consistent with the main strategic policy objectives of the Nixon, Ford, and Carter administrations, it also includes new measures to further those objectives in coming years.

Advocates of this policy approach agree with SDI advocates that we should not count on the threat of assured destruction to prevent a nuclear catastrophe forever. However, they do not believe that the solution to this problem can be found primarily in new military technology. Instead they believe that in the **near term** the best hope lies in early steps to improve strategic stability through arms control agreements, as described below. (Their views on ways to reduce the **long-term** risks inherent in threats of nuclear retaliation are discussed subsequently.)

Under this approach, the United States would make a set of related proposals to the Soviet Union which could include some or all of the following elements:

- Preserving and strengthening the ABM Treaty:
  - —by strong endorsements of its long-term importance by the senior officials of both countries;
  - —by resolving compliance issues in the bilateral Standing Consultative Commission;
  - —by exploring ways to clarify, modify or supplement the Treaty to eliminate troublesome ambiguities or loopholes indicated by events since the Treaty was signed and by new technological developments;[29]
  - —by keeping the U.S. and Soviet BMD research programs at a level and scope no greater than needed to hedge against one another's BMD technology developments. This could include maintaining realistic deployment options and exploring new technologies within the bounds of the ABM Treaty.[30]
- Negotiating a verifiable agreement to ban testing and deployment of anti-satellite weapons and all space-based weapons. Such an agreement would also reinforce the ABM Treaty by eliminating a potential loophole: the testing of devices for potential BMD use under the guise of ASAT testing.
- Negotiating mutual, verifiable limitations on offensive nuclear forces, designed not only to reduce their numbers substantially, but also to decrease counterforce capabilities and strengthen survivability

---

tional Security and Arms Control, Stanford University, 1984); and Richard L. Garwin, Kurt Gottfried, and Henry W. Kendall, *The Fallacy of Star Wars* (New York: Random House, 1984). A list of articles by critics of the Strategic Defense Initiative, including former Secretaries of Defense Harold Brown, Clark Clifford, Robert McNamara, and James Schlesinger, appears in app. J.

[27]For a detailed discussion of these factors, see Thomas K. Longstreth, John E. Pike, and John B. Rhinelander, *The Impact of U.S. and Soviet Ballistic Missile Defense Programs on the ABM Treaty* (Washington, DC: National Campaign to Save the ABM Treaty, March 1985).

[28]For example, in a speech on Nov. 29, 1984, Senator Edward Kennedy said:

> . . . developments in both [the United States and U.S.S.R.] already place the [ABM] Treaty in serious jeopardy. Of particular concern is the development of advanced air defenses that may have a ballistic missile capability. We are now hearing charges and countercharges of Treaty violations. The trends suggest that the superpowers are approaching a point where they must either take concrete measures to renew the Treaty, or risk its abrogation. Congress must act on its own to prevent the Administration from misusing Star Wars to provoke the Soviet Union into abrogating the ABM agreement. We should prohibit the funding of any weapons research or development which could violate that treaty.

[29]For a development of this approach, see Drell, Farley, and Holloway, op. cit. These authors conclude:

> Cooperative action could counter the corrosive effects of unilateral ABM activities which will inevitably cause disputes arising out of deliberate or unintentional divergences in interpretation of the Treaty. More important, it could reinforce confidence that the two nations see the purposes and value of the ABM Treaty in consistent ways, and that each is determined to act separately and jointly toward the fundamental aim of avoiding nuclear war.

[30]The nature of such a research program is discussed in Sidney D. Drell and Thomas H. Johnson (eds.), *Strategic Missile Defense: Necessities, Prospects, and Dangers in the Near Term,* report of a workshop at the Center for International Security and Arms Control (Stanford, CA: Stanford University, Apr. 1985). Also printed in U.S. Senate, Committee on Armed Services, *Hearings of the Subcommittee on Strategic and Theater Nuclear Forces,* Mar. 19, 1985.

of retaliatory forces on both sides.[31] This proposal could include bans on all nuclear missile flight tests and on all nuclear weapon tests.[32] It would also be logical to supplement it with limitations on air defense and on anti-submarine warfare.

Among the proponents of this general policy approach, there is a wide spectrum of views regarding the **long-term** prospects for avoiding nuclear war, either between the United States and the Soviet Union, or between other countries. Some believe that after (but only after) offensive forces have been reduced to a sufficiently low level, deployment of effective nationwide defenses against nuclear attack would help to enable all the countries involved to agree to do away with nuclear weapons altogether.[33] Others believe that for the foreseeable future the most promising way to prevent nuclear war is to maintain a small but invulnerable nuclear retaliatory force in the hands of at least the United States and U.S.S.R., and perhaps Britain, France and China as well.[34] Still others believe that only by improving political relationships between nations or by evolving strong transnational institutions can we hope to banish the long-term threat of nuclear holocaust.[35]

[31]This would be consistent with the conclusion of the Scowcroft Commission that "The central purpose of our arms control efforts should therefore be to enhance U.S. security by increasing strategic stability." (Second report of the President's Commission on Strategic Forces, Mar. 21, 1984, p. 3.)

[32]Proponents of this approach believe that a missile flight test ban would be particularly valuable in preventing the Soviet Union from increasing its destabilizing counterforce capabilities; e.g., by developing more accurate land-based ICBMs, highly accurate SLBMs that could destroy U.S. ICBM silos, or depressed-trajectory SLBMs that could reach U.S. bomber bases before the bombers could escape. See Les AuCoin, "Freeze," *Bulletin of the Atomic Scientists*, November 1984; and "Nailing Shut the 'Window of Vulnerability'," *Arms Control Today*, September 1984. Other arms control advocates propose allowing certain kinds of missile modernization in order to reduce vulnerabilities of both sides' offensive forces. See, for example, Harold Brown and Lynn E. Davis, "Nuclear Arms Control: Where Do We Stand?" *Foreign Affairs*, summer 1984.

[33]For an elaboration of this concept, see Freeman Dyson, *Weapons and Hope* (New York: Harper & Row, 1984), pp. 73 ff. and 280 ff.

[34]See, for example, Richard L. Garwin, "Reducing Dependence on Nuclear Weapons," in David C. Gompert, et al., *Nuclear Weapons and World Politics: Alternatives for the Future* (New York: McGraw-Hill, 1977).

[35]See, for example, Burns H. Weston (ed.) *Toward Nuclear Disarmament and Global Security: A Search for Alternatives* (Boulder, CO: Westview Press, 1984); Randall Forsberg, "Confining the Military to Defense as a Route to Disarmament," *World Policy Journal*, winter 1984; and Robert S. Woito, *To End War: A New Approach to International Conflict* (New York: Pilgrim Press, 1982).

Critical Assumptions for the Non-BMD Policy Approach

This policy approach appears to be based on the following assumptions:

Assumption 1.—The risks of continuing a strategy of deterrence by assured retaliation, stabilized by arms control measures, are less than the risks of introducing BMD into the strategic arms competition.

Advocates of this approach may be unhappy about the U.S. need to rely on assured retaliation to deter Soviet attack, but they do not see a plausible alternative. This assumption is based in part on an assessment that over the next two or three decades we can have higher confidence in our ability to maintain adequately survivable and effective retaliatory forces than in our ability to build BMD systems effective enough to provide "defense dominance." It is also based on the view that the Soviets are more likely to agree to maintaining and stabilizing the current strategic relationship than they are to agree to shifting to a defense-oriented strategic relationship. Most BMD advocates disagree with one or both of these views.

Assumption 2.—It is possible to arrive at, and maintain in force, mutually acceptable, adequately verifiable arms control agreements which will satisfy both sides that neither is deploying significant BMD systems or has a significant lead in BMD break-out capability.

Given the questioned record of Soviet compliance with existing arms control agreements, and given misgivings on each side about new technological developments on the other side, this may be a challenging condition to fulfill.

Some of the critics of arms control believe that it is dangerous to try to cooperate with the Soviet Union. They believe that the Soviets use arms control negotiations solely to attempt to weaken the West. They think any effort to seek mutually advantageous agreements between the superpowers is doomed to failure, because Soviet hostility to the United States will forever dominate any relationship between them. In this view, the Soviets might enter into additional arms control agreements with the United States mainly to limit U.S. progress on BMD while the Soviets prepare to abrogate the treaty, openly or clandestinely, in their own time.

Other observers have a more complex view of Soviet motivations, but nevertheless see significant problems in seeking new arms control agreements. Even given the political will on both sides, there are technical obstacles to effective agreement, especially in the case of space-based or space-attacking weapons. For example, the distinctions between anti-satellite weapons and sensors and potential BMD weapons and sensors is becoming more difficult to draw.

Assumption 3.—A U.S. research program which hedges against Soviet break-out from such arms control agreements will either deter such break-out or provide the United States with an appropriate offensive or defensive response to it.

Many advocates of the non-BMD policy approach assume that the Soviet Union has not already made a firm decision to break out or "creep out" of the ABM Treaty, and that such a U.S. research program could help to deter them from doing so. Some believe the United States could reduce this risk further by developing prototype BMD systems within the bounds of the ABM Treaty, as the Soviets have done, as well as by maintaining a strong capability to overcome potential Soviet BMD systems with penetration aids, enhanced offenses, and other countermeasures.

On the other hand, BMD proponents argue that U.S. BMD research languished under the ABM Treaty regime until the Strategic Defense Initiative began. They contend that a lack of intent to deploy a system might remove incentives for adequate funding for BMD research. A Soviet break-out from an arms control regime limiting BMD could then leave the United States at some disadvantage, at least temporarily.

Assumption 4.—The Soviets can be persuaded to enter into and comply with offensive arms control limitations which would reduce the threat of preemptive nuclear attack, thereby reducing current U.S. incentives to deploy a defense of ICBM silos.

This assumption may not be critical to preserving the non-BMD regime, but offensive arms reductions would be useful, particularly in making continued BMD limitations acceptable in the United States. During the SALT I period and on into the SALT II period, the Soviets continued to add to and improve their ballistic-missile borne hard-target-kill capabilities. The have shown little interest in negotiating away those weapons which the United States finds most destabilizing. Some argue that the current U.S. interest in developing BMD may induce the Soviets to take a serious interest in offensive missile reductions if they believe BMD can be headed off in that way, but seeking such a trade-off does not appear to be current U.S. policy.

Chapter 10

Alternative R&D Programs

Contents

Table

Chapter 10

Alternative R&D Programs

INTRODUCTION

In previous chapters, this report has addressed the potential contributions and liabilities of ballistic missile defenses, and it has primarily discussed the long-term issues associated with developing and deploying BMD. However, technologies now within the state of the art are capable of providing only limited BMD capability. More effective BMD systems cannot be developed without further research and technology development.

This chapter discusses research programs to investigate the possibilities for acquiring more advanced BMD systems. It presents a number of different potential strategies for pursuing BMD research, describes some characteristics by which alternative R&D programs can be compared, and outlines some of the issues Congress must face in the near-term.

There is general agreement that BMD technologies merit investigation. Support for BMD research, however, does not necessarily imply support for the Strategic Defense Initiative (SDI). Possible BMD research programs can differ greatly from the SDI in emphasis, direction, and level of effort. Moreover, research programs having different perceived and intended purposes—even if they have similar technical content—can have very different consequences.

Decisions to be made by Congress in the very near future and in the years to come will have a major impact in ratifying, or in redirecting, major changes which the Reagan Administration has initiated in the U.S. BMD research program. These changes include:

- **Urgency:** Research under the SDI is intended to proceed at a "technology-limited" pace to permit an informed decision to be made at the earliest possible date on whether to enter full-scale engineering development. Proceeding past that point would clearly be inconsistent with ABM Treaty constraints. The pre-SDI program had no such mandate for an early decision on maintaining the ABM Treaty.
- **Visibility:** The SDI has much higher visibility, and a much higher level of Presidential attention, than the previous program of research in BMD-relevant technologies. The decision to spotlight BMD has already been made, and its consequences are already being felt. These consequences certainly include a decision by the Soviets to at least explore their options to respond to the increased probability of a U.S. BMD deployment.
- **Direction:** Under the SDI, emphasis has shifted away from fairly mature technologies, which generally include use of nuclear interceptors, towards nonnuclear defenses which would use much more speculative but potentially more effective technologies.
- **Budget:** Over the next decade, much more is proposed to be spent on ballistic missile defense research than would have been allocated in the absence of the SDI. Large budget increases start with the $3,722 million fiscal year 1986 request, which is almost four times the fiscal year 1984 total and is more than twice what would have been spent within the Department of Defense in fiscal year 1986 under the pre-SDI budget. Subsequent increases proposed for the SDI are even greater, and by fiscal year 1990 are projected to reach a level over eight times the fiscal year 1984 total.
- **Arms Control Policy:** Instead of the pre-SDI approach of seeking deep reductions of offensive forces along with maintenance of the ABM Treaty ban on defenses against ballistic missiles, current arms control policy seeks "greatly reduced

levels of nuclear arms *and an enhanced ability to deter war based upon an increasing contribution of non-nuclear defenses against offensive nuclear arms."*[1]

BMD R&D Options

Near-term decisions by Congress will determine our approach to BMD research. These near-term decisions will not completely determine our longer term policy approach, in part because many factors influencing long-term policy are not under our direct control (e.g., Soviet activities and U.S. progress in technology development). However, decisions made in the short term can significantly affect, or rule out, options for long-term policy. The research options discussed below correspond roughly to the long-run policy approaches discussed in chapter 9, as is shown in table 10-1.

Different approaches that can be taken towards ballistic missile defense research proceed from different sets of basic assumptions about the value and feasibility of BMD and from differing assessments of the consequences of pursuing BMD research. Three such approaches can be distinguished and are presented below. These approaches differ primarily in emphasis and urgency, rather than in which technologies are to be studied. Most BMD-relevant technologies would be investigated, at some level, in all three.

The first approach is the SDI as proposed by the Reagan Administration. The second approach would proceed to BMD deployment faster than the SDI would be able to, and the third approach would conduct BMD research and development at a slower rate than the SDI. Each of the last two approaches is further broken down into two suboptions which differ in the emphasis given to existing versus near-term technologies (in the second approach) or near-term versus far-term technologies (in the third). The five research suboptions are defined as follows:

1. **SDI approach:** Vigorously investigate advanced BMD technologies with the

[1]Quoted from "The U.S. Strategic Concept," enunciated by Paul Nitze in "The Objectives of Arms Control," address before the International Institute of Strategic Studies, London, Mar. 28, 1985. (Emphasis added.)

Table 10-1.—Correlation Between the Near-Term Research Approaches Discussed in This Chapter and the Longer-Term Policy Approaches Discussed in Chapter 9

Near-term R&D approach (ch. 10)	Long-term policy approach (ch. 9)				
	SDI	Early deployment	Intermediate deployment	Silo defense	Non-BMD, arms control
1. SDI	Compatible	Must add near-term deployment	Must commit to deployment	Eventually becomes incompatible	Eventually becomes incompatible
2a. Early deployment	Not very compatible	Compatible	Compatible but not optimal	(see note [a] below)	Incompatible
2b. Intermediate deployment	Conditionally compatible (see note [b] below)	Need to add near-term deployment	Compatible	(see note [a] below)	Incompatible
3a. Funding-limited	Would delay but not rule out	Incompatible	Incompatible	Eventually becomes incompatible	Compatible
3b. Combination	Would delay but not rule out	Incompatible	Incompatible	Compatible	Compatible

[a]Both early deployment and intermediate deployment R&D approaches might be compatible with a "silo defense" long-term policy approach, since defending hardened targets such as missile silos, a technically easier task than defending other types of target, could probably be implemented earlier than other types of BMD. However, to the extent that the early and intermediate deployment R&D approaches are intended to support widespread area defenses, probably including boost-phase weapons, those R&D approaches may be incompatible with the "silo defense" policy approach in which defenses would be limited to specific sites, and where the appearance that nationwide area defenses were being implemented would be avoided.

[b]If the technologies slated for initial deployments in the "intermediate-term" research approach cannot be successfully developed, or if their development triggers offensive countermeasures which render the defense largely ineffective, pursuit of the long-term "SDI" policy approach would be greatly complicated or prevented. Successful deployment of intermediate-term technologies, on the other hand, could then be followed by pursuit of more capable BMD technologies and would be compatible with the goals of the long-term SDI policy approach. This path would take longer and would cost more than pursuit of the "SDI" research approach from the beginning, which would not necessarily include deployments in the intermediate term.

intent to decide in the early 1990s on whether or not to enter full-scale engineering development and subsequent deployment. This approach assumes that while technology now within the state of the art is not good enough to be worth deploying, the long-term potential of advanced BMD technologies is sufficiently promising that a "technology-limited" (i.e., not constrained by lack of funds) effort is warranted to develop that potential. It also assumes that if successfully developed, such technologies could make possible a national security regime (weapon systems and arms control) preferable to the current one and to other alternatives.

2a. **Early deployment approach:** Emphasize early and incremental deployment of currently available BMD technology. This approach places high strategic value on the modest levels of defensive capability which can probably be obtained with existing technology. Although the ABM Treaty permits the United States to defend some ICBMs with a single, highly constrained defensive deployment, most early deployment proposals go well beyond these constraints and could not be pursued under the existing treaty regime.

2b. **Intermediate deployment approach:** Emphasize research on BMD technologies which are beyond the present state of the art, but which, unlike many SDI technologies, might be applicable to deployments in the early to mid-1990s. This approach assumes that investigation of longer run technologies should not delay deployments in the nearer term.

3a. **Funding-limited approach:** Investigate advanced BMD technologies at a funding level well below that requested for the SDI and with a much reduced sense of urgency. Like the SDI, this approach would focus mainly on advanced technologies that may make a highly capable defense possible. Unlike the SDI, however, it does not assume that we will know in a few years whether we can achieve that goal. The program would not aim towards facilitating a development decision at a particular time, nor would it include tests or demonstrations which would raise questions of compliance with the ABM Treaty.

3b. **Combination approach:** Balance research in advanced BMD technologies with the development of near-term deployment options which would include "traditional" BMD technologies (nuclear-armed, radar-guided interceptors) of the sort specifically mentioned in the ABM Treaty. This program, conducted at a funding level well below that requested for the SDI, would aim to deter Soviet abandonment of the ABM Treaty, to hedge against future Soviet BMD developments, to prevent technological surprise, and to investigate the long-term potential of advanced BMD technologies. Like the funding-limited approach, it would not include demonstrations or development work which would raise questions of compliance with the ABM Treaty.

These research options will be described and discussed in detail later in this chapter.

Hedges Against Near-Term Soviet ABM Treaty Breakout

One of the functions of a U.S. BMD research program is to deter or respond to a near-term Soviet ABM Treaty "breakout" (sudden initiation of nationwide BMD deployments) or "creepout" (gradual implementation of nationwide BMD capability without overt Treaty abrogation). A U.S. response to either of these actions would most likely consist of deployment of a near-term U.S. defense, deployment of offensive countermeasures which would ensure that our strategic forces could penetrate Soviet defenses, or some combination of the two.

Near-Term U.S. Defensive Deployment

The *SDI* approach has largely discontinued investigation of "traditional" BMD technologies in favor of nonnuclear technologies which would make intercepts at altitudes high enough to protect soft targets as well as hardened ones.[2] Although protecting soft targets with nonnuclear interceptors is technically much more demanding than defending only hardened targets with nuclear interceptors, advocates of the SDI approach are confident that the technical requirements can be attained within a few years if required. In principle, "traditional" BMD technologies could be restored to the SDI as a hedge against inability to develop other near-term defensive options. However, doing so would require reevaluating SDI's emphasis on nonnuclear technologies, and it would also require additional funds if work on other BMD technologies were not to be impeded.

The *early deployment* and *intermediate deployment* approaches would not wait for a Soviet breakout before deploying defenses. Different versions of these approaches would stress differently the deployment of "traditional" BMD technologies as opposed to nonnuclear ones which have yet to be demonstrated but are nevertheless thought by some to be capable of providing high-confidence deployment options.

The *funding-limited* approach would deemphasize near-term defensive deployments, concentrating on longer term research which could lead to a highly capable defense. It stresses offensive countermeasures (see below), rather than near-term defensive counterdeployments, to respond to near-term Soviet breakout. The *combination* approach would be intended to deter a near-term Soviet breakout by putting more emphasis on improving our ability to deploy a near-term U.S. defense, in addition to developing offensive countermeasures. This approach would pursue research and development of "traditional" BMD technologies (within ABM Treaty constraints) to eliminate the technical risks of depending on yet-to-be demonstrated near-term defensive technologies.

Offensive Countermeasures

The U.S. response to Soviet breakout need not be limited to defense. **Offensive countermeasures intended to penetrate, counter, or evade Soviet defenses are at least as important in deterring or responding to a Soviet defensive deployment as U.S. defensive options are.** Offensive countermeasure research would accompany any of the BMD research options above.

The U.S. program responsible at present for developing offensive BMD countermeasures is the Air Force's Advanced Strategic Missiles Systems (ASMS) Program.[3] These countermeasures include maneuvering reentry vehicles, which evade terminal BMD interceptors by flying unpredictable trajectories, and other penetration aids which would help U.S. warheads defeat Soviet defenses. According to the fiscal year 1982 Arms Control Impact Statement on the ASMS Program,

> Maneuverable re-entry vehicle (MaRV) and penetration aid R&D is expected to provide a high-confidence, low-risk option for timely deployment on current or future ballistic missile systems if needed to offset improved rapidly deployable nationwide Soviet ABM defenses (which would violate the ABM Treaty).
>
> * * *
>
> The present MaRV and penetration aids programs are a hedge against the possibility of such a situation . . .[4]

[2]Even if a defensive interceptor does not use a nuclear warhead, a nuclear explosion can result if the attacking warhead is salvage-fused to detonate when intercepted. Therefore, nonnuclear interceptors (as well as nuclear ones) must intercept at high altitude if soft targets are to be defended.

[3]The U.S. capability to penetrate existing Soviet defenses was not mentioned in the White House January 1985 pamphlet on *The President's Strategic Defense Initiative*. That pamphlet asserts that the Soviets will be able to deploy a nationwide ABM defense system within the next 10 years. Should they decide to do so, it continues, "deterrence would collapse, and we would have no choices between surrender and suicide" (p. 4).

Although any defense deployable by the Soviets in the next 10 years would certainly complicate U.S. targeting, the available offensive countermeasures technologies make it extremely unlikely that we could be forced to choose between "surrender and suicide."

[4]*Fiscal Year 1982 Arms Control Impact Statements*, Statements Submitted to the Congress by the President Pursuant to Section 36 of the Arms Control and Disarmament Act, printed for the use of the Committee on Foreign Affairs and Foreign Relations of the House of Representatives and Senate respectively, Joint Committee Print, 72-434 O, U.S. Government Printing Office, Washington, DC, February 1981, p. 28.

Such offensive countermeasures, of course, would no longer be required were the United States to agree to eliminate its offensive arsenal. However, defenses good enough to permit eliminating offensive nuclear forces are not envisioned for the foreseeable future, even by proponents of strategic defense.[5]

In addition to providing the United States with options to respond to Soviet defenses, investigation of potential offensive countermeasures to BMD systems must also be an integral portion of our own defensive research. Defensive technologies which can be shown to be easily countered will not be as promising as those for which countermeasures cannot be found so readily.

Offensive countermeasure research options differ in their choice of which defense technologies are to be countered and in how far countermeasure and penetration aid research should be taken into advanced development, production, and deployment. Unlike defensive research, there are no treaty constraints banning testing and development of offensive countermeasures.[6]

Soviet BMD Research and Comparison With U.S.

"Traditional" BMD Technologies

The United States and the Soviet Union have conducted research and development activities in BMD both before and after the ABM Treaty was signed. Each has acquired considerable experience with "traditional" BMD technologies, such as the nuclear-armed, radar-guided interceptors of the sort specifically mentioned in the ABM Treaty. **However, although the state of Soviet "traditional" BMD technology probably does not exceed our own, the Soviets are almost certainly better positioned in the near-term to deploy a limited-capability ballistic missile defense system than we are.**

The Soviets have deployed and maintained an ABM system around Moscow utilizing "traditional" BMD technologies. They have also extensively upgraded and modernized that system. Ever since the United States decided that its own similar system was not effective enough to justify maintaining it, the Moscow ABM has been the world's only operational ABM system.

In addition to the Moscow system permitted under the ABM Treaty, the Soviets have built a large radar in Siberia which violates the siting restrictions on such radars in the ABM Treaty. Furthermore, according to the DOD publication *Soviet Military Power, 1985*, the Soviets are "developing a rapidly deployable ABM system to protect important target areas in the U.S.S.R." That report concludes that "the aggregate of [their] ABM and ABM-related activities suggests that the U.S.S.R. may be preparing an ABM defense of its national territory."[7] CIA officials, however, have testified before Congress that they have not judged it likely that the Soviets would in fact move to such a deployment in the near term.[8] They point out that while the Soviets could expand their presently limited ABM system by the early 1990s,

> In contemplating such a deployment . . . [they] will have to weigh the military advantages they would see in such defenses against the disadvantages of such a move,

[5]When asked by Senator Sam Nunn whether they could "envision our having a [defensive] system that would avoid the necessity of deploying our offensive forces," Reagan Administration officials Dr. Robert Cooper (Director of the Defense Advanced Research Projects Agency), Dr. Richard DeLauer (Under Secretary of Defense for Research and Engineering), and Dr. Fred Ikle (Under Secretary of Defense for Policy) responded negatively. They held out the hope that such a condition might someday be achieved, but said that at present there can only be an "optimistic view that that will be possible at some time in the future."

—"Hearings Before the Committee on Armed Services, United States Senate, on Department of Defense Authorization for Appropriations for Fiscal Year 1985," S. Hrg. 98-724, Part 6, Strategic Defense Initiative, Mar. 8, 1984, pp. 2924-2925 and 2939.

[6]However, testing offensive countermeasures which involved nuclear detonations in the atmosphere or in space would violate the Limited Test Ban Treaty and possibly the Outer Space Treaty.

[7]Both quotes from *Soviet Military Power, 1985*, p. 48.

[8]Testimony of National Intelligence Officer Lawrence K. Gershwin before a joint session of the Subcommittee on Strategic and Theater Nuclear Forces of the Senate Armed Services Committee and the Defense Subcommittee of the Senate Committee on Appropriations, June 26, 1985.

particularly the responses by the United States and its Allies.[9]

Advanced Technologies

The Soviets are also undertaking a vigorous research program in advanced BMD (laser and particle beam) technologies.[10] It has been estimated that the total Soviet effort in directed-energy research is larger than that in the United States. However, the **quality** of that work is difficult to determine, and its **significance** is therefore highly controversial. In large part, we are limited to observing what goes *into* their efforts (e.g., the amount of floor space at various Soviet research laboratories, the observable activity at test sites) and what does *not* come out (e.g., absence or cessation of publication on topics known to be under investigation, indicating that the activity has been classified).

In terms of basic technological capabilities, however, the United States remains ahead of the Soviet Union in key areas required for advanced BMD systems, including sensors, signal processing, optics, microelectronics, computers, and software. The United States is roughly equivalent to the Soviets in other relevant areas such as directed energy and power sources. According to the Under Secretary of Defense for Research and Engineering, the Soviet Union does not surpass the United States in any of the 20 "basic technologies that have the greatest potential for significantly improving military capabilities in the next 10 to 20 years."[11]

[9]Written testimony of Robert M. Gates and Lawrence K. Gershwin, op. cit.

[10]*Soviet Military Power, 1985*, U.S. Department of Defense, pp. 43-44.

[11]*The FY 1986 Department of Defense Program for Research, Development, and Acquisition*, Statement by the Under Secretary of Defense, Research and Engineering, 99th Cong., 1st sess., 1985, pp. II-3 and II-4.

ALTERNATIVE R&D PROGRAM DESCRIPTIONS

The alternative R&D program options presented in this chapter are described in terms of basic rationales and objectives, rather than in terms of which technologies would be investigated at what level. The overall effects of conducting BMD research depend on much more than the technical content of the research program—a point which will be returned to below in the section which describes "Political Attributes" (p. 251).

Approach 1: The Strategic Defense Initiative

The goal of the SDI is to advance the state of the art of BMD-relevant technologies to the point where an informed decision could be made on whether to enter full-scale engineering development and subsequent deployment of ballistic missile defenses. The program focuses on resolving those critical technological issues on which a highly effective defense system might rest but which at present are not adequately understood.

The SDI is based on the "technology-limited" research plan formulated by the Defensive Technologies Study Team (DTST), or Fletcher Panel. It is therefore intended to proceed as rapidly as possible, with further progress waiting not for more money but for previous results. Funding requests reached the DTST technology-limited profile with the fiscal year 1986 request.[12]

The SDI research program is intended to comply with all U.S. treaty obligations. However, tests that have been viewed as being ambiguous with respect to treaty compliance are proposed.[13] At any rate, if the research is successful, it would lead to systems for which development and testing would clearly be inconsistent with ABM Treaty constraints.

In addition to developing key BMD technologies, SDI is directed by its charter to "protect U.S. options for near-term deployment of

[12]Report to Congress on the Strategic Defense Initiative, 1985; p. 4.

[13]See app. A.

limited ballistic missile defenses."[14] These options are either to be implemented or held in reserve as a hedge against Soviet defensive breakout. SDI is also specifically instructed to "place principal emphasis on technologies involving nonnuclear intercept and destruction concepts." As a result, development of "traditional" nuclear-armed, radar-guided interceptor technologies has been almost completely discontinued. Research on nuclear directed-energy weapons continues in order to understand their potential and to hedge against Soviet developments in that area.

SDI activities have been grouped into five program elements: Surveillance, Acquisition, Tracking, and Kill Assessment; Directed-Energy Weapons; Kinetic-Energy Weapons; Survivability, Lethality, and Key Technologies; and Systems Concepts/Battle Management Technology (see box). Each program is designed to advance the technology base, to conduct demonstrations that experimentally validate the technology, and to provide direction to focus the technology development on those critical issues which must be resolved before feasibility can be determined.[15]

SDI also attempts to encourage innovation in the U.S. scientific community to aid in identifying new approaches. The Directed-Energy Weapons program element, for example, has set aside 1.5 percent in fiscal year 1985 (1.7 percent is requested for fiscal year 1986) to support high risk, highly innovative approaches which would not otherwise be undertaken. SDI is soliciting advanced technology proposals for these funds from small businesses and the academic community.

Approach 2a: Early Deployment

Advocates of the **early deployment** of strategic defense systems attach a high strategic value to the modest levels of effectiveness that can be provided with presently available BMD technology. They believe that early and incremental deployments are required to address serious problems with existing U.S. strategic capability, and that delaying such deployments until more research has been done may be dangerous. They believe that the ABM Treaty prohibiting such defenses is not in the best interests of the United States, and in embarking on this approach would withdraw from or abrogate it. Although BMD systems effective enough to counter a responsive Soviet threat are not available at present, early deployment advocates are confident that U.S. technological superiority will enable it to prevail should an offensive v. defensive military technology competition ensue.

The research and development program supporting an early deployment approach would have two largely distinct aspects. In the first, those technologies now within the state of the art would have to be engineered into operational status and deployed; in the second, more advanced and presumably more capable concepts would have to be investigated in order to increase defensive capability and to counter responses by the Soviets.

There are many forms that early deployment of BMD technology could take, and their cost and effectiveness depend on what technologies are utilized, what targets are defended, and the nature of the Soviet response. Some have advocated that hardened targets be defended with "traditional" technologies of the type developed prior to the ABM Treaty in 1972, or with more recently proposed nonnuclear, low-altitude interceptors. Others have proposed deployment of space-based chemical infrared lasers or kinetic-kill vehicles. Of these technologies, we have significant experience only with the "traditional" ones, and even their performance in an environment of many nuclear detonations is poorly understood.

The most publicized proposal for early BMD deployment is the one presented by High Frontier.[16] That study recommended near-term deployment of both a terminal defense of hardened targets and a space-based boost-

[14] Caspar Weinberger, "Strategic Defense Initiative Organization (SDIO) Charter," Apr. 24, 1984.

[15] Report to Congress on the Strategic Defense Initiative, 1985; p. 23.

[16] Daniel Graham, *High Frontier: A New National Strategy* (Washington, DC: High Frontier, Inc., 1982).

SDI Program Descriptions (reproduced from Report to the Congress on the Strategic Defense Initiative, 1985, pp. 23-25)

The Surveillance, Acquisition, Tracking, and Kill Assessment (SATKA) Program Element includes a mixture of some of the most and least mature technologies being developed by the SDIO. It includes technology base efforts to support surveillance, acquisition, tracking, and kill assessment that provide: 1) data on the observables from ballistic missiles and their warheads; 2) new radar and optical sensors capable of obtaining detailed imagery of warheads and warhead deployment; and 3) on-board signal and data processing capable of performing necessary computations right at the sensor. The experiments include three general classes: boost-phase surveillance, midcourse tracking, and terminal-phase tracking and discrimination. Space-based surveillance experiments are planned for the early 1990s to demonstrate survivable means of detecting and tracking boosters from very high altitudes in space. Other space-based sensor experiments are to be conducted in the same time frame to explore our ability to track tens of thousands of objects during midcourse flight. Such platforms may ultimately include active sensors to aid in discrimination. A sensor experiment will determine the feasibility of using optical sensors to aid in target discrimination. A terminal imaging radar experiment is planned to demonstrate rapidly evolving ground-based radar capabilities.

The Directed Energy Weapons (DEW) Program Element is advancing the state-of-the-art in the technologies for: 1) high-powered laser and particle beam generation; 2) optics and sensors for correcting and controlling the high power beam; 3) large, lightweight mirrors and lightweight magnets for focusing the beam on the target; 4) precision acquisition, tracking, and pointing to put and hold the beam on target; and 5) fire control to capitalize on those unique features of directed energy weapons such as the ability to measure and control the energy delivered to the target. The DEW technology program includes major experiments at the subcomponent level in the four concepts currently being examined: space-based lasers, ground-based lasers, space-based particle beams, and nuclear-driven directed energy. These concepts are candidates for boost and post-boost phase intercept and for discrimination functions in the other phases. In addition, selected subcomponents for these concepts will be integrated in on-the-ground experiments designed to test interface approaches and resolve technical issues arising from the integration. The work on nuclear-driven directed energy is largely pursued by the Department of Energy and is designed to establish its technical feasibility. Equally important, the work ensures that the U.S. understands the potential impact of these emerging concepts if they were to be used against it by an adversary. It should be reiterated that emphasis in the SDI program is being given to nonnuclear weapons for defense.

The Kinetic Energy Weapons (KEW) Program Element is a collection of related research that would make use of the very high velocity of a small mass to render a ballistic missile or its warhead ineffective. The KEW program contains some of the more mature technology being investigated in the SDI. Efforts include interceptors and hypervelocity gun systems for boost-phase intercept, midcourse intercept, terminal intercept, and defense of space platforms. Both space-based and ground-based kinetic kill vehicles (KKV) are being investigated. The technology thrusts for the space-based KKV include research into a high performance multiple kill vehicle (MKV), fire control/guidance, and booster propulsion. Ground-launched interceptor studies involve both exo- and endo-atmospheric kill. Both space- and ground-based electromagnetic (EM) gun investigations are included. Space-based EM gun investigations include critical technologies such as high-g propulsion, high-g compact structures, long-range high resolution tracking, and multiple MKV tracking. All experiments will be designed and conducted to conform to ABM Treaty constraints.

The Survivability, Lethality, and Key Technologies (SLKT) Program Element provides critical supporting R&T [research and technology]. Understanding the vulnerability of ballistic missiles to the various kill mechanisms is fundamental to assessing their effectiveness against current and responsively hardened targets. Survivability to mission completion, particularly of any defense space assets, is fundamental if defensive options are to be viable. Economical space transportation, on-orbit logistics and maintenance, kilowatt/megawatt sources of power, and multi-megajoule energy storage and conversion are potentially key needs in an affordable defense deployment.

Lethality and target hardening efforts will provide the basic theory underlying kill mechanism/target interactions, the resulting damage and response of the target to damage, and fundamental limitations in hardening countermeasures. The survivability problem includes substantial technology development, particularly in the case of space-based components. It also includes identification and assessment of innovative survivability hardware and tactics and evaluations of the survivability of conceptual designs. Space transportation, logistics, and space power efforts are designed to take advantage of existing DoD and NASA definition efforts and to expand them into the definition phase and satisfaction of the more demanding requirements of a defense-in-depth.

The Systems Concepts/Battle Management Program Element is designed to allow intelligent choices among competing approaches to defense architectures and to develop the technologies necessary to allow eventual implementation of a highly responsive, ultra reliable, survivable, endurable, and cost-effective battle management/command, control, and communications (C^3) system. Threat analyses, mission analyses, conceptual design of defensive architectures and performance requirements definition, and system evaluation for all levels of a layered defense against ballistic missiles will be performed. The battle management/C^3 efforts will provide the tools, methods, and components 1) for development and eventual implementation of the system and 2) to quantify risk and cost of achieving such a system.

Innovative Science and Technology (IS&T)[1] . . . encourages the innovation of the U.S. scientific community to aid SDI research in identifying new approaches. To this end, the Strategic Defense Initiative Organization is soliciting innovative, advanced technology proposals from small businesses and the academic community.

[1]Deleted material was a cross-reference to another section of the SDI Report to Congress not cited above.

phase area defense (Global Ballistic Missile Defense System I) which would use rocket-powered kinetic-kill vehicles. More effective space-based defenses (GBMD II) would be deployed when developed.

Since High Frontier provided a candidate system architecture, including cost estimates and timelines, more detailed analyses can be done on that system than can be performed for other concepts. **However, studies by several groups have shown that High Frontier severely underestimated the cost and overestimated the capability of the GBMD I system.** These studies are discussed in appendix G.

Approach 2b: Intermediate Deployment

Intermediate deployment supporters, like those favoring early deployment, disagree with the SDI premise that a decision to deploy nearer term BMD should be contingent on the successful development of longer term, advanced BMD technologies. They believe that strategic benefits of BMD can be realized in the intermediate-term even without confidence that long-term, highly effective systems can subsequently be deployed. Unlike early deployment supporters, however, backers of intermediate deployment do not believe that deployments should be started now, but rather that those BMD technologies which could lead to deployments in the early to mid-1990s be pursued. They believe that existing technology is inadequate, and they may also seek to leave open the possibility of discussing and negotiating defensive deployments with the Soviets before deployment starts.

The technologies investigated by an intermediate deployment R&D program would be similar to those studied in an early deployment program. In both programs, technologies slated for initial deployments would have to be brought into operational status at the same time that more advanced technologies were investigated.

Approach 3a: Funding-Limited

A BMD research program might share the SDI program's focus on advanced defensive technologies that in the long run may make a thoroughly reliable defense possible, without also sharing the SDI program's premise that we can know in a very few years whether we have any assurance of reaching that goal. A funding-limited approach would conduct a systematic program of laboratory research on technologies which might have potential for leading to highly capable defenses, and it would investigate most of the topics which the SDI proposes to study. This research, however, would not proceed as rapidly as it would under the SDI; it would **investigate** the potential of these technologies without preparing to decide in the near term whether to **exploit** that potential.

Adherents of such an approach would want to maintain the ABM Treaty in the near term. Should a new approach to effective national defense be successfully developed—whether by breakthroughs in defensive technology which made defense-dominant strategies clearly viable, or by implementing constraints in offensive forces so stringent that existing defensive technology could bring about defense dominance—the ABM Treaty would have to be reexamined, and the United States would have to consider an arms control/weapons acquisition approach that integrated offensive and defensive forces. However, advocates of the funding-limited approach would expect such an eventuality to occur well into the future, if at all. Although they might disagree on exactly which criteria should be met before moving defensive research into full-scale engineering development, they would agree that such a decision either should not or cannot be made as early as the SDI approach implies.

In the meantime, the intent of this approach would be to signal that the United States supports the ABM Treaty and wants to deter the Soviet Union from rejecting its own ABM Treaty commitment. The prominence of U.S. BMD research would be reduced from what it would have (and has already had) under the SDI approach, and that research would take on a character more similar to other military R&D programs. Field testing and major technology demonstrations which appeared to be aimed more towards developing a BMD system than towards researching technology would be deferred.

A Soviet near-term ABM Treaty breakout would be deterred primarily by the development of U.S. offensive countermeasures. Unilateral Soviet BMD deployment in the long run would be discouraged by the prospect that the U.S. funding-limited program could be accelerated if the Soviets were to abandon the ABM Treaty regime.

Approach 3b: Combination

The combination approach would balance serious study of advanced BMD technologies with the development of high-confidence, near-term options to deploy BMD systems based on "traditional" technologies. The advanced technologies would be investigated to understand their potential, especially if used against us, and to prevent technological surprise, so that no unanticipated Soviet technological developments would permit them to threaten an ABM Treaty breakout in a way we could not counter. In addition, study of advanced BMD technologies could advance their applications in other military (and possibly civilian) uses.

Near-term deployment options, developed within ABM Treaty constraints, would help to deter a near-term Soviet defensive breakout, and they could provide a response if that contingency occurred. Perhaps more importantly, a prototype "traditional" BMD system would also provide a test-bed for offensive countermeasures which we might have to deploy against nationwide Soviet defenses of the sort already deployed around Moscow.

Like the funding-limited approach, the combination approach would be intended to strengthen rather than to threaten the existing ABM Treaty regime in the near term; it would therefore defer tests or technology demonstrations which appeared to pose questions of compliance with the ABM Treaty. In the long run, advocates of both the combination and the funding-limited approaches believe that it might or might not be desirable to modify the ABM Treaty and our overall strategic arms control approach in the future to incorporate defensive systems, depending on how BMD technology develops and how the U.S.-Soviet strategic relationship evolves.

ALTERNATIVE R&D PROGRAM CHARACTERISTICS

The alternative R&D programs differ in a number of individual characteristics which can be grouped into three categories:

- **Technical attributes** characterize which technologies are to be investigated and which deployment options are to be made available. Each of the alternative R&D programs would produce deployment options, but those options differ widely in what they would consist of, how effective they would be, and how long they would take to implement.
- **Economic attributes** include most directly the cost of a given R&D approach, but also include an R&D program's impact on other activities which compete with it for the same resources (financial, material and facilities, and technical talent).
- **Political attributes** include the effects that a U.S. program to investigate BMD technologies might have on other countries' actions or relationships with us, and the constraints such a program might place on arms control possibilities. They also include the effects that such a program might have on ourselves. In both cases, perceptions may have a greater political impact than either announced intentions or demonstrated capabilities.

Technical Attributes

The technical outputs of a BMD research program will be advances in BMD-relevant technologies which might provide options for deploying ballistic missile defenses. Nearer term options using technologies which are now fairly mature can be developed with relatively high confidence; much more speculative but potentially more powerful technologies might also be developed, but at present there is much less confidence that those technologies will lead to BMD options. If they do, those options would be longer term ones.

None of the alternative R&D approaches described here would abandon research on technologies relevant to long-range BMD concepts. However, the approaches do differ in the relative emphasis put on near-term options as opposed to the longer term ones. The *SDI approach* stresses longer range options. Even though the SDI maintains an active effort in technologies it considers near-term, SDI officials have stated that it could be counterproductive to deploy near-term technologies without also demonstrating that even more effective longer term technologies are feasible and can later be deployed.[17] The experimental technology demonstrations included within the SDI program are not intended to be engineering prototypes of operational BMD components, but they must nevertheless be relevant to the mission and advanced enough to provide a meaningful basis for determining their utility in BMD applications.

Although the *early deployment approach* calls for investigation of longer term possibilities, unlike the SDI it stresses primarily the development of available technologies for near-term deployment. The *intermediate deployment approach* similarly stresses technologies for intermediate-term deployment. The *fund-*

[17] For example, General Abrahamson's testimony before the Strategic and Theater Nuclear Forces subcommittee of the Senate Committee on Armed Services, Feb. 21, 1985.

ing-limited approach would emphasize longer term options; the *combination approach* would develop near-term deployment options in addition to conducting longer range research.

Related to the choice between near- and far-term options is the balance that should be struck between basic research, on the one hand, and development geared towards more immediate application, on the other. The appropriate balance for a given technology depends on the status of the research, the perceived promise of its applications, and the urgency of the task. The first of these depends on the results of the research to date; the latter two also depend in part on the overall approach taken towards BMD research and development.

Prematurely advancing a research program into the development and testing phase can have two major disadvantages: the technologies under investigation can get frozen at an immature level, and the greater expenses of advanced development and testing can absorb resources which would otherwise be devoted to improving the basic technologies or finding better ones. On the other hand, failing to advance a program into development and testing at an appropriate stage delays possible application of newly developed technologies should it become necessary or advantageous to do so.

Economic Attributes

Cost

One obvious characteristic for comparing R&D programs is their respective costs. BMD research will generally yield more results with higher funding until either: 1) the nation's R&D capacity cannot efficiently absorb additional resources, or 2) research reaches a *technology-limited* funding level. Although some of the approaches discussed above will be clearly more expensive than others, in most cases it is not possible to associate a given level of expenditure with a particular approach; most of the approaches are compatible with a range of funding levels.

The *SDI approach*, intended to proceed at a technology-limited pace, will be the most expensive of the research-only approaches—the definition of "technology-limited" means that additional money will not speed up the research, and that therefore budgets larger than those requested for the SDI will not necessarily yield greater results. Furthermore, the SDI approach calls for substantial year-to-year increases, since each year's progress is intended to make possible increasing amounts of follow-up research in subsequent years.[18]

Adding deployment to research would, of course, cost more than research alone. However, limited deployments of existing technologies may be less expensive than ambitious research of more advanced technologies, so approaches which include deployment—the *early deployment* and the *intermediate deployment approaches*—would not necessarily cost more than the SDI approach.[19]

The *funding-limited approach* would probably cost considerably less than the SDI, although in principle it could be carried out at almost any level of funding short of the SDI's technology-limited level. Such a program would also be amenable to a growth rate much slower than that of SDI, in that much follow-on research made possible by technical progress to date would be deferred.[20] The *combination approach* would likewise be compatible with a wide range of funding levels; annual increases would probably also be modest.

Impact on Other R&D

Just as important as the total amount of money which is spent on a research program

[18]Budget requests for the SDI in fiscal year 1986 and future years are presented in app. F, along with the projected requests which the SDI Organization has estimated would have been made for the previously existing BMD programs had SDI not been formed.

[19]The High Frontier study, advocating early BMD deployment, is discussed in app. G. That study contains cost estimates, but others believe that the estimates given for its first space-based deployment should be considerably higher.

[20]A hypothetical range for such a program might be $1.5 to $2.0 billion per year, with annual increases at, or a few percent above, the inflation rate. The modest annual increases, more so than the funding level for any individual year, distinguish the funding-limited approach from the SDI approach.

is the way in which it is spent, and in particular the things on which it is **not** spent. Choices made at the research and development stage constrain the range of possible outcomes, and the **opportunity costs** of forgoing certain investments in order to make others can have a great impact on future BMD developments as well as on other areas. Since resources such as R&D facilities and talent are limited, other military and civilian R&D will suffer to the extent that they are unable to compete against BMD research for those scarce resources. True opportunity costs are difficult to measure, since what would actually have been accomplished had some given amount of money been invested elsewhere cannot be predicted in detail.

"**Spinoffs**" in a sense are the opposite of opportunity costs—they might be described as "opportunity benefits" which have applications in areas other than those of direct interest for BMD research. Spinoffs may be more important in the long run than a research program's direct applications. However, they seldom constitute a justification for pursuing a defined objective that cannot otherwise be supported.

By nature serendipitous, spinoffs are even harder to predict than opportunity costs. However, some generalizations can be made. The more broad-based and basic a research program is, the wider its results are likely to be applied; the further advanced its development, the less its results are likely to be utilized outside of their intended application. For example, basic laser physics has applications throughout the civilian economy as well as in many defense areas; a multi-megawatt, space-qualified chemical laser would have little utility outside a BMD or ASAT system.[21]

Estimates of the impact that a vigorous BMD research and technology development program might have on the civilian sector vary widely. An editorial in an aerospace industry trade journal notes that "there is a school in industry that takes the view that even if the U.S. falls short of its defensive strategy goals, the research program will be the biggest stimulant to technology in this country since the Apollo program."[22] Others are of the opinion that BMD research will be so specialized to military applications that spinoffs for the civilian sector would be better described as "dripoffs."[23]

Political Attributes

The technical and the political aspects of a BMD research program, although related, are quite distinct. The Strategic Defense Initiative, for example, is much more than a cataloged set of technology development programs. Officially described as "*The President's* Strategic Defense Initiative,"[24] it receives unprecedented attention from the highest levels of government. It has been described by the Secretary of Defense as "the only thing that offers any real hope to the world,"[25] and it is meant to set the groundwork for a fundamental shift in national strategy. It featured prominently in the 1984 Presidential election campaign, and it has become the focus of an ideological battle fought in the public pronouncements and private negotiations of the United States, the Soviet Union, our NATO allies, and other nations. Although more difficult to quantify than the technical or even the economic attributes of BMD research and development, the political aspects are nonetheless important and very real.

[21]Plus whatever role it might have for other purposes of war such as space-to-air (e.g., anti-aircraft) or space-to-ground attack, as discussed in the "Non-BMD Applications" section of ch. 7.

[22]William Gregory, "Spark for Technology," *Aviation Week and Space Technology*, May 27, 1985, p. 11.

[23]For example, Lewis Branscomb, vice-president and chief scientist of IBM, and Dieter von Sanden, until recently head of the communications division of Siemens, Germany's largest electronics company, as quoted in "The Diplomatic Round" by John Newhouse, *The New Yorker*, July 22, 1985, p. 49.

[24]White House pamphlet, *The President's Strategic Defense Initiative*, January 1985, GPO : 1985 O-465-450 : QL 3 (emphasis added).

[25]Secretary of Defense Caspar Weinberger; interviewed on ABC TV's "This Week With David Brinkley" (quoted in Cass Peterson, "U.S. Won't Abandon 'Star Wars'," *The Washington Post*, Dec. 21, 1984).

Political Impact on Others

One way in which other countries will respond to our research will be by anticipating its possible outcomes. They cannot predict, any more than we can, how successful our research program will be or what future Congresses and Administrations will decide. They must instead consider a **range** of possibilities—and their reactions may start long before we have decided on or initiated BMD deployment.

Our policy pronouncements, implicitly as well as explicitly stated, will also affect their decisions. Their actions will be based on their **perceptions** of what we might do—perceptions which depend on their analyses of American political processes as well as their estimates of American technological capabilities. Decisions to be made by both the Soviet Union and our NATO allies regarding force modernization (conventional and nuclear), arms control strategy (see the following section), alliance relations, and international affairs will depend in part on their estimates of our own *future* decisions.

Reactions can be stimulated to some extent even when no clear link is drawn between a research program and BMD deployment. However, the likelihood and gravity of allied and Soviet responses will generally depend on how strongly the U.S. BMD research program *appears* to lead to deployment. In this respect, the *SDI approach* presents an ambiguous set of signals. It is intended to proceed as rapidly as possible[26] towards the decision point on whether to enter full-scale development and subsequent deployment, but it is explicitly not committed to crossing that threshold. Regardless of U.S. statements, however, the Soviets may believe that a decision to deploy BMD has already been made provided the technology development proceeds favorably; it would be surprising if they have not already started to analyze their possible options for responding to U.S. defensive deployments.

[26]E.g., at a "technology-limited" pace.

Adoption of an *early deployment* or *intermediate deployment* research strategy makes the connection between research and deployment explicit, and either one would certainly be expected to stimulate prompt Soviet reactions. On the other hand, the *funding-limited* and *combination approaches*, by relaxing the sense of urgency and minimizing the extent to which technology experiments challenged ABM Treaty restrictions, might lessen much of the political impact that would be generated by the other approaches.

Nothing in this section is intended to suggest that the United States should abandon a course of action judged to be in its own best interest because other parties—either allied or adversarial—might misinterpret our purposes. In determining what is in our best interest, however, the reactions of others must be taken fully into account.

Effect on Arms Control

The political impact of the U.S. BMD research program will perhaps be most strongly felt in the area of arms control, where alternative approaches to BMD research can have very different implications. The most direct effect on arms control of conducting BMD research concerns the compatibility of that research with the ABM Treaty. Most BMD *systems* or *components* based on advanced technologies cannot be developed, tested, or deployed under the ABM Treaty regime.[27]

Since the distinction between *technology* development and *component* development is highly controversial, the *SDI approach* raises questions concerning the compatibility of certain technology experiments with Treaty constraints on development and testing.[28] Moreover, with its sense of urgency and its high visibility, the SDI also raises political questions concerning the degree to which the

[27]While laboratory research into any type of BMD system is permitted under the ABM Treaty, there are severe limitations on field testing and development of ABM systems. Only fixed, land-based systems or components can be developed or tested, and only one specified fixed, land-based system can be deployed. See app. A.

[28]These compliance issues are specifically addressed in app. A.

United States is committed to maintaining the ABM Treaty regime. If the Soviets perceive that this U.S. commitment has indeed diminished, the probability that they will act in compliance with the Treaty is reduced.

Although the United States is permitted a very limited BMD deployment under the ABM Treaty, many advocates of the *early deployment approach* find value in going beyond Treaty constraints and favor abandonment of the Treaty. The *intermediate deployment approach*, by deferring deployment for a number of years, provides some time for negotiations between the United States and the U.S.S.R. which could lead to ABM Treaty modifications to permit more extensive, but nevertheless limited, defensive deployments. However, should such an agreement be impossible to reach by the time deployments could be made, intermediate deployment advocates (like early deployment supporters) would probably favor abandoning the Treaty and proceeding with BMD deployments.

On the contrary, advocates of the *funding-limited* and the *combination approaches* would strive not to damage the Treaty regime, at least not until we had identified a preferable alternative that we had confidence could be attained. In their view, mutual U.S. and Soviet adherence to the ABM Treaty would be worth the restrictions that such compliance might impose on our exploitation of BMD technologies. These approaches would relax the urgency of BMD research, easing the political questions; to the extent that technology demonstrations were deemphasized, the questions of treaty compliance would be relaxed as well.

Possible effects of the alternative BMD research approaches on arms control go beyond their impact on the ABM Treaty. These effects on other aspects of arms control are highly controversial, and they may arise even before the ABM Treaty issues do.

Supporters of the *SDI approach* say that the Strategic Defense Initiative has already succeeded in bringing the Soviets back to the bargaining table to discuss offensive arms, and that meaningful reductions in nuclear arsenals can be obtained only after the Soviets come to believe that effective defenses will make offensive forces less useful.

The role of arms control under the SDI approach would be to facilitate a safe transition to a state of highly constrained offenses coupled with effective defenses. However, making BMD deployment contingent on prior agreement with the Soviets in effect gives them a "veto" over U.S. BMD deployment, which the Reagan Administration has emphatically stated will not be permitted. This results in an inherent paradox: U.S. BMD developments will continue even if the Soviets refuse to negotiate a cooperative transition, but such a cooperative agreement is necessary if the long-term SDI goals are to be attained. Moreover, such an agreement would certainly have to be negotiated before deployments start if those deployments are to be regulated in an orderly manner.

The feasibility of any such transition agreement is still very much in question. In addition to regulating offensive and defensive deployments, it might have to regulate offensive and defensive development and testing as well in order to restrict preparations for prohibited deployments. Nobody has yet suggested how the problems of measuring, comparing, and monitoring disparate nuclear forces, problems which have plagued past arms control negotiations, could be satisfactorily resolved in the far more difficult situation in which both offensive and defensive forces are to be closely regulated.

Critics of the SDI point out that the SDI, rather than having driven the Soviets back to arms control negotiations, might instead merely have provided them with a face-saving excuse for reversing their previous decision to walk out—a decision they now regret. The Soviets now say that **reductions** in their offensive forces will be impossible as long as force **increases** might be needed to counter a U.S. defense. These statements may be only propaganda, but they may also accurately describe the initial Soviet reaction to a U.S. defensive deployment. A logical response by the Soviets to a U.S. near-term defense would indeed be the addition of penetration aids and other offensive countermeasures, the proliferation of nuclear warheads, or both. Although the U.S. defensive deployments that such a Soviet decision would anticipate might not be initiated for a number of years, if ever, the consequences of that Soviet decision for the military balance and for arms control prospects would start to be felt immediately. Potential early Soviet reactions therefore affect our choice of near-term BMD research approach, as well as our longer term policy decisions.

By deploying BMD in excess of ABM Treaty limits without waiting for the establishment of a replacement arms control regime, most *early deployment approaches* imply abandonment not only of the ABM Treaty but of the entire strategic arms control process. Not content with the condition of strategic parity prerequisite to arms control, or alternatively believing that the Soviets are not willing to settle for such a state, supporters of this approach would instead attempt to attain and maintain strategic supremacy. *Intermediate deployment approaches* may provide time for negotiation before BMD deployments start. However, should negotiations not be pursued, or should they not be satisfactorily concluded, proceeding to deployment anyway could denote abandonment of the strategic arms control process.

Many supporters of the *funding-limited* and the *combination approaches* believe that long-term improvement of the political relationship between the United States and the Soviet Union, assisted by arms control agreements, would be the most promising way to reduce the risk of nuclear war. They oppose the SDI approach as focusing all U.S. efforts on arms control and conflict avoidance into a single, dubious direction. By lessening the emphasis placed on BMD research, their approaches would leave open other arms control options, such as the ones described in chapter 9. On the other hand, if some SDI supporters are correct in asserting that only U.S. defenses can compel the Soviets to agree to force reductions, then these alternative approaches to offensive arms control will not succeed.

Under the funding-limited and combination approaches, negotiations with the Soviets which attempted to establish the boundaries of permitted versus proscribed BMD research would be desirable for the purposes of clarifying activities by both sides. If the prospect of the United States' developing advanced technologies under the SDI approach sufficiently concerns the Soviets, U.S. proposals for constraining BMD research and technology development by clarifying or extending provisions of the ABM Treaty might have considerable bargaining leverage. Such an agreement would almost certainly have to permit laboratory research, which would be extremely difficult to ban verifiably, but it might constrain more observable activities such as demonstrations of ABM "subcomponents" and other field experiments which the Department of Defense argues are currently not prohibited by the ABM Treaty (see appendix A). Although it might be difficult to construct a verifiable and equitable agreement of this sort, the task would appear easier than reaching agreement on the mutual introduction of strategic defenses.

Political Impact on Ourselves

A multibillion dollar U.S. program to study ballistic missile defense technologies will have a political impact not only on other countries, but also on our own subsequent policy decisions. Creating any large institution also generates constituencies which benefit from that institution's continued existence. This is especially true if the institution, like the Strategic Defense Initiative Organization, exists primarily to spend money for a particular purpose. However, **quantifying** this "institutional momentum" is difficult and controversial. Complex decisions are rarely documented with itemized breakdowns specifying how influential each input criterion was.

Should ballistic missile defense research be greatly accelerated, it would become one of our largest military programs. While some point to precedents for terminating large military programs, such as the cancellation of programs for the DynaSoar lifting body, the Manned Orbiting Laboratory, and the nuclear-powered airplane, others question how easily

Photo credit: Westinghouse

Westinghouse EMACK electromagnetic launcher during assembly and test, February 1982. BMD research conducted with experimental apparatus of this size, easily contained within a building, would be very difficult to control under an arms control agreement.

Congress and the executive branch would be able to terminate a future BMD program, even if the technology advances did not meet initial expectations or requirements.[29]

Since the full effects of "institutional momentum" are poorly understood, it is difficult to predict precisely how much each of the alternative research programs will suffer from it. Relevant factors, however, might include total program budget and the number of people supported by it, along with less tangible items such as visibility and level of attention. The more high-level interest there is in a program, and the more money and prestige that has been committed to it, the harder it will be to make decisions which revise or reverse earlier ones without "losing face."[30]

[29]During a floor debate on the MX missile, Senator Dale Bumpers (D-Ark.) told the Senate that he had been "trying to think . . . when the last time a weapons system was defeated here. Weapons systems have gotten where they are just like Rasputin—you cannot kill one." (*Congressional Record*, Mar. 20, 1985, p. S 3269.)

[30]Along these lines, one observer has noted that "the program manager who will admit that 5 years of research and more than $20 billion have been wasted on an unworkable system probably has yet to be born."

—*William E. Burrows*, "Ballistic Missile Defense: The Illusion of Security," *Foreign Affairs*, spring 1984, p. 855.

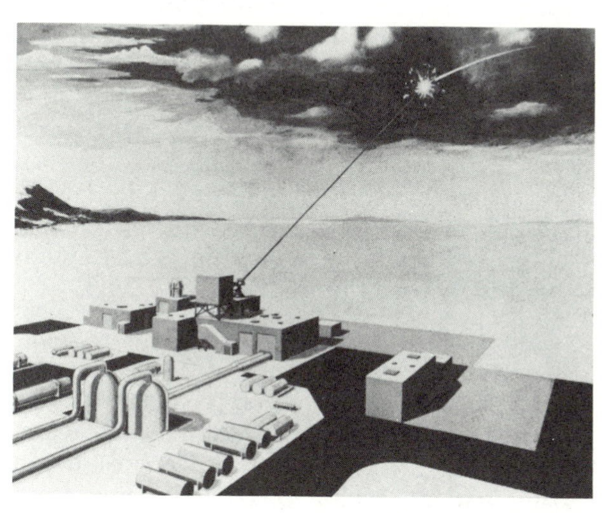

Photo credit: U.S. Air Force

Artist's conception of the existing U.S. High Energy Laser System Test Facility (HELSTF) at White Sands Missile Range, New Mexico.

Photo credit: U.S. Department of Defense

U.S. Defense Department drawing of the Soviet directed-energy research and development site at Sary Shagan proving ground.

Compliance with arms control agreements regulating research, field development, and testing using facilities of the size shown here would be more easily verified than agreements attempting to regulate activities within laboratories.

ISSUES

Other chapters of this report have dealt primarily with issues concerning ballistic missile defense *deployment*, for which decisions are at least several years off (unless the early deployment approach is adopted). Before then, Congress will need to address issues concerning the U.S. program for BMD *research and technology development*.

One set of issues concerns our choice of overall approach to pursuing BMD research. Another group involves specifically those BMD research programs which would prepare options for deployment, or which were intended to permit a decision as to whether deployment options were sufficiently promising to enter full-scale engineering development. A final set of issues pertains to any research and technology development program in areas relevant to ballistic missile defense.

Issues Concerning Choice of Research Approach

The ABM Treaty

Most BMD systems based on advanced technologies which would be investigated by the alternative R&D programs discussed above could not be developed, tested, or deployed under the ABM Treaty regime. One issue is whether or not our program of BMD research is compatible with the ABM Treaty. **A more fundamental issue, however, is whether or not the ABM Treaty continues to be compatible with our national interest.** One's attitude towards that Treaty, or more precisely one's attitude towards the concepts of national security which it embodies, will in large part determine which of the BMD research approaches described above one would choose.

Our current choices are to plan for revision of or withdrawal from the ABM Treaty, to attempt to make it more effective, or to attempt to find a middle ground. That middle ground—bolstering the effectiveness of the ABM Treaty in the short run (thereby preventing near-term Soviet BMD testing and deployment) while explicitly and publicly preparing to decide whether to abandon it later, when we are ready—may be the most difficult to attain.

The testing of new technologies on both sides could, in a few years, undermine the confidence of each that the other was not on the verge of abandoning the Treaty. Therefore, maintaining the BMD limitation regime may require new treaty provisions or other forms of agreement to reduce technical ambiguities (see discussion above on "Effect on Arms Control").

If the Treaty regime is to be sustained, questions of Soviet compliance must be resolved (see discussion of Soviet work on "traditional" BMD technologies, p. 243).

On the other hand, if we decide to revise or withdraw from the ABM Treaty to permit U.S. BMD deployments, our goals should be well-defined and our course of action well-planned. There may be a serious timing problem in carrying out a research program which will not violate the ABM Treaty, but which will give us enough information to decide with confidence that BMD deployment can meet our criteria. If we were to allow the ABM Treaty regime to erode prematurely, and then learn from our BMD research that the new BMD technologies will not fulfill our requirements, we could end up with the worst of both worlds: no arms control to limit Soviet BMD, no effective U.S. BMD, and, quite possibly, proliferated Soviet offensive forces intended to overcome an anticipated U.S. BMD.

An important issue for Congress to consider is how we can carry out our BMD research program so that it does not either prematurely compromise the ABM Treaty through technical ambiguities, or stimulate the Soviets to begin testing and deploying BMD at a time more advantageous to them than to us. At the same time, charges of Soviet noncompliance with the Treaty must be addressed as well. If they cannot be satisfactorily resolved, the United States would effectively have adopted stricter stand-

ards of compliance than those observed by the Soviets, which would put us at a competitive disadvantage.

Congress may wish to review the standards and the procedures by which U.S. and Soviet activities are judged to comply with existing treaty commitments—perhaps by requiring the establishment of an independent, nonpartisan commission to review Soviet activities and to advise Congress and the President on compliance issues associated with tests proposed by the Defense Department.

Anti-Satellite Weapon Arms Control

In the 1985 U.S.-Soviet arms control negotiations in Geneva, the Soviets emphasized the importance they attach to limiting weapons deployed in or directed at space. As both this report and its companion *Anti-Satellite Weapons, Countermeasures, and Arms Control* indicate, anti-satellite weapon technologies are closely related to BMD weapon technologies. Therefore, **those favoring uninhibited research on ballistic missile defense would find arms control measures limiting anti-satellite weapon testing to be highly constrictive.** Indeed, to attempt to remain compliant with the ABM Treaty, some experimental technology demonstrations proposed under the SDI will be conducted as anti-satellite tests. **On the other hand, those interested in strengthening the testing provisions in the ABM Treaty would find anti-satellite weapons test restrictions a useful tool in further constraining BMD development.**

Offensive Weapons Arms Control

The long-term objective of deploying defenses—enabling deep reductions to be made in offensive forces by lessening their utility—directly conflicts with one of the most probable near-term reactions to a defensive deployment—strengthening offensive forces to overwhelm defenses. This strengthening might take the form of adding penetration aids or other countermeasures, deploying additional offensive weapons, or both. Although these changes might turn out to be a waste of resources if the defense could overcome them at lesser cost, the final cost-exchange balance might not be evident in the early stages of deployment, let alone in the research stage. **The effect that our choice of BMD research approach can have on future arms control possibilities is highly significant, as was discussed in the section above on "Effect on Arms Control."**

Near-Term Soviet Breakout Potential

Each of the research approaches needs to account for the possibility of a Soviet breakout or "creepout" from the ABM Treaty. **The major issues in deterring or responding to a Soviet defensive breakout are how important an ability to deploy "traditional" nuclear-armed BMD technologies would be, whether more advanced but still near-term technologies could be relied on, or whether offensive countermeasures alone would suffice.** The *SDI approach* relies on a combination of U.S. ability to penetrate Soviet defenses and an ability to deploy as-yet-untested nonnuclear defense options; it has largely discontinued investigation of the "traditional" ballistic missile defenses of the sort once deployed by ourselves and now deployed by the Soviets. The *early* and *intermediate deployment approaches* handle the threat of Soviet breakout essentially by preempting it. The *funding-limited approach* would emphasize offensive countermeasures to counter a near-term breakout; this approach also holds out the option of accelerating research in advanced technologies up to a technology-limited pace in response to Soviet defensive deployments. In addition to offensive countermeasures and the prospect of acceleration, the *combination approach* would maintain options to deploy a near-term U.S. defense in response to Soviet near-term breakout.

Long-Term Soviet Breakout Potential

The Soviets will almost certainly continue their investigations of advanced BMD technologies. All the U.S. research approaches described here require as a minimum that sufficient U.S. research be done to understand Soviet capabilities. (Some approaches go well beyond that.) **The level of U.S. research in long-term BMD technologies should depend on a decision as to whether understanding potential**

Soviet developments is deemed sufficient, or whether the existing U.S. advantage in advanced BMD technologies can and should be exploited. It should also depend on evaluating the likelihood that valuable capabilities will be forgone by the United States if it does not pursue a more active BMD research program. Giving up what might be valuable options could disadvantage the United States even if the Soviets do not develop those options either.

Issues Concerning Preparation of Deployment Options

R&D/Deployment Coupling

There is an inherent conflict between seeking the ability to make deployment decisions in the near term and seeking to keep control over whether and when such a deployment might be made. Vigorous U.S. R&D programs could lead the Soviets to infer an intent to deploy, and might possibly stimulate them to preempt such a deployment. Therefore, proposals for vigorous R&D programs should demonstrate the ability to cope with a Soviet defensive breakout and associated Soviet offensive actions in a timely way. Offensive countermeasures probably contribute more than defensive actions towards our ability to deter or respond to Soviet defensive breakout.

The Strategic Defense Initiative Organization (SDIO) has the primary responsibility both for directing BMD research and technology development, and for making the case to Congress and to the public that this R&D effort deserves support. It will be the principal source of information about the quality, cost, and adequacy of the technologies which are thought to be ready for full-scale engineering development, but at the same time it will have a large psychological and organizational stake in an affirmative answer to the deployment question. This may create a problem when the time comes for the Secretary of Defense, the President, and Congress to decide whether BMD deployment is appropriate.

There is nothing unusual about this situation, which occurs to a greater or lesser extent when all Department of Defense programs reach their major development milestones. However, even if not peculiar to the SDI, this potential problem may be more acute in the case of SDIO because of the novelty of the technologies involved, the lack of a base of historical experience to serve as a benchmark, the possibility that a "streamlined" program will bypass some of the stages of review that most Defense Department programs must pass through, and the unusual amount of political prestige which both proponents and opponents of SDI will have staked in advance upon the outcome.

If our research program is not to be presumed to be a prelude to deployment, there must be a clearly perceived threshold which requires a positive decision—not merely the lack of a negative one—to cross. The limitations posed by the ABM Treaty provide such a threshold.

Also required, however, is a set of clear decision criteria that must be met before BMD development continues past the point requiring ABM Treaty renegotiation or abrogation. As the level of effort devoted to BMD research increases, a momentum or constituency will be created that will press for continuing and enlarging the research effort, and then for moving from research to demonstrations to deployment. **For this reason, it would be easier to establish clear decision criteria before a few more years of BMD research growth have occurred, and before the time comes to begin the actual decision process.**

Cost Estimates

It is not possible to estimate the cost of BMD deployments in the absence of either a system architecture or cost estimates for candidate system elements. However, **reliable overall cost estimates must exist before an informed development decision can be made.** Cost information, required to determine whether a possible BMD deployment will be affordable, is part of any realistic system design. It is not possible to optimize a system unless there is some way to measure whether a given approach is better or worse than another; the criterion usually utilized for this purpose is minimum cost for various levels of effectiveness.

Any research program leading up to a development or deployment decision must have as a principal priority the determination of credible cost estimates for various levels of defensive capability. Those managing such a program must be able to show whether any proposed defenses can be both affordable and cost-effective.

Relative Pace of Technology and Systems Studies

In an investigation of advanced ballistic missile defense intended to produce deployment options or to facilitate development decisions, technology development and systems studies must proceed in parallel. Without some understanding of technological potential, effective systems cannot be designed. However, without some conception of how it might be applied, technology development may not be effective and may not even be meaningful. **Such a research program needs to decide how to correlate technology development with system studies, and needs to develop a policy regarding how far either should be allowed to progress should unforeseen problems crop up in the other.**

Technology Transfer

The ABM Treaty prohibits the "transfer to other states" of "ABM systems or their components," or of "technical descriptions or blue prints" worked out for their construction.[31] These provisions prohibit the signatory nations from using their allies to circumvent ABM Treaty constraints. As a result, allied participation in a treaty-compliant research program would have to be limited to research which had not reached the "system" or "component" level. Allied participation would also be affected by restrictions which the United States itself might impose, as it does now, on the transfer of military technology to its allies for fear that such technologies may eventually reach the Soviet Union.

[31]Article IX and Agreed Statement G, ABM Treaty. (See app. B.)

In some discussions of BMD research or deployment, it has been suggested that the United States might intentionally transfer BMD technologies to the Soviet Union to prove that the United States did not seek military superiority.[32] **Any such transfer would raise very significant issues. If BMD plans or devices are transferred, potential adversaries might be able to discover vulnerabilities, enabling them to circumvent or destroy our own BMD systems. If technological capability is transferred, rather than specific devices, the American advantage which had enabled us to develop that technology first would necessarily be compromised. Furthermore, many BMD-relevant technologies have applications in other military areas that we may not want to help the Soviets develop.** Approaches towards BMD which assume that we can and should maintain technological supremacy over the Soviets would not be consistent with transfer of U.S. BMD technology to them.

Issues Pertaining to any BMD Research Program

Technology Experiments

Technology demonstration experiments are the most expensive and one of the most controversial aspects of a BMD research program. Demonstrations may be useful to measure technical progress or to provide public evidence that the technology effort in general is succeeding. Moreover, demonstrations are sooner or later needed to determine whether some system components are feasible. On the other hand, advancing our understanding of basic principles and technologies may be preferable to demonstrating the existing state of the art. There is a risk that demonstrations may "lock in" suboptimal levels of technology and divert resources which would otherwise go towards developing improved options.

Demonstrations of BMD technology are particularly complicated by ABM Treaty constraints on developing and testing ABM components or systems. Experiments that raise

[32]For example, see footnote 36, ch. 6.

treaty compliance questions run the risk of provoking a Soviet reaction that could eliminate the option of deferring BMD deployment until technology had advanced further. One possible way to assess whether this risk is worth taking might be to require that before such demonstrations are approved, there should be developed both a plausible system architecture that would use the particular technologies to be demonstrated and a corresponding arms control approach. Congress may wish to satisfy itself beforehand that, if the technologies are proven feasible, such an architecture and arms control regime appear likely to meet satisfactorily whatever criteria are established for proceeding with BMD.

Diversion of Other R&D Efforts

Acceleration of BMD programs affects other military R&D by changing the emphasis of some of those other programs to support the BMD mission. Many BMD programs had originally been pursued for other applications, such as tactical weaponry (particle beams and lasers) or space surveillance (long-wave infrared detection). For example, a system designed to provide early warning of missile launch would be similar in many ways to a system providing coarse pointing information to BMD boost-phase weapons. However, the two will not be identical. If plans to upgrade early warning satellites are subsumed within a longer range effort to develop a BMD tracking system, the original early warning mission may suffer. The alternative, however, would probably be duplication of effort.

Even R&D in nominally unrelated areas can be affected if it competes with BMD research for limited resources, such as highly trained personnel or specialized technical facilities, which cannot be readily increased in the short run.

Allied Relations

Beyond its effects on the ABM Treaty, the U.S. BMD research program can have other foreign policy consequences which should be taken into account in evaluating options. Most of our allies support U.S. BMD research as a counter to Soviet research, and some have inquired how they can participate in this research. However, for the most part they have deep reservations about the wisdom of deploying a strategic defense. **Whether the U.S. BMD research program now, and any BMD deployment in the future, can be conducted so as to avoid endangering the cohesion of our alliances will be an important issue.**[33]

Research and Development of Offensive Forces

There will be a role for U.S. strategic offensive nuclear forces for the foreseeable future in the absence of an agreement to forgo or drastically reduce them. To ensure their effectiveness in the event that the Soviets deploy defenses, the United States will need to continue its development of penetration aids and other countermeasures against defenses. By minimizing the potential effectiveness of Soviet defenses, the existence of such countermeasures would help deter the Soviets from abrogating the ABM Treaty or any subsequent agreement limiting defenses.

However, prudence dictates that we should assume any offensive countermeasure that can be developed by the United States could also be available to the Soviets, and we therefore must consider what such countermeasures would do if deployed against *our* defenses. Development by either side of powerful offensive countermeasures conflicts with the long-term goal of minimizing the role for offenses—a problem which is exacerbated if *defensive* technologies have applications in *offensive* roles (e.g., attacking satellites or aircraft, or particularly attacking enemy defenses).

[33]Alliance issues in particular are discussed in Paul E. Gallis, Mark M. Lowenthal, and Marcia S. Smith, "The Strategic Defense Initiative and United States Alliance Strategy," Congressional Research Service Report No. 85-48 F, Feb. 1, 1985.

Appendixes

Appendix A
Ballistic Missile Defense and the ABM Treaty

Introduction

This appendix examines the provisions of the 1972 Anti-Ballistic Missile Treaty,[1] the limitations these provisions place on development, testing, and deployment of ABM systems, and the sometimes conflicting interpretations that have been applied to the key elements of the treaty. In addition, this appendix discusses the SDI program (as presented in the fiscal year 1986 authorization request) and the issues that this program raises with respect to ABM Treaty compliance.[2] Soviet compliance with the ABM Treaty and Soviet ballistic missile defense programs are not discussed.[3]

This appendix concludes that if one accepts the Defense Department's current interpretation of key terms of the ABM Treaty, one may also conclude that the current SDI program is treaty compliant. Applying a more restrictive interpretation to key treaty terms could have the opposite result.

Treaty Overview

Purpose

The ABM Treaty is an agreement of unlimited duration between the United States and the Soviet Union which places restrictions on the development, testing, and deployment of ballistic missile defense systems. The purposes of this treaty, as stated in Article I, are to "limit anti-ballistic missile (ABM) systems,"[4] and to prevent either party from deploying "ABM systems for a defense of the territory of its country."[5] Although the treaty does allow limited ABM deployments, such deployments are restricted so that they could neither provide a nationwide ABM defense nor serve as the basis for deploying one. The effect of the ABM Treaty is to leave essentially unimpaired the penetration capability of either side's ballistic missile forces.

Major Provisions

Article III of the ABM Treaty prohibits all ABM deployments except those which are explicitly permitted. This article, as amended,[6] allows one fixed, land-based ABM site in each country to be located either at the nation's capital or at an ICBM field. No more than 100 interceptor missiles and 100 launchers can be deployed at the allowed site. If the national capital is chosen as the ABM site, no more than six radar complexes—each having a radius of no more than 3 kilometers—are allowed. A site to defend ICBM fields may have 2 large ABM radars and 18 smaller ABM radars. These provisions were designed to accommodate existing U.S. and Soviet ABM systems.

The United States originally elected to deploy its ABM system at the ICBM field at Grand Forks, North Dakota. This system is no longer operational, although the acquisition radar is still used for early warning purposes. The Soviets elected to deploy their ABM system around Moscow. This system is operational and is being modernized within the limits of the treaty.

Article IV permits testing, at designated test sites, of certain systems not deployable under Article III. However, systems permitted at test sites, as well as deployments, are severely constrained by **Article V**, in which "each party undertakes not to develop, test, or deploy ABM systems or components which are sea-based, air-based, space-

[1] Treaty Between the United States of America and the Union of Soviet Socialist Republics on the Limitation of Anti-Ballistic Missile Systems, which entered into force Oct. 3, 1972. App. B contains the full texts of the Treaty, its agreed interpretations, and its 1976 Protocol.

[2] The Reagan Administration's view on compliance of the SDI program with the ABM Treaty is described in detail on pp. B-1 to B-9 of *Report to the Congress on the Strategic Defense Initiative, 1985*, issued by the Department of Defense in April 1985.

Other views on this issue are discussed in:
- Abram Chayes, Antonia Chayes, and Eliot Spitzer, "Space Weapons and International Law," *Daedalus*, summer 1985.
- Thomas K. Longstreth, John E. Pike, and John B. Rhinelander, *The Impact of U.S. and Soviet Ballistic Missile Defense Programs on the ABM Treaty* (Washington, DC: National Campaign to Save the ABM Treaty, March 1985).
- Alan B. Sherr, *Legal Issues of the "Star Wars" Defense Program* (Boston, MA: Lawyers Alliance for Nuclear Arms Control, Inc., June 1984).
- R. Jeffrey Smith, " 'Star Wars' Tests and the ABM Treaty," *Science*, July 5, 1985, pp. 29-31.

[3] For two different views on these subjects, see: *Soviet Military Power, 1985*, U.S. Department of Defense (Washington, DC: U.S. Government Printing Office, April 1985); and Longstreth, et al., op. cit. Soviet BMD research is also discussed briefly in chs. 3 and 10 of this study.

[4] Ibid., Article I (1).

[5] Ibid., Article I (2).

[6] Originally, the treaty had allowed each side one ABM site to defend its capital and another site to defend one ICBM field. The treaty was amended by a 1976 Protocol to allow only one ABM site on each side.

based, or mobile land-based." Only fixed, land-based systems can be developed or tested, and only the fixed, land-based systems specified in Article III can be deployed. The second part of Article V prohibits launchers capable of firing more than one interceptor as well as launchers capable of being rapidly reloaded. **Agreed Statement E**, approved by U.S. and U.S.S.R. delegation heads at the same time that the treaty was signed, makes clear that Article V prohibits development, testing, or deployment of ABM interceptor missiles carrying more than one independently guided warhead.

Giving non-ABM systems ABM capabilities is prohibited in **Article VI(a)**, as is the testing of non-ABM systems "in an ABM mode."[7] **Part (b)** of Article VI restricts ABM battle management radars by requiring early warning radars to be on the periphery of the country and oriented outward. **Agreed Statement F** excludes radars used "for the purposes of tracking objects in outer space or for use as national technical means of verification" from the location and orientation restrictions in Article VI(b).

Article XII prohibits interference with verification of the treaty, both by banning interference with the national technical means used for verification and by prohibiting "deliberate concealment measures" which would impede verification by national technical means.

Article XIII establishes the Standing Consultative Commission (SCC) to handle questions relating to treaty compliance, to consider possible amendments, and to consider proposals for further limiting strategic arms.

Agreed Statement D reaffirms the parties' intentions not to deploy ABM systems or components except those specifically allowed in Article III. The Statement notes that ABM components based on "other physical principles" and capable of substituting for interceptors, launchers, or radars would be "subject to discussion" in the Standing Consultative Commission. "Specific limitations" on such new systems and their components would require amendment of the treaty. In the absence of amendment, Article III of the Treaty would prohibit the deployment of such new components. Article V would prohibit their development, test or deployment if they were to be space-, air-, sea-, or mobile land-based.

[7]Although the treaty does not define "non-ABM systems," these could include air defense systems, anti-tactical ballistic missile systems, strategic offensive missiles, or anti-satellite weapons.

Definitions

Ballistic missile defense involves a complicated and rapidly evolving set of technologies. Recognizing this, the drafters of the ABM Treaty tried to use language that was precise enough to effectively limit then-existing ABM systems, yet flexible enough to constrain technologies which might be developed in the future. This attempt to control potential ABM systems unavoidably introduces an element of ambiguity. The treaty language discussed below has been the focus of continued legal and technical scrutiny since the ABM Treaty was drafted; however, recent interest in advanced ABM systems has caused these discussions to take on increased significance. The relationship between these terms and the current SDI research program is discussed in the following section.

The drafters of the ABM Treaty recognized that ambiguities would arise, particularly with regard to new technologies (the so-called "other physical principles" mentioned in Agreed Statement D), but they assumed that such ambiguities would be dealt with in the context of the SCC or through treaty amendment. The reason for this assumption is a practical one. Treaty language is the expression of the agreed expectations of the parties. Put simply, a treaty means what the parties have agreed that it means. Unilateral determinations of compliance—although essential to the domestic political debate—do not bind other parties. To the extent that such determinations are inconsistent with the expectations of other parties to a treaty, then the basis of the treaty is eroded. This issue of compliance is, of course, separate from broader considerations such as the U.S. determination of the present and future value of the ABM Treaty.

"ABM Systems"

Article II of the ABM Treaty defines an anti-ballistic missile system as "a system to counter strategic ballistic missiles or their elements in flight trajectory." This definition is followed by the words "currently consisting of" and then a list of three items: ABM interceptor missiles, ABM launchers, and ABM radars. However, the treaty is *not* restricted to these specific systems. This subject is discussed in greater detail below.

The ABM system definition is limited to *strategic* weapons. Systems to counter *tactical* missiles are not covered at all. It is important to note that the treaty defines an ABM as a system *to counter* strategic weapons. It does not say "system designed to counter," as the Soviets would have liked, nor does it read "system capable of countering," which was the United States' preferred wording. The United States was concerned that, by upgrading surface-to-air missiles (SAMs), the U.S.S.R. would be able to deploy a considerable ABM capability. The Soviet Union, on the other hand, was concerned that it would be forced to classify some 10,000 SAMs as ABM interceptors.[8] The current treaty language is, therefore, a compromise between the Soviet and U.S. positions. The treaty lists the components of a then-existing ABM system but is silent on the question of how to characterize future technologies as ABM systems or components. Neither the U.S. "capabilities" test nor the Soviet "intentions" test is sanctioned by Article II of the Treaty.[9]

Some of the problems caused by the lack of a clear definition in Article II are solved by the prohibition in Article VI against giving non-ABM systems ABM capabilities. As a result, all systems which are ABM-capable, whether or not they were designed for that purpose, are either considered ABM systems under Article II or else are in violation of Article VI(a), which prohibits giving ABM capability to non-ABM systems.

Testing "in an ABM Mode"

Although Article VI prohibits the testing of non-ABM components "in an ABM mode," the ABM treaty does not define these terms. The United States, in a unilateral statement attached to the treaty, provided its interpretation of this phrase.[10] By the U.S. definition, a launcher was tested "in an ABM mode" if it was "used to launch an ABM interceptor missile"; a missile was "tested in an ABM mode" if it was "... flight tested against a target vehicle which has a flight trajectory with characteristics of a strategic ballistic missile flight trajectory ..."; and a radar was tested "in an ABM mode" if it "makes measurements on a cooperative target vehicle [with a strategic ballistic missile flight trajectory] ... or makes measurements in conjunction with the test of an ABM interceptor missile or an ABM radar at the same test range."

In 1978, the United States and the Soviet Union reached an agreement in the SCC regarding the interpretation of the phrase "in an ABM mode";[11] however, the text of the 1978 Agreed Statement remains classified.

"Development"

Because the path between research and deployment of any sophisticated weapon system is long and complicated, considerable effort has gone into determining precisely what is meant by the treaty's ban on specific types of ABM development. Perhaps the clearest definition of the words "development" and "develop," as referred to in Articles IV and V of the ABM Treaty, was provided by Gerard C. Smith, the chief U.S. negotiator of the ABM Treaty. In testimony before the Senate Armed Services Committee in 1972, Ambassador Smith stated:

> The obligation not to develop [ABM] systems, devices or warheads would be applicable only to that stage of development which follows laboratory development and testing. The prohibitions on development contained in the ABM Treaty would start at that part of the development process where field testing is initiated on either a prototype or breadboard model. It was understood by both sides that the prohibition on 'development' applies to activities involved after a component moves from the laboratory development and testing stage to the field testing stage, wherever performed. The fact that early stages of the development process, such as laboratory testing, would pose problems for verification by National Technical Means is an important consideration in reaching this definition. Exchanges with the Soviet Delegation made clear that this definition is also the Soviet interpretation of the term 'development'.[12]

[8]U.S. Congress, Office of Technology Assessment, *Arms Control in Space: Workshop Proceedings* (Washington, DC: U.S. Government Printing Office, May 1984), OTA-BP-ISC-28, p. 33.

[9]The compromise language of the treaty does not resolve this still current and controversial issue. The *Report to Congress on the Strategic Defense Initiative, 1985*, op. cit., states on p. B-2 that "Compliance [with the ABM Treaty] must be based on **objective assessments of capabilities** which support a single standard for both sides and not on **subjective judgments as to intent** which could lead to a double standard of compliance." (Emphasis added.)

[10]On Apr. 7, 1972, the U.S. Delegation made the following statement:

> ... To clarify our interpretation of "tested in an ABM mode," we note that we would consider a launcher, missile or radar to be "tested in an ABM mode" if, for example, any of the following events occur: (1) a launcher is used to launch an ABM interceptor missile, (2) an interceptor missile is flight tested against a target vehicle which has a flight trajectory with characteristics of a strategic ballistic missile flight trajectory, or is flight tested in conjunction with the test of an ABM interceptor missile or an ABM radar at the same test range, or is flight tested to an altitude inconsistent with interception of targets against which air defenses are deployed, (3) a radar makes measurements on a cooperative target vehicle of the kind referred to in item (2) above during the reentry portion of its trajectory or makes measurements in conjunction with the test of an ABM interceptor missile or an ABM radar at the same test range. Radars used for purposes such as range safety or instrumentation would be exempt from application of these criteria.

[11]U.S. Senate Committee on Foreign Relations, *SALT II Treaty: Background Documents*; "Miscellaneous Agreements Relating to the Standing Consultative Commission" forwarded from J. Brian Atwood, Department of State, to Senator Frank Church, Nov. 13, 1979.

[12]Senate Armed Services Committee, July 18, 1972.

ABM "Component"

The limitations of the ABM Treaty apply to "ABM systems or their components" and, under the terms of Agreed Statement D, to future systems and components which might be substituted for these. This raises two related questions. First, how does one distinguish between an ABM component, the testing or deployment of which is prohibited, and a subcomponent or adjunct, the testing or deployment of which is allowed? Second, how does one determine whether a system, component, or subcomponent is capable of substituting for a missile, a launcher, or a radar? The treaty language and the Agreed Statements which accompany the Treaty are silent on this point.

It is the Defense Department's position that the entire SDI research program as submitted in the fiscal year 1986 authorization request is treaty compliant. In its 1985 Report to Congress on the Strategic Defense Initiative, DOD acknowledges that Ambassador Smith's definition of "development," combined with the limitations of Article V, would prohibit the "field testing" of "ABM systems" and "components," or their "prototypes" and "breadboard models," which are other than fixed land-based.[13] However, the Defense Department maintains that the experiments currently planned for the SDI program "are designed to demonstrate technical feasibility that can be established without involving ABM systems or components or devices with their capabilities."[14] DOD is arguing that since they are testing subcomponents and not components, and since the specific systems they are testing cannot be substituted for an ABM missile, launcher, or radar, then this research is allowed under Ambassador Smith's interpretation of the Treaty.

Others disagree with DOD's interpretation. They argue that this line of reasoning ignores the history of the treaty negotiations which clearly suggests that the individual parts of an ABM system need not perform the complete range of battle functions to be considered an "ABM component." A report by the National Campaign to Save the ABM Treaty recently made the following argument:

> [The] early *Nike-Zeus* [U.S. ABM] system had not one or two, but *four* separate types of radars, for target acquisition, decoy discrimination, target tracking and interceptor tracking. Under . . . [the DOD] . . . interpretation of the difference between a "component" and an "adjunct," all of these radars would be considered to be adjuncts to one another, and none of them would be considered to be a component.[15]

The debate on this issue reflects disagreement as to whether the classification of something as an ABM system or component should be based solely on its capabilities in isolation, or whether other factors should be examined, such as its capability when combined with other devices or the apparent intentions of the parties (whether declared or evidenced by a clear pattern of activities). DOD is arguing that one looks to the capabilities of the tested systems alone to determine whether they can substitute for ABM systems or components; if they can, then they are banned by the Treaty, if they cannot, then they are allowed. Others maintain that this view is too restrictive. They argue that although capabilities are important, one must also examine the apparent intended application of a technology. Standing alone, individual technologies may have no ABM capability; however, in combination, they may have a significant ABM potential.

In addition, the tested capabilities of specific systems may not always be an adequate measure of potential. Lack of ABM capability may result from true technological limitations, or from "treaty compliant" design features that could be easily altered (e.g., putting on wheels, inserting a few additional electronic devices, or readjusting some control parameters). The distinction between these two cases must ultimately be made by the other side with the help of its national technical means of verification. It is unlikely that either side will be content to rely on the word of the other that a given experiment is treaty compliant; presence or absence of ABM capability must be manifested in ways which are amenable to verification. According to the report of the National Campaign to Save the ABM Treaty:

> The clear intention of Article V was to limit the development of new types of ABM technology at the earliest possible stage, that is, at the time that they would become detectable by national technical means.[16]

---

[13] *Report to Congress on the Strategic Defense Initiative, 1985*, op. cit., p. B-4.

[14] Ibid., p. B-2.

[15] Longstreth, et al., op. cit., p. 29.

[16] Ibid., p. 30.

U.S. Research Programs and the ABM Treaty

The SDI Program

The purpose of this section is to examine specific elements of the current U.S. BMD research programs and to determine whether they raise important questions of ABM Treaty compliance. However, there is no simple formula for deciding what is and what is not banned by the 1972 ABM Treaty. Previous sections have examined the language of the treaty and described the controversy surrounding such terms as "ABM system," "component," and "capable of substituting for." As this discussion makes clear, the inherent limitations of language and the rapid pace of technology make it impossible to develop clear, unambiguous, and objective standards by which to measure all possible research programs. As noted earlier, the general conclusion of this appendix is that if one accepts the Defense Department's current interpretation of key treaty terms, one may also reasonably accept the conclusion that the current SDI program is treaty compliant. Applying a different interpretation to these key terms could have the opposite result.

With these caveats in mind, it is useful to examine the actual elements of the SDI program. Current SDI program plans call for 15 major experiments designed to demonstrate technologies which may eventually have ABM applications. Three of the experiments will examine sensor technologies, four will involve directed-energy technologies, three will study kinetic-energy technologies, and five will involve the testing of fixed, ground-based ABM components.

Sensor Programs:

Boost Surveillance and Tracking System (BSTS).—BSTS is a space-based experiment to demonstrate technology for upgrading the current satellite early warning system. If successful, the experiments will permit a decision to proceed with similar but more advanced technologies for ABM purposes. BSTS will be capable of performing early warning functions; however, DOD asserts that it "will be limited in capability so that it cannot substitute for an ABM component." In particular, it will not be given the capability to process launch detection data in real time. For this reason, DOD claims that this system does not violate Article V(1) of the ABM Treaty which bans the development, testing, or deployment of space-based ABM components.[17]

[17] *Report to Congress on the Strategic Defense Initiative, 1985*, op. cit., p. B-6.

DOD is correct in arguing that the currently proposed BSTS system would be limited to an early warning role. However, the issue of BSTS Treaty compliance stems not only from the system's capabilities, but also from the changing nature of early warning systems. When the ABM Treaty was drafted, early warning satellites were not considered to be ABM components, or part of an ABM system, because the satellites had limited capabilities and BMD weapon systems had not yet been conceived which could use the boost-phase data these satellites produced.[18] BSTS, like its predecessors, is an early warning system; however, unlike its predecessors, BSTS might eventually contribute to the effectiveness of a layered ABM system. Assuming the existence of BMD weapons which could use BSTS data to provide acquisition and tracking information, BSTS would have to be given closer scrutiny than it would if it could only serve as an advanced early warning system.

Space Surveillance and Tracking System (SSTS).—Originally designed as an upgrade to the ground-based *Spacetrack* satellite tracking network, SSTS will demonstrate the space-based technology necessary to track and identify objects already in space.[19] SSTS technology, if perfected, could be used to support the U.S. ASAT weapon or to provide information for midcourse ABM interceptors. DOD maintains that the SSTS program is ABM Treaty compliant because the "capabilities of any demonstration satellites will be significantly less than those necessary to achieve ABM performance levels or substitute for an ABM component."[20]

If developed as originally conceived—i.e., as a component of our satellite tracking network—SSTS would probably not have raised serious ABM compliance issues even though such a system could have supplied information useful to BMD research. However, now that SSTS is part of the SDI program, DOD's assessment that it is not an ABM component will probably need to be periodically reexamined as more specific information on testing procedures and system capabilities becomes available.

Airborne Optical Adjunct (AOA).—The AOA experiment will demonstrate the technical feasibility of using optical sensors on an airborne platform for BMD applications. As part of its feasibility demonstration, AOA will observe ballistic missile

[18] Early warning *radars*, on the other hand, being similar in capability to ABM battle management radars, are specifically limited by the Treaty.

[19] SSTS tracks and identifies objects in space; BSTS identifies launches and objects entering space.

[20] *Report to Congress on the Strategic Defense Initiative, 1985*, op. cit., p. B-7.

tests at agreed ABM Test Ranges. DOD maintains that because of limitations on sensor and platform performance, the AOA could not substitute for an ABM component and therefore does not violate the ban in Article V(1) against developing air-borne ABM components.

Clearly, if AOA were designated as a "component" rather than as an "adjunct," the planned tests "in an ABM mode" would violate Article V(1) of the ABM Treaty. Here, as in other SDI programs, the distinction between an adjunct or subcomponent and an ABM component depends less on objective determinations of capability than on how one defines those terms.

Directed-Energy Programs:

ALPHA/LODE/LAMP.—ALPHA is a ground-based laser designed to explore the potential of chemical lasers in space-based BMD applications. LODE (Large Optics Demonstration Experiment) and LAMP (LODE Advanced Mirror Program) are experiments to demonstrate critical beam control and optics. In the late 1980s, the LODE/LAMP mirror is to be integrated with a high-power chemical laser using LODE beam control technology. DOD reports that "All of these tests are under-roof experiments using devices incapable of achieving ABM performance levels."[21]

The ALPHA/LODE/LAMP series of tests, if conducted in the laboratory, would seem to be consistent with the generally accepted view that the ABM Treaty's prohibitions on development only apply "to that stage of development which follows laboratory development and testing."

Acquisition, Tracking and Pointing (ATP).—For the near term, ATP[22] experiments will concentrate on ground-based, laboratory-level experiments on the technology required for space- and ground-based weapon sensors. In the future, "the measurement of booster plumes from space is a distinct possibility,"[23] as are "experiments with passive sensors in the Shuttle bay."[24] The Shuttle may also be used in follow-on experiments "to explore pointing and tracking technology."[25] It is DOD's position that "If conducted these experiments will use technologies which are only part of the set of technologies ultimately required for an ABM component. These devices will also not be capable of achieving ABM performance levels."

As long as the ATP tests remain in the laboratory there would be no violation of the ABM Treaty. The proposed space-based tests would violate Article V(1)'s prohibition against testing space-based ABM components if they were considered as "components" or as being able to substitute for ABM components. Administration officials have argued that these are generic experiments investigating pointing and tracking technologies which would have many applications and could not substitute for ABM components.

Integration of High-Powered Laser and Optical Devices.—The Defense Department eventually plans to integrate ALPHA/LODE/LAMP, ATP, and perhaps other laser and optical subsystems into one "experimental device." This "experimental device" will be used for "ground-based testing against ground-based static targets." DOD claims that these "important subsystems . . . (separate or in whole) are not ABM components or prototypes." This position rests on three arguments: 1) this "experimental device" is not capable of being based in space; 2) the power, optics, and laser wavelength are not compatible with atmospheric propagation at ranges useful for ABM applications; and 3) tests are not planned against missiles or their elements in flight.

This argument rests on the assumption that the "experimental device" in question here, although more than a subsystem or adjunct, is still less than a component or prototype. The ultimate credibility of this assumption probably cannot be assessed until more precise information becomes available on the nature of the "experimental device" and its tests.

Ground-Based Laser Uplink.—These experiments will use a ground-based laser to examine the effects of the atmosphere on beam propagation. DOD maintains that the tests are treaty compliant because "the testing mode and capabilities are below the power level and beam quality required for a ground-based laser ABM weapon, and testing will not include strategic ballistic missiles or their elements in flight."[26]

The testing of ground-based lasers at agreed ranges would not violate the terms of the ABM Treaty. The testing of mirrors in space to redirect the beam of a ground-based laser would raise compliance questions.

[21] Ibid.

[22] The ATP program is a replacement for the now-canceled *TALON GOLD* tracking telescope. Originally, *TALON GOLD* was to have flown on the Shuttle to test the technology necessary to ensure that a laser was properly aimed at its target.

[23] *Report to Congress on the Strategic Defense Initiative, 1985*, op. cit., p. B-6.

[24] Ibid., p. B-7.

[25] Ibid.

[26] Ibid.

Kinetic-Energy Programs:

Space-Based Kinetic-Kill Vehicles.—This program will be designed to prove the feasibility of rocket-propelled projectile launch and guidance. If successful, this technology might be used as an anti-satellite weapon or to defend against such weapons. In a more advanced form, space-based kinetic-kill vehicles might have applications as ABM interceptors. To attempt to ensure that this program does not violate the ABM Treaty, DOD intends to limit the performance of the demonstration hardware to satellite defense missions. Testing may include "intercepts of certain orbital targets simulating anti-satellite weapons."

The ABM Treaty does not ban anti-satellite weapons or weapons used for satellite defense, unless those weapons are tested "in an ABM mode," or could substitute for ABM systems or components. However, it should be noted that the trajectory of a ballistic missile in flight—although not orbital—resembles in many ways that of a satellite. Anti-satellite weapons and other "gray area" systems will be discussed in a later section.

Land-Based Electromagnetic Railgun.—This program will demonstrate the capability to launch unguided and guided projectiles from an electromagnetic accelerator know as a "railgun." DOD claims that test devices will not be ABM components, will not be tested "in an ABM mode," and will not have ABM capabilities.

Testing a railgun in the laboratory or in a fixed, ground-based mode at an ABM test range would not violate the terms of the ABM Treaty.

Space-Based Electromagnetic Railgun.—This program would investigate the feasibility of space-based railgun operation. DOD claims that the program would "demonstrate a capability to defend against anti-satellite interceptors and will also permit a decision to be made on the applicability of more advanced technology for ABM purposes." However, "specific performance parameters . . . will be established to satisfy Treaty compliant guidelines."

As with space-based kinetic-kill vehicles, space-based railguns might be tested as ASAT weapons or satellite defense weapons without violating the ABM Treaty. However, as discussed below, the distinctions between ASAT and BMD technologies and applications become less clear as the systems become more capable.

ABM Systems or Components:

Fixed, Ground-Based ABM Launchers.—SDI also plans to conduct tests of "ABM components" at designated ABM test ranges. Two such tests, the *High Endoatmospheric Defense Interceptor (HEDI)* and the *Exoatmospheric Reentry-Vehicle Interceptor Subsystem (ERIS)*, will demonstrate the capability to intercept strategic ballistic missile warheads within and above the atmosphere. Since such tests will be at agreed test ranges, using fixed, ground-based launchers which cannot be rapidly reloaded, and since each interceptor missile is not intended to deliver more than one independently targetable warhead, these two programs are permitted by the ABM Treaty.

Terminal Imaging Radar (TIR).—TIR is a radar that will be tested "in an ABM mode." This radar will be used to discriminate between reentry vehicles and transfer this information to interceptor missiles. DOD has announced that since the TIR tests will be conducted at a designated ABM test range from a fixed, land-based platform, they are treaty compliant.

If TIR were mobile, testing it "in an ABM mode" would violate Article V(1) of the Treaty. As this and similar technologies are developed, it will be necessary to distinguish between those systems which are incapable of operation except when fixed and land-based and those which are designed to be fixed and land-based but could operate in a mobile mode with little or no redesign.

Long Wavelength Infrared (LWIR) Probe.—The LWIR probe appears to be designed to provide a data base with which to evaluate optical system sensors. It is conceivable that this technology might also eventually substitute for current ABM radars. Even if operated as a "pop-up radar," systems based on the LWIR probe would not seem to violate Article V(1)'s prohibition against sea-, air-, space-, and mobile land-based ABM systems and components. In any case, since DOD plans to conduct the LWIR tests from fixed, land-based launchers at agreed test ranges, this program does not seem to raise treaty compliance issues.[27]

Integrated Demonstration.—DOD will eventually wish to test the HEDI and ERIS interceptors with the Terminal Imaging Radar and associated command, control, and communication systems to perform terminal defense engagements. If conducted at agreed test ranges with fixed, ground-based launchers and radars, and assuming no rapidly reloadable launchers or multiple independently guided warheads, then such tests would be allowed under the treaty.

[27] *Report to Congress on the Strategic Defense Initiative, 1985*, op. cit., p. B-9.

Other "Gray Area" Programs

In addition to the questions raised by current and proposed BMD programs, research into anti-satellite weapons, anti-tactical ballistic missiles, and large phased-array radars also pose ABM Treaty questions. In certain cases, parts of these technologies could also function as components or adjuncts to BMD systems; in other cases, research essential for non-ABM systems will supply information critical to BMD research.

Anti-Satellite Weapons.—There is great overlap between BMD and ASAT technologies. In general, even a poor anti-ballistic missile could be an excellent ASAT. The trajectory of a missile reentry vehicle while outside the atmosphere—peak altitude on the order of 1,000 km and velocity slightly suborbital—is similar to that of a satellite. The Soviet GALOSH ABM system was not designed as an ASAT but may have ASAT capability for satellites in orbits similar to ICBM trajectories. The U.S. miniature homing vehicle ASAT weapon evolved from a design originally intended for mid-course BMD.

Conversely, since technologies investigated for ASAT may also be useful in a BMD role, aggressive ASAT development could aid in the development of advanced BMD systems. Technology development ostensibly for advanced ASAT systems might provide a loophole for undertaking BMD research which might otherwise be considered a violation of the ABM treaty.

Developing an ASAT system which had BMD capability, or upgrading one to give it BMD capability, would be a violation of either Article V or VI of the ABM Treaty. Nonetheless, since information valuable to ABM research could be obtained from tests "in an ASAT mode" even before an ABM capability was achieved, ASAT weapon development could help to erode the ABM Treaty.

Large Phased-Array Radars.—Another relevant connection between ASAT systems and the ABM Treaty involves the large phased-array radars required for ASAT space surveillance and battle management. Space-track radars may be hard for an adversary to distinguish from the early-warning radars and ABM battle management radars which are currently limited by the ABM treaty. In addition to their space surveillance and tracking role, such radars can also provide early warning of missile and bomber attack and would be essential components of any ABM system. Such systems may also be used to observe missile tests in order to assist verifying compliance with treaty obligations.

Agreed Statement F in the ABM Treaty exempts space-track radars, and radars used for national technical means of verification, from the siting restrictions on ABM and early-warning radars. ASAT development will certainly stimulate development and deployment of space monitoring radars and sensors. To the extent that the distinction between an early warning radar and a space track radar is ambiguous, confusion can result which raises additional ABM Treaty compliance questions.

Anti-Tactical Ballistic Missiles.—Since the ABM Treaty prohibits defenses only against *strategic* missiles, anti-tactical ballistic missiles (ATBM) systems are not prohibited. Anti-tactical ballistic missiles were not included in the ABM Treaty because the United States wished to protect SAM-D, a surface-to-air missile then under development.[28] Since the treaty was signed, the Soviets have developed and deployed a weapon similar to the original SAM-D.

Aggressive ATBM development and deployment might affect the continuing viability of the ABM Treaty. Missiles deployed under the rubric of anti-tactical ballistic missiles could have an impact on the penetrativity of both sides' SLBMs. Eventually, ATBM systems could become so capable as to completely undercut the provisions of the ABM Treaty which prevent the development and deployment of systems to defend against ICBMs.

SDI and the Allies

Under Article IX of the ABM Treaty, the United States and the Soviet Union each agree not to "transfer to other States, and not to deploy outside its national territory, ABM systems or their components limited by [the ABM] Treaty." Agreed Statement G of the Treaty declares the intention of the signatories that Article IX's provisions should extend to "technical descriptions or blue prints specially worked out for the construction of ABM systems and their components . . . "

The Reagan Administration has stated its intention to "proceed with cooperative research with the Allies in areas of technology that could contribute to the SDI research program."[29] However, the Administration has assured Congress that

[28]SAM-D was intended to have some capability against short-range tactical ballistic missiles as well as against aircraft. However, as SAM-D developed (changing its name to 'Patriot'), its anti-tactical missile capabilities were not pursued.

[29]*Report to Congress on the Strategic Defense Initiative, 1985*, op. cit., p. A-4.

such research will be "consistent with existing international obligations including the ABM Treaty,"[30] and that "the United States will not seek to arrange for the Allies to do for the United States what it cannot do under the Treaty."[31]

Attempts to define the precise nature of Article IX's prohibitions encounter many of the difficulties already discussed (e.g., how to define an ABM system or component or how to characterize advanced ATBMs). The ABM Treaty does not constrain cooperative laboratory research efforts. The Treaty would, however, prevent joint development, testing, production, or deployment of ABM systems or components, including those—e.g., fixed, land-based launchers and interceptors—which the United States, acting alone, could legally develop, test, produce, and deploy.

[30]Ibid.
[31]Ibid.

Appendix B

Texts of the 1972 ABM Treaty, Its Agreed Interpretations, and Its 1976 Protocol*

Treaty Between the United States of America and the Union of Soviet Socialist Republics on the Limitation of Anti-Ballistic Missile Systems

Signed at Moscow May 26, 1972
Ratification advised by U.S. Senate August 3, 1972
Ratified by U.S. President September 30, 1972
Proclaimed by U.S. President October 3, 1972
Instruments of ratification exchanged October 3, 1972
Entered into force October 3, 1972

The United States of America and the Union of Soviet Socialist Republics, hereinafter referred to as the Parties,

Proceeding from the premise that nuclear war would have devastating consequences for all mankind,

Considering that effective measures to limit anti-ballistic missile systems would be a substantial factor in curbing the race in strategic offensive arms and would lead to a decrease in the risk of outbreak of war involving nuclear weapons,

Proceeding from the premise that the limitation of anti-ballistic missile systems, as well as certain agreed measures with respect to the limitation of strategic offensive arms, would contribute to the creation of more favorable conditions for further negotiations on limiting strategic arms,

Mindful of their obligations under Article VI of the Treaty on the Non-Proliferation of Nuclear Weapons,

Declaring their intention to achieve at the earliest possible date the cessation of the nuclear arms race and to take effective measures toward reductions in strategic arms, nuclear disarmament, and general and complete disarmament,

Desiring to contribute to the relaxation of international tension and the strengthening of trust between States,

Have agreed as follows:

Article I

1. Each party undertakes to limit anti-ballistic missile (ABM) systems and to adopt other measures in accordance with the provisions of this Treaty.

2. Each Party undertakes not to deploy ABM systems for a defense of the territory of its country and not to provide a base for such a defense, and not to deploy ABM systems for defense of an individual region except as provided for in Article III of this Treaty.

Article II

1. For the purpose of this Treaty an ABM system is a system to counter strategic ballistic missiles or their elements in flight trajectory, currently consisting of:

(a) ABM interceptor missiles, which are interceptor missiles constructed and deployed for an ABM role, or of a type tested in an ABM mode;

*Taken from U.S. Arms Control and Disarmament Agency, *Arms Control and Disarmament Agreements: Texts and Histories of Negotiations,* 1982 edition, pp. 139-147 and 162-163.

(b) ABM launchers, which are launchers constructed and deployed for launching ABM interceptor missiles; and

(c) ABM radars, which are radars constructed and deployed for an ABM role, or of a type tested in an ABM mode.

2. The ABM system components listed in paragraph 1 of this Article include those which are:

(a) operational;
(b) under construction;
(c) undergoing testing;
(d) undergoing overhaul, repair or conversion; or
(e) mothballed.

Article III

Each Party undertakes not to deploy ABM systems or their components except that:

(a) within one ABM system deployment area having a radius of one hundred and fifty kilometers and centered on the Party's national capital, a Party may deploy: (1) no more than one hundred ABM launchers and no more than one hundred ABM interceptor missiles at launch sites, and (2) ABM radars within no more than six ABM radar complexes, the area of each complex being circular and having a diameter of no more than three kilometers; and

(b) within one ABM system deployment area having a radius of one hundred and fifty kilometers and containing ICBM silo launchers, a Party may deploy: (1) no more than one hundred ABM launchers and no more than one hundred ABM interceptor missiles at launch sites, (2) two large phased-array ABM radars comparable in potential to corresponding ABM radars operational or under construction on the date of signature of the Treaty in an ABM system deployment area containing ICBM silo launchers, and (3) no more than eighteen ABM radars each having a potential less than the potential of the smaller of the above-mentioned two large phased-array ABM radars.

Article IV

The limitations provided for in Article III shall not apply to ABM systems or their components used for development or testing, and located within current or additionally agreed test ranges. Each Party may have no more than a total of fifteen ABM launchers at test ranges.

Article V

1. Each Party undertakes not to develop, test, or deploy ABM systems or components which are sea-based, air-based, space-based, or mobile land-based.

2. Each Party undertakes not to develop, test, or deploy ABM launchers for launching more than one ABM interceptor missile at a time from each launcher, not to modify deployed launchers to provide them with such a capability, not to develop, test, or deploy automatic or semi-automatic or other similar systems for rapid reload of ABM launchers.

Article VI

To enhance assurance of the effectiveness of the limitations on ABM systems and their components provided by the Treaty, each Party undertakes:

(a) not to give missiles, launchers, or radars, other than ABM interceptor missiles, ABM launchers, or ABM radars, capabilities to counter strategic ballistic missiles or their elements in flight trajectory, and not to test them in an ABM mode; and

(b) not to deploy in the future radars for early warning of strategic ballistic missile attack except at locations along the periphery of its national territory and oriented outward.

Article VII

Subject to the provisions of this Treaty, modernization and replacement of ABM systems or their components may be carried out.

Article VIII

ABM systems or their components in excess of the numbers or outside the areas specified in this Treaty, as well as ABM systems or their components prohibited by this Treaty, shall be destroyed or dismantled under agreed procedures within the shortest possible agreed period of time.

Article IX

To assure the viability and effectiveness of this Treaty, each Party undertakes not to transfer to other States, and not to deploy outside its national territory, ABM systems or their components limited by this Treaty.

Article X

Each Party undertakes not to assume any international obligations which would conflict with this Treaty.

Article XI

The Parties undertake to continue active negotiations for limitations on strategic offensive arms.

Article XII

1. For the purpose of providing assurance of compliance with the provisions of this Treaty, each Party shall use national technical means of verification at its disposal in a manner consistent with generally recognized principles of international law.

2. Each Party undertakes not to interfere with the national technical means of verification of the other Party operating in accordance with paragraph 1 of this Article.

3. Each Party undertakes not to use deliberate concealment measures which impede verification by national technical means of compliance with the provisions of this Treaty. This obligation shall not require changes in current construction, assembly, conversion, or overhaul practices.

Article XIII

1. To promote the objectives and implementation of the provisions of this Treaty, the Parties shall establish promptly a Standing Consultative Commission, within the framework of which they will:

(a) consider questions concerning compliance with the obligations assumed and related situations which may be considered ambiguous;

(b) provide on a voluntary basis such information as either Party considers necessary to assure confidence in compliance with the obligations assumed;

(c) consider questions involving unintended interference with national technical means of verification;

(d) consider possible changes in the strategic situation which have a bearing on the provisions of this Treaty;

(e) agree upon procedures and dates for destruction or dismantling of ABM systems or their components in cases provided for by the provisions of this Treaty;

(f) consider, as appropriate, possible proposals for further increasing the viability of this Treaty; including proposals for amendments in accordance with the provisions of this Treaty;

(g) consider, as appropriate, proposals for further measures aimed at limiting strategic arms.

2. The Parties through consultation shall establish, and may amend as appropriate, Regulations for the Standing Consultative Commission governing procedures, composition and other relevant matters.

Article XIV

1. Each Party may propose amendments to this Treaty. Agreed amendments shall enter into force in accordance with the procedures governing the entry into force of this Treaty.

2. Five years after entry into force of this Treaty, and at five-year intervals thereafter, the Parties shall together conduct a review of this Treaty.

Article XV

1. This Treaty shall be of unlimited duration.

2. Each Party shall, in exercising its national sovereignty, have the right to withdraw from this Treaty if it decides that extraordinary events related to the subject matter of this Treaty have jeopardized its supreme interests. It shall give notice of its decision to the other Party six months prior to withdrawal from the Treaty. Such notice shall include a statement of the extraordinary events the notifying Party regards as having jeopardized its supreme interests.

Article XVI

1. This Treaty shall be subject to ratification in accordance with the constitutional procedures of each Party. The Treaty shall enter into force on the day of the exchange of instruments of ratification.

2. This Treaty shall be registered pursuant to Article 102 of the Charter of the United Nations.

DONE at Moscow on May 26, 1972, in two copies, each in the English and Russian languages, both texts being equally authentic.

FOR THE UNITED STATES OF AMERICA

FOR THE UNION OF SOVIET SOCIALIST REPUBLICS

President of the United States of America

General Secretary of the Central Committee of the CPSU

Agreed Statements, Common Understandings, and Unilateral Statements Regarding the Treaty Between the United States of America and the Union of Soviet Socialist Republics on the Limitation of Anti-Ballistic Missiles

1. Agreed Statements

The document set forth below was agreed upon and initialed by the Heads of the Delegations on May 26, 1972 (letter designations added);

AGREED STATEMENTS REGARDING THE TREATY BETWEEN THE UNITED STATES OF AMERICA AND THE UNION OF SOVIET SOCIALIST REPUBLICS ON THE LIMITATION OF ANTI-BALLISTIC MISSILE SYTEMS

[A]

The Parties understand that, in addition to the ABM radars which may be deployed in accordance with subparagraph (a) of Article III of the Treaty, those non-phased-array ABM radars operational on the date of signature of the Treaty within the ABM system deployment area for defense of the national capital may be retained.

[B]

The Parties understand that the potential (the product of mean emitted power in watts and antenna area in square meters) of the smaller of the two large phased-array ABM radars referred to in subparagraph (b) of Article III of the Treaty is considered for purposes of the Treaty to be three million.

[C]

The Parties understand that the center of the ABM system deployment area centered on the national capital and the center of the ABM system deployment area containing ICBM silo launchers for each Party shall be separated by no less than thirteen hundred kilometers.

[D]

In order to insure fulfillment of the obligation not to deploy ABM systems and their components except as provided in Article III of the Treaty, the Parties agree that in the event ABM systems based on other physical principles and including components capable of substituting for ABM interceptor missiles, ABM launchers, or ABM radars are created in the future, specific limitations on such systems and their components would be subject to discussion in accordance with Article XIII and agreement in accordance with Article XIV of the Treaty.

[E]

The Parties understand that Article V of the Treaty includes obligations not to develop, test or deploy ABM interceptor missiles for the delivery by each ABM interceptor missile of more than one independently guided warhead.

[F]

The Parties agree not to deploy phased-array radars having a potential (the product of mean emitted power in watts and antenna area in square meters) exceeding three million, except as provided for in Articles III, IV and VI of the Treaty, or except for the purposes of tracking objects in outer space or for use as national technical means of verification.

[G]

The Parties understand that Article IX of the Treaty includes the obligation of the US and the USSR not to provide to other States technical descriptions or blue prints specially worked out for the construction of ABM systems and their components limited by the Treaty.

2. Common Understandings

Common understanding of the Parties on the following matters was reached during the negotiations:

A. Location of ICBM Defenses

The U.S. Delegation made the following statement on May 26, 1972:

Article III of the ABM Treaty provides for each side one ABM system deployment area centered on its national capital and one ABM system deployment area containing ICBM silo launchers. The two sides have registered agreement on the following statement: "The Parties understand that the center of the ABM system deployment area centered on the national capital and the center of the ABM system deployment area containing ICBM silo launchers for each Party shall be separated by no less than thirteen hundred kilometers." In this connection, the U.S. side notes that its ABM system deployment area for defense of ICBM silo launchers, located west of the Mississippi River, will be centered in the Grand Forks ICBM silo launcher deployment area. (See Agreed Statement [C].)

B. ABM Test Ranges

The U.S. Delegation made the following statement on April 26, 1972:

Article IV of the ABM Treaty provides that "the limitations provided for in Article III shall not apply to ABM systems or their components used for development or testing, and located within current or additionally agreed test ranges." We believe it would be useful to assure that there is no misunderstanding as to current ABM test ranges. It is our understanding that ABM test ranges encompass the area within which ABM components are located for test purposes. The current U.S. ABM test ranges are at White Sands, New Mexico, and at Kwajalein Atoll, and the current Soviet ABM test range is near Sary Shagan in Kazakhstan. We consider that non-phased array radars of types used for range safety or instrumentation purposes may be located outside of ABM test ranges. We interpret the reference in Article IV to "additionally agreed test ranges" to mean that ABM components will not be located at any other test ranges without prior agreement between our Governments that there will be such additional ABM test ranges.

On May 5, 1972, the Soviet Delegation stated that there was a common understanding on what ABM test ranges were, that the use of the types of non-ABM radars for range safety or instrumentation was not limited under the Treaty, that the reference in Article IV to "additionally agreed" test ranges was sufficiently clear, and that national means permitted identifying current test ranges.

C. Mobile ABM Systems

On January 29, 1972, the U.S. Delegation made the following statement:

Article V(1) of the Joint Draft Text of the ABM Treaty includes an undertaking not to develop, test, or deploy mobile land-based ABM systems and their components. On May 5, 1971, the U.S. side indicated that, in its view, a prohibition on deployment of mobile ABM systems and components would rule out the deployment of ABM launchers and radars which were not permanent fixed types. At that time, we asked for the Soviet view of this interpretation. Does the Soviet side agree with the U.S. side's interpretation put forward on May 5, 1971?

On April 13, 1972, the Soviet Delegation said there is a general common understanding on this matter.

D. Standing Consultative Commission

Ambassador Smith made the following statement on May 22, 1972:

The United States proposes that the sides agree that, with regard to initial implementation of the ABM Treaty's Article XIII on the Standing Consultative Commission (SCC) and of the consultation Articles to the Interim Agreement on offensive arms and the Accidents Agreement,[1] agreement establishing the SCC will be worked out early in the follow-on SALT negotiations; until that is completed, the following arrangements will prevail: when SALT is in session, any consultation desired by either side under these Articles can be carried out by the two SALT Delegations; when SALT is not in session, *ad hoc* arrangements for any desired consultations under these Articles may be made through diplomatic channels.

Minister Semenov replied that, on an *ad referendum* basis, he could agree that the U.S. statement corresponded to the Soviet understanding.

E. Standstill

On May 6, 1972, Minister Semenov made the following statement:

In an effort to accommodate the wishes of the U.S. side, the Soviet Delegation is prepared to proceed on the basis that the two sides will in fact observe the obligations of both the Interim Agreement and the ABM Treaty beginning from the date of signature of these two documents.

In reply, the U.S. Delegation made the following statement on May 20, 1972:

---

[1]See Article 7 of Agreement to Reduce the Risk of Outbreak of Nuclear War Between the United States of America and the Union of Soviet Socialist Republics, signed Sept. 30, 1971.

The U.S. agrees in principle with the Soviet statement made on May 6 concerning observance of obligations beginning from date of signature but we would like to make clear our understanding that this means that, pending ratification and acceptance, neither side would take any action prohibited by the agreements after they had entered into force. This understanding would continue to apply in the absence of notification by either signatory of its intention not to proceed with ratification or approval.

The Soviet Delegation indicated agreement with the U.S. statement.

3. Unilateral Statements

The following noteworthy unilateral statements were made during the negotiations by the United States Delegation:

A. Withdrawal from the ABM Treaty

On May 9, 1972, Ambassador Smith made the following statement:

The U.S. Delegation has stressed the importance the U.S. Government attaches to achieving agreement on more complete limitations on strategic offensive arms, following agreement on an ABM Treaty and on an Interim Agreement on certain measures with respect to the limitation of strategic offensive arms. The U.S. Delegation believes that an objective of the follow-on negotiations should be to constrain and reduce on a long-term basis threats to the survivability of our respective strategic retaliatory forces. The USSR Delegation has also indicated that the objectives of SALT would remain unfulfilled without the achievement of an agreement providing for more complete limitations on strategic offensive arms. Both sides recognize that the initial agreements would be steps toward the achievement of more complete limitations on strategic arms. If an agreement providing for more complete strategic offensive arms limitations were not achieved within five years, U.S. supreme interests could be jeopardized. Should that occur, it would constitute a basis for withdrawal from the ABM Treaty. The U.S. does not wish to see such a situation occur, nor do we believe that the USSR does. It is because we wish to prevent such a situation that we emphasize the importance the U.S. Government attaches to achievement of more complete limitations on strategic offensive arms. The U.S. Executive will inform the Congress, in connection with Congressional consideration of the ABM Treaty and the Interim Agreement, of this statement of the U.S. position.

B. Tested in ABM Mode

On April 7, 1972, the U.S. Delegation made the following statement:

Article II of the Joint Text Draft uses the term "tested in an ABM mode," in defining ABM components, and Article VI includes certain obligations concerning such testing. We believe that the sides should have a common understanding of this phrase. First, we would note that the testing provisions of the ABM Treaty are intended to apply to testing which occurs after the date of signature of the Treaty, and not to any testing which may have occurred in the past. Next, we would amplify the remarks we have made on this subject during the previous Helsinki phase by setting forth the objectives which govern the U.S. view on the subject, namely, while prohibiting testing of non-ABM components for ABM purposes: not to prevent testing of ABM components, and not to prevent testing of non-ABM components for non-ABM purposes. To clarify our interpretation of "tested in an ABM mode," we note that we would consider a launcher, missile or radar to be "tested in an ABM mode" if, for example, any of the following events occur: (1) a launcher is used to launch an ABM interceptor missile, (2) an interceptor missile is flight tested against a target vehicle which has a flight trajectory with characteristics of a strategic ballistic missile flight trajectory, or is flight tested in conjunction with the test of an ABM interceptor missile or an ABM radar at the same test range, or is flight tested to an altitude inconsistent with interception of targets against which air defenses are deployed, (3) a radar makes measurements on a cooperative target vehicle of the kind referred to in item (2) above during the reentry portion of its trajectory or makes measurements in conjunction with the test of an ABM interceptor missile or an ABM radar at the same test range. Radars used for purposes such as range safety or instrumentation would be exempt from application of these criteria.

C. No-Transfer Article of ABM Treaty

On April 18, 1972, the U.S. Delegation made the following statement:

In regard to this Article [IX], I have a brief and I believe self-explanatory statement to make. The U.S. side wishes to make clear that the provisions of this Article do not set a precedent for whatever provision may be considered for a Treaty on Limiting Strategic Offensive Arms. The question of transfer of strategic offensive arms is a far more complex issue, which may require a different solution.

D. No Increase in Defense of Early Warning Radars

On July 28, 1970, the U.S. Delegation made the following statement:

Since Hen House radars [Soviet ballistic missile early warning radars] can detect and track ballistic missile warheads at great distances, they have a significant ABM potential. Accordingly, the U.S. would regard any increase in the defenses of such radars by surface-to-air missiles as inconsistent with an agreement.

Protocol to the Treaty Between the United States of America and the Union of Soviet Socialist Republics on the Limitation of Anti-Ballistic Missile Systems

Signed at Moscow July 3, 1974
Ratification advised by U.S. Senate November 10, 1975
Ratified by U.S. President March 19, 1976
Instruments of ratification exchanged May 24, 1976
Proclaimed by U.S. President July 6, 1976
Entered into force May 24, 1976

The United States of America and the Union of Soviet Socialist Republics, hereinafter referred to as the Parties,

Proceeding from the Basic Principles of Relations between the United States of America and the Union of Soviet Socialist Republics signed on May 29, 1972,

Desiring to further the objectives of the Treaty between the United States of America and the Union of Soviet Socialist Republics on the Limitation of Anti-Ballistic Missile Systems signed on May 26, 1972, hereinafter referred to as the Treaty,

Reaffirming their conviction that the adoption of further measures for the limitation of strategic arms would contribute to strengthening international peace and security,

Proceeding from the premise that further limitation of anti-ballistic missile systems will create more favorable conditions for the completion of work on a permanent agreement on more complete measures for the limitation of strategic offensive arms,

Have agreed as follows:

Article I

1. Each Party shall be limited at any one time to a single area out of the two provided in Article III of the Treaty for deployment of anti-ballistic missile (ABM) systems or their components and accordingly shall not exercise its right to deploy an ABM system or its components in the second of the two ABM system deployment areas permitted by Article III of the Treaty, except as an exchange of one permitted area for the other in accordance with Article II of this Protocol.

2. Accordingly, except as permitted by Article II of this Protocol: the United States of America shall not deploy an ABM system or its components in the area centered on its capital, as permitted by Article III(a) of the Treaty, and the Soviet Union shall not deploy an ABM system or its components in the deployment area of intercontinental ballistic missile (ICBM) silo launchers as permitted by Article III(b) of the Treaty.

Article II

1. Each Party shall have the right to dismantle or destroy its ABM system and the components thereof in the area where they are presently deployed and to deploy an ABM system or its components in the alternative area permitted by Article III of the Treaty, provided that prior to initiation of construction, notification is given in accord

with the procedure agreed to in the Standing Consultative Commission, during the year beginning October 3, 1977 and ending October 2, 1978, or during any year which commences at five year intervals thereafter, those being the years for periodic review of the Treaty, as provided in Article XIV of the Treaty. This right may be exercised only once.

2. Accordingly, in the event of such notice, the United States would have the right to dismantle or destroy the ABM system and its components in the deployment area of ICBM silo launchers and to deploy an ABM system or its components in an area centered on its capital, as permitted by Article III(a) of the Treaty, and the Soviet Union would have the right to dismantle or destroy the ABM system and its components in the area centered on its capital and to deploy an ABM system or its components in an area containing ICBM silo launchers, as permitted by Article III(b) of the Treaty.

3. Dismantling or destruction and deployment of ABM systems or their components and the notification thereof shall be carried out in accordance with Article VIII of the ABM Treaty and procedures agreed to in the Standing Consultative Commission.

Article III

The rights and obligations established by the Treaty remain in force and shall be complied with by the Parties except to the extent modified by this Protocol. In particular, the deployment of an ABM system or its components within the area selected shall remain limited by the levels and other requirements established by the Treaty.

Article IV

This Protocol shall be subject to ratification in accordance with the constitutional procedures of each Party. It shall enter into force on the day of the exchange of instruments of ratification and shall thereafter be considered an integral part of the Treaty.

DONE at Moscow on July 3, 1974, in duplicate, in the English and Russian languages, both texts being equally authentic.

For the United States of America:

RICHARD NIXON

President of the United States of America

For the Union of Soviet Socialist Republics:

L. I. BREZHNEV

General Secretary of the Central Committee of the CPSU

Appendix C

Effects of BMD Deployment on Existing Arms Control Treaties

The arms control treaties which are most directly relevant to BMD deployment are the 1972 ABM Treaty, the 1967 Outer Space Treaty, the 1963 Limited Test Ban Treaty, the 1974 Threshold Test Ban Treaty, and the 1970 Non-Proliferation Treaty.[1] The ABM Treaty is discussed extensively in chapters 6, 9, and 10, and in appendix A. The others are discussed briefly in this appendix.

Outer Space Treaty

Article IV of the Outer Space Treaty[2] begins:

> States Parties to the Treaty undertake not to place in orbit around the Earth any objects carrying nuclear weapons or any other kinds of weapons of mass destruction, install such weapons on celestial bodies, or station such weapons in outer space in any other manner . . .

Article IX includes the following provision:

> If a State Party to the Treaty has reason to believe that an activity or experiment planned by it or its nationals in outer space . . . would cause potentially harmful interference with activities of other States Parties in the peaceful exploration and use of outer space, . . . it shall undertake appropriate international consultations before proceeding with any such activity or experiment.

Depending on the specific nature of a BMD system deployment which utilizes space-based components, there may be a conflict with one or both of these provisions. For example, Article IV would prohibit placing in orbit a BMD satellite which contains a directed-energy weapon that is powered by a nuclear explosive device.

Limited Test Ban Treaty

Article I of the Limited Test Ban Treaty[3] prohibits each Party from carrying out any kind of nuclear explosion in outer space, in the atmosphere, or under water. Although underground nuclear explosions are permitted, it is very unlikely that the United States or the Soviet Union would deploy a BMD system which relies on space-based directed-energy weapons powered by nuclear explosions without having tested them in space. Thus it is very likely that BMD deployments of that type would require withdrawal from the Limited Test Ban Treaty.

Threshold Test Ban Treaty

The Reagan Administration has reported to Congress that directed-energy weapons driven by nuclear explosions may require nuclear explosive devices on the order of 1,000 kilotons or higher.[4] This would be far above the 150-kiloton limit imposed by the Threshold Test Ban Treaty[5] on tests of such devices. The Administration also stated in 1984 that "at this time there is no indication of a need to test above 150 kt."[6]

Non-Proliferation Treaty

While Article II of the Non-Proliferation Treaty[7] obliges the nonnuclear-weapon parties to refrain from acquiring nuclear weapons, Article VI obliges the parties which possess nuclear weapons to " . . . pursue negotiations in good faith on effective measures relating to cessation of the nuclear arms race at an early date . . . " Most of the nonnuclear-weapon parties believe that these two obligations constitute a balanced deal.[8] In recent years they have been complaining strongly in international fora that they have been keeping their side of the

[1]The texts of these treaties and histories of their negotiations appear in *Arms Control and Disarmament Agreements*, U.S. Arms Control and Disarmament Agency, Washington, DC, 1982.

[2]"Treaty on Principles Governing the Activities of States in the Exploration and Use of Outer Space, Including the Moon and Other Celestial Bodies," which entered into force Oct. 10, 1967. It has over 80 parties, including the United States, the U.S.S.R., and the People's Republic of China.

[3]"Treaty Banning Nuclear Weapon Tests in the Atmosphere, in Outer Space, and Under Water," which entered into force Oct. 10, 1963. It has about 100 parties, including the United States and U.S.S.R.

[4]"Fiscal Year 1985 Arms Control Impact Statements," issued March 1984. U.S. Senate Print 98-149, p. 253.

[5]"Treaty Between the United States of America and the Union of Soviet Socialist Republics on the Limitation of Underground Nuclear Weapon Tests." This treaty was signed by President Nixon on July 3, 1974. Although it has not been ratified, both Parties have announced their intention to observe its 150-kiloton limitation pending ratification.

[6]Ibid.

[7]"Treaty on the Non-Proliferation of Nuclear Weapons," which entered into force March 5, 1970. It has about 120 parties, including the United States and U.S.S.R.

[8]See, for example, *Nuclear Proliferation Factbook*, U.S. Congress, Congressional Research Service, Washington, DC, September 1980, pp. 459-496; Coit D. Blacker and Gloria Duffy (eds.), *International Arms Control: Issues and Agreements* (Stanford, CA: Stanford University Press, 1984), pp. 153-159 and 169-172.

treaty's bargain, but that there has been insufficient progress toward ending the superpowers' nuclear arms race.[9]

From an American viewpoint, the spread of nuclear weapons to many additional countries would not only constitute a serious threat to U.S. national security, but would also threaten the security of all states. Hence U.S. representatives have argued, in the U.N. and elsewhere, that mutual abstinence from acquisition of nuclear weapons is in the self interest of states not now possessing them, regardless of when or whether the superpowers succeed in their efforts to halt and reverse their nuclear arms competition. To date no party has withdrawn from the Non-Proliferation Treaty, perhaps because the parties agree with that argument or because they still hope for progress toward nuclear disarmament.

If the U.S. Strategic Defense Initiative should lead to a U.S.-Soviet agreement to reduce offensive nuclear forces and to amend the ABM Treaty to permit deployment of nonnuclear defenses on an agreed schedule, the nonnuclear-weapon states might well consider that a step toward "cessation of the nuclear arms race." On the other hand, if either the United States or the Soviet Union should abrogate the ABM Treaty before a U.S.-Soviet agreement is concluded on a new strategic arms control regime, the nonnuclear-weapon states would probably perceive little hope for progress toward nuclear arms reductions. In that case, there would be a substantially increased risk that some parties would withdraw from the Non-Proliferation Treaty, and it would become much more difficult to persuade additional states to adhere to it.

[9]For example, in October 1984 the Nigerian delegate stated to the U.N. General Assembly, "The Non-Proliferation Treaty will continue to be a cornerstone of the nonproliferation regime only if all parties assume their responsibilities and obligations with sincerity. As long as the nuclear Powers continue with their vertical proliferation of nuclear weapons, [nuclear weapon] threshold States will consider it their right and duty to keep their options open, and non-nuclear-weapon states will doubt the wisdom of continued adherence to the Treaty."

Appendix D

Defense Requirements for Assured Survival

Opinion varies greatly on how much damage the United States could sustain from a Soviet nuclear attack and still survive. Opinion also differs on what is important in determining whether the United States has survived. Some believe that what matters is how well society would survive and reconstitute itself. Others argue that the nation will have survived if it recovers its superpower status and its economy in some specified number of years. Still others argue that survival is assured only if the number of casualties can confidently be kept below some "limited" number. However, within this group opinions differ on what that number is. Some believe that the nation can sustain 10 million casualties or more, while others believe that if the nation suffered hundreds of thousands of civilian deaths in a short period of time it would be a catastrophe without precedent, and the nation could certainly not be said to have survived.[1]

This appendix illustrates how the number of casualties might be related to defense capability if the Soviets were to attempt to maximize U.S. casualties. Most observers would probably agree that an extremely capable defense would be required to keep casualties low if the Soviets decided to attack in an effort to maximize casualties. Because of the great destructive power of nuclear weapons and the concentration of U.S. population in major urban areas, a small number of nuclear weapons detonating over populous areas would cause large numbers of casualties. Planners seeking a defense to assure survival would most likely make "defense conservative" estimates. They would give the offense a great benefit of the doubt and estimate the capability of their defense very conservatively in order to minimize the likelihood that casualties would exceed their expectations.

This appendix illustrates how such worst-case estimates of casualties might be made and how they would be related to defense capability. It is illustrative of an approach to the problem of determining requirements for assured survival. It is not a prediction of casualties that would result from an attack on the United States. Casualties need not be as high as shown here, and they might be considerably lower. We assume that the Soviets attack to maximize casualties and that no civil defense measures are taken. Different Soviet attack tactics, evacuation of cities, and preferential defense of the most heavily populated areas might all contribute to reducing casualties. On the other hand, long-term nuclear effects spreading far beyond the immediate blast area might increase casualties.

OTA does not predict either that the required defense levels are achievable or that they are not achievable.

This appendix presents a rather rudimentary calculation in order to illustrate the problem. We recognize that the results can be refined substantially by taking advantage of detailed, sophisticated information on population distribution, aimpoint uncertainties, and nuclear weapon effects.

It is assumed that since the United States has extremely capable defenses, the Soviets are denied the capability for a meaningful attack on U.S. military assets, and they hence concentrate their forces to produce the greatest number of casualties they can. A force of 9,000 RVs, roughly equal to the current Soviet force, is assumed. Each RV is assumed to have a 750 kiloton (kt) yield. For simplicity, the attacking weapons are all ballistic missile RVs and the defense is BMD only.

The basic scenario is as follows. The Soviets know about how capable the U.S. defense is. They prepare a list of aimpoints such that the first is the most densely populated part of the United States, the second is the second most densely populated part, and so on. They allocate their weapons against the most populous part of this list in a manner to be described and do not attack the rest of the United States.

To illustrate some of the uncertainties in this calculation, four cases have been examined. In two cases, the worst for the United States, the Soviets are assumed to know *exactly* how good the U.S. defense is, and they allocate their weapons to achieve an expected one weapon penetrating to each aimpoint. The number of aimpoints is equal to 9,000(1-P_k), where P_k is the probability that the

[1] Some people believe that even if many tens of millions of Americans died, society would remain intact (or rapidly reconstitute itself) and the nation would have survived. Others believe that society can be destroyed even if casualties are relatively low.

defense kills an RV attempting to penetrate it.[2] As a worst case, we assume that they hit each aimpoint. In actuality, some aimpoints would survive. In the other two cases, the Soviets only know roughly how good the defense is, so they target 100 RVs on each of 90 aimpoints. Their hits are distributed randomly among the aimpoints. In each case, they begin with the most populous aimpoint and allocate weapons in descending order until all their weapons are allocated.

For each of the two cases described in the preceding paragraph, we use two different kill criteria, for a total of four cases. In two of the cases, the Soviets distribute their weapons to produce 3 pounds per square inch (psi) overpressure over the entire area attacked. In the other two cases, they distribute their weapons to put 5 psi over the area attacked. We assume in each case that everyone living in the attacked area is killed. It is beyond the scope of this appendix to determine the minimum overpressure that would kill everyone subjected to it, although it seems likely that the answer is between 3 and 5 psi.[3] A 1978 ACDA report[4] says that 3 to 5 psi would cause total burn out in urban areas. A 1979 OTA report[5] estimated that most of those exposed to 5 psi would be killed immediately or seriously wounded, and that half of those exposed to 2 to 5 psi would be killed or seriously wounded. Many of the wounded would eventually die for lack of care. A 750 kt weapon detonated at 2,000 feet above the ground would produce 5 psi or more overpressure over about 24 square miles and 3 psi or more over 50 square miles.

In order to understand how U.S. population is distributed among the most populous parts of the nation we examined both the most populous cities and the most densely populated counties and cities. These are listed in tables D-1 and D-2, respectively. The distributions of cumulative population as a function of total area occupied obtained from these were reasonably similar, despite the fact that there were many areas that appeared on one list but not on the other.

Figure D-1 shows the number of people living in the most populous parts of the United States. It is arrived at by summing down tables D-1 and D-2 in rank order, beginning with number 1. If the Soviets wanted to maximize casualties, they would begin by allocating their weapons against the most heavily populated areas, and work their way up the cumulative curves until they ran out of weapons. Figure D-2 repeats figure D-1, but also shows the number of detonations required to produce 5 psi over a given area and the number of weapons required to produce 3 psi. For example, 40 deto-

[2] $(1-P_k)$ is the probability that an RV gets through the defense, so $9{,}000(1-P_k)$ is the number of RVs they expect to get through the defense. Hence, they aim at that number of aimpoints. An actual calculation of the number of aimpoints would probably be more sophisticated than this, since some of the intended aimpoints will receive more than one detonation while others will be successfully defended. The worst they can do, from the U.S. perspective, is to hit each intended aimpoint.

[3] Some maintain that in either case the number of casualties is likely to exceed the population of the area attacked, because effects such as fallout and groundwater contamination, as well as destruction of vital services, would kill far beyond the blast area.

[4] "An Analysis of Civil Defense in Nuclear War" December 1978.

[5] U.S. Congress, Office of Technology Assessment, *The Effects of Nuclear War* (Washington, DC: U.S. Government Printing Office, May 1979), OTA-NS-89

Table D-1.—Population, Area, and Population Density of the Most Populous U.S. Cities

Rank	City	Population (thousands)	Area (square miles)	Population per square mile
1	New York	7,072	301.5	23,455
2	Chicago	3,005	228.1	13,174
3	Los Angeles	2,967	464.7	6,384
4	Philadelphia	1,688	136.0	12,413
5	Houston	1,595	556.4	2,867
6	Detroit	1,203	135.6	8,874
7	Dallas	904	333.0	2,715
8	San Diego	876	320.0	2,736
9	Phoenix	790	324.0	2,437
10	Baltimore	787	80.3	9,798
11	San Antonio	786	262.7	2,992
12	Indianapolis	701	352.0	1,991
13	San Francisco	679	46.4	14,633
14	Memphis	646	264.1	2,447
15	Washington	638	62.7	10,181

SOURCE: Statistical Abstract of the United States, 1984. Department of Commerce, Bureau of the Census. Populations are based on the 1980 census.

Table D-2.—The Most Densely Populated Counties and Independent Cities in the United Statess

Rank	Name	Population per square mile	Area (square miles)	Population (thousands)
1	New York, NY	64,395	22	1,428
2	Kings, NY	31,762	70	2,231
3	Bronx, NY	28,006	42	1,169
4	Queens, NY	17,411	109	1,891
5	San Francisco, CA	14,636	46	679
6	Philadelphia, PA	12,413	136	1,688
7	Hudson, NJ	11,993	46	557
8	Suffolk, MA	11,472	57	650
9	Washington, DC	10,181	63	638
10	Baltimore, MD (city)	9,793	80	787
11	St. Louis, MO (city)	7,379	61	453
12	Alexandria, VA (city)	6,867	15	103
13	Essex, NJ	6,696	127	851
14	Richmond, NY	5,995	59	352
15	Arlington, VA	5,878	26	153
16	Cook, IL	5,485	958	5,254
17	Norfolk, VA (city)	5,037	53	267
18	Union, NJ	4,886	103	504
19	Falls Church, VA (city)	4,830	2	10
20	Nassau, NY	4,610	287	1,322
21	Denver, CO	4,452	111	492
22	Milwaukee, WI	3,997	241	965
23	Charlottsville, VA	3,827	10	40
24	Wayne, MI	3,801	615	2,338
25	Richmond, VA (city)	3,650	60	219

SOURCE: *County and City Data Book*, 1983. Department of Commerce, Bureau of the Census. Population Data based on 1980 census.

nations would be required to produce 3 psi overpressure over a total area of 2,000 square miles, and 80 would be required to produce 5 psi over the same area. The most populous 2,000 square miles contains about 17 million people.

We can now calculate the expected number of casualties from an attack on our population, as a function of the effectiveness of our BMD as measured by the probability that an RV is killed by it, P_k. In the worst case, the Soviets use all their weapons against a number of aimpoints equal to the number of RVs they expect to penetrate the defense, $9{,}000(1\text{-}P_k)$, and their weapons detonate successfully at all of them. They pick the most populous aimpoints. This provides an upper bound on the number of prompt casualties. In the other case, they allocate 100 of their weapons against each of the 90 most lucrative aimpoints. In this case the probability that any aimpoint is destroyed is given by the expression $1\text{-}P_k^{100}$. The number of prompt casualties is this multiplied by the total population at those aimpoints, which is about 25 million for 3 psi coverage and about 20 million for 5 psi coverage.

Figures D-3 and D-4 show the results for the four cases. They show the number of casualties as a function of the effectiveness of the U.S. BMD system. Figure D-3 shows what the number of casualties would be if one believes that 3 psi is sufficient to kill almost everyone, and figure D-4 shows what the results would be if one believes that 5 psi is necessary.

Basic Observations

If the Soviets were intent on killing Americans, it would require an extremely capable defense to keep casualties "low." A defense that permitted 1 percent of the Soviet weapons through might result in casualties well in excess of 10 million. It would appear that keeping casualties below 1 million would require a defense that could stop in excess of 99.9 percent of the Soviet attack. While we would need defenses with these capabilities to be confident that we could keep casualties low, lesser (but still quite capable) defenses *might* result in casualties much lower than what is indicated in this worst case analysis. Soviet weapons might not be so heavily concentrated on a few cities, and populations might evacuate or take other protective measures. Finally, not everyone agrees that assured survival requires guaranteeing very low expected casualties. By some definitions, the nation would survive even if millions of Americans did not.

Figure D-1.—Cumulative U.S. Population Beginning With the Most Populous Areas

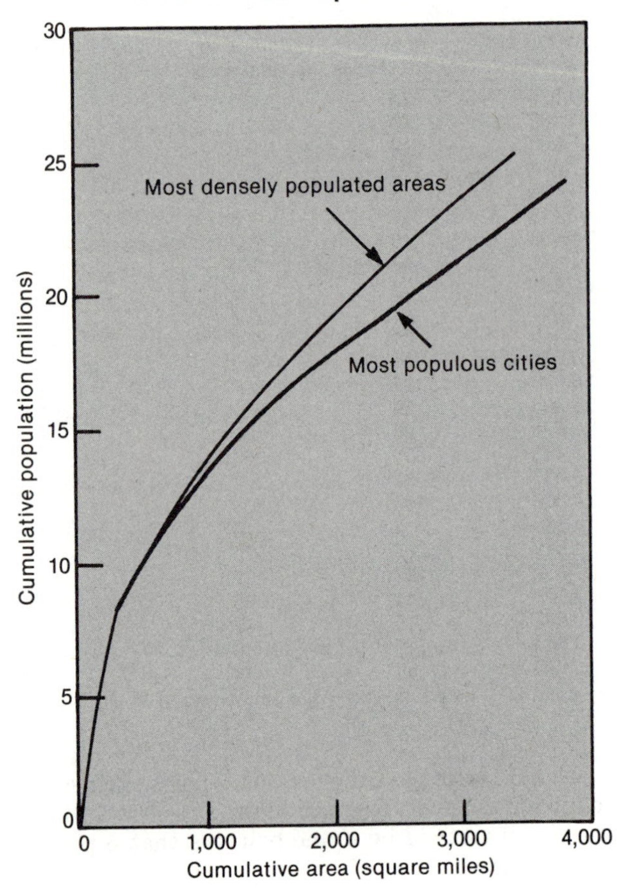

Note: Curves have been smoothed.

SOURCE: Office of Technology Assessment.

Figure D-2.—U.S. Population at Risk to a Soviet Attack Designed to Maximize Casualties

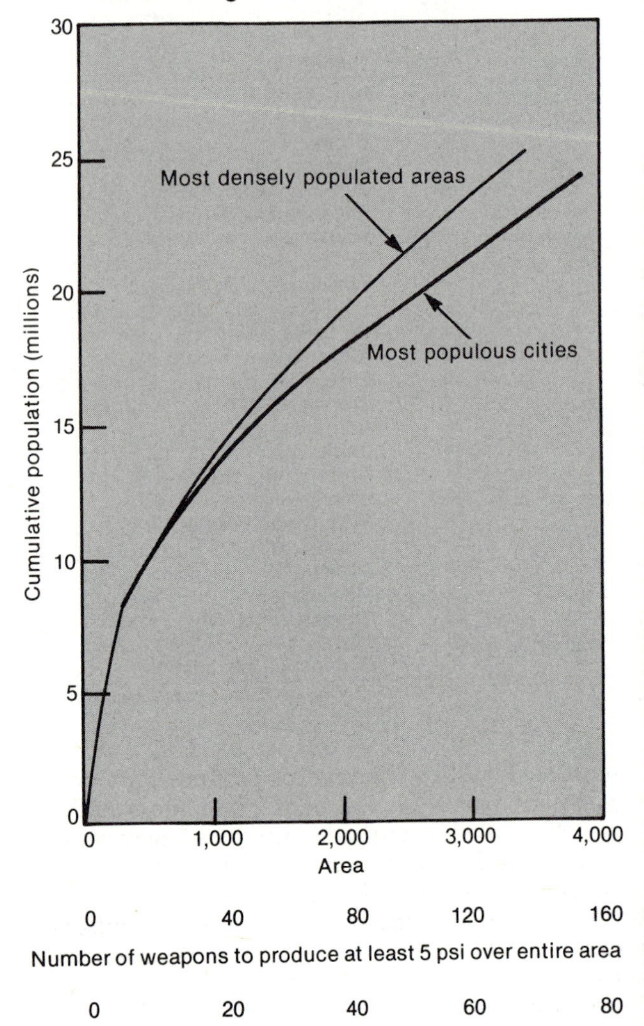

SOURCE: Office of Technology Assessment.

Figure D-3.—Potential Casualties From a 9,000 RV Soviet Attack Designed to Maximize Casualties: Attack Designed for 3 PSI Over Target Areas

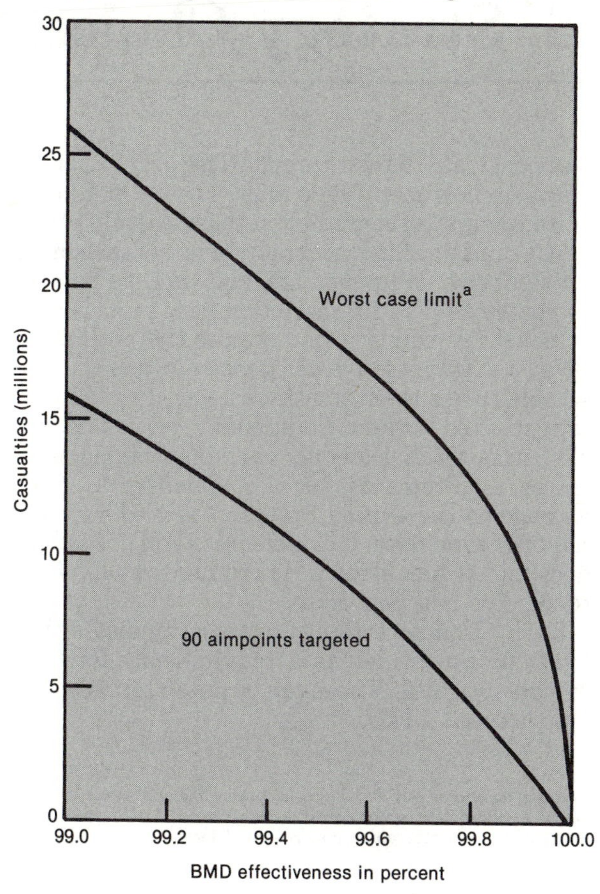

[a]Soviets know exactly how capable U.S. defense is and target accordingly. They hit every city they target.

SOURCE: Office of Technology Assessment.

Figure D-4.—Potential Casualties From a 9,000 RV Soviet Attack Designed to Maximize Casualties: Attack Designed for 5 PSI Over Target Areas

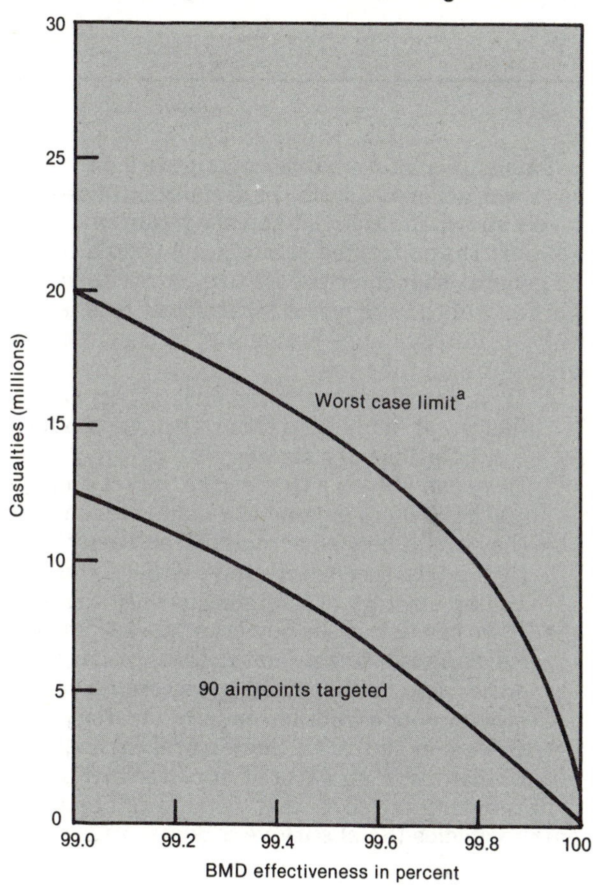

[a]Soviets know exactly how capable U.S. defense is and target accordingly. They hit every city they target.

SOURCE: Office of Technology Assessment.

Appendix E

Defense Capability Levels and U.S. Strategy Choices

This appendix describes how figure 5-4 in chapter 5 was generated from the definitions of defense levels shown in table E-1 and the requirements to support the suggested strategies as listed below. In general, the requirements to underwrite or support any of the suggested strategies—retaliation-only, countervailing, prevailing, or assured survival—are as follows:

- If we can absorb a Soviet first strike and inflict great damage on them, then we can have a **retaliation-only strategy**.
- If we can absorb a first strike, inflict damage on the Soviets beyond the value of whatever they might hope to accomplish and deny them their goals, then we can have either a **countervailing strategy** or a **retaliation-only strategy**.
- If we can defeat the Soviets while keeping our losses at a "tolerable" level, then we can adopt either a **prevailing strategy**, a **countervailing strategy**, or a **retaliation-only strategy**
- If we can survive a Soviet first strike, then we can have an **assured survival strategy**.

Retaliating requires that some number of reentry vehicles (RVs) survive a Soviet first strike and penetrate to their targets. There are different views on how many RVs must survive and penetrate to support a credible retaliation-only strategy. Countervailing generally requires that more RVs survive and penetrate, and that we be able to use those RVs for more than just punishment attacks. Prevailing would require that still more RVs be able to survive and penetrate, and that we be able to use them to attack a variety of important selected targets. Additionally, prevailing, unlike either retaliation-only or countervailing, generates requirements for U.S. defenses to limit damage to the United States.[1] Assured survival requires even more U.S. defenses than prevailing does, but it has little or no requirement for RVs to survive and penetrate.

Each of these strategy choices implies either limits on Soviet defenses, requirements for U.S. defenses, or both. These can be put in terms of the four defense levels.

[1]In the absence of Soviet defense, countervailing and retaliation-only do not require defenses, but do not exclude them either. If the Soviets have defense, countervailing may require defense.

Table E-1.—Levels of Defense Capability

Region	Level	Description
Offense-dominated	0 no defense	
	1 "some ICBMs"*	A defense capable of ensuring the survival of a useful fraction of the ICBMs, but not capable of protecting cities
Transition	2 "either/or"	A defense (including BMD) that can ensure the survival of most ICBMs or a high degree of urban survival against a follow-on (or simultaneous) attack, but not both
Defense-dominated	3 "most ICBMs/some cities"	A defense that ensures a high level of survival of military targets. Massive damage can only be obtained by concentrating the entire offense against cities
	4 "extremely capable"	Ensures a high level of urban survival against a full attack. The attacker cannot have high confidence that any cities can be destroyed

*Terms in quotes are a shorthand used to identify the levels.

NOTE: For simplicity the chapter often divides targets into ICBMs and cities. There are, of course, many other types of targets that might be attacked, but discussing them all in each case would greatly expand the text. ICBMs are representative of strategic military targets (although by no means an accurate model of them all). "Cities" is typically used as a short hand for people, economic assets, and social structure. A level 1 defense, for example, might be used to defend the C^3 system rather than the ICBMs.

The requirements for assured survival are the simplest to specify. We would need a level 4 defense regardless of what defense the Soviets built. Furthermore, if we had a level 4 defense that supported an assured survival strategy, no Soviet defense could undermine that strategy.

Prevailing would require either a level 4 defense, or, if we were to plan to strike first to reduce the Soviet offensive forces, a level 3 defense.[2] However, this U.S. defense alone would not assure an option to prevail. If the Soviet defense were sufficiently capable, it could keep us from having enough RVs surviving and penetrating to support a prevailing strategy. Level 4 Soviet defenses would certainly keep us from having a prevailing strategy. Very few U.S. RVs would reach their targets. A level 3 Soviet defense would keep us from attacking military targets, which would prevent us from satisfying the definition of prevailing. However, if the Soviets had level 3 and we had level 4, we would have a large, possibly exploitable, advantage. This might be called an opportunity to prevail despite not being able to destroy military targets. What we would call our strategy would not be as significant as the large advantage.

If we had a level 3 defense we could strike first against a range of targets and defend against the ragged retaliation, thereby limiting damage to ourselves, perhaps enough to prevail. However, if the Soviets had a level 2 defense, they might prevent us from destroying enough of their forces to keep our losses to their retaliation "tolerable." Therefore, if we had level 3 and they had level 2, prevailing might not be a practical option. We could, however, countervail since they could not destroy our retaliatory forces in a first strike.

With the exceptions noted above, a Soviet defense at level 3 or above would limit our strategy choice to retaliation-only, unless we had a level 4 defense that would support an assured survival strategy. The only targets we could expect to destroy would be cities. A Soviet level 4 defense would call into question our ability to retaliate. We could not be certain that we could inflict great damage on their cities. Therefore, if they had a level 4 defense and we did not have assured survival, our only option would be a retaliation-only strategy, but it might not have much prospect of being successful. Of course, if we also had a level 4 defense, neither side could be certain of its ability to damage the other, and the Soviets would have no advantage over us.

Currently, we have a countervailing strategy. If we add defense and the Soviets have none, we could certainly continue to have this option. If the Soviets add a level 1 defense while the United States has no defense, they could use it to deny us the ability to retaliate against some military targets, although we could still attack their cities. We might no longer be able to countervail, although we could certainly retaliate. A U.S. level 1 defense would restore our ability to countervail by ensuring the survival of RVs to replace the ones the Soviet defense might destroy. Higher levels of Soviet defense, however, could deny us the option to countervail by protecting a range of military and civilian targets.

[2]If we had level 2 and the Soviets had no defense, we might attempt to prevail by striking first. In order to defend our cities, we would have to leave military targets undefended against a Soviet retaliation.

Appendix F

BMD and the Military R&D Budget

Funding Levels

Military research and development currently constitutes about two-thirds of all Federal spending on R&D. In fiscal year 1985, $34.7 billion of the $52.0 billion total appropriated by Congress for R&D went to defense-related activities in the Departments of Defense and Energy.[1]

Ballistic missile defense technologies have been investigated with Department of Defense R&D funds since the 1950s. BMD funding (including deployment costs) reached its highest level (in real dollars) in fiscal year 1972. After the ABM Treaty was signed, overall BMD funding dropped.

In the fiscal year 1984 budget, the most recent one submitted before the Strategic Defense Initiative Organization was formed, just under $1 billion was appropriated for BMD research. According to Defense Department funding projections at that time, BMD programs within DOD would have been allocated about $12 billion for fiscal years 1984 through 1989 had the SDI not been formed, with about $2 billion more to have been spent by the Department of Energy. Under the SDI funding profile, DOD projections for the same period totaled $26 billion.[2] A more recent projection, which included an estimate for fiscal year 1990 and also accounted for congressional action on the fiscal year 1985 request, gave a $33 billion total for SDI for the seven fiscal years between fiscal years 1984 and 1990, inclusive.[3]

However, the periods covered by the above estimates are artifacts of the Pentagon's planning process. The budget for each fiscal year includes projections for the four subsequent years as well, but provides no information beyond that. Ambassador Paul Nitze, special advisor to the President and Secretary of State on arms control, has estimated that it will take "at least 10 years" (e.g., not before 1995) to determine whether a ballistic missile defense can meet the tests of survivability and cost-effectiveness.[4] The program would take even longer if slowed by unanticipated difficulties or congressional budget cuts. The total 10-year SDI cost from fiscal year 1984 through 1993 has been estimated to be $70 billion in nominal (uncorrected for inflation) dollars if it were to proceed on budget and on schedule; delays and overruns would further increase this total.[5]

Defense Department R&D Budget Categories

The Department of Defense places R&D programs in five categories ranging from basic research to engineering and operational development. *Basic research*, category 6.1, is scientific

---

[1]Aggregate figures from table I of Willis Shapley, Albert Teich, and Jill Pace, *Congressional Action on R&D in the FY 1985 Budget* (Washington, DC: R&D Budget and Policy Project of the American Association for the Advancement of Science in cooperation with the Intersociety Working Group, November 1984). The R&D budget as requested by the President, before being acted on by Congress, is analyzed in *AAAS Report IX: Research and Development, FY 1985 Budget*, by the Intersociety Working Group, 1984. Data for this publication is drawn from *Special Analysis K*, prepared by the Office of Management and Budget, and from *Federal Funds*, prepared by the National Science Foundation.

[2]"Analysis of the Costs of the Administration's Strategic Defense Initiative, 1985-1989," Congressional Budget Office, May 1984. For planning purposes, the Department of Defense annually prepares Five-Year Defense Plans which project future spending levels. However, the CBO report based its projections beyond fiscal year 1986 on press reports, since DOD typically does not provide its "out-year" projections to Congress.

[3]Departing from usual practice, in 1985 the Office of the Secretary of Defense (Comptroller) released out-year SDI budget projections to Congress. The Strategic Defense Initiative Organization has also compiled a corresponding breakdown of what would have been requested over that time period had the SDI not been formed. The two budget requests for fiscal years 1984 through 1990, in millions of dollars, are:

	SDI	Pre-SDI	
Fiscal year	*(DOD)*	*(DOD)*	*(DOE)*
1984	991	991	NA
1985	1,397	1,527	210
1986	3,722	1,802	295
1987	4,908	2,181	365
1988	6,165	2,699	439
1989	7,300	2,982	505
1990	8,634	NA	NA
Total	33,122	12,182	1,814

NA = not available. Note that the aggregate SDI figure is a 6-year sum whereas the aggregate pre-SDI figure is only over 5 years. Fiscal year 1984 and 1985 figures for the SDI budget represent appropriated, not requested, funds.

[4]Quoted by Walter Pincus in "Decade of Study Seen for 'Star Wars'," *The Washington Post*, Apr. 27, 1985.

[5]John Pike, *The Strategic Defense Initiative: Budget and Program* (Washington, DC: Federation of American Scientists, Feb. 10, 1985), pp. 81-83. The FAS projection of $69 billion fell in between a $66 billion estimate by the Electronic Industries Association and an $80 billion estimate by DMS, a private market research firm. Budget estimates given in constant fiscal year 1986 dollars are 86 percent of the those cited here, which are in current-year (nominal) dollars.

FAS cites Norman Augustine, President of Martin Marietta Denver Aerospace, as having frequently noted that defense development projects typically take one-third longer and cost one-third more than initially estimated. This factor moves the FAS research and technology development estimate of $70 billion closer to $100 billion.

study and experimentation directed toward acquiring knowledge in those fields relating to long-term national security needs. Category 6.2, *exploratory development*, refers to research directed towards solving specific military problems, short of major development projects. It includes development of "brass-board" level hardware, intended to validate design concepts, which need not be to scale and which do not meet operational specifications.

Advanced development, category 6.3, consists of projects which have moved into the development of prototypes in field configuration for technical or operational testing. Projects at this stage are evaluated for their suitability for military use. Category 6.4 is *engineering development*, and refers to systems meeting all military specifications for operational use which are destined for production in the very near term. (If approved for production, weapons systems leave the R&D budget.) The fifth category, 6.5, is for management support, and it funds installations required for general R&D use (i.e., testing ranges). Operational systems development, or development conducted on systems which have already been deployed, does not appear as a "6" category but is funded by line items elsewhere in the budget.

The five program elements constituting the Strategic Defense Initiative were created by consolidating portions of 27 previous program elements spanning the range from 6.1 to 6.5.[6] The resulting aggregates are therefore difficult to categorize, and they have been placed by DOD more or less arbitrarily in category 6.3, advanced development.

For the Department of Defense as a whole, only a small fraction of R&D funds (2.5 percent in the fiscal year 1985 request) was requested for basic research programs in category 6.1. Another 6.5 percent was requested for applied research, which is considered to be category 6.2 plus a portion "6.3A" of category 6.3 programs in the early stages of advanced development. The overwhelming majority of DOD's R&D funds, 90 percent in fiscal year 1985, were requested for development activities in category 6.4 and the remainder of category 6.3, with the final 1 percent being requested for R&D facilities.[7]

Department of Energy

Within the Department of Energy, $2.2 billion was allocated for defense-related R&D programs in fiscal year 1985. These programs include all development of nuclear weapons, along with other items such as naval nuclear reactor development.[8] (Dual military/civilian nuclear power programs, such as nuclear power for space systems, are funded elsewhere in the DOE budget and are not included in this total.) While the Department of Energy conducts research relevant to strategic defense, those programs are not formally part of the Strategic Defense Initiative.

[6] "Analysis of the Costs of the Administration's Strategic Defense Initiative, 1985-1989," Congressional Budget Office, May 1984

[7] *AAAS Report IX*, note 1 above, table II-4. These figures are for the budget requested by the President; similar breakdowns for the budget as enacted by Congress were not available but would differ only marginally.

[8] DOE does not use the 6.1 to 6.5 budget categories used by the Department of Defense, and a breakdown of DOE R&D funds equivalent to the one above for DOD is not available. However, comparing table I-7 (total defense expenditures) of AAAS Report IX (*supra*, note 1 above) with table II-4 (Department of Defense expenditures) yields a somewhat comparable analysis. These figures are for the fiscal year 1985 budget as submitted by the President; no such figures were compiled for the budget as enacted by Congress:

Basic Research	$0
Applied Research	0.7 billion
Development	1.3 billion
Facilities	0.5 billion

Although this analysis shows that no DOE funds were attributed for basic research in defense-related programs, $0.9 billion was allocated for basic research elsewhere within DOE (table I-9 of the AAAS Report).

Appendix G

Studies of the High Frontier Global Ballistic Missile Defense I

Deployment Costs

High Frontier[1] asserted that its GBMD I constellation of BMD satellites carrying kinetic-kill vehicles could be built using "off-the-shelf" technology and could be "fully deployed in five or six years at a minimum cost of some $10-$15 billion."[2]

However, the Department of Defense obtained a much higher estimate. Shortly after the High Frontier report was published, Dr. Robert Cooper, director of the Defense Advanced Research Projects Agency, commented on the High Frontier proposal for a subcommittee of the Senate Committee on Armed Services:

> . . . The DOD has worked with the High Frontier analysts throughout the development of their concept and supports the basic Damage-Denial goal. However, as hardware developers of war fighting systems, we do not share their optimism in being able to develop and field such a capability within their timeframe and cost projections. We have conducted several in-house analyses and have experienced some difficulties in ratifying the existence of "off-the-shelf components or technologies" to provide the required surveillance, command and control, and actually perform the intercepts within the orbital and physical conditions described. Our understanding of the systems implications and costs would lead us to project expenditures on the order of $200 to $300 billion in acquisition costs alone for the proposed system.[3]

A year later, John Gardner, Director for Defensive Systems in the Office of the Under Secretary of Defense for Research and Engineering, described some of the DOD analyses before a subcommittee of the Senate Committee on Armed Services:

> We have conducted studies, both in the Army and in the Air Force, on the High Frontier concept as we understand it. Generally, those studies were associated with understanding the concept, identifying the technical issues and risks; doing work to optimize the system and estimating the cost for a deployed system as well as its survivability.
>
> ***
>
> While we believe that the technical capabilities of the system are certainly appropriately described by the High Frontier, we do have some reservations about the survivability of a system of the kind that has been described.
>
> ***
>
> We have looked at system components in some detail. I would say that of the various elements of the system—the spacecraft, the search and acquisition system, and the interception system—I believe that the judgment is that the highest risk would exist in the interception system and in the command and control that would be required to drive and control the whole system.
>
> ***
>
> We have made estimates of the cost of such a system, using the costing techniques that are common to the Department of Defense for both defensive systems, space launches, and satellite systems. It is on the basis of the cost estimates that estimates have been made ranging from $50 to $60 billion, and to numbers considerably in excess of that . . .[4]

R&D Costs

Since the initial High Frontier deployments were assumed not to require much further technical development, High Frontier estimated that research and development of the entire GBMD I system would cost only $1 billion.[5] This estimate for developing the **entire** GBMD I system can be compared with $1.275 billion (in 1982 dollars) that the Air Force plans to have allocated over the 19 fiscal years from 1972 to 1990 to develop the air-launched and infrared-guided Miniature Vehicle

[1]Daniel O. Graham, *High Frontier, A New National Strategy* (Washington, DC: High Frontier, 1982).

[2]Ibid., p. 8. Estimates by High Frontier of the cost of its entire program, including terminal defense, two layers of space-based defense, improved space transportation, a space station, a high performance spaceplane, and a satellite power system, total $24 billion in the first 5 years and $40 billion for the first 8 years. The GBMD I portion along, the first space-based layer, is estimated to cost $10 to $15 billion.

[3]Response to question submitted for the record following Dr. Cooper's Mar. 10, 1982, appearance before the Subcommittee on Strategic and Theater Nuclear Forces of the Senate Committee on Armed Services. Hearings on S. 2248, DOD Authorization for Appropriations for Fiscal Year 1983, Part 7, p. 4635.

[4]Hearings before the Strategic and Theater Nuclear Forces subcommittee of the Senate Committee on Armed Services, Mar. 23, 1983, printed in S. Hrg. 98-49, Part 5, Department of Defense Authorization for Appropriations for Fiscal Year 1984, Senate Committee on Armed Services, p. 2668-9.

[5]Graham, op cit., p. 128.

ASAT weapon.[6] However, the ASAT weapon is roughly equivalent to the GBMD I **kill vehicle alone,**[7] and its development would not address the other requirements of the GBMD I system: the carrier satellites, the system wide surveillance, acquisition, and kill assessment sensors, overall command and control, battle management software and hardware, on-orbit transportation and logistical support, and system survivability.

Moreover, technology developed for the MV ASAT would most likely not be sufficient even for the GBMD kill vehicle it corresponds to. The ASAT is designed to find satellites against the cold background of space; GBMD kill vehicles must operate looking down against a warm earth. The booster's exhaust plume, of course, is much hotter than either the booster or the earth background, but a booster cannot be killed by attacking its exhaust. Either the kill vehicle will need to track some part of the booster itself, or it will need to know where the booster is relative to the exhaust—both of which are more difficult tasks than locating an isolated satellite with nothing behind it. In many cases, the kill vehicle would not be able to reach the booster before burnout, obviating plume tracking.

Area of Coverage

In a background paper done for OTA,[8] Ashton Carter analyzed the High Frontier GBMD I system. He states that, due to the slow speed (1 km/sec) of the individual kill vehicles, they would not be able to travel very far during the boost phase of a Soviet ICBM. "[T]housands of satellites would be needed worldwide for continuous coverage of Soviet ICBM fields," wrote Carter. "The High Frontier concept with only 432 satellites would therefore have meager coverage of Soviet ICBM fields."[9] He noted that in the only example of boost-phase intercept given in the High Frontier report,[10] the kill vehicle would have been launched 53 seconds **before** its target ICBM was launched, with no explanation of how the defense would know **in advance** when that launch would occur. Although the High Frontier study also discusses interceptions during the post-boost period, the kill vehicle sensors postulated for the initial deployment (GBMD I) would not be appropriate for phases following boost phase. Carter's overall conclusion was that

> It would therefore appear that the technical characteristics of the High Frontier scheme result in a defensive system of extremely limited capability for boost phase intercept of present Soviet ICBMs and with no capability against future MX-like Soviet boosters, even with no Soviet effort to overcome the defense.[11]

Overall Capability

On February 21, 1985, 2 years after John Gardner had testified about High Frontier before a Senate Armed Services subcommittee, Strategic Defense Initiative Organization Director James A. Abrahamson appeared before the same subcommittee. Senator Sam Nunn asked Abrahamson about a recent High Frontier claim that a 95 percent-effective defense could be built using off-the-shelf technology in 5 to 6 years.

Abrahamson did not substantiate that assertion.[12] He stated that the kinetic-kill vehicle concept adopted by High Frontier is considered by SDI to be one of the more mature BMD technologies, and that it could provide part of an initial, partial capability for boost-phase intercept against current ballistic missiles. However, later on in his testimony, Abrahamson indicated that the kinetic-kill vehicles considered by the Strategic Defense Initiative would be solid-propelled rockets traveling five to eight times faster than the High Frontier design. Abrahamson also emphasized that it takes more than a weapons concept to make an overall ballistic missile defense. The full job requires tracking, surveillance, and command and control, and is a more complex issue than was implied by the High Frontier publication that Senator Nunn referred to.

In his testimony, Abrahamson highlighted the basic difference between the High Frontier approach and that being pursued by the SDI. He said

[6]From the Dec. 31, 1983 Comprehensive Selected Acquisition Report (SAR) on Space Defense and Operations (ASAT). Figures are corrected from the total given in the SAR (in 1977 dollars) to 1982 dollars using the DOD Air Force RDT&E deflator.

[7]According to John Gardner's testimony, "the space-based component of [the High Frontier GBMD I defense] does postulate the use of defensive interceptors that take advantage of the technology that is currently being developed as part of the anti-satellite program."

—Senate Armed Services Committee Hearings, Mar. 23, 1983, p. 2667.

[8]Ashton Carter, *Directed Energy Missile Defense in Space,* background paper prepared under contract for the Office of Technology Assessment, April 1984.

[9]Ibid., p. 35.

[10]Graham, op cit., p. 122.

[11]Carter, op cit., p. 35. According to the Department of Defense, the Soviets are currently developing an ICBM, the SS-X-24, which is similar in many characteristics to the MX. (*Soviet Military Power, 1985.* See pp. 29-30 for data on U.S. and Soviet ICBMs.)

[12]Testimony given below is paraphrased from General Abrahamson's spoken testimony.

that there were two dangerous consequences that could happen should the United States deploy only a partial BMD capability and then stop. First, it might drive the Soviets in precisely the wrong direction, stimulating them to build up offensive forces if they thought that they could overwhelm the defense. Second, if the U.S. system were based on a single concept, the entire system would be vulnerable should the Soviets discover a countermeasure to that concept.

Alluding to the survivability problems mentioned about High Frontier by Gardner 2 years earlier, Abrahamson explained that the SDI did not yet know enough to be confident that a High Frontier-type system could not be countered or easily knocked out. He noted that General Graham did not have the resources available to him to investigate all the countermeasure problems and the command and control difficulties. The High Frontier program could be a good start, Abrahamson said, but he did not know if it would be the best start. At present, he would not recommend that the United States proceed to deploy it.

Appendix H

Excerpts From Statements on BMD by Reagan Administration Officials

The conclusion of President Reagan's March 23, 1983, speech on Defense Spending and Defensive Technology.

Weekly Compilation of

Presidential Documents

Now, thus far tonight I've shared with you my thoughts on the problems of national security we must face together. My predecessors in the Oval Office have appeared before you on other occasions to describe the threat posed by Soviet power and have proposed steps to address that threat. But since the advent of nuclear weapons, those steps have been increasingly directed toward deterrence of aggression through the promise of retaliation.

This approach to stability through offensive threat has worked. We and our allies have succeeded in preventing nuclear war for more than three decades. In recent months, however, my advisers, including in particular the Joint Chiefs of Staff, have underscored the necessity to break out of a future that relies solely on offensive retaliation for our security.

Over the course of these discussions, I've become more and more deeply convinced that the human spirit must be capable of rising above dealing with other nations and human beings by threatening their existence. Feeling this way, I believe we must thoroughly examine every opportunity for reducing tensions and for introducing greater stability into the strategic calculus on both sides.

One of the most important contributions we can make is, of course, to lower the level of all arms, and particularly nuclear arms. We're engaged right now in several negotiations with the Soviet Union to bring about a mutual reduction of weapons. I will report to you a week from tomorrow my thoughts on that score. But let me just say, I'm totally committed to this course.

If the Soviet Union will join with us in our effort to achieve major arms reduction, we will have succeeded in stabilizing the nuclear balance. Nevertheless, it will still be necessary to rely on the specter of retaliation, on mutual threat. And that's a sad commentary on the human condition. Wouldn't it be better to save lives than to avenge them? Are we not capable of demonstrating our peaceful intentions by applying all our abilities and our ingenuity to achieving a truly lasting stability? I think we are. Indeed, we must.

After careful consultation with my advisers, including the Joint Chiefs of Staff, I believe there is a way. Let me share with you a vision of the future which offers hope. It is that we embark on a program to counter the awesome Soviet missile threat with measures that are defensive. Let us turn to the very strengths in technology that spawned our great industrial base and that have given us the quality of life we enjoy today.

What if free people could live secure in the knowledge that their security did not rest upon the threat of instant U.S. retaliation to deter a Soviet attack, that we could intercept and destroy strategic ballistic missiles before they reached our own soil or that of our allies?

I know this is a formidable, technical task, one that may not be accomplished before the end of this century. Yet, current technology has attained a level of sophistication where it's reasonable for us to begin this effort. It will take years, probably decades of effort on many fronts. There will be failures and setbacks, just as there will be successes and breakthroughs. And as we proceed, we must remain constant in preserving the nuclear deterrent and maintaining a solid capability for flexible response. But isn't it worth every investment necessary to free the world from the threat of nuclear war? We know it is.

In the meantime, we will continue to pursue real reductions in nuclear arms, negotiating from a position of strength that can be ensured only by modernizing our strategic forces. At the same time, we must take steps to reduce the risk of a conventional military conflict escalating to nuclear war by improving our non-nuclear capabilities.

America does possess—now—the technologies to attain very significant improvements in the effectiveness of our conventional, non-nuclear forces. Proceeding boldly with these new technologies, we can significantly reduce any incentive that the Soviet Union may have to threaten attack against the United States or its allies.

As we pursue our goal of defensive technologies, we recognize that our allies rely upon our strategic offensive power to deter attacks against them. Their vital interests and ours are inextricably linked. Their safety and ours are one. And no change in technology can or will alter that reality. We must and shall continue to honor our commitments.

I clearly recognize that defensive systems have limitations and raise certain problems and ambiguities. If paired with offensive systems, they can be viewed as fostering an aggressive policy, and no one wants that. But with these considerations firmly in mind, I call upon the scientific community in our country, those who gave us nuclear weapons, to turn their great talents now to the cause of mankind and world peace, to give us the means of rendering these nuclear weapons impotent and obsolete.

Tonight, consistent with our obligations of the ABM treaty and recognizing the need for closer consultation with our allies, I'm taking an important first step. I am directing a comprehensive and intensive effort to define a long-term research and development program to begin to achieve our ultimate goal of eliminating the threat posed by strategic nuclear missiles. This could pave the way for arms control measures to eliminate the weapons themselves. We seek neither military superiority nor political advantage. Our only purpose—one all people share—is to search for ways to reduce the danger of nuclear war.

My fellow Americans, tonight we're launching an effort which holds the promise of changing the course of human history. There will be risks, and results take time. But I believe we can do it. As we cross this threshold, I ask for your prayers and your support.

Thank you, good night, and God bless you.

Note: The President spoke at 8:02 p.m. from the Oval Office at the White House. The address was broadcast live on nationwide radio and television.

Following his remarks, the President met in the White House with a number of administration officials, including members of the Cabinet, the White House staff, and the Joint Chiefs of Staff, and former officials of past administrations to discuss the address.

In a speech on March 29, 1985, President Reagan said:[1]

. . . Two years ago, I challenged our scientific community to use their talents and energies to find a way that we might eventually rid ourselves of the need for nuclear weapons—starting with ICBMs. We seek to render obsolete the balance of terror—or Mutual Assured Destruction, as it's called—and replace it with a system incapable of initiating armed conflict or causing mass destruction, yet effective in preventing war. Now, this is not and should never be misconstrued as just another method of protecting missile silos.

. . . The means to intercept ballistic missiles during their early-on boost phase of trajectory would enable us to fundamentally change our strategic assumptions, permitting us to shift our emphasis from offense to defense.

. . . We're not discussing a concept just to enhance deterrence, but rather a new kind of deterrence; not just an addition to our offensive forces, but research to determine the feasibility of a comprehensive nonnuclear defensive system—a shield that could prevent nuclear weapons from reaching their targets.

[1]Speech to the National Space Club, Mar. 29, 1985.

The Administration has not presented in detail its view of how it thinks the U.S./U.S.S.R. strategic relation would evolve as BMD developments proceed and efforts are made to manage the evolution. Administration spokesmen have, however, given broad descriptions of the major parts of that evolution and the reasons why they believe it to be plausible. Some of these are excerpted in this appendix. For a deeper understanding, the reader should read the sources in their entirety.[2]

In a statement released January 3, 1985, and published in a White House pamphlet *The President's Strategic Defense Initiative*, President Reagan said:[3]

> . . . The SDI research program will provide to a future President and a future Congress the technical knowledge required to support a decision on whether to develop and later deploy advanced defensive systems.
>
> At the same time, the United States is committed to the negotiation of equal and verifiable agreements which bring real reductions in the power of the nuclear arsenals of both sides. To this end, my Administration has proposed to the Soviet Union a comprehensive set of arms control proposals. We are working tirelessly for the success of these efforts, but we can and must go further in trying to strengthen the peace.
>
> Our research under the Strategic Defense Initiative complements our arms reduction efforts and helps to pave the way for creating a more stable and secure world. That the research we are undertaking is consistent with all of our treaty obligations, including the 1972 Anti-Ballistic Missile Treaty.
>
> In the near term, the SDI research program also responds to the ongoing and extensive Soviet anti-ballistic missile (ABM) effort, which includes actual deployments. It provides a powerful deterrent to any Soviet decision to expand its ballistic missile defense capability beyond that permitted by the ABM Treaty. **And, in the long-term, we have confidence that SDI will be a crucial means by which both the United States and the Soviet Union can safely agree to very deep reductions, and eventually, even the elimination of ballistic missiles and the nuclear weapons they carry.** [emphasis added]

The White House publication which accompanied this statement elaborated on the arms control implications of the Strategic Defense Initiative as follows:[4]

> The United States does not view defensive measures as a means of establishing military superiority. Because we have no ambitions in this regard, deployments of defensive systems would most usefully be done in the context of a cooperative, equitable, and verifiable arms control environment that regulates the offensive and defensive developments and deployments of the United States and Soviet Union. Such an environment could be particularly useful in the period of transition from a deterrent based on the threat of nuclear retaliation, through deterrence based on a balance of offensive and defensive forces, to the period when adjustments to the basis of deterrence are complete and advanced defensive systems are fully deployed. During the transition, arms control agreements could help to manage and establish guidelines for the deployment of defensive systems.
>
> The SDI research program will complement and support U.S. efforts to seek equitable, verifiable reductions in offensive nuclear forces through arms control negotiations. Such reductions would make a useful contribution to stability, whether in today's deterrence environment or in a potential future deterrence environment in which defenses played a leading role.
>
> A future decision to develop and deploy effective defenses against ballistic missiles could support our policy of pursuing significant reductions in ballistic missile forces. To the extent that defensive systems could reduce the effectiveness and, thus, value of ballistic missiles, they also could increase the incentives for negotiated reductions. Significant reductions in turn would serve to increase the effectiveness and deterrent potential of defensive systems.

This prediction has been explained by George A. Keyworth II, science advisor to President Reagan, in the following terms:[5]

> Strategic defenses of the type we can reasonably project—even in their early modes—can be vital catalysts for arms control . . . In fact, early and intermediate defenses will undoubtedly be imperfect, and any nuclear weapon that makes it through to its target will be devastating. While hardened military assets can be very successfully defended by these transition systems, civilian population centers will still be hostage to a determined adversary. Critics cite this as a major failing. In fact, it is crucial to stability during those transition years, because as long as there is some leakage in those transition defense technologies, there remains a retaliatory deterrent against first strike.
>
> . . . But we will once again have a common ground for negotiating real weapons reductions. After all, realistic, survivable, retaliatory arsenals do not have to be enormous, not nearly as large

[2]A list of statements and articles on BMD by Administration spokesmen appears in app. I.

[3]*The President's Strategic Defense Initiative* (Washington, DC: The White House, January 1985), GPO:1985 O-465-450:QL 3, p. i.

[4]Ibid., pp. 5-6.

[5]George A. Keyworth II, "The Case For: An Option for a World Disarmed," *Issues in Science and Technology*, fall 1984, pp. 42-44.

as the arsenals we now require to survive preemptive strikes (or in the Soviet case, to launch them). With the preemptive option clouded, or even removed, we would have an opportunity to negotiate major arms reductions that would still leave each side with a strong retaliatory deterrent.

At that point we would have accomplished two things, two goals that have eluded us for 20 years. We would have reduced both nations' perceptions that the other could launch a successful disarming first strike, and we would have drastically reduced the size of the arsenals.

. . . These options will probably become available when the strategic nuclear forces we must build today to maintain our near-term deterrence reach the limits of their operational lifetimes. We then have a new option: rather than replace them, let each side retain only token nuclear forces for their sole remaining purpose—restricted retaliation.

It is only at this point, in the presence of near-zero arsenals, that arms control begins to have any real meaning in the minds of ordinary people. Only when the prospect of final world holocaust reverts to "mere" catastrophe—that is, when the stockpiles can be measured in the dozens, rather than in the tens of thousands—can we once again depend on the sun coming up the next day.

. . . Soviet habits, attitudes, and policies are the product of a thousand years of brutal historical experience. There is no reason to believe that the Soviet Union will suddenly become a country that we would trust to respect the legal requirements of a near-total disarmament treaty.

. . . Strategic defense provides the option to break this cycle. Although we cannot disinvent nuclear weapons, and although nations will continue to distrust one another, heavily defended countries could nonetheless realistically enter into treaties to reduce nuclear forces to near zero. The scale of cheating necessary to provide an arsenal capable of successfully engaging several layers of active defenses would be so large as to be impractical within the context of normal intelligence-gathering capabilities.

Strategic defense therefore provides an option for a world effectively disarmed of nuclear weapons, yet still retaining national sovereignty and security. In fact, deployment of strategic defense is the only way in which the superpowers will be able to achieve these very deep arms reductions.

In another article[6] he wrote:

When [the Soviets] look seriously at the loss of utility of their ICBMs as a preemptive force, they will have no choice but to admit that the age of the ICBM as *the* dominant weapon is passing. They, and we, will no doubt begin to replace ICBMs with other weapons, but in so doing we will be phasing out the most feared and most destabilizing of the nuclear weapons. This is the key issue and, to my mind, the strongest reason we have to pursue the strategic defense initiative. With the ICBM tarnished and with the need to look to other options to preserve national security, both the Soviets and we will have a mutual basis to negotiate reductions in ICBM forces. If ICBMs serve only to retaliate in case the other side does attack first, then both sides can consider truly massive reductions in ICBM warheads. Ten or twenty nuclear weapons are virtually all the retaliatory deterrent that any country needs—and those are the levels of weapons that arms controls ought to be aiming for.

On February 13, 1985, the Director of the U.S. Arms Control and Disarmament Agency, Kenneth Adelman, told the International Institute for Strategic Studies:

[If SDI succeeds in making defenses more cost-effective than offenses], SDI can then prove a real incentive to deep reductions in offensive nuclear systems through arms control. We hope for that kind of incentive from SDI.

We must scrupulously guard against a vicious cycle of defensive efforts—even research for defense—spurring the other side onto more offensive weapons in order to saturate prospective defenses, and so on, and so on. That snowball effect would undercut stability and weaken deterrence.

That risk can be reduced and managed through the kind of overall strategic discussions Secretary Shultz launched in Geneva last month and that Ambassador Kampelman will take up further when the arms talks begin again next month. This type of exchange with the Soviet Union—an in-depth dialog about critical strategic relationships, strategic concepts, strategic stability—is indispensable to an effective SDI approach.

No one has a crystal ball in this complicated business. We need data to provide a sound basis for decisions several years off on whether or not to pursue strategic defensive systems further . . . [a] managed evolution—one involving the Soviets and the Allies intimately all along the way—could lead to a safer world.

***

Most broadly, we will be going 'back to basics' in looking at the relationship between offensive and defensive forces. We will be describing to the Soviets, in some detail and with some care, the kind of strategic concept that will guide us in the period ahead. We envision it as falling into three phases.

During the first phases, deterrence will continue to rest almost exclusively on offensive nuclear retaliatory capabilities. We believe that this can

[6]George A. Keyworth II, "The Case for Arms Control and the Strategic Defense Initiative," *Arms Control Today*, April 1985, p. 8.

be done at greatly reduced levels of nuclear forces and with full compliance with the ABM Treaty, and we will seek both. We hope the Soviets believe and will act likewise. This period could last ten or fifteen years, or longer or even indefinitely, depending largely on the progress and results of the on-going SDI research.

The second phase will be one of transition. During this period, and assuming successful development of some effective non-nuclear defensive systems, we would begin to move towards a strategic posture with ever-greater reliance on defense, rather than offense. A transition of indefinite duration, this period will help lay the technical and political groundwork necessary for the ultimate goal of eventually eliminating nuclear arms completely.

The last period is one with its hallmark being the complete elimination of nuclear arms. The technical knowledge of how to make these weapons and the danger of cheating would persist. These risks, unfortunately, can never be eliminated, but effective defenses would give insurance against them. The enormous and depressing nuclear threat hanging over the world could be lifted.

These three stages have to evolve gradually and, as I have said, depend critically upon a cooperative effort between the United States, in consultation with its key Allies, and the Soviet Union.

This theme was elaborated on by Ambassador Paul H. Nitze in a speech to the Philadelphia World Affairs Council on February 20, 1985. He summarized the strategic basis for the upcoming talks in Geneva as follows:

During the next ten years, the U.S. objective is a radical reduction in the power of existing and planned offensive nuclear arms, as well as the stabilization of the relationship between offensive and defensive nuclear arms, whether on earth or in space. We are even now looking forward to a period of transition to a more stable world, with greatly reduced levels of nuclear arms and an enhanced ability to deter war based upon an increasing contribution of non-nuclear defenses against offensive nuclear arms. This period of transition could lead to the eventual elimination of all nuclear arms, both offensive and defensive. A world free of nuclear arms is an ultimate objective to which we, the Soviet Union, and all other nations can agree.

He then went on to say:

It would be worthwhile to dwell on this concept in some detail. To begin with, it entails three time phases: the near term, a transition phase, and an ultimate phase.

The Near Term: For the immediate future—at least the next ten years—we will continue to base deterrence on the ultimate threat of nuclear retaliation. We have little choice; today's technology provides no alternative.

That being said, we will press for radical reductions in the number and power of strategic and intermediate-range nuclear arms. Offensive nuclear arsenals on both sides are entirely too high and potentially destructive, particularly in the more destabilizing categories such as the large MIRVed [multiple independently-targeted reentry vehicles] Soviet ICBM [intercontinental ballistic missile] and SS-20 forces.

At the same time, we will seek to reverse the erosion that has occurred in the Anti-Ballistic Missile (ABM) Treaty regime—erosion that has resulted from Soviet actions over the last ten years. These include the construction of a large phased-array radar near Krasnoyarsk in central Siberia in violation of the ABM Treaty's provisions regarding the location and orientation of ballistic missile early warning radars.

For the near term, we will be pursuing the SDI research program—in full compliance with the ABM Treaty, which permits such research. Likewise, we expect the Soviets will continue their investigation of the possibilities of new defensive technologies, as they have for many years.

We have offered to begin discussions in the upcoming Geneva talks with the Soviets as to how we might together make a transition to a more stable and reliable relationship based on an increasing mix of defensive systems.

The Transition Period: Should new defensive technologies prove feasible, we would want at some future date to begin such a transition, during which we would place greater reliance on defensive systems for our protection and that of our allies.

The criteria by which we will judge the feasibility of such technologies will be demanding. The technologies must produce defensive systems that are survivable; if not, the defenses would themselves be tempting targets for a first strike. This would decrease rather than enhance stability.

New defensive systems must also be cost effective at the margin—that is, it must be cheap enough to add additional defensive capability so that the other side has no incentive to add additional offensive capability to overcome the defense. If this criterion is not met, the defensive systems could encourage a proliferation of countermeasures and additional offensive weapons to overcome deployed defenses, instead of a redirection of effort from offense to defense.

As I said, these criteria are demanding. If the new technologies cannot meet these standards, we are not about to deploy them. In the event, we would have to continue to base deterrence on the ultimate threat of nuclear retaliation. However, we hope and have expectations that the scientific community can respond to the challenge.

We would see the transition period as a cooperative endeavor with the Soviets. Arms control would play a critical role. We would, for example, envisage continued reductions in offensive nuclear arms.

Concurrently, we would envisage the sides beginning to test, develop, and deploy survivable and cost-effective defenses at a measured pace, with particular emphasis on non-nuclear defenses. Deterrence would thus begin to rely more on a mix of offensive nuclear and defensive systems instead of on offensive nuclear arms alone.

The transition would continue for some time—perhaps for decades. As the U.S. and Soviet strategic and intermediate-range nuclear arsenals declined significantly, we would need to negotiate reductions in other types of nuclear weapons and involve, in some manner, the other nuclear powers.

The Ultimate Period: Given the right technical and political conditions, we would hope to be able to continue the reduction of nuclear weapons down to zero.

The global elimination of nuclear weapons would be accompanied by widespread deployments of effective non-nuclear defenses. These defenses would provide assurance that were one country to cheat—for example, by clandestinely building ICBMs or shorter-range systems, such as SS-20s—it would not be able to achieve any exploitable military advantage. To overcome the deployed defenses, cheating would have to be on such a large scale that there would be sufficient notice so that countermeasures could be taken.

Were we to reach the ultimate phase, deterrence would be based on the ability of the defense to deny success to a potential aggressor's attack. The strategic relationship could then be characterized as one of mutual assured security.

Ambassador Nitze then went on to say:

We would have to avoid a mix of offensive and defensive systems that, in a crisis, would give one side or the other incentives to strike first. That is precisely why we would seek to make the transition a cooperative endeavor with the Soviets. . .

In an interview with *U.S. News and World Report* printed March 18, 1985, National Security Adviser Robert McFarlane said:[7]

Now, there is a relationship between reductions of offensive systems and the integration of defensive systems because of the potentially destabilizing effect of either side achieving a first-strike capability through possession of both.

So our policy must be to first establish agreement between ourselves and the Russians on the value of defensive systems. Once we have reached agreement on that, then we must establish a path for the integration of these defensive systems into the force structure that will be stable.

In an interview on ABC Network television broadcast June 6, 1985, Secretary of Defense Caspar Weinberger said:

We're working for a program that could be a thoroughly reliable defense that could indeed give us the confidence that all of these missiles could be destroyed. But if we get only a partial result, it still will be very worthwhile.

In a speech on March 29, 1985, George Keyworth described the goal of the SDI as follows:

Is the SDI the means to protect people or to protect weapons? Protecting weapons represents **no** change in present policy. It simply strengthens—**entrenches**—the doctrine of Mutual Assured Destruction. Protecting people, on the other hand, holds out the promise of dramatic change.[8]

This clear purpose of the President has been repeated time and time again by Cap Weinberger, Bud McFarlane, and myself. But the ambiguity over SDI's real goal remains. It is fostered by three main tenets: First is the assertion, embraced by those anxious to protect both past strategic doctrine and future nuclear systems, that "strengthening deterrence" must be the **primary** goal for SDI. Second is that protecting weapons, especially ICBM silos, is the nearer-term and **most likely** goal for SDI. And third is that defense of European military targets against tactical ballistic missiles is the most **politically attractive** near-term goal for SDI.

If these arguments continue to be used as the basis to achieve Congressional and Allied support, I believe the opportunity for strategic change—and the President's objective—is lost.

***

Terminal defenses **within** the SDI also can play a very **real** part in an overall "layered" defense. But attempts to make terminal defense our first move, within the **SDI,** does **not** start us in the direction of the President's objective.

Following is the text of a "Fact Sheet" on the Strategic Defense Initiative, issued by The White House on June 1, 1985, and published by the Department of State:

[7] "Prospects Are Good for Arms Pact—But Not Soon," *U.S. News and World Report,* Mar. 18, 1985, pp. 24-25.

[8] "The President's Strategic Defense Initiative," remarks to the SDIO University Review Forum, Mar. 29, 1985.

Special Report No. 129

The Strategic Defense Initiative

June 1985

United States Department of State
Bureau of Public Affairs
Washington, D.C.

In his speech of March 23, 1983, President Reagan presented his vision of a future in which nations could live secure in the knowledge that their national security did not rest upon the threat of nuclear retaliation but rather on the ability to defend against potential attacks. The Strategic Defense Initiative (SDI) research program is designed to determine whether and, if so, how advanced defensive technologies could contribute to the realization of this vision.

The Strategic Context

The U.S. SDI research program is wholly compatible with the Anti-Ballistic Missile (ABM) Treaty, is comparable to research permitted by the ABM Treaty which the Soviets have been conducting for many years, and is a prudent hedge against Soviet breakout from ABM Treaty limitations through the deployment of a territorial ballistic missile defense. These important facts deserve emphasis. However, the basic intent behind the Strategic Defense Initiative is best explained and understood in terms of the strategic environment we face for the balance of this century and into the next.

The Challenges We Face. Our nation and those nations allied with us face a number of challenges to our security. Each of these challenges imposes its own demands and presents its own opportunities. Preserving peace and freedom is, and always will be, our fundamental goal. The essential purpose of our military forces, and our nuclear forces in particular, is to deter aggression and coercion based upon the threat of military aggression. The deterrence provided by U.S. and allied military forces has permitted us to enjoy peace and freedom. However, the nature of the military threat has changed and will continue to change in very fundamental ways in the next decade. Unless we adapt our response, deterrence will become much less stable and our susceptibility to coercion will increase dramatically.

Our Assumptions About Deterrence. For the past 20 years, we have based our assumptions on how deterrence can best be assured on the basic idea that if each side were able to maintain the ability to threaten retaliation against any attack and thereby impose on an aggressor costs that were clearly out of balance with any potential gains, this would suffice to prevent conflict. Our idea of what our forces had to hold at risk to deter aggression has changed over time. Nevertheless, our basic reliance on nuclear retaliation provided by offensive nuclear forces, as the essential means of deterring aggression, has not changed over this period.

This basic idea—that if each side maintained roughly equal forces and equal capability to retaliate against attack, stability and deterrence would be maintained—also served as the foundation for the U.S. approach to the strategic arms limitation talks (SALT) process of the 1970s. At the time that process began, the United States concluded that deterrence based on the capability of offensive retaliatory forces was not only sensible but necessary, since we believed at the time that neither side could develop the technology for defensive systems which could effectively deter the other side.

Today, however, the situation is fundamentally different. Scientific developments and several emerging technologies now do offer the possibility of defenses that did not exist and could hardly have been conceived earlier. The state of the art of defense has now progressed to the point where it is reasonable to investigate whether new technologies can yield options, especially non-nuclear options, which could permit us to turn to defense not only to enhance deterrence but to allow us to move to a more secure and more stable long-term basis for deterrence.

Of equal importance, the Soviet Union has failed to show the type of restraint, in both strategic offensive and defensive forces, that was hoped for when the SALT process began. The trends in the development of Soviet strategic offensive and defensive forces, as well as the growing pattern of Soviet deception and of noncompliance with existing agreements, if permitted to continue unchecked over the long term, will undermine the essential military balance and the mutuality of vulnerability on which deterrence theory has rested.

Soviet Offensive Improvements. The Soviet Union remains the principal threat to our security and that of our allies. As a part of its wide-ranging effort further to increase its military capabilities, the Soviet Union's improvement of its ballistic missile force, providing increased prompt, hard-target kill capability, has increasingly threatened the survivability of forces we have deployed to deter aggression. It has posed an especially immediate challenge to our land-based retaliatory forces and to the leadership structure that commands them. It equally threatens many critical fixed installations in the United States and in allied nations that support the nuclear retaliatory and conventional forces which provide our collective ability to deter conflict and aggression.

Improvement of Soviet Active Defenses. At the same time, the Soviet Union has continued to pursue strategic advantage through the development and improvement of active defenses. These active defenses provide the Soviet Union a steadily increasing capability to counter U.S. retaliatory forces and those of our allies, especially if our forces were to be degraded by a Soviet first strike. Even today, Soviet active defenses are extensive. For example, the Soviet Union possesses the world's only currently deployed antiballistic missile system, deployed to protect Moscow. The Soviet Union is currently improving all elements of this system. It also has the world's only deployed antisatellite (ASAT) capability. It has an extensive air defense network, and it is aggressively improving the quality of its radars, interceptor aircraft, and surface-to-air missiles. It also has a very extensive network of ballistic missile early warning radars. All of these elements provide them an area of relative advantage in strategic defense today and, with logical evolutionary improvement, could provide the foundation of decisive advantage in the future.

Improvement in Soviet Passive Defenses. The Soviet Union is also spending significant resources on passive defensive measures aimed at improving the survivability of its own forces, military command structure, and national leadership. These efforts range from providing rail and road mobility for its latest generation of ICBMs [intercontinental ballistic missiles] to extensive hardening of various critical installations.

Soviet Research and Development on Advanced Defenses. For over two decades, the Soviet Union has pursued a wide range of strategic defensive efforts, integrating both active and passive elements. The resulting trends have shown steady improvement and expansion of Soviet defensive capability. Furthermore, current patterns of Soviet research and development, including a longstanding and intensive research program in many of the same basic technological areas which our SDI program will address, indicate that these trends will continue apace for the foreseeable future. If unanswered, continued Soviet defensive improvements will further erode the effectiveness of our own existing deterrent, based as it is now almost exclusively on the threat of nuclear retaliation by offensive forces. Therefore, this longstanding Soviet program of defensive improvements, in itself, poses a challenge to deterrence which we must address.

Soviet Noncompliance and Verification. Finally, the problem of Soviet noncompliance with arms control agreements in both the offensive and defensive areas, including the ABM Treaty, is a cause of very serious concern. Soviet activity in constructing either new phased-array radar near Krasnoyarsk, in central Siberia, has very immediate and ominous consequences. When operational, this radar, due to its location, will increase the Soviet Union's capability to deploy a territorial ballistic missile defense. Recognizing that such radars would make such a contribution, the ABM Treaty expressly banned the construction of such radars at such locations as one of the primary mechanisms for ensuring the effectiveness of the treaty. The Soviet Union's activity with respect to this radar is in direct violation of the ABM Treaty.

Against the backdrop of this Soviet pattern of noncompliance with existing arms control agreements, the Soviet Union is also taking other actions which affect our ability to verify Soviet compliance. Some Soviet actions, like their increased use of encryption during testing, are directly aimed at degrading our ability to monitor treaty compliance. Other Soviet actions, too, contribute to the problems we face in monitoring Soviet compliance. For example, Soviet increases in the number of their mobile ballistic missiles, especially those armed with multiple, independently-targetable reentry vehicles, and other mobile systems, will make verification less and less certain. If we fail to respond to these trends, we could reach a point in the foreseeable future where we would have little confidence in our assessment of the state of the military balance or imbalance, with all that implies for our ability to control escalation during crises.

Responding to the Challenge

In response to this long-term pattern of Soviet offensive and defensive improvements, the United States is compelled to take certain actions designed both to maintain security and stability in the near term and to ensure these conditions in the future. We must act in three main areas.

Retaliatory Force Modernization. First, we must modernize our offensive nuclear retaliatory forces. This is necessary to reestablish and maintain the offensive balance in the near term and to create the strategic conditions that will permit us to pursue complementary actions in the areas of arms reduction negotiations and defensive research. For our part, in 1981 we embarked on our strategic modernization program aimed at reversing a long period of decline. This modernization program was specifically designed to preserve stable deterrence and, at the same time, to provide the incentives necessary to cause the Soviet Union to join us in negotiating significant reductions in the nuclear arsenals of both sides.

In addition to the U.S. strategic modernization program, NATO is modernizing its longer range intermediate-range nuclear forces (LRINF). Our British and French allies also have underway important programs to improve their own national strategic nuclear retaliatory forces. The U.S. SDI research program does not negate the necessity of these U.S. and allied programs. Rather, the SDI research program depends upon our collective and national modernization efforts to maintain peace and freedom today as we explore options for future decision on how we might enhance security and stability over the longer term.

New Deterrent Options. However, over the long run, the trends set in motion by the pattern of Soviet activity, and the Soviets' persistence in that pattern of activity, suggest that continued long-term dependence on offensive forces may not provide a stable basis for deterrence. In fact, should these trends be permitted to continue and the Soviet investment in both offensive and defensive capability proceed unrestrained and unanswered, the resultant condition could destroy the theoretical and empirical foundation on which deterrence has rested for a generation.

Therefore, we must now also take steps to provide future options for ensuring deterrence and stability over the long term, and we must do so in a way that allows us both to negate the destabilizing growth of Soviet offensive forces and to channel longstanding Soviet propensities for defenses toward more stabilizing and mutually beneficial ends. The Strategic Defense Initiative is specifically aimed toward these goals. In the near term, the SDI program also responds directly to the ongoing and extensive Soviet antiballistic missile effort, including the existing Soviet deployments permitted under the ABM Treaty. The SDI research program provides a necessary and powerful deterrent to any near-term Soviet decision to expand rapidly its antiballistic missile capability beyond that contemplated by the ABM Treaty. This, in itself, is a critical task. However, the overriding, long-term importance of SDI is that it offers the possibility of reversing the dangerous military trends cited above by moving to a better, more stable basis for deterrence and by providing new and compelling incentives to the Soviet Union for seriously negotiating reductions in existing offensive nuclear arsenals.

The Soviet Union recognizes the potential of advanced defense concepts—especially those involving boost, postboost, and mid-course defenses—to change the strategic situation. In our investigation of the potential these systems offer, we do not seek superiority or to establish a unilateral advantage. However, if the promise of SDI technologies is proven, the destabilizing Soviet advantage can be redressed. And, in the process, deterrence will be strengthened significantly and placed on a foundation made more stable by reducing the role of ballistic missile weapons and by placing greater reliance on defenses which threaten no one.

Negotiation and Diplomacy. During the next 10 years, the U.S. objective is a radical reduction in the power of existing and planned offensive nuclear arms, as well as the stabilization of the relationship between nuclear offensive and defensive arms, whether on earth or in space. We are even now looking forward to a period of transition to a more stable world, with greatly reduced levels of nuclear arms and an enhanced ability to deter war based upon the increasing contribution of non-nuclear defenses against offensive nuclear arms. A world free of the threat of military aggression and free of nuclear arms is an ultimate objective to which we, the Soviet Union, and all other nations can agree.

To support these goals, we will continue to pursue vigorously the negotiation of equitable and verifiable agreements leading to significant reductions of existing nuclear arsenals. As we do so, we will continue to exercise flexibility concerning the mechanisms used to achieve reductions but will judge these mechanisms on their ability to enhance the security of the United States and our allies, to strengthen strategic stability, and to reduce the risk of war.

At the same time, the SDI research program is and will be conducted in full compliance with the ABM Treaty. If the research yields positive results, we will consult with our allies about the potential next steps. We would then consult and negotiate, as appropriate, with the Soviet Union, pursuant to the terms of the ABM Treaty, which provide for such consultations, on how deterrence might be strengthened through the phased introduction of defensive systems into the force structures of both sides. This commitment does not mean that we would give the Soviets a veto over the outcome anymore than the Soviets have a veto over our current strategic and intermediate-range programs. Our commitment in this regard reflects our recognition that, if our research yields appropriate results, we should seek to move forward in a stable way. We have already begun the process of bilateral discussion in Geneva needed to lay the foundation for the stable integration of advanced defenses into the forces of both sides at such time as the state of the art and other considerations may make it desirable to do so.

The Soviet Union's View of SDI

As noted above, the U.S.S.R. has long had a vigorous research, development, and deployment program in defensive systems of all kinds. In fact, over the last two decades the Soviet Union has invested as much overall in its strategic defenses as it has in its massive strategic offensive buildup. As a result, today it enjoys certain important advantages in the area of active and passive defenses. The Soviet Union will certainly attempt to protect this massive, long-term investment.

Allied Views Concerning SDI

Our allies understand the military context in which the Strategic Defense Initiative was established and support the SDI research program. Our common understanding was reflected in the statement issued following President Reagan's meeting with Prime Minister Thatcher in December, to the effect that:

First, the U.S. and Western aim was not to achieve superiority but to maintain the balance, taking account of Soviet developments;

Second, that SDI-related deployment would, in view of treaty obligations, have to be a matter for negotiations;

Third, the overall aim is to enhance, and not to undermine, deterrence; and,

Fourth, East-West negotiations should aim to achieve security with reduced levels of offensive systems on both sides.

This common understanding is also reflected in other statements since then—for example, the principles suggested recently by the Federal Republic of Germany that:

- The existing NATO strategy of flexible response must remain fully valid for the alliance as long as there is no more effective alternative for preventing war; and,
- The alliance's political and strategic unity must be safeguarded. There must be no zones of different degrees of security in the alliance, and Europe's security must not be decoupled from that of North America.

SDI Key Points

Following are a dozen key points that capture the direction and scope of the program:

1. The aim of SDI is not to seek superiority but to maintain the strategic balance and thereby assure stable deterrence.

A central theme in Soviet propaganda is the charge that SDI is designed to secure military superiority for the United States. Put in the proper context of the strategic challenge that we and our allies face, our true goals become obvious and clear. Superiority is certainly not our purpose. Nor is the SDI program offensive in nature. The SDI program is a research program aimed at seeking better ways to ensure U.S. and allied security, using the increased contribution of defenses—defenses that threaten no one.

2. Research will last for some years. We intend to adhere strictly to ABM Treaty limitations and will insist that the Soviets do so as well.

We are conducting a broad-based research program in full compliance with the ABM Treaty and with no decision made to proceed beyond research. The SDI research program is a complex one that must be carried out on a broad front of technologies. It is not a program where all resource considerations are secondary to a schedule. Instead, it is a responsible, organized research program that is aggressively seeking cost-effective approaches for defending the United States and our allies against the threat of nuclear-armed and conventionally armed ballistic missiles of all ranges. We expect that the research will proceed so that initial development decisions could be made in the early 1990s.

3. We do not have any preconceived notions about the defensive options the research may generate. We will not proceed to development and deployment unless the research indicates that defenses meet strict criteria.

The United States is pursuing the broadly based SDI research program in an objective manner. We have no preconceived notions about the outcome of the research program. We do not anticipate that we will be in a position to approach any decision to proceed with development or deployment based on the results of this research for a number of years.

We have identified key criteria that will be applied to the results of this research whenever they become available. Some options which could provide interim capabilities may be available earlier than others, and prudent planning demands that we maintain options against a range of contingencies. However, the primary thrust of the SDI research program is not to focus on generating options for the earliest development/deployment decision but options which best meet our identified criteria.

4. Within the SDI research program, we will judge defenses to be desirable only if they are survivable and cost effective at the margin.

Two areas of concern expressed about SDI are that deployment of defensive systems would harm crisis stability and that it would fuel a runaway proliferation of Soviet offensive arms. We have identified specific criteria to address these fears appropriately and directly.

Our survivability criterion responds to the first concern. If a defensive system were not adequately survivable, an adversary could very well have an incentive in a crisis to strike first at vulnerable elements of the defense. Application of this criterion will ensure that such a vulnerable system would not be deployed and, consequently, that the Soviets would have no incentive or prospect of overwhelming it.

Our cost-effectiveness criterion will ensure that any deployed defensive system would create a powerful incentive not to respond with additional offensive arms, since those arms would cost more than the additional defensive capability needed to defeat them. This is much more than an economic argument, although it is couched in economic terms. We intend to consider, in our evaluation of options generated by SDI research, the degree to which certain types of defensive systems, by their nature, encourage an adversary to try simply to overwhelm them with additional offensive capability while other systems can discourage such a counter effort. We seek defensive options which provide clear disincentives to attempts to counter them with additional offensive forces.

In addition, we are pressing to reduce offensive nuclear arms through the negotiation of equitable and verifiable agreements. This effort includes reductions in the number of warheads on ballistic missiles to equal levels significantly lower than exist today.

5. It is too early in our research program to speculate on the kinds of defensive systems—whether ground-based or space-based and with what capabilities—that might prove feasible and desirable to develop and deploy.

Discussion of the various technologies under study is certainly needed to give concreteness to the understanding of the research program. However, speculation about various types of defensive systems that might be deployed is inappropriate at this time. The SDI is a broad-based research program investigating many technologies. We currently see real merit in the potential of advanced technologies providing for a layered defense, with the possibility of negating a ballistic missile at various points after launch. We feel that the possibility of a layered defense both enhances confidence in the overall system and compounds the problem of a potential aggressor in trying to defeat such a defense. However, the paths to such a defense are numerous.

Along the same lines, some have asked about the role of nuclear-related research in the context of our ultimate goal of non-nuclear defenses. While our current research program certainly emphasizes non-nuclear technologies, we will continue to explore the promising concepts which use nuclear energy to power devices which could destroy ballistic missiles at great distances. Further, it is useful to study these concepts to determine the feasibility and effectiveness of similar defensive systems that an adversary may develop for use against future U.S. surveillance and defensive or offensive systems.

6. The purpose of the defensive options we seek is clear—to find a means to destroy attacking ballistic missiles before they can reach any of their potential targets.

We ultimately seek a future in which nations can live in peace and freedom, secure in the knowledge that their national security does not rest upon the threat of nuclear retaliation. Therefore, the SDI research program will place its emphasis on options which provide the basis for eliminating the general threat posed by ballistic missiles. Thus, the goal of our research is not, and cannot be, simply to protect our retaliatory forces from attack.

If a future president elects to move toward a general defense against ballistic missiles, the technological options that we explore will certainly also increase the survivability of our retaliatory forces. This will require a stable concept and process to manage the transition to the future we seek. The concept and process must be based upon a realistic treatment of not only U.S. but Soviet forces and out-year programs.

7. U.S. and allied security remains indivisible. The SDI program is designed to enhance allied security as well as U.S. security. We will continue to work closely with our allies to ensure that, as our research progresses, allied views are carefully considered.

This has been a fundamental part of U.S. policy since the inception of the Strategic Defense Initiative. We have made a serious commitment to consult, and such consultations will precede any steps taken relative to the SDI research program which may affect our allies.

8. If and when our research criteria are met, and following close consultation with our allies, we intend to consult and negotiate, as appropriate, with the Soviets pursuant to the terms of the ABM Treaty, which provide for such consultations, on how deterrence could be enhanced through a greater reliance by both sides on new defensive systems. This commitment should in no way be interpreted as according the Soviets a veto over possible future defensive deployments. And, in fact, we have already been trying to initiate a discussion of the offense-defense relationship and stability in the defense and space talks underway in Geneva to lay the foundation to support such future possible consultations.

If, at some future time, the United States, in close consultation with its allies, decides to proceed with deployment of defensive systems, we intend to utilize mechanisms for U.S.-Soviet consultations provided for in the ABM Treaty. Through such mechanisms, and taking full account of the Soviet Union's own expansive defensive system research program, we will seek to proceed in a stable fashion with the Soviet Union.

9. It is our intention and our hope that, if new defensive technologies prove feasible, we (in close and continuing consultation with our allies) and the Soviets will jointly manage a transition to a more defense-reliant balance.

Soviet propagandists have accused the United States of reneging on commitments to prevent an arms race in space. This is clearly not true. What we envision is not an arms race; rather, it is just the opposite—a jointly managed approach designed to maintain, at all times, control over the mix of offensive and defensive systems of both sides and thereby increase the confidence of all nations in the effectiveness and stability of the evolving strategic balance.

10. SDI represents no change in our commitment to deterring war and enhancing stability.

Successful SDI research and development of defense options would not lead to abandonment of deterrence but rather to an enhancement of deterrence and an evolution in the weapons of deterrence through the contribution of defensive systems that threaten no one. *We would deter a potential aggressor by making it clear that we could deny him the gains he might otherwise hope to achieve rather than merely threatening him with costs large enough to outweigh those gains.*

U.S. policy supports the basic principle that our existing method of deterrence and NATO's existing strategy of flexible response remain fully valid, and must be fully supported, as long as there is no more effective alternative for preventing war. It is in clear recognition of this obvious fact that the United States continues to pursue so vigorously its own strategic modernization program and so strongly supports the efforts of its allies to sustain their own commitments to maintain the forces, both nuclear and conventional, that provide today's deterrence.

11. For the foreseeable future, offensive nuclear forces and the prospect of nuclear retaliation will remain the key element of deterrence. Therefore, we must maintain modern, flexible, and credible strategic nuclear forces.

This point reflects the fact that we must simultaneously use a number of tools to achieve our goals today while looking for better ways to achieve our goals over the longer term. It expresses our basic rationale for sustaining the U.S. strategic modernization program and the rationale for the critically needed national modernization programs being conducted by the United Kingdom and France.

12. Our ultimate goal is to eliminate nuclear weapons entirely. By necessity, this is a very long-term goal, which requires, as we pursue our SDI research, equally energetic efforts to diminish the threat posed by conventional arms imbalances, both through conventional force improvements and the negotiation of arms reductions and confidence-building measures.

We fully recognize the contribution nuclear weapons make to deterring conventional aggression. We equally recognize the destructiveness of war by conventional and chemical means, and the need both to deter such conflict and to reduce the danger posed by the threat of aggression through such means. ■

Published by the United States Department of State • Bureau of Public Affairs
Office of Public Communication • Editorial Division • Washington, D.C. • June 1985

Appendix I

List of Reagan Administration Statements on BMD

Abrahamson, James A., Director, Strategic Defense Initiative Organization, Statement before the U.S. Senate Armed Services Subcommittee on Strategic and Theater Nuclear Forces, Apr. 24, 1984.

Abrahamson, James A., Statement before the U.S. Senate Committee on Foreign Relations, Apr. 25, 1984.

Abrahamson, James A., Statement before the U.S. House of Representatives Foreign Affairs Subcommittee on International Security and Scientific Affairs, July 26, 1984, and subsequent responses to additional questions on pages 441-466.

Abrahamson, James A., "The Strategic Defense Initiative," *Defense/84*, August 1984.

Adelman, Kenneth L., Director, Arms Control and Disarmament Agency, "What's Next for Strategic Stability and Arms Control," Speech to the International Istitute for Strategic Studies in London, Feb. 13, 1985.

Cooper, Robert S., Director, Defense Advanced Research Projects Agency, Statement before the U.S. Senate Armed Services Subcommittee on Strategic and Theater Nuclear Forces, May 2, 1983.

DeLauer, Richard D., Under Secretary of Defense for Research and Engineering, "The President's Strategic Defense Initiative," Statement before the U.S. House of Representatives Armed Services Subcommittee on Research and Development, Mar. 1, 1984.

DeLauer, Richard D., Statement before the U.S. Senate Committee on Armed Services, Mar. 8, 1984.

Department of Defense, *Overview of Strategic Defense Initiative: Fact Sheet*, Mar. 9, 1984.

Department of Defense, *The Strategic Defense Initiative: Defensive Technologies Study*, March 1984.

Department of Defense, *Defense Against Ballistic Missiles: An Assessment of Technologies and Policy Implications*, April 1984.

Department of Defense, *Report to the Congress on the Strategic Defense Initiative*, April 1985.

Ikle, Fred C., Under Secretary of Defense for Policy, "Nuclear Strategy: Can There be a Happy Ending?" *Foreign Affairs*, spring 1985.

Keyworth, George A., II, Science Adviser to the President, Speech to the Armed Forces Communications and Electronics Association, Washington, DC, Oct. 13, 1983.

Keyworth, George A., II, "Reassessing Strategic Defense," Speech to the Council on Foreign Relations, Washington, DC, Feb. 15, 1984.

Keyworth, George A., II, "Ballistic Missile Defense: Current Issues," Speech to the Brookings Forum on the Future of Ballistic Missile Defense, Washington, DC, Feb. 29, 1984.

Keyworth, George A., II, "A Sense of Obligation—the Strategic Defense Initiative," *Aerospace America*, April 1984.

Keyworth, George A., II, Statement before the U.S. Senate Committee on Foreign Relations, Apr. 25, 1984.

Keyworth, George A., II, "Strategic Defense Initiative: The Rational Route to Effective Nuclear Arms Control," *Government Executive*, June 1984.

Keyworth, George A., II, "Strategic Defense: A Catalyst for Arms Reductions," in *Proceedings of the Third Annual Seminar of the Center for Law and National Security*, University of Virginia, Charlottesville, Va., June 23, 1984.

Keyworth, George A., II, "The Case for Strategic Defense: An Option for a World Disarmed," *Issues in Science and Technology*, fall 1984.

Keyworth, George A., II, "New Technologies: The Catalysts for Reductions in Nuclear Arms," Speech at the University of Maryland, College Park, Md., Sept. 5, 1984.

Keyworth, George A., II, "Today's Big Stick: Technology," Speech to the Air Force Association, Washington, DC, Sept. 17, 1984.

Keyworth, George A., II, "The President's Strategic Defense Initiative," Speech to the SDIO University Review Forum, Mar. 29, 1985.

Keyworth, George A., II, "The Case for Arms Control and the Strategic Defense Initiative," *Arms Control Today*, April 1985.

Keyworth, George A., II, "Security and Stability: The Role for Strategic Defense," Speech at the

University of California, San Diego, May 1, 1985.

Keyworth, George A., II, "The President's Strategic Defense Initiative," Speech to the Aerospace Association, May 21, 1985.

Keyworth, George A., II, "SDI—Where Does the United States Stand Now?" Speech to the American Security Council Foundation, June 4, 1985.

McFarlane, Robert C., Assistant to the President for National Security Affairs, "Strategic Defense Initiative," Speech to the Overseas Writers Association, Washington, DC, Mar. 7, 1985.

McFarlane, Robert C., Interview, *U.S. News and World Report*, Mar. 18, 1985, pp. 25-26.

Nitze, Paul H., Special Adviser to the President and the Secretary of State for Arms Reductions, "On the Road to a More Stable Peace," Speech to the Philadelphia, Pennsylvania, World Affairs Council, Feb. 20, 1985.

Nitze, Paul H., "The Objectives of Arms Control," Speech to the International Institute for Strategic Studies in London, Mar. 28, 1985.

Reagan, Ronald, "Defense Spending and Defensive Technology," Speech on nationwide television, Mar. 23, 1983.

Reagan, Ronald, *The President's Strategic Defense Initiative*, White House pamphlet issued Jan. 3, 1985.

Reagan, Ronald, Speech to the National Space Club, Washington, DC, Mar. 29, 1985.

Shultz, George, Secretary of State, "Arms Control: Objectives and Prospects," Speech to the Austin, Texas, Council on Foreign Relations, Mar. 28, 1985.

Weinberger, Caspar W., Secretary of Defense, News conference on the Strategic Defense Initiative, Mar. 27, 1984.

Weinberger, Caspar W., Speech at the Foreign Press Center, Washington, DC, Dec. 19, 1984.

Weinberger, Caspar W., Speech to the American Society of Newspaper Editors, Washington, DC, Apr. 11, 1985.

White House, *Fact Sheet on The Strategic Defense Initiative*, issued June 1, 1985, 11 pp.

Yonas, Gerold, Chief Scientist, Strategic Defense Initiative Organization, "The Reagan Strategic Defense Initiative," *Daedalus*, spring 1985.

Yonas, Gerold, "Strategic Defense Initiative: The politics and science of weapons in space," *Physics Today*, June 1985, pp. 24-32

Appendix J

Articles by Critics of the Strategic Defense Initiative

Ball, George, "The War for Star Wars," *The New York Review of Books*, Apr. 11, 1985, pp. 38-44.

Bethe, Hans A., et al., "Space-Based Ballistic Missile Defense," *Scientific American*, October 1984, pp. 39-49.

Boutwell, Jeffrey, and Richard A. Scribner, *The Strategic Defense Initiative: Some Arms Control Implications*, American Association for the Advancement of Science, Washington, DC, May 1985, 44 pp.

Brown, Harold, "The Strategic Defense Initiative: Defensive Systems and the Strategic Debate," *Survival*, March/April 1985, pp. 55-64. Also printed in Discussion Paper No. 104, California Seminar on International Security and Foreign Policy, Santa Monica, Calif., March 1985.

Bundy, McGeorge, et al., "The President's Choice: Star Wars or Arms Control," *Foreign Affairs*, 63:264-278, winter 1984/85.

Burrows, William E., "Ballistic Missile Defense: The Illusion of Security," *Foreign Affairs*, 62:843-856, spring 1984.

Chayes, Abram, "Treaties and Legal Issues," Speech on ballistic missile defense at Symposium on Space, National Security and C-cubed-I, The Mitre Corporation, Bedford, Mass., Oct. 25, 1984. Mitre Document M85-3, pp. 29-32.

Clausen, Peter A., "SDI in Search of A Mission," *World Policy Journal*, spring 1985, pp. 249-303.

Clifford, Clark, Testimony before the Committee on Foreign Affairs, U.S. House of Representatives, May 1, 1985.

Drell, Sidney, and Thomas H. Johnson, eds., *Strategic Missile Defense: Necessities, Prospects, and Dangers in the Near Term*, Report of a workshop at the Center for International Security and Arms Control, Stanford University, April 1985, 23 pp. Also printed in Hearings of the Subcommittee on Strategic and Theater Nuclear Forces, Committee on Armed Services, U.S. Senate, March 19, 1985.

Drell, Sidney, Philip J. Farley, and David Holloway, *The Reagan Strategic Defense Initiative: A Technical, Political, and Arms Control Assessment,* Center for International Security and Arms Control, Stanford University, 1984, 142 pp.

Drell, Sidney, Philip J. Farley, and David Holloway, "Preserving the ABM Treaty: A Critique of the Reagan Strategic Defense Initiative," *International Security*, 9(2):51-91, fall, 1984.

Drell, Sidney, and Wolfgang K.H. Panofsky, "The Case Against Strategic Defense: Technical and Strategic Realities," *Issues in Science and Technology*, fall 1984.

Garwin, Richard L., "Countermeasure: Defeating Space-based Defense," *Arms Control Today*, May 1985.

Garwin, Richard L., "How Many Orbiting Lasers for Boost-Phase Intercept?" *Nature*, May 23, 1985, pp. 286-290.

Garwin, Richard L., et al., *The Fallacy of Star Wars,* Random House, 1984.

Garwin, Richard, et al., "Space Weapons," *Bulletin of the Atomic Scientists*, May 1984.

Glaser, Charles L., "Why Even Good Defenses May Be Bad," *International Security* 9(2):92-123, fall 1984. Abridged version appears in *Bulletin of the Atomic Scientists*, March 1985, pp. 13-16.

Hartung, William D., et al., *The Strategic Defense Initiative: Costs, Contractors and Consequences,* Council on Economic Priorities, New York, 1985.

Hoopes, Townsend, "The Star Wars Proposal," *CNS Reports*, The Committee for National Security, Washington, DC, winter 1985.

Jacky, Jonathan, "The 'Star Wars' Defense Won't Compute," *The Atlantic Monthly*, June 1985, pp. 18-30.

Kaiser, Robert G., "A Disarming Lack of Candor," *The Washington Post*, Mar. 10, 1985.

Krauthammer, Charles, "The Illusion of Star Wars," *The New Republic*, May 14, 1984, pp. 13-17.

League of Women Voters, "Space-Age Defense: Pipe Dream or Protection?" *National Voter*, League of Women Voters, Washington, DC, spring 1985, pp. 1-8.

Longstreth, Thomas K., John E. Pike, and John B. Rhinelander, *The Impact of U.S. and Soviet Ballistic Missile Defense Programs on the ABM Treaty*, National Campaign to Save the ABM Treaty, Washington, DC, March 1985, 99 pp.

McNamara, Robert S., and Hans A. Bethe, "Reducing the Risk of Nuclear War," *The Atlantic Monthly*, July 1985, pp. 43-51.

McNamara, Robert S., Testimony before the Committee on Foreign Affairs, U.S. House of Representatives, May 1, 1985.

Paine, Christopher, "The ABM Treaty: Looking for Loopholes," *Bulletin of the Atomic Scientists*, August/September 1983, pp. 13-16.

Pike, John, *The Strategic Defense Initiative Budget and Program*, Federation of American Scientists, Washington, DC, July 1985.

Panofsky, Wolfgang K. H., "The Strategic Defense Initiative: Perception vs Reality," *Physics Today*, June 1985, pp. 34-45.

Rathjens, George, and Jack Ruina, "The Uncertainty of Ballistic Missile Defense," *Daedalus*, summer 1985.

Rathjens, George, "The Strategic Defense Initiative: The Imperfections of 'Perfect Defense'," *Environment*, June 1984, pp. 6-13.

Rhinelander, John B., "How to Save the ABM Treaty," *Arms Control Today*, May 1985.

Schlesinger, James R., Speech on ballistic missile defense at Symposium on Space, National Security and C-cubed-I, The Mitre Corporation, Bedford, Mass., Oct. 25, 1984. Mitre Document M85-3, pp. 55-62. See also pp. 959-961 in Mr. Schlesinger's article "The Eagle and the Bear," *Foreign Affairs*, summer 1985.

Sedacca, Sandra, and Robert DeGrasse, *Star Wars: Questions and Answers on the Space Weapons Debate*, Common Cause, Washington, DC, February 1985, 36 pp.

Sewall, Sarah, "Militarizing the Last Frontier: The Space Weapons Race," *Defense Monitor*, Vol. 12, No. 5, 1983, Center for Defense Information, Washington, DC, 8 pp.

Sherr, Alan B., *Legal Issues of the "Star Wars" Defense Program*, Lawyers Alliance for Nuclear Arms Control, Inc., Boston, Mass., June 1984.

Stein, Jonathan, *From H-Bomb to Star Wars: The Politics of Strategic Decision Making*, Lexington Books, 1984.

Stone, Jeremy, "The Four Faces of Star Wars: Anatomy of a Debate," *F.A.S. Public Interest Report: Journal of the Federation of American Scientists*, March 1985,

Articles by Richard L. Garwin, Spurgeon M. Keeny, Jr., and George Rathjens in Ashton B. Carter and David N. Schwartz, eds., *Ballistic Missile Defense,* The Brookings Institution, 1984.

Appendix K
Excerpts From Soviet Statements on BMD

The Soviet reaction to President Reagan's March 23, 1983, speech was prompt and strongly negative. Four days after the speech was given, the Soviet President, Yuri Andropov, denounced President Reagan's proposal to develop new types of BMD systems. Andropov said the idea of defensive measures might seem attractive to the uninformed, but:

> . . . In fact, the strategic offensive forces of the United States will continue to be developed and upgraded at full tilt and along quite a definite line at that, namely that of acquiring a nuclear first strike capability. Under these conditions, the intention to secure itself the possibility of destroying, with the help of the ABM defenses, the corresponding strategic systems of the other side, that is, of rendering it unable to deal a retaliatory strike, is a bid to disarm the Soviet Union in the face of the U.S. nuclear threat. . . [It] is only mutual restraint in the field of ABM defenses that will allow progress in limiting and reducing offensive weapons, that is in checking and reversing the strategic arms race as a whole. Today, however, the United States intends to sever this interconnection. Should this conception be converted into reality, this would actually open the floodgates of a runaway race of all types of strategic arms, both offensive and defensive. Such is the real purport, the seamy side, so to say, of Washington's "defensive conception."[1]

These themes have been reiterated vigorously and persistently ever since by Soviet newspaper commentators, scientists, diplomats, and senior officials.

In an interview with *U.S. News and World Report* in April 1984, the Director of the Soviet Institute of Space Research, Roald Sagdeyev, commented on the U.S. Strategic Defense Initiative as follows:

> We have made a detailed analysis. We believe that even if it would be possible to build such a system—a very expensive system—it would not prove to be an absolute shield. Its penetrability would remain quite high.
>
> It will always be possible—and at a lower cost—to interfere with such a system or to foil it by increasing the number of attacking weapons.
>
> A space-based defense system would prove to be extraordinarily destabilizing. When those who command such a system understand that it does not provide 100 percent protection, they might be seduced by the idea of attempting a first strike. Our conclusion is quite pessimistic; It will lead to a new round in the arms race and will increase the emphasis on developing first-strike weapons.[2]

Following are excerpts from an article by another prominent Soviet scientist, Yevgeny P. Velikhov, a vice-president of the Soviet Academy of Sciences:

> . . . [The deployment of a BMD system] would significantly complicate the maintenance of deterrence, making it highly unstable, for it would stimulate the illusion of advantages (damage limitation and even a chance for surviving nuclear war) associated with a first strike. . . . [I]f both sides possessed space-based [BMD] systems the destabilizing effect would be much greater than if such systems were available to only one side. In the context of strategic logic (without considering psychological and political aspects) this thinking arises from the fact that if both sides had these systems, their impetus for a preemptive first strike would be greater, since each side could hope to secure an advantage by striking first.
>
> . . . The development of [space-based anti-missile systems] could stimulate an increase in the arsenals of strategic delivery vehicles and nuclear warheads, for example, strategic cruise missiles, including sea- and ground-launched cruise missiles . . . If tests of the space-based systems were to begin, to say nothing of their actual deployment, the permanent ABM Treaty, signed on May 26, 1972, would be threatened. . . . It is hard to overestimate the importance of this U.S.-Soviet Treaty today for it remains the only ratified and acting agreement in the area of strategic arms limitation. . . .
>
> Abrogation of the ABM Treaty would in turn undoubtedly lessen chances for reaching mutually beneficial strategic arms limitation and reduction agreements in the near future. The stabilizing regime created by the 1972 ABM Treaty could be strengthened significantly by agreements on the non-deployment in space of any weapons and the non-use of force in space.
>
> ***
>
> [Space-based BMD systems] would inevitably become a serious obstacle for U.S.-Soviet cooperation in the peaceful uses of space. Yet the potential value of such cooperation is important from economic, scientific and technological points of view, because of the many mutually complementary characteristics of the Soviet and U.S. space programs. Cooperation in this area could be a very positive factor, politically and psychologically, in improving U.S.-Soviet relations in general, and in

[1] From *Pravda*, Mar. 27, 1983.

[2] *U.S. News and World Report*, Apr. 23, 1984, p. 50.

strengthening the confidence between the peoples and the leaders of the two great powers.

The potential impact of a large-scale space-based anti-missile program on the strategic balance would be to substantially increase both the risk of a preemptive strike and the likelihood of wrong and fatal decisions in crises. Hence even if rough parity in strategic forces were preserved, strategic stability would be seriously undermined.[3]

In August 1983 the Soviet Union formally proposed at the United Nations General Assembly a revised draft treaty on "The Prohibition of the Use of Force in Outer Space and From Space Against the Earth."[4] The provisions of this draft would ban space-based weapons, anti-satellite systems, and military use of manned spacecraft.

Former Soviet President Konstantin Chernenko issued several statements on ballistic missile defense. Following are illustrative excerpts:

We are resolutely against the development of broad-scale antimissile defense systems, which cannot be viewed in any other way than as aimed at the unpunished perpetration of nuclear aggression. There is an indefinite Soviet-American treaty on antimissile defense, prohibiting the creation of such systems. It must be rigorously observed.[5]

Today, no limitation and, all the more, no reduction of nuclear arms can be attained without effective measures that would prevent the militarization of outer space... Using the term "defense" is juggling with words. In its substance, this is an ... aggressive concept. The aim is to try to disarm the other side and deprive it of a capability to retaliate in the event of nuclear aggression against it.

To put it simply, the aim is to acquire a capability to deliver a nuclear strike, counting on impunity with an anti-ballistic missile shield to protect oneself from retaliation.... [U.S. BMD deployment not only would mean] the end of the process of nuclear arms limitation and reduction, but [it] would become a catalyst of an uncontrolled arms race in all fields.[6]

In a lengthy interview on Moscow television January 13, 1985, then Soviet Foreign Minister Andrei Gromyko discussed the results of his January 7-8 meeting with Secretary of State George Shultz. The following excerpts refer to space weapons:

...[It] is impossible to examine either the question of strategic armaments or the question of intermediate-range nuclear weapons without examining the question of ... averting an arms race in space. In the end the American side agreed to adopt such a viewpoint. This fact is a positive one.

... [Preventing "militarization of space" means] arms intended for use against targets in space should be banned categorically and also that arms intended for use from space against... targets on the ground, in the sea and in the air should be banned categorically.

***

[If] accords [on preventing militarization of space] became clear, then it would be possible to move forward also on questions of strategic armaments. The Soviet Union would be willing not only to examine this problem of strategic armaments but would also be willing to reduce them sharply. ... And on the contrary, if there were no movement forward in space questions, then it would be superfluous even to speak about the possibility of a reduction in strategic armaments.

***

We are told this: After all, the United States does not have the intention of striking a blow at the Soviet Union. We tell them: Well, then, it follows that the Soviet Union must rely on your conscience, on the conscience of Washington. Well, first of all, we are not very convinced that Washington has very great reserves of this merchandise.... And second, if we were to mentally trade places with you, ... if we were trying to create such a system, corresponding statements, statements to the effect that: You should rely on our conscience. Would they be sufficient for you? Silence. Silence.

***

[The] chief barrier that separates the policy of the Soviet Union from that of the United States is atomic weapons.... [Reaching agreement at the Geneva negotiations] would therefore undoubtedly denote a big step forward in matters relating to improving bilateral Soviet- U.S. relations, especially if one takes account of the fact that both sides are major powers with broad-ranging international interests.

The June 4, 1985, issue of *Pravda* contained a long article on the ABM Treaty by Marshal Sergei F. Akhromeyev, Chief of the Soviet General Staff. Excerpts follow:

The limitation, still more the reduction, of nuclear arms is inconceivable in conditions of the militarization of space. The creation and deployment in space of strike arms will inevitably lead to an increase in the quantity of, and to the qualitative improvement of, strategic nuclear arms.... The creation of the large-scale space ABM system contemplated in the United States has a clear aggressive point: This system is a most important element in the integrated offensive potential of the

[3] Yevgeny P. Velikhov, "Effect on Strategic Stability," *Bulletin of the Atomic Scientists*, May 1984.

[4] U.N. General Assembly Document No. A/38/194.

[5] TASS, Dec. 20, 1984.

[6] "Chernenko Again Warns U.S. on Space Plan," *New York Times*, Feb. 1, 1985, p. 3.

side that has created it, undermines strategic equilibrium, and provides the opportunity for the United States to deliver a first strike in the hope that the retaliatory strike against U.S. territory can be averted.

How is the other side, the Soviet Union, supposed to behave under these conditions? It is left with no choice; it will be forced to ensure the restoration of the strategic balance and to build up its own strategic offensive forces, supplementing them with means of defense. Therefore, any attempts to limit strategic offensive armaments while creating space strike means are futile.

... The military-political significance of the Soviet-U.S. ABM Treaty is extremely great. This treaty is one of the foundations on which relations between the sides are based. By signing it the Soviet Union and the United States recognized that in the nuclear age only mutual restraint in the sphere of ABM systems will make it possible to advance along the path of limiting and reducing nuclear arms, that is, to curb the strategic arms race as a whole.

... If the [ABM Treaty] were to lapse for any reason, the foundation on which talks between the sides on nuclear arms limitation could be based and conducted would disappear. This would effectively mean the collapse of talks and an uncontrolled arms race for decades.

... The U.S. Administration's actions in creating a new class of weapons—space strike means—are incompatible with the principles forming the foundation of the ABM Treaty. By proclaiming the "Strategic Defense Initiative" and embarking on the practical implementation of a large-scale anti-ballistic missile system with space-based elements, Washington is effectively working directly to undermine the treaty.

***

[U.S. leaders] are saying that the U.S. actions running counter to the treaty can somehow be legitimized, for instance, by revising this document and making amendments to it agreed with the Soviet side....

All this is merely an unworthy ploy aimed at reassuring public opinion.... The United States is working toward changing the meaning of the Treaty itself and emasculating it of its main content—the ban on the deployment of an ABM defense of the country's territory.

The Soviet Union, of course, will not countenance the Treaty on the Limitation of ABM Systems being transformed into a cover for U.S. policy aimed at ensuring an arms race in the sphere of space anti-ballistic missile systems.

... If space strike arms are banned, and preparations for their creation are halted at the stage of scientific research work, broad opportunities will be opened up for a radical reduction of nuclear arms. The Soviet Union has already proposed a reduction of strategic offensive arms by one-fourth. Given the non-militarization of space, it is possible to carry out even more profound reductions.... For its part, the Soviet Union will persistently seek in Geneva specific, mutually acceptable agreements that would make it possible to put an end to the arms race and carry forward the cause of disarmament.

In a speech[7] in May 1985, Soviet General Chairman Mikhail S. Gorbachev said:

There are no people in the world who are not worried by the U.S. plans to militarize space. This worry is well grounded. Let us take a realistic view of matters: the implementation of these plans would thwart disarmament talks.

Moreover, it would dramatically increase the threat of a truly global, all-destroying military conflict. Anyone capable of an unbiased analysis of the situation and sincerely wishing to safeguard peace cannot help opposing "star wars."

In a nationally televised speech June 26, 1985, Chairman Gorbachev said:

We are prepared to seek accord not only about ending the arms race, but about the greatest of arms reductions—right up to general and complete disarmament. At present, as you know, we are holding talks with the United States in Geneva. The task before them, as the Soviet leadership understands it, is to end the arms race on earth and prevent one in space. We embarked upon the negotiations in order to achieve these aims in practice. But all the indications are that this is precisely what the U.S. Administration, and the military-industrial complex which it serves, do not want. The attainment of serious accords evidently does not enter into their plans. They are continuing to implement their gigantic program of forcing through the production of more and more new types of weapons of mass destruction in the hope of achieving superiority over the countries of socialism, and dictating their will to them. The Americans have not only failed to put forward any serious proposals in Geneva for curtailing the arms race, but on the contrary, are taking steps that make such a curtailment impossible. I am thinking of the so-called "star wars" program to create offensive space weapons. Talk of its supposed defensive nature is, of course, a fairy tale for the gullible. The idea is to attempt to paralyze the Soviet Union's strategic arms and guarantee the opportunity of an unpunished nuclear strike against our country.

This is the essence of the matter, and one which we cannot fail to take into account. If the Soviet Union is faced with a real threat from space, it will find a way to effectively counter it. Let no one, and

[7]TASS, May 27, 1985.

I say this quite definitely, doubt this. For the time being, one thing is clear—that is, that the American program for the militarization of space plays the role of a blank wall, barring the way to the achievement in Geneva of the relevant accords.

By its militarist policy the U.S. Administration is assuming a grave responsibility to mankind. If our partners at the Geneva talks continue with their line of playing for time at the meetings of the delegations, avoiding a solution of the questions for which they have assembled and using this time to push ahead with their military programs in space, on the ground, and at sea, we shall then of course have to assess the whole situation anew. We simply cannot allow the talks to be used again to divert attention and to cover up military preparations, whose purpose is to secure U.S. strategic superiority and achieve world dominance. In rebuffing these schemes, I am confident that we will be supported by the really peace-loving forces throughout the whole world and that we will be supported by the Soviet people.

In a letter[8] sent July 5, 1985, to American scientists, Chairman Gorbachev said:

. . . On behalf of the Soviet leadership I want to state in all definiteness that the Soviet Union will not be the first to make a step into outer space with weapons. We shall make every effort to convince other countries, and above all the United States of America, not to make such a fatal step which would inevitably increase the threat of nuclear war and would give an impetus to the uncontrolled arms race in all directions.

Proceeding from this goal, the Soviet Union, as you evidently know, has made a radical proposal in the United Nations organization, tabling a draft treaty on the prohibition of the use of force in space and from space against earth. If the United States joined the vast majority of states that have supported this initiative, the issue of space weapons could be closed once and for all.

At the Soviet-American talks on nuclear and space arms in Geneva we are seeking to come to terms on a full ban on the development, testing, and deployment of space attack systems. Such a ban would make it possible not only to preserve outer space for peaceful development, research, and scientific discoveries, but also to launch the process of sharply reducing, then eliminating nuclear weapons.

We have also repeatedly taken unilateral steps which have been called upon to set a good example to the United States. It is for two years now that the Soviet Union has maintained its moratorium on the placement of anti-satellite weapons in outer space, and it will continue abiding by it for as long as the other states will be acting in the same way. Lying on the table in Washington is our proposal for both sides to put a total end to efforts to develop new anti-satellite systems and for such systems already possessed by the U.S.S.R. and the United States, including those whose testing has not yet been completed, to be scrapped. The actions of the American side will show already in the near future which decision the U.S. Administration will prefer.

Strategic stability and trust would, no doubt, be strengthened if the United States agreed together with the U.S.S.R. in a binding form to reaffirm commitment to the regime of the Treaty on the Limitation of Anti-Ballistic Missile Systems, a treaty of unlimited duration.

The Soviet Union is not developing attack space weapons or a large-scale ABM system, just as it is not laying the foundation for such a defense. It strictly adheres to its obligations under the treaty as a whole and, in its particular aspects, unswervingly observes the spirit and the letter of that document of paramount importance. We invite the American leadership to join us in that undertaking, [and] renounce the plans of space militarization that are now in the making, plans which would invariably lead to the breakup of that document—the key link of the entire process of nuclear arms limitations.

The U.S.S.R. proceeds from the premise that the practical fulfillment of the task of preventing an arms race in space and terminating it on earth is possible given the political will and sincere desire of both sides to work toward attaining that historic goal. The Soviet Union has such a desire and such a will. . . .

[8]"Outer Space Should Serve Peace," article giving text of Mikhail Gorbachev's reply to a message from the Union of Concerned Scientists, TASS, July 5, 1985.

Appendix L

References on Strategic Nuclear Policy

This list represents an effort to select a representative sampling of the voluminous literature on this subject. Lawrence Freedman's book contains an extensive bibliography.

Books

Allison, Graham T., Albert Carnesale, and Joseph S. Nye, Jr., eds., *Hawks, Doves, and Owls: An Agenda for Avoiding Nuclear War* (New York: W. W. Norton, 1985).

Berman, Robert P., and John C. Baker, *Soviet Strategic Forces: Requirements and Responses* (Washington: The Brookings Institution, 1982).

Blair, Bruce G., *Strategic Command and Control: Redefining the Nuclear Threat* (Washington: The Brookings Institution, 1985).

Blechman, Barry M., ed., *Rethinking the U.S. Strategic Posture* (Cambridge, MA: Ballinger, 1982).

Boutwell, Jeffrey, Donald Hafner, and Franklin A. Long, eds., *Weapons in Space: The Politics and Technology of Ballistic Missile Defense and Anti-Satellite Weapons* (New York: W.W. Norton, 1985). (Also published in the spring 1985 and summer 1985 issues of *Daedalus*.)

Bracken, Paul, *The Command and Control of Nuclear Forces* (New Haven, CT: Yale University Press, 1983.

Brodie, Bernard, *Escalation and the Nuclear Option* (Princeton, NJ: Princeton University Press, 1966).

Brown, Harold, *Thinking About National Security —Defense and Foreign Policy in a Dangerous World* (Boulder, CO: Westview Press, 1983).

Brown, Harold, and Lynn E. Davis, *Nuclear Arms Control Choices* (Boulder, CO: Westview Press, 1984).

Carnesale, Albert, et al., *Living with Nuclear Weapons* (New York: Bantam Books, 1983).

Carter, Ashton B., and David N. Schwartz, eds., *Ballistic Missile Defense* (Washington: The Brookings Institution, 1984).

Chayes, Abram, amd Jerome B. Wiesner, eds., *ABM* (New York: Harper and Row, 1969).

Davis, Jacquelyn K., et al, *The Soviet Union and Ballistic Missile Defense* (Cambridge, MA: Institute for Foreign Policy Analysis, 1980).

Drell, Sidney D., Philip J. Farley, and David Holloway, *The Reagan Strategic Defense Initiative: A Technical, Political, and Arms Control Assessment* (Stanford, CA: Stanford University Center for International Security and Arms Control, 1984.)

Dyson, Freeman, *Weapons and Hope* (New York: Harper and Row, 1984).

Ford, Daniel, *The Button: The Pentagon's Strategic Command and Control System* (New York: Simon and Schuster, 1985).

Freedman, Lawrence, *The Evolution of Nuclear Strategy* (New York: St. Martin's Press, 1981).

Frei, Daniel, and Christian Catrina, *Risks of Unintentional Nuclear War* (London: Taylor and Francis, 1982).

Gallis, Paul E., Mark M. Lowenthal, and Marcia S. Smith, *The Strategic Defense Initiative and United States Alliance Strategy*, Report No. 85-48-F (Washington: Congressional Research Service, U.S. Congress, February 1, 1985).

Garwin, Richard L., Kurt Gottfried, and Henry W. Kendall, *The Fallacy of Star Wars* (New York: Random House, 1984).

Gompert, David C., et al., *Nuclear Weapons and World Politics: Alternatives for the Future* (New York: McGraw-Hill, 1977).

Graham, Daniel O., *High Frontier: A New National Strategy* (Washington: High Frontier, 1982).

Graham, Daniel O., and Gregory A. Fossedal, *A Defense That Defends; Blocking Nuclear Attack* (Greenwich, CT: Devin-Adair, 1983).

Gray, Colin S., *Nuclear Strategy and Strategic Planning* (Philadelphia: Foreign Policy Research Institute, 1984.)

Green, Philip, *Deadly Logic: The Theory of Nuclear Deterrence* (Columbus, OH: Ohio State University Press, 1966).

Hartung, William D., et al, *The Strategic Defense Initiative: Costs, Contractors and Consequences* (New York: Council on Economic Priorities, 1985).

Herspring, Dale, and Robin Laird, *The Soviet Union and Strategic Arms* (Boulder, CO: Westview Press, 1984).

Holloway, David, *The Soviet Union and the Arms Race* (New Haven, CO: Yale University Press, 1983).

Huntington, Samuel P., ed., *The Strategic Imperative: New Policies for American Security* (Cambridge, MA: Ballinger, 1982).

Jasani, Bhupendra, ed., *Space Weapons—The*

Arms Control Dilemma (London: Taylor and Francis, 1984).

Jasani, Bhupendra, and Christopher Lee, *Countdown to Space War* (London: Taylor and Francis, 1984).

Jastrow, Robert, *How to Make Nuclear Weapons Obsolete* (Boston, MA: Little, Brown & Co., 1985).

Jervis, Robert, *The Illogic of American Nuclear Strategy* (Ithaca, NY: Cornell University Press, 1984).

Kahan, Jerome H., *Security in the Nuclear Age—Developing U.S. Strategic Arms Policy* (Washington: The Brookings Institution, 1975).

Kegley, Charles W., Jr., and Pat McGowan, eds., *Foreign Policy: USA/U.S.S.R.* (Beverly Hills, CA: Sage Publications, 1982).

Kennan, George F., *The Nuclear Delusion* (New York: Pantheon, 1982).

Krepon, Michael, *Strategic Stalemate: Nuclear Weapons and Arms Control in American Politics* (New York: St. Martin's Press, 1984).

Kolkowicz, Roman, and Neil Joeck, eds., *Arms Control and International Security* (Boulder, CO: Westview Press, 1984).

Lebow, Richard Ned, *Between Peace and War: The Nature of International Crisis* (Baltimore, MD: Johns Hopkins University Press, 1981).

Leebaert, Derek, ed., *Soviet Military Thinking* (London: George Allen and Unwin, 1981.

Mandelbaum, Michael, *The Nuclear Question: The United States and Nuclear Weapons, 1946-1976* (New York: Cambridge University Press, 1979).

Martin, Laurence, ed., *Strategic Thought in the Nuclear Age* (Baltimore, MD: Johns Hopkins University Press, 1979).

McNamara, Robert S., *The Essence of Security: Reflections in Office* (New York: Harper and Row, 1968).

Nacht, Michael, *The Age of Vulnerability: Threats to Nuclear Stalemate* (Washington: The Brookings Institution, 1985).

National Academy of Sciences, *Nuclear Arms Control: Background and Issues* (Washington: National Academy Press, 1985).

Nincic, Miroslav, *The Arms Race: The Political Economy of Military Growth* (New York: Praeger, 1982).

O'Neill, Robert and D.M. Horner, eds., *New Directions in Strategic Thinking* (London: George Allen and Unwin, 1981).

Palme, Olof, et al., *Common Security: A Programme for Disarmament* (London: Pan Books Ltd., 1982).

Payne, Keith B., *Nuclear Deterrence in U.S.-Soviet Relations* (Boulder, CO: Westview Press, 1982).

Schneider, William, Jr., et al., *U.S. Strategic-Nuclear Policy and Ballistic Missile Defense: The 1980s and Beyond* (Cambridge, MA: Institute for Foreign Policy Analysis, 1980).

Smith, Gerard, *Doubletalk: The Story of the First Strategic Arms Limitation Talks* (New York: Doubleday, 1980).

Smoke, Richard, *National Security and the Nuclear Dilemma* (Reading, MA: Addison-Wesley, 1984).

Smoke, Richard, *War: Controlling Escalation* (Cambridge, MA: Harvard University Press, 1977).

Stein, Jonathan B., *From H-Bomb to Star Wars: The Politics of Strategic Decision Making* (Lexington, MA: Lexington Books, 1984).

Stockholm International Peace Research Institute, *The Arms Race and Arms Control, 1984* (London: Taylor and Francis, 1984.

Ury, William L., and Richard Smoke, *Beyond the Hotline: Controlling a Nuclear Crisis* (Cambridge, MA: Nuclear Negotiation Project, Harvard Law School, 1984).

Weinberger, Caspar W., *Report of the Secretary of Defense to the Congress* (Washington: Department of Defense, February 4, 1985).

Weston, Burns H., ed., *Toward Nuclear Disarmament and Global Security: A Search for Alternatives* (Boulder, CO: Westview Press, 1984).

White, Ralph K., *Fearful Warriors: A Psychological Profile of U.S.-Soviet Relations* (New York: The Free Press, 1984).

Woito, Robert S., *To End War: A New Approach to International Conflict* (New York: Pilgrim Press, 1982).

Yanarella, Ernest J., *The Missile Defense Controversy: Strategy, Technology, and Politics, 1955-1972* (Lexington, KY: University Press of Kentucky, 1977).

Yost, David S., *NATO's Strategic Options: Arms Control and Defense* (Elmsford, NY: Pergamon Press, 1982).

Articles

Abrahamson, James A., "The Strategic Defense Initiative," *Defense/84*, August 1984.

AuCoin, Les, "Nailing Shut the Window of Vulnerability," *Arms Control Today*, September 1984.

Ball, Desmond, "Can Nuclear War Be Controlled?" *Adelphi Paper No. 169*, International Institute for Strategic Studies, London, 1981.

Barkenbus, Jack N., and Alvin M. Weinberg, "Defense-Protected Build-Down," *Bulletin of the Atomic Scientists*, October 1984, pp. 18-23.

Bethe, Hans A., et al, "Space-Based Ballistic Missile Defense," *Scientific American*, October 1984, pp. 39-49.

Broad, William J., "Allies in Europe Are Apprehensive About Benefits of 'Star Wars' Plan," *New York Times*, May 13, 1985, pp. A-1, A-6.

Brown, Harold, "The Strategic Defense Initiative: Defensive Systems and the Strategic Debate," *Survival*, March/April 1985, pp. 55-64. Also printed in Discussion Paper No. 104, California Seminar on International Security and Foreign Policy, Santa Monica, Calif., March 1985.

Brown, Harold, and Lynn E. Davis, "Nuclear Arms Control: Where Do We Stand?" *Foreign Affairs*, 62:1145-1160, summer 1984.

Brzezinski, Zbigniew, Robert Jastrow, and Max M. Kampelman, "Defense in Space is Not Star Wars," *New York Times Magazine*, Jan. 27, 1985.

Bundy, McGeorge, et al, "The President's Choice: Star Wars or Arms Control," *Foreign Affairs*, 63:264-278, winter 1984/85.

Bundy, McGeorge, et al, "Nuclear Weapons and the Atlantic Alliance," *Foreign Affairs*, 60:753-768, spring 1982.

Burrows, William E., "Ballistic Missile Defense: The Illusion of Security," *Foreign Affairs*, 62:843-856, spring 1984.

Clifford, Clark M., Testimony before the Committee on Foreign Affairs, U.S. House of Representatives, May 1, 1985.

Drell, Sidney D., and Wolfgang K.H. Panofsky, "The Case Against Strategic Defense: Technical and Strategic Realities," *Issues in Science and Technology*, fall 1984, pp. 45-65.

Drell, Sidney D., Philip J. Farley, and David Holloway, "Preserving the ABM Treaty: A Critique of the Reagan Strategic Defense Initiative," *International Security*, 9(2):51-91, fall 1984.

Forsberg, Randall, "Confining the Military to Defense as a Route to Disarmament," *World Policy Journal*, 1:285-318, winter 1984.

Garthoff, Raymond L., "Mutual Deterrence and Strategic Arms Limitation in Soviet Policy," *International Security*, 3(1):112-147, summer 1978.

Garwin, Richard, et al, "Space Weapons," *Bulletin of the Atomic Scientists*, May 1984.

George, Alexander L., "Crisis Management: The Interaction of Political and Military Considerations," *Survival*, 26:223-235, September/October 1984.

Glaser, Charles L., "Why Even Good Defenses May Be Bad," *International Security* 9(2):92-123, fall 1984. Abridged version appears in *Bulletin of the Atomic Scientists*, March 1985, pp. 13-16.

Gray, Colin S., "Strategic Stability Reconsidered," *Daedalus (Journal of the American Academy of Arts and Sciences)*, 109 (4):135-154, fall 1980.

Hafner, Donald, "The Strategic Defense Initiative and Nuclear Deterrence," *Daedalus*, spring 1985.

Hart, Douglas M., "Soviet Approaches to Crisis Management: The Military Dimension," *Survival*, 26:214-223, September/October 1984.

Howe, Sir Geoffrey, "Defence and Security in the Nuclear Age," Speech to the Royal United Services Institute in London, Mar. 15, 1985.

Ikle, Fred C., "Nuclear Strategy: Can There be a Happy Ending?" *Foreign Affairs*, spring 1985, pp. 810-826.

Ikle, Fred C., "Can Nuclear Deterrence Last Out the Century?" *Foreign Affairs*, winter 1973.

Keyworth, George A., II, "A Sense of Obligation—The Strategic Defense Initiative," *Aerospace America*, April 1984, pp. 56-62.

Keyworth, George A., II, "Strategic Defense Initiative: The Rational Route to Effective Nuclear Arms Control," *Government Executive*, June 1984.

Keyworth, George A., II, "The Case For Strategic Defense: An Option For a World Disarmed," *Issues in Science and Technology*, fall 1984.

Krauthammer, Charles, "The Illusion of Star Wars," *The New Republic*, May 14, 1984, pp. 13-17.

Kupperman, Robert H., and Harvey A. Smith, "Strategies of Mutual Deterrence," *Science*, 176:18-23, Apr. 7, 1972.

McNamara, Robert S., and Hans A. Bethe, "Reducing the Risk of Nuclear War," *The Atlantic Monthly*, July 1985, pp. 43-51.

McNamara, Robert S., Testimony before the Committee on Foreign Affairs, U.S. House of Representatives, May 1, 1985.

McNamara, Robert S., "The Military Role of Nuclear Weapons: Perceptions and Misperceptions," *Foreign Affairs*, fall 1983.

Murray, Russell, 2nd, "Consequences of Nuclear Warfare," Testimony before the Joint Economic Committee, U.S. Congress, July 11, 1984.

Panofsky, Wolfgang K.H., "The Mutual Hostage Relationship between America and Russia," *Foreign Affairs*, 52 (1):109-118, October 1973.

Payne, Keith B., et al, "The Strategic Defense Initiative," *Orbis*, summer 1984, pp. 215-256.

Rathjens, George, and Jack Ruina, "The Uncertainty of Ballistic Missile Defense," *Daedalus*, Summer 1985.

Rathjens, George, "The Strategic Defense Initiative: The Imperfections of 'Perfect Defense'," *Environment*, June 1984, pp. 6-13.

Schlesinger, James R., Speech on ballistic missile defense at Symposium on Space, National Security and C-cubed-I, The Mitre Corporation, Bedford, Mass., October 25, 1984. Mitre Document M85-3, pp. 55-62.

Scowcroft, Brent, et al, "Report of the President's Commission on Strategic Forces," April 6, 1983.

Scowcroft, Brent, et al, "Second Report of the President's Commission on Strategic Forces," March 21, 1984.

Shulman, Marshall D., "SALT and the Soviet Union," in *SALT: The Moscow Agreements and Beyond*, ed. by Mason Willrich and John B. Rhinelander, (New York: The Free Press, 1974).

Steinbruner, John D., and Thomas M. Garwin, "Strategic Vulnerability: The Balance Between Prudence and Paranoia," *International Security*, 1(1):138-170, Summer 1976.

Steinbruner, John D., "National Security and the Concept of Strategic Stability," *Journal of Conflict Resolution*, 22:411-428, September 1978.

Toomay, John, "The Utility of Ballistic Missile Defense," *Daedalus*, Summer 1985.

Warnke, Paul C., "Consequences of Nuclear Warfare," Testimony before the Joint Economic Committee, U.S. Congress, July 11, 1984.

Yost, David S., "Ballistic Missile Defense and the Atlantic Alliance," *International Security*, Fall 1982, pp 143-174.

Appendix M

References on Soviet Strategic Policy

Berman, Robert P., and John C. Baker, *Soviet Strategic Forces: Requirements and Responses* (Washington: The Brookings Institution, 1982).

Davis, Jacquelyn K., et al, *The Soviet Union and Ballistic Missile Defense* (Cambridge, MA: Institute for Foreign Policy Analysis, 1980).

Deane, Michael J., *The Role of Strategic Defense in Soviet Strategy* (Coral Gables, FL: Advanced International Studies Institute in association with the University of Miami, 1980).

Ermarth, Fritz W., "Contrasts in American and Soviet Strategic Policies," *International Security*, fall 1978.

Garthoff, Raymond L., *Detente and Confrontation: American-Soviet Relations from Nixon to Reagan* (Washington: The Brookings Institution, 1985).

Garthoff, Raymond L., "BMD and East-West Relations," in Ashton B. Carter and David N. Schwartz, eds., *Ballistic Missile Defense* (Washington: The Brookings Institution, 1984).

Garthoff, Raymond L., "Mutual Deterrence and Strategic Arms Limitation in Soviet Policy," *International Security*, 3(1):112-147, summer 1978.

Goure, Daniel, and Gordon H. McCormick, "Soviet Strategic Defense: The Neglected Dimension of the U.S.-Soviet Balance," *Orbis*, spring 1980.

Graybeal, Sidney, and Daniel Goure, "Soviet Ballistic Missile Defense Objectives: Past, Present, and Future," in *U.S. Arms Control Objectives and the Implications for Ballistic Missile Defense*, Proceedings of a symposium at Harvard University in November, 1979 (Cambridge, MA: Harvard University Center for Science and International Affairs, 1980).

Green, William C., *Soviet Nuclear Weapons Policy: A Research Guide* (Boulder, CO: Westview Press, 1985).

Hart, Douglas M., "Soviet Approaches to Crisis Management: The Military Dimension," *Survival*, 26:214-223, September/October 1984.

Herspring, Dale, and Robin Laird, *The Soviet Union and Strategic Arms* (Boulder, CO: Westview Press, 1984).

Holloway, David, "The Soviet Union and the SDI," *Daedalus*, summer 1985.

Holloway, David, *The Soviet Union and the Arms Race* (New Haven, CN: Yale University Press, 1983).

Leebaert, Derek, ed., *Soviet Military Thinking* (London: George Allen and Unwin, 1981).

Lockwood, Jonathan S., *The Soviet View of U.S. Strategic Doctrine: Implications for Decision Making* (New Brunswick, NJ: Transaction Books, 1983).

Meyer, Stephen M., "Soviet Military Programs and the New High Ground," *Survival*, September 1983, pp. 204-215.

Shenfield, Stephen, "Soviets May Not Imitate Star Wars," *Bulletin of the Atomic Scientists*, June/July 1985, pp. 38-39.

Shulman, Marshall D., "U.S.-Soviet Relations and the Control of Nuclear Weapons," in *Rethinking the U.S. Strategic Posture*, ed. by Barry M. Blechman (Cambridge, MA: Ballinger, 1982).

Shulman, Marshall D., "SALT and the Soviet Union," in *SALT: The Moscow Agreements and Beyond*, ed. by Mason Willrich and John B. Rhinelander (New York: The Free Press, 1974).

Strode, Dan L., and Rebecca V. Strode, "Diplomacy and Defense in Soviet National Security Policy," *International Security*, 8(2):91-116, fall 1983.

Walt, Stephen M., *Interpreting Soviet Military Statements: A Methodological Analysis*, Report No. CNA-81-0260.10, Center for Naval Analyses, Alexandria, Va., December, 1983.

Appendix N
Glossary of Acronyms and Terms

List of Acronyms

ABM —anti-ballistic missile
ALCM —air-launched cruise missile
ASAT —anti-satellite
BMD —ballistic missile defense
C^3I —command, control, communications, and intelligence
CONUS —continental United States
DEW —directed-energy weapon
DSAT —defensive satellite
GLCM —ground-launched cruise missile
ICBM —intercontinental ballistic missile
IR —infrared
IRBM —intermediate-range ballistic missile
KEW —kinetic-energy weapon
KKV —kinetic-kill vehicle
LWIR —long-wave infrared
MaRV —maneuverable reentry vehicle
MIRV —multiple independently targeted reentry vehicle
MILSAT—military satellite
MPS —multiple protective shelters, once to be used for basing MX
MWIR —medium-wave infrared
MX —experimental missile, newest addition to U.S. ICBM arsenal, also called "Peacekeeper"
PBV —post-boost vehicle
RV —reentry vehicle
SDI —Strategic Defense Initiative
SDIO —Strategic Defense Initiative Organization
SLBM —submarine-launched ballistic missile
SLCM —sea-launched cruise missile
SWIR —short-wave infrared
UV —ultraviolet

Definitions of Terms

Ablative Shield: A shield that evaporates when heated, absorbing laser energy and protecting the object which is behind it from heat damage.

ABM Treaty: A Treaty of 1972, signed and ratified by the Soviet Union and the United States, prohibiting development of many types of anti-ballistic missile systems and limiting deployments on each side to a specified number of land-based units, which use only rocket interceptors and ground-based radar.

Acquisition: Detection of a potential target by the sensors of a weapons system.

Active Sensor: One that illuminates a target, producing return secondary radiation, which is then detected in order to track and/or identify the target. An example is **ladar** (cf.).

Adaptive Optics: Optical systems which can be modified (e.g., by controlling the shape of a mirror) to compensate for distortions. An example is the use of information from a beam of light, passing through the atmosphere to compensate for the distortion suffered by another beam of light on its passage through the atmosphere. Used to eliminate the "twinkling" of stars in observational astronomy and to reduce the dispersive effect of the atmosphere on laser beam weapons.

Air-breathing: Describing a flying weapon that travels through the atmosphere and uses air in its propulsion system. Examples are jet aircraft and cruise missiles. Specifically does not include ballistic missiles.

Analog Processing: Problem solving in a computer by means of direct manipulation of the magnitudes of a physical quantity. For example, the sizes of different voltage pulses may be compared, added, subtracted, etc., in the course of solving a problem (cf. **digital processing**).

Anti-satellite Weapon (ASAT): A weapon to destroy satellites in space.

Anti-simulation: Deceiving adversary sensors by making a strategic target look like a decoy.

Area Defense: An ABM defense covering a large area. Ususally implies the capability to protect "soft" (i.e. not hardened missile silos or bunkers) targets.

Ballistic Missile Defense (BMD): A defense system that is designed to protect territory from attacking ballistic missiles. Usually conceived as having several independent layers.

Battle Management: The set of instructions and rules and the corresponding hardware controlling the operation of a BMD system. Sensors and interceptors are allocated by the system, and the updated battle results are presented to the (human) command for analysis and possible intervention.

Birth-to-death Tracking: The tracking of objects from the time that they are deployed from a booster or post-boost vehicle until they are killed or detonated.

Bistatic Radar: Radar systems in which the receiver and transmitter are separated.

Blackout: The disabling of radar by means of a nuclear explosion. The intense electromagnetic energy released generates a large background that obscures signals and renders many types of radar useless for minutes or longer.

Boost Phase: The phase of a missile trajectory from launch to burnout of the final stage. For ICBMs, this phase typically lasts from 3 to 5 minutes, but studies indicate that reductions to the order of 1 minute could be possible.

Brightness: In this report, the amount of power that can be delivered per unit solid angle by a directed-energy weapon.

Coherence: The matching, in space and time, of the wave structure of different parallel rays of a single frequency of electromagnetic radiation. This results in the mutual reinforcing of the energy of these different components of a larger beam. Lasers can produce coherent radiation.

Command Guidance: The steering and control of a missile by transmitting commands to it.

Common Mode Failure: Refers to a type of system failure in which diverse components are disabled by the same single cause.

Constellation Size: The number of defensive weapon satellites placed in orbit about the Earth as part of a BMD system.

Counter-countermeasures: In this report, measures taken by the defense to defeat offensive countermeasures.

Countermeasures: In this report, measures taken by the offense to overcome aspects of a BMD system.

Cruise Missile: A missile traveling within the atmosphere at aircraft speeds and, usually, low altitude, whose trajectory is preprogrammed. It is capable of achieving high accuracy in striking a distant target. It is maneuverable during flight, is constantly propelled, and therefore does not follow a ballistic trajectory. Cruise missiles may be nuclear armed, but do not have to be.

Dazzling: In this report, the temporary blinding of a sensor by overloading it with an intense signal of electromagnetic radiation, e.g., from a laser or a nuclear explosion.

Decoy: An object that is designed to make an observer believe that the object is more valuable than is actually the case. Usually, in this report, a decoy refers to a light object, not containing a warhead, designed to look like a nuclear-armed reentry vehicle.

Defensive Satellite Weapon (DSAT): A device that is intended to defend satellites in space by destroying attacking ASAT weapons.

Defensive Technologies Study Team (DTST): A committee, generally known as the "Fletcher Panel," after its Chair, appointed by President Reagan to investigate the technologies of potential BMD systems.

Diffraction: The spreading out of electromagnetic radiation as it leaves an aperture, such as a mirror. The angle of spread, which cannot be eliminated by focusing, is proportional to the ratio of the wavelength of radiation to the diameter of the aperture.

Digital Processing: The most familiar type of computing, in which problems are solved through the mathematical manipulation of streams of numbers.

Directed-Energy Weapon: A weapon that kills its target by delivering energy to it at or near the speed of light. Includes lasers and particle beam weapons.

Discrimination: The ability of a defensive system to differentiate decoys or other nonthreatening objects from targets, e.g., a threatening booster rocket, post-boost vehicle, or RV.

Early Warning: In this report, early detection of an enemy ballistic missile launch, usually by means of surveillance satellites and long-range radar.

Electromagnetic Radiation: A form of propagated energy, arising from electric charges in motion, that produces a simultaneous wavelike variation of electric and magnetic fields in space. The highest frequencies (or shortest wavelengths) of such radiation are possessed by gamma rays, which originate from processes within atomic nuclei. As one goes to lower frequencies, the electromagnetic spectrum includes X-rays, ultraviolet light, visible light, infrared light, microwaves, and radio waves.

Electron-volt: The energy gained by an electron in passing through a potential difference of one volt.

Endoatmospheric: Within the atmosphere; an endoatmospheric interceptor reaches its target within the atmosphere.

Exoatmospheric: Outside the atmosphere; an exoatmospheric interceptor reaches its target in space.

Fast-Burn Booster: A ballistic missile that can burnout much more quickly than current versions, possibly before exiting the atmosphere entirely. Such rapid burnout complicates a boost-phase defense.

Fission: The breaking apart of the nucleus of an

atom, usually by means of a neutron. For very heavy elements, such as uranium, a significant amount of energy is produced by this process. When controlled, this process yields energy which may be extracted for civilian uses, such as commercial electric generation. When uncontrolled, energy is liberated very rapidly: such fission is the energy source of uranium- and plutonium-based nuclear weapons; it also provides the trigger for fusion weapons.

Fratricide: The destructive effect of the earlier-detonating weapons in a barrage on those weapons which arrive later.

Functional Kill: The destruction of a target by disabling vital components in a way not immediately detectable, but nevertheless able to prevent the target from functioning properly. An example is the destruction of electronics in a guidance system by a neutral particle beam.

Fusion: The fusing of two atomic nuclei, usually of light elements, such as hydrogen. For light elements, energy is liberated by this process. Hydrogen bombs produce most of their energy through the fusion of hydrogen into helium.

Geosynchronous Orbit: An orbit about 35,800 km above the Equator. A satellite placed in such an orbit revolves around the Earth once per day, maintaining the same position relative to the surface of the Earth. It then appears to be stationary, and is useful as a communications relay or as a surveillance post.

Hard Kill: Destruction of a target in such a way as to produce unambiguous visible evidence of its neutralization.

Hardness: In this report, a property of a target, measured by the power needed per unit area to destroy the target by means of a directed-energy weapon. A hard target is more difficult to kill than a soft target.

Homing Device: A device, mounted on a missile, that uses sensors to detect the position or to help predict the future position of a target, and then directs the missile to intercept the target. It usually updates frequently during the flight of the missile.

Impulse Kill: The destruction of a target, using directed energy, by ablative shock. The intensity of directed energy is such that the surface of the target violently and rapidly boils off, delivering a mechanical shock wave to the rest of the target and causing structural failure.

Inverse Synthetic Aperture Radar (ISAR): A type of radar similar to **synthetic aperture radar** (cf.), but which uses information from the motion of targets in order to provide high resolution.

Ionization: The removal or addition of one or more electrons to a neutral atom, forming a charged ion.

Keep-out Zone: A volume around a space asset, off limits to parties not owners of the asset. Keep-out zones could be negotiated or unilaterally declared. The right to defend such a zone by force and the legality of unilaterally declared zones under the Outer Space Treaty remain to be determined.

Kill Assessment: The detection and assimilation of information indicating the destruction of an object under attack. Kill assessment is one of the many functions to be performed by a battle management system.

Kinetic-Energy Weapon: A weapon that uses kinetic energy, or energy of motion, to kill an object. Weapons that use kinetic energy are a rock, a bullet, a nonexplosively armed rocket, and an electromagnetic railgun.

Ladar: A technique analogous to radar, but which uses laser light rather than radio or microwaves. The light is bounced off a target and then detected, with the return beam providing information on the distance and velocity of the target.

Laddering Down: A hypothetical technique for overcoming a terminal phase missile defense. Successive salvos of **salvage-fused** (cf.) RVs attack. The detonations of one salvo disable local ABM abilities so that following salvos are able to approach the target more closely before being, in turn, intercepted. Eventually, by repeating the process, the target is reached and destroyed.

Lasant: A material that can be stimulated to produce laser light.

Laser: A device that produces a narrow beam of coherent radiation through a physical process known as stimulated emission. Lasers are able to focus large quantities of energy at great distances, and are among the leading candidates for BMD weapons.

Layered Defenses: The use of several layers of BMD at different phases of the missile trajectory. Each layer is designed to be as independent as possible of the others, and each would probably use its own, distinctive set of missile defense technologies.

Leverage: In this report, refers to the advantage gained by boost-phase intercept, when a single booster kill may eliminate many RVs and decoys before they are deployed. This could provide a favorable cost-exchange ratio for the defense, and would reduce stress on later layers of the defense system.

Limited Test Ban Treaty: The multilateral Treaty signed and ratified by the United States and the U.S.S.R. in 1963 which prohibits nuclear tests in all locations except underground.

Megawatt: One million watts; a unit of power. A typical commercial electric plant generates about 500 to 1,000 megawatts.

Mev: One million electron-volts. A unit of energy usually used in reference to nuclear processes. It is equivalent to the energy that an electron gains in crossing a potential of 1 million volts.

Micron: One-millionth of a meter (equivalently, one-thousandth of a millimeter). Roughly twice the wavelength of visible light.

Midcourse Phase: The phase of a ballistic missile trajectory in which the RVs travel through space on a ballistic course towards their targets. This phase lasts up to 20 minutes.

Military Satellite (MILSAT): A satellite used for military purposes, such as navigation or intelligence gathering.

Monostatic Radar: A radar system in which the receiver and transmitter are colocated.

Multiple Independently-targetable Reentry Vehicle (MIRV): One of several RVs on the same post-boost vehicle that can be independently placed on a ballistic course towards a target after completion of the boost phase.

Multiple Phenomenology: A system using repeated observations of potential targets by means of different physical principles and different sensor systems. In the case of sensor systems, the use of multiple phenomenology makes it more difficult for an adversary to deceive them.

Multistatic Radar: A radar system with a transmitter and several receivers, all separated.

Optical Processing: A type of analog processing (q.v.) in which the behavior of light beams, passed through optical systems, is used in problem solving.

Outer Space Treaty of 1967: A signed and ratified agreement between the Soviet Union, the United States, and other nations, forbidding the basing of nuclear or other weapons of mass destruction in space.

Parallel Processing: The use of different paths in a computer to work simultaneously on different calculations needed to solve a single problem, thus reducing the time needed for the overall calculation.

Passive Sensor: One that detects naturally occurring emissions from a target for tracking and/or identification purposes.

Penetration Aid: In this report, a device mounted on a post-boost vehicle with RVs, that is used to confuse defenses. It may be a decoy or anything else that renders more difficult the defense's job of detecting and killing the RVs or the PBV.

Phased-Array Radar (PAR): A radar with elements that are physically stationary, but with a beam that is electronically steerable and can switch rapidly from one target to another. Used for tracking many objects, often at great distances.

Pointing: The aiming of sensors or defense weapons at a target with sufficient accuracy either to track the target or to aim with sufficient accuracy to destroy it.

Post-boost Phase: The phase of a missile trajectory, after the booster's stages have finished firing, in which the various RVs are independently placed on ballistic trajectories towards their targets. In addition, **penetration aids** (cf.) are dispensed from the post-boost vehicle. The length of this phase is typically 3 to 5 minutes, but could be drastically reduced.

Power Supply: In this report, a source of energy for a BMD component. It may be ground- or space-based, and may range from commercial electric plants to space-based nuclear reactors.

Preferential Defense: The concentration of (usually limited) defensive assets on a subset of sites in order to assure the survival of some of them.

Preferential Offense: The concentration of offensive assets on a subset of targets.

Pumping: In this report, the raising of the molecules or atoms of a **lasant** (cf.) to an energy state above the normal lowest state, in order to produce laser light. This results when they fall back to a lower state. Pumping may be done using electrical, chemical, or nuclear energy.

Redout: The blinding or dazzling of infrared detectors due to high levels of infrared radiation produced in the upper atmosphere by a nuclear explosion.

Reentry: The return of objects, originally launched from Earth, into the atmosphere.

Reentry Vehicle (RV): As used in this report, reentry vehicles are small containers containing nuclear warheads. They are released from the last stage of a booster rocket or from a **post-boost vehicle** (cf.) early in the ballistic trajectory. They are thermally insulated to survive rapid heating during the high velocities of reentry into the atmosphere, and are designed to protect their contents until detonation at their targets.

Responsive Threat: The **threat** (cf.) after taking

into account modernization and BMD countermeasures.

Robust: In this report, describing a system, indicating its ability to endure and perform its mission against a reactive adversary. Also used to indicate ability to survive under direct attack.

Safeguard: A U.S. midcourse and terminal-phase defense for ICBMs, deployed in 1975 and deactivated in 1976 due to its limited cost-effectiveness.

Salvage-fused: Describing a warhead that is set to detonate when it is attacked. Usually refers to a nuclear warhead.

Selectivity: In this report, refers to choosing a subset of targets, either for attack or defense. See preferential defense and preferential offense.

Semi-active Sensor: One that does not generate radiation itself, but that detects radiation reflected by targets when they are illuminated by other BMD components. Such devices are used for tracking and identification and can operate without revealing their own locations.

Sensors: Electronic instruments that can detect radiation from objects at great distances. The information can be used for tracking, aiming, discrimination, attacking, kill assessment, or all of the above. Sensors may detect any type of electromagnetic radiation or several types of nuclear particles.

Sentinel: ABM system designed for light area defense against a low-level ballistic missile attack on the United States. Developed into the Safeguard (cf.) system in late 1960s.

Shoot-back: In this report, the technique of defending a space asset by shooting at an attacker.

Signature: Distinctive type of radiation emitted or reflected by a target, which can be used to identify that target.

Simulation: The art of making a decoy look like a more valuable strategic target (cf. **anti-simulation**).

Slew Time: The time needed for a weapon to reaim at a new target after having just fired at a previous one.

Soft Kill: Same as **functional kill.**

Space Mines: Hypothetical devices that can track and follow a target in orbit, with the capability of exploding on command or by pre-program, in order to destroy the target.

Spartan: Nuclear-armed long-range midcourse interceptor used in Safeguard/Sentinel systems (cf.).

Sprint: Nuclear-armed short-range interceptor used in Safeguard/Sentinel systems (cf.).

SS-18: Largest ICBM in current Soviet inventory, credited with carrying 10 RVs, but capable of holding many more.

Stimulated Emission: Physical process by which an excited molecule is induced by incident radiation to emit radiation at an identical frequency and in phase with the incident radiation. Lasers operate by stimulated emission.

Structured Attack: An attack in which the arrival of warheads on their diverse targets is precisely timed for maximum strategic impact.

Synthetic Aperture Radar (SAR): A radar technique that processes echoes of signals emitted at different points along a satellite's orbit. The highest resolution achievable by such a system is theoretically equivalent to that of a single large antenna as wide as the distance between the most widely spaced points along the orbit that are used for transmitting positions.

Terminal Phase: The final phase of a ballistic missile trajectory, lasting about a minute or less, in which the RVs reenter the atmosphere and detonate at their targets.

Thermal Kill: The destruction of a target by heating it, using directed energy, to the degree that structural components fail.

Threat: The anticipated inventory of enemy weapons. In the context of this report, the inventory is of nuclear weapons and their delivery systems, as well as of decoys, penetration aids, and other BMD countermeasures.

Track File: Information stored in computer memory containing position coordinates and velocity components of a target. In this report, refers to such information concerning offensive weapons during their trajectories: e.g., boosters, RVs, decoys.

Tracking: The monitoring of the course of a moving target. Ballistic objects may have their tracks predicted by the defensive system, using several observations and physical laws.

Transition: In this report, the period in which the world strategic balance would shift from offense-dominance to defense-dominance.

Warhead: A weapon, usually a nuclear weapon, contained in the payload of a missile.

Anti-Satellite Weapons, Countermeasures, and Arms Control

Foreword

At the requests of the House Armed Services Committee and the Senate Foreign Relations Committee, OTA undertook an assessment of the opportunities and risks involved in an accelerated program of research on new ballistic missile defense technologies, including those that might lead to deployment of weapons in space. The resulting report, *Ballistic Missile Defense Technologies*, is being published concurrently with this volume. This report on *Anti-Satellite Weapons, Countermeasures, and Arms Control* discusses additional implications of the same or similar technologies.

Closely related to BMD technology, system survivability, and arms control issues are questions about the development and deployment of anti-satellite weapons. Whether or not the United States decides to deploy BMD systems in space, other military uses of space will continue to grow in importance. How can the United States respond to the potential threat to its military capabilities posed now and in the future both by Soviet military satellites and by Soviet anti-satellite weapons (ASAT)? This report examines U.S. options for countering Soviet military satellite capabilities and explores both unilateral and cooperative measures for limiting the ASAT threat. Possible unilateral steps include active and passive countermeasures as well as deterrence; possible cooperative steps include a variety of arms control agreements. The report examines the pros and cons of several illustrative "arms control regimes" for space weapons, ranging from lesser to greater limitations than now exist. It suggests that some combinations of unilateral and cooperative measures might provide more military security than either type alone.

It should be recognized that the relative roles of anti-satellite weapons, countermeasures, and arms control will be strongly affected by the course followed in the development and deployment of space-based BMD systems.

OTA gratefully acknowledges the contributions of the many individuals, firms, laboratories, and government agencies who assisted its research and writing for this report.

JOHN H. GIBBONS
Director

Glossary of Acronyms and Terms

Glossary of Acronyms

ABM —anti-ballistic missile
ASAT —anti-satellite
ASATCC —ASAT Control Center
BMD —ballistic missile defense
CMC —Cheyenne Mountain Complex
DSAT —defensive satellite
DEW —directed-energy weapon
FEL —free-electron laser
FOBS —Fractional-Orbit Bombardment System
GEO —geosynchronous Earth orbit
GRASER —gamma-ray amplification by stimulated emission of radiation
GSO —geostationary Earth orbit
HEO —high-Earth orbit
INW —isotropic nuclear weapon
°K —degrees Kelvin
KEW —kinetic-energy weapon
LASER —Light amplification by stimulated emission of radiation; the acronym "laser" is no longer capitalized in current usage
LEO —low-Earth orbit
LIDAR —light detection and ranging
LTBT —Limited Test Ban Treaty
MW —megawatt
MeV —million electron-volts
MHV —miniature homing vehicle
MILSAT —military satellite
MV —miniature vehicle
NSSC —National Space Surveillance Center
PMOC —Prototype Mission Operations Center
RADAR —radio detection and ranging; the acronym "radar" is no longer capitalized in current usage
ROCC —Regional Operations Control Center
SAR —synthetic aperture radar
SPADATS—Space Detection and Tracking System
SPADOC —Space Defense Operations Center
WWMCCS—World-Wide Military Command and Control System

Glossary of Terms

Ablative Shield: A shield that evaporates when heated, absorbing laser energy and protecting the object which is behind it from heat damage.

Ablative Shock: Generation of a mechanical shock wave at the surface of an object exposed to intense pulsed electromagnetic radiation. A thin layer of the objects surface violently and rapidly boils off; the resulting vapor suddenly exerts pressure against the surface, generating a mechanical shock wave at the surface. This shock wave then propagates deeper into the object and can cause melting, vaporization, and spallation of surface material and structural failure of the object.

ABM Treaty: A Treaty of 1972, signed and ratified by the Soviet Union and the United States, prohibiting development of many types of anti-ballistic missile systems and limiting deployments on each side to a specified number of land-based units, which use only rocket interceptors and ground-based radar.

Acquisition: Detection of a potential target by the sensors of a weapons system.

Active Sensor: One that illuminates a target, producing return secondary radiation, which is then detected in order to track and or identify the target. An example is *LIDAR*.

Adaptive Optics: Optical systems which can be modified (e.g., by controlling the shape of a mirror) to compensate for distortions. An example is the use of information from a beam of light, passing through the atmosphere to compensate for the distortion suffered by another beam of light on its passage through the atmosphere. Used to eliminate the "twinkling" of stars in observational astronomy and to reduce the dispersive effect of the atmosphere on laser beam weapons.

Amplified Spontaneous Emission: See *Superradiance*.

Anti-satellite Weapon (ASAT): A weapon to destroy satellites in space.

Anti-simulation: Deceiving adversary sensors by making a strategic target look like a decoy.

Apogee: The maximum altitude attained by an Earth satellite.

Ballistic Missile Defense (BMD): A defense system that is designed to protect a territory from attacking ballistic missiles.

Birth-to-death Tracking: The tracking of space objects—e.g., satellites, reentry vehicles, or decoys which simulate these—from the time that they are deployed from a booster or post-boost vehicle until they are destroyed.

Bistatic Radar: A radar system which has transmitters and receivers stationed at two locations; a special case of multistatic radar.

Boost Phase: The phase of a missile trajectory from launch to burnout of the final stage. For ICBMs, this phase typically lasts from 3 to 5 minutes, but studies indicate that reductions to the order of 1 minute are possible.

Brightness: In this report, the amount of power that can be delivered per unit solid angle by a directed-energy weapon.

Capital Satellite: A highly valued or costly satellite, as distinct from an inexpensive decoy satellite. Some decoys might be so expensive as to be considered capital satellites.

Chaff: Confetti-like metal foil ribbons which can be ejected from spacecraft (or terrestrial vehicles) to reflect enemy radar signals, thereby creating false targets or screening actual targets from the "view" of radar.

Coherence: The matching, in space (transverse) or time (temporal coherence), of the wave structure of different parallel rays of a single frequency of electromagnetic radiation. This results in the mutual reinforcing of the energy of these different components of a larger beam. Lasers and radar systems produce partially coherent radiation.

Command Guidance: The steering and control of a missile by transmitting commands to it.

Counter-countermeasures: Measures taken to defeat countermeasures.

Countermeasures: In this report, measures taken by the offense to overcome aspects of a BMD system.

Dazzling: In this report, the temporary blinding of a sensor by overloading it with an intense signal of electromagnetic radiation, e.g., from a laser or a nuclear explosion.

Decoy: An object that is designed to make an observer believe that the object is more valuable than is actually the case. Usually, in this report, a decoy refers to a light object designed to look like a satellite.

Deep Space: The region of outer space at altitudes greater than 3,000 nautical miles (about 5,600 kilometers) above the Earth's surface.

Defensive Satellite (DSAT) Weapon: A space-based ASAT weapon that is intended to defend satellites by destroying attacking ASAT weapons.

Defensive Technologies Study Team (DTST): A committee, generally known as the "Fletcher Panel," after its Chair, appointed by President Reagan to investigate the technologies of potential BMD systems.

Delta-V: A numerical index of the maneuverability of a satellite or rocket. It is the maximum change in velocity which a spacecraft could achieve in the absence of a gravitational field.

Diffraction: The spreading out of electromagnetic radiation as it leaves an aperture, such as a mirror. The degree of spread, which cannot be eliminated by focusing, is proportional to the ratio of the wavelength of radiation to the diameter of the aperture.

Digital Processing: The most familiar type of computing, in which problems are solved through the mathematical manipulation of streams of numbers.

Directed-Energy Weapon: A weapon that kills its target by delivering energy to it at or near the speed of light. Includes lasers and particle beam weapons.

Discrimination: The ability of a surveillance system to distinguish decoys from intended targets, e.g., certain types of satellites.

Early Warning: In this report, early detection of an enemy ballistic missile launch, usually by means of surveillance satellites and long-range radar.

Electromagnetic Radiation: A form of propagated energy, arising from electric charges in motion, that produces a simultaneous wavelike variation in electric and magnetic fields in space. The highest frequencies (or shortest wavelengths) of such radiation are possessed by gamma rays, which originate from processes within atomic nuclei. As one goes to lower frequencies, the electromagnetic spectrum includes X-rays, ultraviolet light, visible light, infrared light, microwave, and radio waves.

Electron-volt: The energy gained by an electron in passing through a potential difference of 1 volt. 6.25 quintillion electron-volts equals 1 joule; 22.5 billion trillion electron-volts equals 1 kilowatt-hour.

Elliptical Orbit: A noncircular *Keplerian orbit.*

Endoatmospheric: Within the atmosphere; an endoatmospheric interceptor intercepts its target within the atmosphere.

Ephemeris: A collection of data about the predicted positions (or apparent positions) of celestial objects, including artificial satellites, at various times in the future. A satellite ephemeris might contain the orbital elements of satellites and predicted changes in these.

Equatorial Orbit: An orbit above the Earth's Equator.

Excimer: A contraction for "excited dimer"; a type of lasant. A dimer is a molecule consisting of two atoms. Some dimers—e.g., xenon chloride and krypton fluoride—are molecules which cannot exist under ordinary conditions of approximate thermal equilibrium but must be created in an "excited"—i.e., energized—condition by special "pumping" processes in a laser.

Exoatmospheric: Outside the atmosphere; an exoatmospheric interceptor intercepts its target in space.

Fission: The breaking apart of the nucleus of an atom, usually by means of a neutron. For very heavy elements, such as uranium, a significant amount of energy is produced by this process. When controlled, this process yields energy which may be extracted for civilian uses, such as commercial electric generation. When uncontrolled energy is liberated very rapidly: such fission is the energy source of uranium- and plutonium-based nuclear weapons; it also provides the trigger for fusion weapons.

Fratricide: In this report, the unintended destruction of some of a nation's weapons or other military systems (e.g., satellites) by others.

Free-electron Laser: A type of laser which does not use ordinary matter as a lasant but instead generates radiation by the interaction of an electron beam with a static magnetic or electric field. Loosely speaking, free-electron laser technology resembles and evolved from that used by particle accelerators ("atom smashers"). Lasers which are not free-electron lasers are bound-electron lasers.

Functional Kill: The destruction of a target by disabling vital components in a way not immediately detectable, but nevertheless able to prevent the target from functioning properly. An example is the destruction of electronics in a guidance system by a neutral particle beam.

Fusion: More specifically, nuclear fusion: The fusing of two atomic nuclei, usually of light elements, such as hydrogen. For light elements, energy is liberated by this process. Hydrogen bombs produce most of their energy through the fusion of hydrogen into helium.

Graser: See *Gamma-ray Laser*.

Gamma-ray Laser: A laser which generates a beam of *gamma rays*; also called a *graser*. A gamma-ray laser, if developed, would be a type of X-ray laser; although it would employ nuclear reactions, it need not (but might) employ nuclear fission or fusion reactions or explosions.

Gamma Rays: X-rays emitted by the nuclei of atoms.

Geostationary Orbit: An orbit at an altitude of 35,800 kilometers above the Earth's Equator. A satellite placed in such an orbit revolves around the Earth once per day, maintaining the same position relative to the surface of the Earth. It then appears to be stationary, and can be used as a communications relay or as a surveillance post.

Geosynchronous Orbit: An orbit about 35,800 km above the Equator. A satellite placed in such an orbit revolves around the Earth once per day. See *Geostationary Orbit*.

Gray: The Systême International unit of absorbed dose of ionizing radiation. One gray (abbreviated 1 Gy) is 1 joule of absorbed energy per kilogram of matter.

Hard Kill: Destruction of a target in such a way as to produce unambiguous visible evidence of its neutralization.

Hardness: A property of a target; measured by the power needed per unit area to destroy the target by means of a directed-energy weapon. A hard target is more difficult to kill than a soft target.

High-Earth Orbit: An orbit about the Earth at an altitude greater than 3,000 nautical miles (about 5,600 kilometers).

Homing Device: A device, mounted on a missile, that uses sensors to detect the position or to help predict the future position of a target, and then directs the missile to intercept the target. It usually updates frequently during the flight of the missile.

Impulse: A mechanical jolt delivered to an object. Physically, impulse is a force applied for a period of time, and the Systême Internationale unit of impulse is the newton-second (abbreviated N-s). See *Impulse Intensity*.

Impulse Intensity: Mechanical impulse per unit area. The Systême Internationale unit of impulse intensity is the pascal-second (abbreviated Pa-s) A conventionally used unit of impulse intensity is the "tap," which is one dyne-second per square centimeter; hence 1 tap = 0.1 Pa-s.

Impulse Kill: The destruction of a target, using directed energy, by ablative shock. The intensity of directed energy may be so great that that the surface of the target violently and rapidly boils off delivering a mechanical shock wave to the rest of the target and causing structural failure.

Inclination: The inclination of an orbit is the (dihedral) angle between the plane containing the orbit and the plane containing the Earth's Equator. An equatorial orbit has an inclination of 0° for a satellite traveling eastward or 180° for a satellite traveling westward. An orbit having an inclination between 0° and 90° and in which a satellite is traveling generally eastward is called a prograde orbit. An orbit having an inclination of 90° passes above the north and south poles and is called a polar orbit. An orbit having an inclination of more than 90° is called a retrograde orbit.

Ionization: The removal or addition of one or more electrons to a neutral atom, forming a charged ion.

Isotropic: Independent of direction; referring to

the radiation of energy, it means "with equal intensity in all directions," i.e., omnidirectional.

Isotropic Nuclear Weapon (INW): A nuclear explosive which radiates X-rays and other forms of radiation with approximately equal intensity in all directions. The term "isotropic" is used to distinguish them from nuclear directed-energy weapons.

Joule: The Système Internationale unit of energy. One kilowatt-hour is 3.6 million joules.

Keep-out Zone: A volume around a space asset, off limits to parties not owners of the asset. Keep-out zones could be negotiated or unilaterally declared. The right to defend such a zone by force and the legality of unilaterally declared zones under the Outer Space Treaty remain to be determined.

Kelvin Temperature: A scale of temperature on which zero degrees Kelvin (abbreviated 0° K) corresponds to "absolute zero." Temperature in degrees Kelvin equals temperature in degrees Celsius plus 273.16, thus ice melts at 273.16° K, and water boils at 373.16° K.

Keplerian Orbit: The orbit which a satellite would follow if the Earth were a uniform sphere with no atmosphere, and if other simplifying assumptions were valid. Such an orbit would be an ellipse having the center of the Earth as one focus. A special case of such an orbit is a circular orbit about the center of the Earth.

Kill Assessment: The detection and assimilation of information indicating the destruction of an object under attack. Kill assessment is one of the many functions to be performed by a battle management system.

Kinetic-Energy Weapon: A weapon that uses kinetic energy, or energy of motion, to kill an object. Weapons that use kinetic energy are a rock, a bullet, a nonexplosively armed rocket, and an electromagnetic railgun.

Lasant: A material that can be stimulated to produce laser light. Many materials can be used as lasants; these can be in solid, liquid, or gaseous form (consisting of molecules—including excimers— or atoms) or in the form of a plasma (consisting of ions and electrons). Lasant materials useful in high-energy lasers include carbon dioxide, carbon monoxide, deuterium fluoride, hydrogen fluoride, iodine, xenon chloride, krypton fluoride, and selenium, to mention but a few.

Laser: A device that produces a narrow beam of coherent radiation through a physical process known as stimulated emission. Lasers are able to focus large quantities of energy at great distances, and are among the leading candidates for BMD weapons.

LIDAR: A technique analogous to radar, but which uses laser light rather than radio or microwaves. The light is bounced off a target and then detected, with the return beam providing information on the distance and velocity of the target.

Limited Test Ban Treaty: The multilateral Treaty signed and ratified by the United States and the U.S.S.R. in 1963 which prohibits nuclear tests in all locations except underground.

Megawatt: One million watts; a unit of power. A typical commercial electric plant generates about 500 to 1,000 megawatts.

MeV: One million electron-volts. A unit of energy usually used in reference to nuclear processes. It is equivalent to the energy that an electron gains in crossing a potential of 1 million volts.

Micron: One-millionth of a meter (equivalently, one-thousandth of a millimeter). Roughly twice the wavelength of visible light.

Midcourse Phase: The phase of a ballistic missile trajectory in which the RVs travel through space on a ballistic course towards their targets. This phase lasts up to 20 minutes.

Military Satellite (MILSAT): A satellite used for military purposes, such as navigation or intelligence gathering.

Miniature Homing Vehicle (MHV)/Miniature Vehicle (MV): An air-launched direct-ascent ("pop-up") kinetic-energy ASAT weapon currently being developed and tested by the U.S. Air Force.

Monostatic Radar: A radar system in which the receiver and transmitter are colocated.

Multistatic Radar: A radar system that has transmitters and receivers stationed at multiple locations; typically, a radar system with a transmitter and several receivers, all of which are geographically separated. A special case is bistatic radar. An advantage of multistatic radar over monostatic radar is that even if transmitters—which might be detected by the enemy when operating—are attacked, receivers in other locations might not be noticed and might thereby escape attack.

Obscurant: A material—e.g., smoke or chaff—used to conceal an object from observation by a radio or optical sensor. Smoke may be used to conceal an object from observation by an optical sensor, and chaff may be used to conceal an object from observation by a radio sensor (e.g., radar).

On-line: Operating, as distinct from dormant.

Orbital Elements: Any set of several parameters (e.g., apogee, perigee, inclination, etc.) used to specify a Keplerian orbit and the position of a satellite in such an orbit at a particular time. Seven independent orbital elements are required to umambiguously specify the position of a satellite in a Keplerian orbit at a particular time.

Outer Space Treaty of 1967: A multilateral treaty

signed and ratified by both the United States and the Soviet Union. Article IV of the Outer Space treaty forbids basing nuclear weapons or other weapons of mass destruction in space.

Passive Sensor: One that detects naturally occurring emissions from a target for tracking and/or identification purposes.

Perigee: The minimum altitude attained by an Earth satellite.

Phased-Array Radar (PAR): A radar with elements that are physically stationary, but with a beam that is electronically steerable and can switch rapidly from one target to another. Used for tracking many objects, often at great distances.

Pointing: The aiming of sensors or defense weapons at a target with sufficient accuracy either to track the target or to aim with sufficient accuracy to destroy it.

Polar Orbit: An orbit having an inclination of 90°.

Prograde Orbit: An orbit having an inclination of between 0° and 90°. See *Retrograde Orbit*.

Pumping: In this report, the raising of the molecules or atoms of a lasant to an energy state above the normal lowest state, in order to produce laser light. This results when they fall back to a lower state. Pumping may be done using electrical, chemical, or nuclear energy.

Rad: A unit of absorbed dose of ionizing radiation. One rad is 0.001 *gray*.

Radar: A technique for detecting targets in the atmosphere or in space by transmitting radio waves (e.g., microwaves) and sensing the waves reflected by objects. The reflected waves (called "returns" or "echos") provide information on the distance to the target and the velocity of the target and may also provide information about the shape of the target. (Originally an acronym for "RAdio Detection And Ranging.")

Radian: A unit of angular measure. One radian is about 57.3°. One microradian (0.000001 radian) is the angle subtended by an object 1 meter across at a distance of 1,000 kilometers.

Reaction Decoy: A decoy deployed only upon warning or suspicion of imminent attack.

Reentry: The return of objects, originally launched from Earth, into the atmosphere.

Retrograde Orbit: An orbit having an inclination of more than 90°. See *Prograde Orbit*.

Robust: In this report, describing a system, indicating its ability to endure and perform its mission against a reactive adversary. Also used to indicate ability to survive under direct attack.

Salvage-fused: Describing a warhead, that is set to detonate when it is attacked. Usually refers to a nuclear warhead.

Sensors: Electronic instruments that can detect radiation from objects at great distances. The information can be used for tracking, aiming, discrimination, attacking, kill assessment, or all of the above. Sensors may detect any type of electromagnetic radiation or several types of nuclear particles.

Shoot-back: In this report, the technique of defending a space asset by shooting at an attacker.

Signature: Distinctive type of radiation emitted or reflected by a target, which can be used to identify that target.

Simulation: The art of making a decoy look like a more valuable strategic target See *Anti-simulation*.

Slew Time: The time needed for a weapon to reaim at a new target after having just fired at a previous one.

Smoke: An obscurant which may be used in the atmosphere or in space to conceal an object from observation by an optical sensor.

Soft Kill: Same as *functional kill*.

Space Detection And Tracking System (SPADATS): A network of space surveillance sensors operated by the U.S. Air Force.

Space Mines: Hypothetical devices that can track and follow a target in orbit, with the capability of exploding on command or by pre-program, in order to destroy the target.

Stimulated Emission: Physical process by which an excited molecule is induced by incident radiation to emit radiation at an identical frequency and in phase with the incident radiation. Lasers operate by stimulated emission.

Superfluorescence: See *Superradiance*.

Superradiance: The process used by a superradiant laser to generate or amplify a laser beam in a single pass through a lasant material, or—in the case of a free-electron laser—through an electric or magnetic field in the presence of an electron beam. Superradiance is actually a form of stimulated emission. Also known as superfluorescence, or amplified spontaneous emission.

Superradiant Laser: A laser in which the beam passes through the lasant only once; mirrors are not required for the operation of such a laser, as they are with more conventional lasers which are sometimes called "cavity lasers" to distinguish them from superradiant lasers. Free-electron lasers may also be superradiant; the laser beam of a superradiant free-electron laser would pass once through an electric or magnetic field (instead of a lasant) in the presence of an electron beam.

Synthetic Aperture Radar (SAR): A radar system which correlates the echoes of signals emitted at different points along a satellite's orbit or an airplane's flight path. The highest resolution achievable by such a system is theoretically equivalent to that of a single large antenna as wide as the distance between the most widely spaced points along the orbit that are used for transmitting positions. In practice, resolution will be limited and by the radar receiver's signal-processing capability or by the limited coherence of the radio signal emitted by the radar transmitter.

Thermal Kill: The destruction of a target by heating it, using directed energy, to the degree that structural components fail.

Threat: The anticipated inventory of enemy weapons and method of using them.

Tracking: The monitoring of the course of a moving target. Ballistic objects may have their tracks predicted by the defensive system, using several observations and physical laws.

Warhead: A weapon, usually a nuclear weapon, contained in the payload of a missile.

World-Wide Military Command and Control System (WWMCCS): A communications network linking U.S. forces.

X-ray Laser: A laser which generates a beam or beams of *X-rays*. Also called an "x-raser" or "XRL."

X-rays: Electromagnetic radiation having wavelengths shorter than 10 nanometers (10 billionths of a meter).

Contents

Chapter 1

Executive Summary

Contents

Tables

Chapter 1

Executive Summary

INTRODUCTION

For over two decades the United States and the Soviet Union have used satellites for military purposes. As a result of recent technological advances, military satellites will soon be able to play a more significant role in terrestrial conflicts. These space assets will be able to supply more types of information, more rapidly, to more diverse locations. Some will carry out target acquisition, tracking, and kill assessment functions, thus operating more directly than before as components of weapons systems.

This growing military utility also makes satellites attractive targets for opposing military forces. Both the Soviet Union and the United States have been developing anti-satellite (ASAT) weapons. These weapons could weaken the opponent's military capabilities by depriving his forces of the services of some satellites. The existing Soviet anti-satellite weapons—and future, potentially more effective ASATs—pose a growing defense problem for the United States.

A variety of unilateral measures, passive and active, may improve the survivability of U.S. military satellites. At present, it is unclear whether such survivability measures will be adequate to guard against the highly developed ASAT threats of the future. Another possible contributor to satellite survivability is mutually agreed arms control. A judicious combination of certain arms control measures and unilateral satellite survivability measures might provide more security to U.S. military satellites than either type of measure alone.

At the same time, however, arms control measures which constrained the threat to U.S. satellites would also constrain the ability of the United States to weaken Soviet military capabilities by attacking their satellites in time of war. In addition, limits on ASATs would severely limit the kinds of ballistic missile defense weapons that might be deployed in the future. (The subject of ballistic missile defense is dealt with in a companion OTA report, *Ballistic Missile Defense Technologies*.)

This report explains the dilemmas facing U.S. policymakers and assesses the pros and cons of some options for dealing with the challenge of anti-satellite weapons, particularly in the light of projected future weapons technology.

PRINCIPAL FINDINGS

Current Soviet military satellites pose only a limited threat to U.S. military capabilities, but future space systems will pose a greater threat.

The Soviet Union currently uses satellites to perform a wide variety of tasks including missile launch detection, communications, navigation, meteorological surveillance, photographic and radar reconnaissance, and collection of electromagnetic intelligence (e.g., radar emissions). Many of these satellites, although not "weapons" themselves, support and enhance the effectiveness of terrestrial Soviet forces that would engage in direct combat. For example, if navigation satellites improve munition delivery accuracies, then fewer munitions are required to accomplish a given objective. The growing military utility of satellites has rekindled U.S. interest in ASAT weapons.

Some Soviet satellites already supply limited targeting information to other terrestrial assets. The Administration has expressed its concern about:

> . . . present and projected Soviet space systems which, while not weapons themselves, are designed to support directly the U.S.S.R.'s terrestrial forces in the event of a conflict. These include ocean reconnaissance satellites which use radar and electronic intelligence in efforts to provide targeting data to Soviet weapon platforms which can quickly attack U.S. and allied surface fleets.[1]

At present Soviet radar (RORSAT) and electronic intelligence (EORSAT) ocean reconnaissance satellites pose only a limited threat to U.S. and allied surface fleets. RORSATs and EORSATs are typically deployed at altitudes and inclinations which offer limited observation range. Although the observation "swath" of these satellites will eventually cover most of the Earth, if only one or two of these satellites are operational—as has been customary in peacetime—then a ship would be exposed to observation only intermittently and might successfully evade the satellite. The Soviet Union could increase the number of deployed RORSATs and EORSATs, thereby making evasion more difficult. Other countermeasures exist which could further reduce the threat posed by these satellites, but such measures might not be available to merchant resupply vessels operating during a protracted nonnuclear conflict.

In the future, sophisticated communication, navigation, and surveillance satellites are likely to play a greater role in all levels of terrestrial conflict. This will increase the incentive for both the United States and the Soviet Union to develop and deploy ASAT weapons.

Possible responses to the threat posed by Soviet military satellites are numerous and diverse.

A variety of options are available to mitigate the threat to U.S. and allied security posed by Soviet military satellites (MILSATs). These options include nondestructive as well as destructive measures; those presented below are not mutually exclusive.

- Possible nondestructive responses to Soviet MILSATs:
  - —*Force Augmentation:* U.S. combat or support forces could be increased to counter the increase in effectiveness which Soviet forces could derive from use of military satellites. Force augmentation is often, but not always, more costly than other means of mitigating the threat posed by Soviet military satellites.
  - —*Passive Countermeasures:* By using passive measures to conceal or disguise their identity and nature, U.S. forces could reduce the utility of Soviet reconnaissance satellites. For example, assets now detectable by radar might be redesigned to reflect radar signals only weakly in order to evade detection by radar satellites, or radio silence might be practiced, or covert signaling techniques used to prevent detection by satellites that collect signals intelligence.
  - —*Electronic Countermeasures and Electro-optical Countermeasures:* Electronic countermeasures such as "jamming" (i.e., overloading enemy receivers with strong signals) and "spoofing" (i.e., sending deceptive signals) could be used to interfere with satellite functions. Electro-optical countermeasures such as "dazzling" (temporary "blinding") or spoofing optical sensors are also available. However, these countermeasures —especially spoofing—require detailed knowledge of the satellite systems (e.g., operational frequencies, receiver sensitivity, etc.) against which they are directed.
- Possible Destructive Responses to Soviet MILSATs:
  - —*Inadvertent But Inherent ASAT Capabilities:* The inherent ASAT capabilities of nuclear weapons such as ICBMs and SLBMs could be used to destroy low-altitude Soviet satellites; with some modifications, these weapons might

[1] President Ronald Reagan, *Report to the Congress: U.S. Policy on ASAT Arms Control*, Mar. 31, 1984.

also be used to attack satellites at higher altitudes. Some types of non-nuclear interceptors (e.g., that demonstrated in the U.S. Army's 1984 Homing Overlay Experiment (HOE)) which might eventually be developed and deployed for BMD purposes, would have some inherent ASAT capability. Finally, any highly maneuverable spacecraft capable of noncooperative rendezvous—e.g., the U.S. Space Shuttle—has some ASAT potential.

—*Planned ASAT Weapons:* When operational, the current USAF MV ASAT weapon will be able to destroy Soviet military satellites in low-Earth orbit.

—*Advanced ASAT Weapons:* Space- or ground-based directed-energy weapons or advanced kinetic-energy weapons could be developed that would be able to destroy Soviet satellites beyond the range of existing or planned U.S. ASAT weapons.

The United States is now more dependent on satellites to perform important military functions than is the Soviet Union.

In choosing between ASAT weapon development and arms control, one wishes to pursue that course which makes the greater contribution to U.S. national security. This is often characterized as a choice between developing a capability to destroy Soviet satellites while assuming U.S. satellites will also be at risk, or protecting U.S. satellites to some extent through arms control while forfeiting effective ASAT weapons. The better choice could, in principle, be identified by comparing the utility which the United States expects to derive from its military satellites with the disutility which the United States would expect to suffer from Soviet MILSATs during a conflict. Such a comparison—although possible in principle—is made exceedingly difficult by the number of conflict scenarios which must be considered and by the lack of consensus or official declaration about the relative likelihood and undesirability of each scenario.

Although national utility for space system support is difficult to assess precisely and meaningless to compare between nations, it is apparent that the United States is more dependent on MILSATs *to perform important military functions* than is the Soviet Union. The United States has global security commitments and force deployments, while the Soviet Union has few forces committed or deployed outside the borders and littoral waters of members of the Warsaw Treaty Organization and Cuba. The United States has corresponding requirements for global and oceanic command and control communications (C^3) capabilities and relies largely on space systems to provide these requirements. The Soviet Union, on the other hand, can rely on landline communications systems and over-the- horizon radio links for many of its C^3 needs. Satellite communications links are used by the Soviet Union but are not as essential as those of the United States. In addition, the Soviet Union has greater capability to reconstitute satellites which are lost in action; hence even to the extent the Soviet Union is dependent on space system support, it is less dependent on individual satellites for some functions. The United States also has fewer alternative terrestrial means for collecting intelligence than does the Soviet Union, which can exploit the freedom and openness of U.S. society for this purpose.

Soviet ASAT capabilities threaten U.S. military capabilities to some extent now and potentially to a much greater extent in the future.

The Soviet Union tested a coorbital satellite interceptor system from 1968 until its self-imposed moratorium of August 1983. The Reagan Administration considers this ASAT system to be operational. The interceptors are believed to be capable of attacking satellites at altitudes of up to 5,000 kilometers, depending on their orbital inclination. At present there appear to be only two launchpads for Soviet coorbital interceptors, both located at Tyuratam.

Artist's conception of U.S. MV anti-satellite weapon attacking a satellite. This weapon is still being tested.

Photo credits: U.S. Department of Defense

Artist's conception of a Soviet coorbital anti-satellite weapon attacking a satellite. This weapon is considered by the U.S. Department of Defense to be operational.

The growing military importance of satellites has made them attractive targets. Both the United States and the Soviet Union have been developing anti-satellite (ASAT) weapons.

The existing Soviet ASAT weapon may be effective for negating low-altitude U.S. military satellites, such as are used for navigation (Transit), meteorological surveillance (Defense Meteorological Support Program satellites), and other purposes. Assistant Secretary of Defense Richard Perle has stated:

> We believe that this Soviet anti-satellite capability is effective against critical U.S. satellites in relatively low orbit, that in wartime we would have to face the possibility, indeed the likelihood, that critical assets of the United States would be destroyed by Soviet anti-satellite systems. . . . If, in wartime, the Soviet Union were to attack critical satellites on which our knowledge of the unfolding conventional war depended, . . . we would have little choice but . . . to deter continuing attacks on our eyes and ears, without which we could not hope to prosecute successfully a conventional war.[2]

[2]Statement of The Honorable Richard Perle, Assistant Secretary of Defense (International Security Policy), in *Hearings before the Subcommittee on Strategic and Theater Nuclear Forces of the Senate Committee on Armed Services: Review of the FY 1985 Defense Authorization Bill*, Mar. 15, 1984 [S.Hrg. 98-724, Pt. 7., p. 3452].

The current Soviet interceptor and the booster that it has been tested with cannot reach critical U.S. early warning and communication satellites in high orbits. If the Soviet ASAT weapon were mated with a larger booster—a procedure which has yet to be tested—it might be able to reach these U.S. satellites.

In addition to the coorbital interceptor, the Soviet Union is testing ground-based lasers which the Reagan Administration believes have ASAT capabilities. The U.S. Department of Defense estimates that the U.S.S.R. could test a space-based laser within the decade.[3] Advanced directed-energy weapons such as lasers and particle beam weapons—if developed and deployed—could give the Soviets an "all altitude," "instantaneous kill" capability. As the United States increases its reliance on space systems to perform vital military functions (e.g., the MILSTAR communication satellite system), an increase in Soviet ASAT capabilities could create a significant threat to U.S. national security.

[3]U.S. Department of Defense, *Soviet Military Power*, 1985, p. 44.

Aside from its intentional ASAT capabilities, the Soviet Union could currently attack low-altitude satellites with its nuclear ABMs, ICBMs, and SLBMs. With some modification, these nuclear assets might also be used to attack satellites in higher orbits. Current Soviet spacecraft (i.e., Soyuz, Salyut), because of their limited maneuver and rendezvous capabilities, do not have a significant ASAT potential. Future Soviet spacecraft, such as the expected Soviet "Shuttle" and space plane, will have greater inherent ASAT capabilities. The Soviets also have the technological capability to conduct electronic warfare against space systems.

Several technologies on the horizon could lead to a new generation of highly capable ASATs.

The following advanced ASATs could be developed and deployed by either the United States or the Soviet Union:

- *Space Mines:* These would be deployed within lethal range and would continuously trail their target. Using a conventional or nuclear explosive charge, a space mine would destroy its quarry almost instantly on command or (if salvage-fused) when attacked or disturbed.
- *High-Power Radio-Frequency Weapons:* These would be devices capable of producing intense, damaging beams of electromagnetic radiation that could be used to jam communication and radar systems at low power levels or to overload and burn out satellite electronics at higher power levels;
- *High-Energy Laser Weapons:* High-energy lasers may eventually be capable of producing intense, damaging beams of electromagnetic radiation that could jam optical communication and sensor systems at low power levels or cause permanent damage at higher power levels. Ground-based lasers would have infrequent opportunities to attack satellites but, unless attacked themselves, could shoot inexpensively and repeatedly. Space-based reflectors could also be used to relay laser beams from ground-based lasers to their targets. Space-based lasers

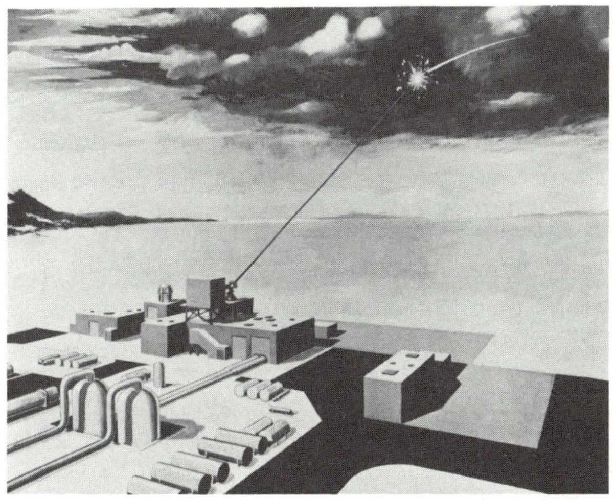

Photo credit: U.S. Air Force

Artist's conception of the High Energy Laser Test Facility, currently under construction at White Sands Missile range, New Mexico.

Photo credit: U.S. Department of Defense

Artist's conception of high-energy laser facility at the Sary Shagan test facility in the Soviet Union.

High-energy lasers may eventually be effective ASAT weapons. Ground-based lasers would have infrequent opportunities to attack satellites, but, unless attacked themselves, could shoot inexpensively and repeatedly.

might be able to attack several satellites in quick succession; space-based X-ray lasers might be able to attack several satellites instantly and simultaneously.

- *Neutral Particle Beam Weapons:* Powerful particle accelerators, similar to those now used in scientific research, might eventually be developed which could destroy the hardened electronics of a spacecraft.
- *Kinetic-Energy Weapons:* Space- or ground-based kinetic-energy weapons (similar to the current U.S. MV ASAT) would probably be small, homing vehicles that destroy their target by colliding with it at extremely high velocities.

Possible U.S. responses to the Soviet ASAT threat are numerous and diverse.

The United States could respond to the threat posed by the Soviet ASAT threat in several ways; both unilateral and diplomatic options are available.

- Possible unilateral responses to Soviet ASATs:
  —*Reduce Dependence on Military Satellites:* No matter what satellite survivability or arms control measures are taken, there will always be some risk that critical satellites can be destroyed or rendered inoperable. The United States must exercise caution in the extent of its reliance on space assets to perform tasks essential to the national security. Nonetheless, some space systems perform vital military functions which cannot be duplicated—or can be duplicated only imperfectly—by terrestrial systems.
  —*Passive Countermeasures:* Passive countermeasures such as hiding, deception (use of decoys), evasion (maneuvering), hardening (making satellites more durable), and proliferation (adding more satellites) all offer significant protection from the current and perhaps future Soviet ASAT weapons. Decoys would probably be effective against a wide variety of ASAT weapons and will be particularly economical for the protection of small satellites capable of being imitated by small, cheap decoys. Combinations of these passive responses—e.g., decoys for "dark" spare satellites—could offer even greater protection than individual measures alone.
  —*Active Countermeasures:* Active countermeasures may be destructive or nondestructive. Destructive countermeasures could include giving satellites a self-defense capability or providing critical satellites with an escort defense. Nondestructive countermeasures might include electronic countermeasures and electro-optical countermeasures such as jamming. Attacking Soviet ASAT control facilities is also a potential—though dangerous—active countermeasure.
  —*Deterrence:* The Soviets might be deterred from attacking U.S. satellites if the United States declared its willingness to retaliate for attacks on U.S. space assets. Such retaliation could be against Soviet space assets, in which case the United States would need a capable ASAT weapon, or it could be against Soviet terrestrial assets. The former alternative assumes that the Soviets value the preservation of their satellites at least as highly as they value the destruction of U.S. satellites. The latter alternative, of course, carries a greater risk of uncontrolled escalation if deterrence should fail.
  —*Keep-Out Zones:* The United States could declare and defend protective zones around critical satellites. Defended keep-out zones could offer significant protection against current ASAT weapons for some satellites. This subject is discussed in detail below.
- Possible diplomatic responses to Soviet ASATs:
  —*Arms Control:* The United States, the Soviet Union, and other spacefaring nations could negotiate limitations on the testing, deployment, or hostile use of anti-satellite systems.

—*Rules of the Road:* The United States, the Soviet Union, and other spacefaring nations could negotiate restrictions on potentially provocative activities in space, such as unexplained close approaches to foreign satellites or irradiation of foreign satellites with low-power directed-energy beams. With such agreed restrictions in force, these activities would justify defensive or retaliatory measures.

Of the future ASAT weapons now foreseeable, those which would be most effective if used in a preemptive or aggressive surprise attack would be space-based and therefore subject to attack by similar weapons.

Preemptive attack would be an attractive countermeasure to space-based ASAT weapons. If each side feared that only a preemptive attack could counter the risk of being defeated by enemy preemption, then a crisis situation could be extremely unstable. While salvage-fusing, if it proved practicable, would diminish this risk, it would create a risk of space war breaking out by accident. For example if a meteoroid destroyed a satellite, it might set off a chain reaction of salvage fusing which would destroy all satellites. To the extent that protection was sought through "shoot-back" rather than "shoot-first" tactics, a premium would be placed on having the biggest and best ASATs deployed, which could lead to an intense arms race.

Foreseeable passive or active countermeasures may be inadequate to guarantee the survival of large military satellites attacked by advanced ASAT weapons.

Passive or active countermeasures might have only limited effectiveness against very advanced ASAT or BMD weapons. For example, it might be uneconomical to rely on passive measures to protect large and expensive satellites from a powerful neutral particle beam weapon. Shielding satellites against a neutral particle beam weapon could cost more than it would to scale up the weapon to penetrate the shielding, and such a weapon could slew its beam quickly enough to make evasion infeasible. With such a weapon it might be as economical to damage spare satellites as they are brought "on line" as it would be to damage initially operational satellites.

Active measures, such as "shoot-back" with a weapon of longer effective range, could provide protection against some ASAT weapons but not against weapons such as space mines or single-pulse lasers which could destroy satellites instantly and without warning. However, it might not be economical to attack satellite systems composed of many small, cheap satellites—and possibly decoys as well—with expensive advanced ASAT weapons. Such satellites could perform a number of important functions (e.g., communication or navigation) without encouraging a proliferation of advanced weapons to attack them. Therefore, although it would be difficult to protect *individual* satellites, satellite *systems* performing some critical functions might retain a fair degree of survivability.

A commitment to satellite survivability is important whether or not ASAT development, or arms control, or both, are pursued.

The United States should place more emphasis on means to ensure the survivability of critical military satellites and, particularly, their associated ground stations and data links, regardless of whether ASAT limitations are agreed upon. The existence of non-ASAT weapons (e.g., ICBMs, ABMs) and space systems (e.g., maneuverable spacecraft) with some inherent ASAT capability makes it impossible to ban the ability to attack satellites. Therefore, even under the most restrictive ASAT arms control regime, programs for satellite survivability and countermeasures must be pursued. In the absence of arms control limitations on ASATs, ensuring satellite survivability will be a more demanding task since highly capable directed-energy ASAT weapons or space mines could be deployed.

In the absence of restrictions on the development or deployment of ASAT weapons, satellite survivability can be enhanced if the United States is willing to negotiate or declare keep-out zones and is able thereafter to defend such zones against unauthorized penetration by foreign spacecraft.

Although passive or active countermeasures alone may be insufficient to protect satellites, if combined with keep-out zones they could offer a significant degree of protection from certain ASAT weapons. Without keep-out zones space mines could be predeployed next to all critical military space systems. A keep-out zone of sufficient size would reduce the effectiveness of such weapons. However, advanced, directed-energy or kinetic energy ASAT weapons may be able to function effectively even outside very large keep-out zones.

Should ASAT development be pursued, the United States will need to formulate an employment policy.

At present, no clear consensus exists among those Administration military space policy analysts and executives interviewed by OTA on the conditions under which the United States would attack foreign satellites or on the manner in which it would retaliate for an attack on U.S. satellites. If the United States continues with its ASAT development and deployment plans, it will be necessary to formulate an employment policy.

If the United States wishes to enhance the deterrent value of its ASAT weapons, it may choose to publicize certain aspects of its employment policy. It might, for example, promise that the United States would not use its ASAT capabilities in an aggressive or preemptive first strike but might use them in a defensive or retaliatory reaction to an attack against the United States or its allies, even if U.S. satellites were not attacked. That is, the United States might announce a "no first strike but possible first use" policy for the employment of ASAT weapons as it has for the employment of nuclear weapons.

If defensive satellites (DSATs) are deployed by the United States to defend its satellites, or if certain satellites are given self-defense capabilities, the United States would have to decide under what circumstances it would use these assets and whether it wished to publicly announce this employment policy. The United States might declare in advance that it would fire at satellites suspected of being ASATs if they approached U.S. satellites within possibly lethal range. However, such a declaration would have an uncertain legal status and might generate considerable political opposition from both spacefaring and non-spacefaring nations.

Certain arms control provisions would reduce the probability that advanced ASATs will be developed or deployed. However, arms control could not guarantee the survival of U.S. satellites attacked by residual or covert Soviet ASAT weapons.

Arms control provisions, such as a ban on all testing of all systems "in an ASAT mode," would reduce the likelihood that the Soviet Union could successfully develop and deploy advanced, highly capable ASAT weapons. The categories of weapons eliminated might include space mines capable of "shadowing" valuable military assets in any orbit, directed-energy weapons with kill radii of hundreds to thousands of kilometers, and advanced kinetic-energy weapons. In the absence of an agreement limiting their development, each side would have a strong incentive to seek continually more effective means to attack threatening satellites and to defend valuable assets. In the absence of adequate countermeasures, the "instantaneous kill" capability of some advanced ASATs might be destabilizing in a crisis, because they would give each side an incentive to "shoot first" or else risk the loss of its space assets.

Photo credit: Lockheed

Launch of U.S. Homing Overlay Experiment (HOE) to test nonnuclear anti-ballistic missile technology.

Photo credit: U.S. Department of Defense

Artist's conception of the Soviet GALOSH ABM interceptor currently deployed around Moscow.

ABM interceptors may have some inherent ASAT capabilities. ABMs, as well as ICBMs and SLBMs, present a threat which may not be easily resolved through arms control.

A ban on testing weapons in an ASAT mode would be less effective at reducing the threat posed by weapon systems with inherent ASAT capability and by the existing Soviet ASAT weapon. ICBMs, SLBMs, and ABM interceptors with nuclear payloads are examples of systems with residual ASAT capability. Although these systems lack the kind of precision guidance necessary to actually collide with a satellite, the long-range destructiveness of their nuclear payloads makes them potentially effective ASATs. The Shuttle's recent success at retrieving satellites strongly suggests the ASAT potential of future maneuverable spacecraft, although costly vehicles like the Shuttle would probably not risk approaching satellites which might be booby-trapped. However, the range, effectiveness, and reaction time of even advanced maneuverable spacecraft not designed as weapons would be substantially less than those of intentional ASAT weapons. Although the development of maneuverable spacecraft would not be inhibited by most ASAT testing limitations, some limits could be placed on operating them "in an ASAT mode."

Arms control provisions might substitute for passive and active countermeasures in reducing the threat posed by ASAT weapons, but arms control would be more effective if combined with countermeasures.

U.S. satellites can be protected either by increasing their survivability or by reducing the threat posed by Soviet ASAT weapons. Passive or active countermeasures are designed to do the former while arms control provisions would hopefully do the latter. In considering the advantages and disadvantages of ASAT arms control, ASAT development, and countermeasures, it is important to consider them in packages. A combination of arms control provisions and passive countermeasures, for example, or of passive countermeasures and active countermeasures, could provide greater security than each component of such a package might provide alone.

The benefits of most ASAT limitations conflict with the benefits of ASAT exploitation.

Although ASAT arms control might prevent the Soviet Union from developing advanced, highly capable ASAT weapons, it would also place a similar restriction on the United States. Therefore, although U.S. satellites might be less vulnerable to ASAT attack, the United States would have to give up the ability to strike at Soviet satellites which threaten U.S. and allied forces. Although arms control might prevent an expensive and potentially destabilizing arms race in space, it would also limit the ability of the United States to use its comparative advantage in advanced technology to protect U.S. satellites and place threatening Soviet satellites at risk.

Not all arms control regimes are inconsistent with ASAT development or deployment. "Rules of the road" for space—e.g., negotiated keep-out zones—might be pursued simultaneously with ASAT research, testing, and deployment.

Effective ASAT arms control would likely place significant restrictions on the testing and deployment of future ballistic missile defense systems.

There is considerable overlap between BMD and ASAT technologies. Since even a poor ballistic missile defense system would probably have excellent ASAT capabilities, any ASAT limitation or test ban would almost certainly impede BMD development. Conversely, technology development ostensibly for advanced ASAT systems might provide some limited BMD capabilities, or, at minimum, information useful in BMD research.

Some available unilateral actions have clear benefits:

- *Deployment of attack sensors* on valuable satellites in order to provide information to support a retaliation decision;
- *Deployment of a space-based, space surveillance system* in order to provide information to support verification of compliance with future arms control agreements; to provide warning information required for effective evasion or dispensing of decoys; to support a decision to retaliate in the event of an attack; and to provide information required for the targeting of ASAT/DSAT weapons[4];
- *Hardening of military satellites against nuclear effects* to a modest degree in order to preclude "cheap kills" by nuclear-armed ICBMs, SLBMs, or ABMs;
- *Development or maintenance of electronic countermeasures and electro-optical countermeasures,* which would be relatively cheap, useful at all conflict levels, and unlikely to be prohibited by arms control agreements.

The ASAT weapon under development in the United States is sufficient to meet the threat posed by current, low-orbit Soviet military satellites.

Should the United States decide that it is in our national security interest to deploy ASAT weapons, the current MV program and, potentially, interceptors based on the recently tested HOE technology, are sufficient to respond to the threat posed by existing Soviet military satellites in low orbit. The current U.S. ASAT could be made even more effective by the addition of a space-based space surveil-

[4]Because of its usefulness, such a surveillance system would be an attractive target for a Soviet ASAT attack. The ultimate utility of such a system, therefore, hinges on its survivability.

lance system to aid the process of targeting and/or additional basing facilities.

Many of the functions performed by Soviet military satellites may eventually be performed by more capable satellites orbiting at altitudes out of reach of the U.S. MV. Should this happen, it would provide a strong incentive to extend the MV's capabilities by adding a larger booster or to develop a newer, more capable ASAT system.

COMPARISON OF POTENTIAL ASAT AND ARMS CONTROL REGIMES

There are widely varying views about the wisdom of deploying weapons which are to operate in space or against space objects. This fact, combined with more general concerns about the Soviet military threat and the dangers of the U.S./Soviet arms race, has made it difficult to forge a national consensus on the subject of ASAT weapons. Some people oppose ASAT weapons as a matter of principle because these weapons would operate in space or because such developments would contribute to the arms race. Others believe the benefits of ASAT weapons are outweighed by the risk they pose to current U.S. space systems, which are seen as essential for maintaining U.S./Soviet strategic stability. Still others see the development of ASAT and BMD weapons as a means to exploit U.S. technological advantages to enhance U.S. power, reduce the threat of conflict and global nuclear war, and reduce the damage done by such a war should it ever occur.

In its analysis, OTA has attempted to take into consideration this range of viewpoints and, to the greatest extent possible, show how it leads to a range of policy options. Many of the choices that will be made over the next several years will require a delicate balancing of strategic, economic, and political considerations. There is little doubt that reasonable persons can and will disagree as to the most appropriate nature of this balance.

Seven international legal regimes and corresponding military postures are considered critically below. Each of these regimes is intended to facilitate assessment of the effectiveness and desirability of different combinations of ASAT and BMD technology development, satellite survivability, and arms control. Each regime is constructed so that it is different from the other regimes and so that it contains elements which might reasonably be expected to co-exist in the same proposal.

1. Existing Constraints

The first regime is defined by treaties and agreements presently in force; these are the Limited Test Ban Treaty, the Outer Space Treaty, and the Anti-Ballistic Missile Treaty. The existing international legal regime prohibits the *use* of ASAT capabilities except in self-defense, the *testing or deployment* of space-based weapons with BMD capability, and the *testing or deployment in space* of nuclear space mines or ASATs that would require a nuclear detonation as a power source.

Table 1-1.—Effect of Regimes on ASAT Development and Arms Control

	Restrict with arms control	Develop ASAT weapons
Existing constraints	No	Yes
Comprehensive ASAT and space-based weapon ban	Yes	No
Test ban and space-based weapon ban	Yes	Yes/No[a]
One each/no new types	Yes	Yes[b]
Rules of the road	Yes	Yes[c]
Space sanctuary	Yes	Yes[c]
Ballistic missile defense	No	Yes

[a]In this regime ASAT weapons could be developed, tested, and deployed on Earth but not in space. The United States could pursue ASAT development within the bounds of the treaty, or it could forego ASAT development entirely.
[b]All ASAT weapons other than "current types" could not be tested or deployed in space.
[c]Development and deployment optional but strongly supported by advocates of this regime.

With these few exceptions, all other ASAT weapon development and deployment activities would be allowed. It is, therefore, permissible for the United States and the Soviet Union to develop and deploy coorbital interceptors (like the current Soviet system), direct-ascent interceptors (like the current U.S. system), terrestrial or space-based lasers, space-based neutral particle beam weapons, and weapons based on maneuvering spacecraft.

In the current regime both the United States and the Soviet Union could develop, test, deploy, and use such passive countermeasures as hiding, deception, evasion, hardening, and proliferation. Active, nondestructive defenses, such as electronic or electro-optical countermeasures, would also be allowed. Active, destructive defenses, such as shoot-back or DSATs, would be allowed as long as they did not violate any of the treaties enumerated above.

The primary advantage of the current regime is that it allows the almost unrestrained application of U.S. technology to the twin problems of protecting U.S. satellites and placing threatening Soviet satellites at risk. Under this regime, the United States would be free to use its comparative advantage in advanced technology to keep pace with expected developments in Soviet ASATs and other military satellites. Advanced U.S. ASATs might discourage the development of more capable Soviet space systems designed to place U.S. terrestrial assets at risk. In addition, the United States would be free to respond to Soviet ASAT weapons with increasingly sophisticated defensive weapons and countermeasures, thereby reducing the probability that the Soviets would ever use their intentional or inherent ASAT capabilities.

In the existing regime, research and development on new ballistic missile defense technologies can also proceed without the constraints that might be imposed by certain ASAT arms control regimes. Testing of advanced ASATs could provide valuable information that would contribute to the development of very capable BMD systems. Therefore, some types of generic space-weapon research could be conducted without first having to modify or withdraw from the ABM Treaty.

Some view advanced ASAT research as dangerous for this very reason. They argue that such research will gradually erode the usefulness of the ABM Treaty, thereby precipitating a defensive and offensive arms competition on Earth and in space. Rather than protecting satellites, a competition in space weapons might severely reduce their military utility. Under conditions of unrestrained competition, security might be purchased, if at all, only at the price of a substantial and sustained commitment to the development of increasingly sophisticated offensive and defensive space weapons. In such an environment, ensuring the survivability of satellites would require more than simple hardening or evasion. Costly measures might have to be taken such as the deployment of precision decoys, predeployed spares, or acquiring the ability to quickly reconstitute space assets. Satellites capable of defending themselves or a companion satellite might also have to be developed and deployed.

Should space-based weapons such as space mines or directed-energy weapons be deployed, these might be capable of the almost instantaneous destruction of a large number of critical satellites and ASATs. This could force nations into a situation in which they must "use or lose" their own pre-deployed space weapons. This might supply the incentive to escalate an otherwise manageable crisis.

2. Comprehensive ASAT and Space-Based Weapon Ban

A comprehensive ASAT and space-based weapon ban would require the United States and the Soviet Union to agree to forgo the possession of specialized ASAT weapons, the testing—on Earth or in space—of specialized ASAT capabilities, the testing in an "ASAT mode"[5] of systems (e.g., ICBMs or ABMs)

[5]Testing in an "ASAT mode" would include tests of ground-, air-, sea- or space-based systems against targets in space or against points in space.

which have inherent ASAT capabilities, and the deployment in space of any weapon. Such a regime would require the U.S.S.R. to destroy all of its coorbital interceptors and the United States to destroy all of its direct-ascent interceptors.[6]

Although this regime contains the most far-reaching arms control provisions, it would have the disadvantage of being the most difficult to verify. Unlike an ASAT test ban and space-based weapons ban regime, a comprehensive ban would prohibit *possession and testing* of ASAT weapons *on Earth*.[7] Because the current Soviet coorbital interceptor is a relatively small spacecraft launched on a much larger, general-purpose booster, the Soviet Union could maintain and perhaps even expand its ASAT force without the the United States gaining unambiguous evidence of a violation.

Since the United States might agree to a comprehensive ASAT ban only after considerable domestic political friction over questions of compliance and verification, it is important to consider how such a ban might make a greater contribution to U.S. national security than a ban on ASAT testing and space-based weapon deployment (discussed below). The purpose of both bans would be to prevent the use of ASATs, or, at minimum, to reduce the probability that an ASAT attack would be effective. An ASAT test ban would primarily affect weapons reliability, while an ASAT possession ban, if observed, would affect both availability and reliability. It is conceivable that the risk posed by possible illegal Soviet use of ASAT weapons might be somewhat lower in a régime in which the Soviets could not lawfully possess ASAT weapons. Presumably, the inability to overtly possess ASAT weapons would diminish one's ability to use them effectively. Furthermore, an absolute ban on possession might make it less likely that the current generation of ASAT weapons could be upgraded and held in readiness in significant numbers.

However, if the United States can only be confident that the Soviets are complying with a treaty to the extent we can verify compliance, then the United States would not have confidence that this regime offered any greater protection to our satellites than would a test ban and space deployment ban.

3. ASAT Weapon Test Ban and Space-Based Weapon Deployment Ban

In this regime, in addition to adhering to treaties and agreements presently in force, the Soviet Union and the United States would agree to forgo all testing in an "ASAT mode" and the deployment of any weapon in space. Such a ban would not only prohibit the testing of both current and future ASAT systems but would also place similar restrictions on BMD systems with ASAT capabilities. This regime would not ban terrestrial research on ASAT or space-based weapons and would not attempt to ban their possession. Therefore, if it were judged to be desirable, ASAT and BMD weapons could be developed (though not tested in space) and held in readiness on Earth.

In a test ban regime, the passive countermeasures and nondestructive active countermeasures that were discussed in the "existing regime" could still be developed and employed. Destructive active countermeasures such as "shoot-back" or DSATs could not be tested or deployed but could be developed and held in readiness.

Although a ban on testing in an "ASAT mode" would not eliminate all threats to satellites, it would reduce the cost and complexity of ensuring a reasonable level of satellite survivability. The United States would still benefit from "hardening" its satellites and deploying spares and decoys, but the more elaborate, expensive, precaution of developing and deploying DSATs would be prohibited and, indeed, less attractive. In the absence of reliable, effective ASATs, satellites would pre-

[6]Such an agreement might resemble the draft treaty proposed to the United Nations by athe U.S.S.R. in August of 1983, except that draft also bans the testing or use of manned spacecraft for military purposes. See U.N. Document A/38/194, Aug. 23, 1983.

[7]A comprehensive ban would not ban systems with inherent ASAT capabilities, such as ICBMs, ABMs, and maneuverable spacecraft.

Table 1-2.—Sensor Technology for Compliance Monitoring

Prohibitable action	Observables	Sensors
ASAT attack:		Attack sensors:
KEW[a] impact	acceleration	accelerometers
Pulsed HEL[b] irradiation	acceleration	accelerometers
Continuous HEL irradiation	heating	thermistors
NPB[c] irradiation	ionization	ionization detectors
Keep-out zone penetration	position of thermal radiation source (ASAT)	space-based LWIR[d] thermal imager[e f]
Interception test	positions of thermal radiation sources (ASAT and target)	space-based LWIR thermal imager[e f]
NPB ASAT operation	thermal radiation from ASAT	space-based LWIR thermal imager[e f]
HEL ASAT operation	thermal radiation from ASAT	space-based LWIR thermal imager[e f]
Irradiation of target with NPB	gamma radiation from target	gamma-ray spectrometer[g]
Irradiation of target with pulsed HEL	thermal radiation from target	space-based LWIR thermal imager
Irradiation of target with pulsed HEL	reflected radiation from target	space-based multispectral imager
Irradiation of target with continuous HEL	position of thermal radiation source (target)	space-based LWIR thermal imager
Irradiation of target with continuous HEL	reflected radiation from target	space-based multispectral imager
Nuclear explosive aboard satellite	gamma radiation from fissile or fusile nuclei activated by cosmic radiation or by particle beams	gamma-ray spectrometer (and optional particle beam generator)

[a]Kinetic-energy weapon.
[b]High-energy laser.
[c]Neutral-particle beam.
[d]Long-wavelength infrared.
[e]The LWIR telescope on the Infrared Astronomical Satellite (IRAS) exemplifies demonstrated space-based thermal imager technology; this instrument is described in *Astrophysical Journal,* 278 (1, Pt. 2); L1-L85, Mar. 1, 1984 (Special Issue on the Infrared Astronomical Satellite).
[f]Radar and passive radio direction-finding methods could also be useful for tracking, if hiding measures are not employed by the penetrating spacecraft. LWIR tracking is emphasized here because it is difficult to counter by such measures.
[g]A target irradiated by a high-energy neutral particle beam will emit gamma rays, neutrons, and other observable particles, just as it will, at a slower rate, when bombarded by natural cosmic rays. These gamma rays could be detected by a gamma-ray spectrometer such as those which have been carried by Soviet Venusian and lunar landers and by U.S. NASA Ranger and Apollo spacecraft. (NASA report SP-387, pp. 3-20.)

sumably be of greater utility since the United States might have higher confidence that they would be available when needed.

Relative to the existing regime, the primary advantage of a regime banning testing of ASAT capabilities and deployment of space-based weapons would be that highly valued U.S. satellites in higher orbits—e.g., the future MILSTAR system—could be protected with some confidence from advanced ASAT weapons, especially if protected as well by passive countermeasures. The fact that advanced ASATs could not be overtly tested would reduce the probability that they would be developed and deployed. If they were developed and used without prior testing, a test ban would reduce the probability that they would be successful.

As in the existing regime, the United States could retain a capability to attempt to negate low-altitude Soviet satellites with its MV ASAT (or, possibly, with interceptors based on the HOE technology) since a "no test" ban would not prohibit ASAT possession. However, confidence in the operational capability of both the U.S. and Soviet ASAT systems would degrade over time without continued operational testing.

There would be two important disadvantages to this regime. First a ban on testing in an "ASAT mode" and deploying space-based weapons would not offer absolute protection for satellites; there would remain some possibility that an untested—or covertly tested—advanced ASAT, if suddenly deployed and used, might actually work well enough to overcome passive countermeasures. Second, without an ASAT weapon the United States would lack a fully tested means to attack threatening satellites. The United States would, therefore, have to place greater reliance on countermeasures to protect its terrestrial assets. It is unclear whether countermeasures alone will be able to keep pace with the threat posed by advances in military satellites.

Depending on one's viewpoint, an additional advantage or disadvantage of this regime is that the testing of some types of advanced BMD weapons would be prohibited. This prohibition might even include some ground-based BMD weapons such as the U.S. HOE (Homing Overlay Experiment) ABM interceptor, which is currently allowed under the ABM Treaty. Although such limitations would only be slightly more restrictive than those of the ABM Treaty, they would be very restrictive when compared to a regime in which the ABM Treaty was no longer in force.

4. One Each/No New Types

Regime four would include arms limitation provisions which would restrict the United States and the Soviet Union to their current ASATs and prohibit the testing in an "ASAT mode" and deployment, in space, of more advanced systems. Existing treaties and agreements would remain in force. In addition to banning the testing or deployment in space of new types of ASATs, the "no new types" agreement would prohibit making current systems more capable so they could attack targets at higher altitudes. BMD systems would also be banned if they had ASAT capabilities.

In a "no new types" regime, the passive countermeasures and nondestructive active countermeasures that were discussed in the "existing regime" could still be developed and employed. Destructive active countermeasures such as "shoot-back" or DSATs could not be tested or deployed but could be developed and held in readiness. Current ASATs—should they already possess the capability when the treaty is signed—could be used to attack other ASATs.

The primary advantage of a "no new types" regime, relative to the existing regime, would be that, by prohibiting the testing of advanced ASAT weapons, highly valued U.S. satellites in higher orbits could be protected with some confidence. In addition, the United States could retain a capability to negate low-altitude Soviet satellites (e.g., RORSAT) in the event of war and to respond in kind to a Soviet ASAT attack.

A primary disadvantage of a "no new types" regime would be that allowed (i.e., tested, nonnuclear) U.S. ASAT weapons would be inadequate to negate threatening Soviet satellites if such satellites were moved to higher orbits—a feasible but technologically difficult and costly Soviet countermeasure. An additional disadvantage to this regime is that attempts to define "new types" of ASATs would be likely to result in the same ambiguity and distrust that resulted from attempts to define "new types" of ICBMs in the SALT II negotiations. Finally, the degree of protection afforded high-altitude satellites by a ban on testing "new types" would be uncertain; there would remain some probability that an untested advanced ASAT, if suddenly deployed and used, might actually work. Systems with inherent ASAT capabilities (ICBMs, ABMs, maneuvering spacecraft) would also still exist.

As in the test ban and space-based weapon ban regime, a "no new types" regime would limit the testing of some types of advanced BMD weapons which are currently allowed under the ABM Treaty.

5. "Rules of the Road" for Space

Whether or not the United States and the Soviet Union agree to restrict ASAT weapons, they might negotiate a set of "rules of the road" for space operations. These rules could serve the general purpose of reducing suspicion and encouraging the orderly use of space, or they could be designed specifically to aid in the defense of space assets. Examples of general rules might include agreed limits on minimum separation distance between satellites or restrictions on very low-altitude overflight by manned or unmanned spacecraft. These general rules might also be used to establish new, stringent requirements for advance notice of launch activities. Specific rules for space defense might include agreed and possibly defended "keep-out zones," grants or restrictions on the rights of inspection, and limitations on high-velocity fly-bys or trailing of foreign satellites. It might also be desirable to establish a means by which to obtain timely information and consult concerning ambiguous or threatening activities.

The "rules of the road" discussed above—if implemented in the absence of restrictions on ASAT weapon development—would not remove the threat of ASAT attack. The primary purpose of such a regime would not be to restrict substantially the activities of the parties, but rather, to make the intentions behind these activities more transparent. Although the degree of protection for U.S. space assets to be gained from a "rules of the road" agreement would be less than from other arms limitation regimes, the costs would also be correspondingly less if other nations failed to comply with such rules. One must assume that in the absence of ASAT arms control, both ASAT development and satellite survivability programs will be given high priority. This being the case, offensive and defensive measures would be available to respond to violations of "rules of the road."

If they were defended, "keep-out zones" would probably offer the closest thing to security in a "rules of the road" regime. Space mines designed to shadow satellites and detonate on command would lose a great deal of their utility if held at bay by a defended keep-out zone. Nonetheless, there are a number of difficulties with trying to implement this regime, not the least of which would be the reaction of other space-faring nations.

ASAT weapons such as nuclear interceptors must be kept at a range of several hundred kilometers from moderately hardened satellites in order to protect such satellites; advanced directed-energy ASAT weapons might have to be kept much farther away. Given the number of satellites currently in orbit, this would present several problems. Satellites in geostationary orbit are already so closely spaced that a keep-out zone sufficiently large to protect satellites from a nuclear weapon would displace other satellites. It is possible that critical strategic warning and communications satellites could function in supersynchronous orbits.[8] If so, there would be adequate room to accommodate large keep-out zones around satellites in such orbits.

[8] I.e. higher than geosynchronous orbital altitude.

There are too many satellites in low-Earth orbit to accommodate large keep-out zones. However, it might be feasible to establish smaller keep-out zones around such satellites and, in addition, to specify a minimum angular separation between orbital planes to prevent continuous trailing.

6. Space Sanctuary

Regime six would establish altitude limits above which military satellites could operate but where the testing or deployment of weapons would be forbidden. A "space sanctuary" regime would not constrain ASAT weapon development, testing, or deployment in space but would attempt to enhance security by prohibiting these activities in deep space (i.e., above 3,000 nautical miles, or about 5,600 kilometers) where critical strategic satellites are based. At present, the altitude of these strategic satellites makes them invulnerable to attack by the current Soviet and U.S. ASATs.

In a "space sanctuary" regime, the passive countermeasures and nondestructive active countermeasures that were discussed in the "existing regime" could still be developed and employed. Unlike the "test ban" or "no new types" regime, destructive, active countermeasures such as DSATs could be tested and deployed, but not in deep space. Deployment in deep space of "shoot-back" capabilities or DSATs would probably be prohibited since it might be impossible to differentiate these weapons from offensive ASATs.

The primary advantage of this regime would be that it could protect satellites in high orbits from the current generation of ASAT weapons. In addition, a deep-space sanctuary regime would constrain ASAT development less than would a comprehensive test ban regime or a no-new-types regime. However, should the United States and the Soviet Union choose to pursue advanced ASAT weapons, a space sanctuary might offer only limited protection.

The greatest risks in a space sanctuary regime would be posed by advanced directed-energy weapons which could be tested and deployed at low altitudes. Such testing and deployment would probably be adequate to guarantee effectiveness against targets at higher altitudes. Satellites at very high, supersynchronous altitudes might still derive some protection from this regime, but violation of the sanctuary by highly maneuverable kinetic-energy weapons or by satellites covertly carrying powerful nuclear or directed-energy weapons would remain a risk. For this reason, sanctuaries might provide less security than would keep-out zones (discussed above), because *any* foreign satellite entering an agreed keep-out zone could be fired upon, while a satellite entering a sanctuary could be lawfully fired upon only if it could be proven that it was, or carried, a weapon.

7. Space-Based BMD

The seventh regime might result from U.S. or Soviet withdrawal from the ABM Treaty followed by the deployment of space-based BMD systems. Since even a modest BMD system would make a very capable ASAT weapon, in a "space-based BMD" regime there could be no attempt to restrain ASAT development. Moreover, each side would probably want the freedom to develop new ASAT weapons capable of destroying the opponent's space-based BMD systems.

The ASAT weapons allowable under the "space-based BMD" regime would include all of those in the "existing regime," plus weap-

Photo credit: U.S. Department of Energy, Los Alamos National Laboratory

Artist's conception of the Space Shuttle deploying a neutral particle beam weapon.

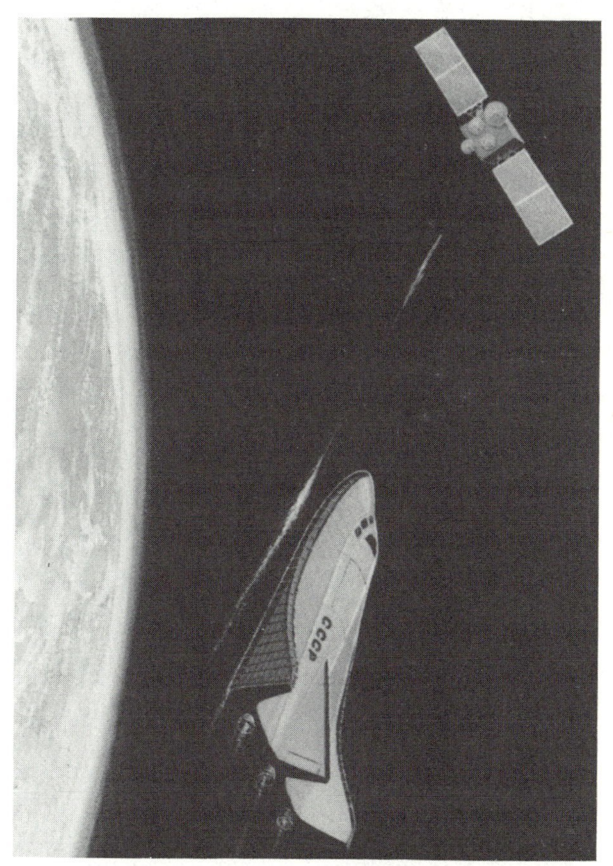

Photo credit: U.S. Department of Defense

Artist's conception of a manned Soviet space plane attacking a satellite.

Neither of the weapons illustrated here exists today, but in the absence of agreed limitations, both the United States and the Soviet Union will probably develop a wide range of advanced space weapons.

ons capable of countering ballistic missiles in flight. Defensive measures would be less constrained and more essential than in the "existing regime." In particular, advanced space-based weapons such as neutral particle beam weapons could be deployed at low altitudes and then used as ASAT or DSAT weapons. Passive countermeasures and nondestructive active countermeasures like those discussed in the "existing regime" would be developed and employed.

Depending on one's viewpoint, the principal advantage, or disadvantage, of a space-based BMD regime would be that it would allow the United States and the Soviet Union to deploy highly capable weapons in space. On March 23, 1983, President Reagan called for a vigorous research program to determine the feasibility of advanced BMD systems, suggesting that the deployment of such systems, if feasible, could offer an alternative to the current stalemate in strategic nuclear weapons. Given the inherent ASAT capabilities of advanced BMD weapons, satellites would be most vulnerable in a space-based BMD regime. Before the United States deployed space-based BMD systems it would have to determine, first, that the contribution that such systems made to U.S. security was great enough to compensate for the threat which similar opposing systems would pose to U.S. satellites; and, second, that space-based BMD components could be protected at competitive cost against advanced ASAT weapons.

ASAT countermeasures must prove to be effective for space-based BMD platforms if a decision to deploy them is to make sense. Perhaps large improvements in the effectiveness or economy of passive countermeasures such as combinations of hardening, deception, and proliferation would provide the needed protection. Alternatively, the superior fire-power or massive shielding of BMD weapons might give them a degree of protection unattainable by smaller, less capable satellites.

With respect to other military satellites, the expense of equipping them with countermeasures to insure some level of survivability against advanced BMD systems would be considerable. However if, as some argue, space-based missile defenses could make us more secure and encourage the Soviets to make real reductions in offensive missiles, this would reduce the threat of U.S./Soviet conflict. In a world where conflict was less likely, satellite vulnerability would be less important.

Others, of course, disagree strongly with this argument. They claim that space-based missile defenses will decrease our security by encouraging greater competition in both offensive and defensive weapons. In a world of space-based weapons and higher U.S./Soviet tension, satellite vulnerability would be a critical and potentially destabilizing factor.

TREATIES OF LIMITED DURATION

Each of the regimes examined above could be negotiated as a treaty of indefinite or limited duration or, alternatively, as one which remained in force as long as periodic reviews were favorable. Each of these alternatives would have its advantages and disadvantages. Treaties of indefinite duration are more effective at discouraging the pursuit of banned activities, yet require a greater degree of foresight regarding the long-term interests of the signatories and can foreclose technological options for the indefinite future.[9] Treaties of limited duration allow parties to take advantage of future technological options, yet can

[9]Treaties of unlimited duration usually contain a clause which states that if a country's "supreme national interests" are threatened, then that country may withdraw from the treaty. In addition to "supreme national interest clauses," treaties may also contain specific unilateral or agreed statements regarding specific understandings about related events. For example, The 1972 ABM Treaty contains a unilateral statement by the United States which links the continued viability of the treaty to "more complete limitations on strategic arms."

encourage aggressive development programs designed to reach fruition at the termination of the designated period. Treaties which call for a periodic reassessment of agreed limitations in theory have great flexibility, yet, in practice, often result in a strong presumption that they should be continued.

The United States might, for example, enter into a treaty limiting ASATs with the explicit and public reservation that we would withdraw from this treaty if and when we were ready to test and deploy a ballistic missile defense system in ways that the ASAT Treaty would forbid. Alternatively, we might take the public position that we intended to restrict our BMD activities so as to remain within the limits of an ASAT Treaty. While the former position would suggest a treaty of limited duration and the latter a treaty of unlimited duration, this need not be the case. It would be perfectly possible to sign a treaty of unlimited duration, with the standard provision allowing for withdrawal, accompanied by a clear statement of some of the conditions under which we intended to withdraw.

From one point of view, the exact language in a treaty regarding its duration would be less important than the intentions of the parties. After all, there have been numerous examples of treaties of unlimited duration that were violated soon after they were signed and examples of treaties of limited duration that continued in force after they had expired (e.g., the "Interim Offensive Agreement" signed at SALT I). The real issue would be whether the parties believe that adherence to the treaty in question continued to be in their national security interest.

The Reagan Administration has recently indicated that it intends to conduct ASAT tests to gather information useful in advanced BMD research.[10] Given the close connection between these two technologies, an ASAT treaty of even limited duration would require modification of current SDI program plans. Thus, to the extent that the United States wished to maintain the most rapid pace of advanced BMD research within the bounds of the ABM Treaty, such a treaty would not be desirable. Conversely, to the extent that the United States wished to slow the pace of Soviet BMD research and would be willing to defer decisions regarding the testing of space-based or space-directed weapons, an ASAT treaty of limited duration could contribute to that result.

[10]The purpose of tests "in an ASAT mode" would be to investigate advanced technologies without violating the ABM Treaty. The Department of Defense recently told Congress that, "To ensure compliance with the ABM Treaty the performance of the demonstration hardware will be limited to the satellite defense mission. Intercepts of certain orbital targets simulating anti-satellite weapons can clearly be compatible with this criteria." [*Report to the Congress on the Strategic Defense Initiative*, Department of Defense, 1985, app. B, p. 8.]

Chapter 2

Introduction

Contents

Chapter 2
Introduction

OVERVIEW

This report examines the issues raised by the development of weapons capable of attacking objects stationed in space. It analyzes the military utility of space systems, describes the technical characteristics and military value of anti-satellite (ASAT) weapons, and discusses the effectiveness of a number of satellite defenses and technical countermeasures. Finally, the report examines how various levels of ASAT arms control might contribute to U.S. national security when combined with various survivability measures and various levels of ASAT development and deployment.

Believing that the development of weapons capable of attacking missiles in flight or objects in space would likely have a strong effect on "deterrence, crisis stability, arms control and . . . national security policy," the House Committee on Armed Services[1] and the Senate Committee on Foreign Relations[2] asked the Office of Technology Assessment to prepare this report. The committees requested that the report should, among other things, assess:

- the feasibility, effectiveness, and cost of various space-based or space-directed concepts;[3]
- the relationship between capabilities that can reasonably be expected and the impact of the technology exploitation effort on the overall strategic policy of the United States;[4]
- the implications of anti-satellite weapons and space-based or space-directed missile defense concepts for standing arms control agreements, particularly the Anti-Ballistic Missile, Outer Space, and Limited Test Ban Treaties;[5] and
- the prospects for future space-related arms control agreements, including an assessment of advantages, disadvantages, and verifiability.[6]

The subject of ballistic missile defense (BMD)—particularly space-based BMD—was of special interest to both the House Armed Services and Senate Foreign Relations Committees. This subject is dealt with in a companion OTA report, *Ballistic Missile Defense Technologies.*

There is a strong relationship between ASAT and BMD technologies and the technical, political, and diplomatic actions taken in one sphere will almost certainly affect the other. For this reason, OTA assessed the two subjects at the same time, with a single staff, and with the advice of a single advisory panel. In each of these reports, OTA has endeavored to make clear the relationship between these two sets of technologies, and where appropriate has provided cross-references to further assist the reader.

In producing this unclassified report, OTA was able to draw on a wide range of classified material. Appendices of classified notes on this report are available to individuals having appropriate security clearances and who require access to that material.

[1]Letter from Melvin Price, Chairman, William L. Dickinson, Ranking Minority Member, and Les Aspin of the House Armed Services Committee, to John H. Gibbons, Director, Office of Technology Assessment, Mar. 5, 1984.

[2]Letter from Charles H. Percy, Chairman, Claiborne Pell, Ranking Member, Larry Pressler, and Paul E. Tsongas of the Senate Foreign Relations Committee to John H. Gibbons, Director, Office of Technology Assessment, Mar. 20, 1984.

[3]Ibid.

[4]Supra, note 1.

[5]Supra, note 2.

[6]Ibid.

SPACE WEAPONS: ATTITUDES AND CONTROVERSY

Assuming that highly capable and militarily useful ASAT weapons can be built at an acceptable cost, then why not proceed with development and deployment? Why should the U.S. Congress give more attention to ASATs than it gives to other new terrestrial weapon systems (e.g., anti-ship or anti-aircraft weapons)?

ASATs and BMD

Going forward with ASAT weapon development or, alternatively, agreeing to restrict such development through arms control measures, could have important consequences for advanced, space-based BMD technologies. Over the past several years a major debate on strategic defense has been taking place in the United States. Some believe that ballistic missile defenses can be developed that may eventually allow the United States to abandon the current policy of deterrence through assured retaliation. Others believe that even increased research on BMD alternatives might precipitate an offensive arms race with each side hastening to counter possible defenses with more and better offensive arms. This debate was intensified by President Reagan's March 23, 1983, speech which outlined what was later to become the Strategic Defense Initiative.

Since the debate over ballistic missile defense involves a fundamental reassessment of this country's strategic policy, decisionmakers are reluctant to proceed with ASAT weapon development, deployment, or arms control decisions that may tie their hands with respect to future technologies or that may commit them irrevocably to a course with unforeseen consequences. Some people believe that ASAT weapon development programs will be used to accomplish BMD research, thereby avoiding the strictures of the ABM Treaty and the scrutiny of Congress. Others believe that ASAT arms control restrictions would impede future BMD research and development programs. Given these opposing viewpoints, the decision to go forward with or, alternatively, to restrict ASAT development must be made in the broader context of this country's reassessment of its strategic posture and the military utility of space.

Attitudes Toward the Military Uses of Space

In addition to understanding the complex relationship between ASAT and BMD technologies, one must also recognize that people think about the military use of space in radically different ways. There are a great many views—both pro and con—regarding weapons that would operate in or from space; it is useful to examine several of the more frequently stated positions.[7]

Opposition to Space Weapons

Some people oppose the development of weapons that would operate in or from space because they feel such activities run counter to the legal and political history of space. They point to the many examples of successful international cooperation in space science, commerce, law, and politics and see these activities as reducing international tension and contributing broadly to peace and development. Space weapons are seen as violating the spirit and, in some cases, the letter of the treaties and agreements to which the United States is a party. They point to the language of the 1967 Outer Space Treaty, which states that space activities should be conducted "in the interest of maintaining international peace and security and promoting international cooperation and understanding."[8] Adherents of this viewpoint emphasize that every American President since Eisenhower has stated

[7]For a discussion of various "space doctrines," see: "Space Doctrines," Lt. Col. D. Lupton, USAF (Ret.), *Strategic Review*, fall 1983, pp. 36-47; see also: Lt. Col. D. Lupton, USAF (Ret.), *On Space Warfare: A Spacepower Doctrine*, U.S. Air Force, Air University Command, Center for Aerospace Doctrine, Research, and Education, Maxwell Air Force Base, Alabama, 1985.

[8]"Treaty on Principles Governing the Activities of States in the Exploration and Use of Outer Space, Including the Moon and Other Celestial Bodies," 18 U.S.T. 2410, T.I.A.S. 6347, Article III.

support for the idea that space should not be an arena of conflict and that space exploration should contribute to peace. The web of commitments that the United States has fashioned over the past 25 years through its agreements and unilateral declarations is seen as imposing a positive burden on the United States to support the broad ideals stated in the Outer Space Treaty.

Others—although acknowledging the importance of the laws and the history of space—base their opposition to space weapons on the belief that the deployment of such weapons in space, if not halted now, will be impossible to reverse. Since neither the United States nor the Soviet Union now has weapons that are *based* in space, they feel that it is both possible and desirable to prevent the arms race from extending to this new environment. This view is widely held in countries other than the United States and the Soviet Union. Over the last several years, the Soviet Union has made a strong effort to place the blame for the militarization of space on the United States. The American point of view is that the arms race is a burden imposed on the United States by the inordinate military preparations of the Soviet Union. Nonetheless, many nonallied governments, as well as important segments of the populations of even our allies, view the superpower arms race as a dangerous and destabilizing activity.[9] Those who see the superpower arms race as a dangerous process which the protagonists are doing little to halt are likely to see military development in space as an integral part of that process.

Some opposition to space weapons derives from the fact that such weapons would place at risk critical communication and information-gathering satellites that contribute to the stability of the U.S./Soviet relationship.[10] Space weapons are seen as destabilizing and likely to increase the possibility that a nuclear war might occur either through accident or intention. At present, nations can use space to peer within the boundaries of other sovereign states to obtain otherwise inaccessible information and early warning of attack. For this reason, many believe that space is of greater value to the United States than to the Soviet Union, since the Soviet Union has other means of gathering information in the open U.S. society. Adherents to this position maintain that, though there are many potential military uses of space, the communication and information-gathering activities are the most important. They argue that these benefits will be jeopardized by U.S./Soviet military space activities such as ASAT weapons development or space-based BMD. Although these latter activities also have military utility, they are not seen as outweighing the risk that such systems would create.

Support for Space Weapons

Those who support the development of weapons that would operate in or from space generally emphasize the importance of being able to exert military power in space. Some supporters view space as merely another sphere of military activity; others feel that military space activities might offer a means by which to fundamentally alter the U.S./Soviet strategic balance. Advocates of the former viewpoint emphasize that the increase in the number of Soviet military space systems with enhanced capabilities creates a threat to which the the United States must be prepared to respond. In particular, supporters of this position stress the importance of being able to destroy satellites which assist the Soviets in targeting U.S. terrestrial forces. They believe that in order to deter or, alternatively, to prevail in terrestrial conflicts, the United States must be able to operate in, and respond to threats from, space just as it does on land, at sea, or in the air.[11]

[9]U.S. Congress, Office of Technology Assessment, *Unispace '82: A Context for Cooperation and Competition—A Technical Memorandum*, OTA-TM-ISC-26 (Washington, DC: U.S. Government Printing Office, March 1983).

[10]See: R. Garwin, K. Gottfried, and D. Hafner, "Anti-Satellite Weapons," *Scientific American*, vol. 250, No. 6, June 1984, pp. 45 ff.

[11]Colin Gray, "Why an ASAT Treaty Is a Bad Idea," *Aerospace America*, April 1984, pp. 70 ff.; and R. F. Futrell, *Ideas, Concepts, and Doctrine: A History of Basic Thinking in the United States Air Force, 1907-1964*, U.S. Air Force, Air University Command, Maxwell Air Force Base, Alabama, 1974, pp. 279-282, summarizing views of Gen. Thomas D. White, USAF.

Some space weapons advocates see space as more than just another theater of military operation; they see it as a solution to the current stalemate in offensive nuclear weapons. They argue that space-based ballistic missile defenses can provide the opportunity for the United States to abandon its current doctrine of assured retaliation. Should both the United States and the Soviet Union possess space-based defensive forces, then more desirable offensive-defensive or purely defensive strategies can be developed. Other space power advocates see space weapons as a means to capture the "high ground."[12] The current U.S. lead in military space technology is seen as granting a military advantage over the Soviet Union—an advantage which, if not seized, will soon be lost.

Because views about the military uses of space vary so widely, it has been difficult to forge a national consensus on the subject of ASAT weapons. Some people oppose ASAT weapons as a matter of principle because these weapons would operate in space, others oppose ASAT weapons because they believe the benefits of such weapons are outweighed by the risk they pose to current U.S. space systems. Some people support ASAT weapons simply because they feel the United States must be able to respond to Soviet threats from any theater. Other supporters see space as a means to project U.S. power, reduce the threat of conflict and global nuclear war, and reduce the damage done by such a war should it ever occur.

In its analysis, OTA has attempted to take into consideration this range of viewpoints and, to the greatest extent possible, show it leads to a variety of policy options. As this report demonstrates, the opportunities and risks that might result from developing or not developing ASAT weapons or from pursuing or not pursuing ASAT arms control cannot be simply stated. Many of the choices that will be made over the next several years will require a delicate balancing of strategic, economic, and political interests. There is little doubt that reasonable persons can and will disagree as to the most appropriate nature of this balance.

[12]Lt. Gen. Daniel O. Graham, USA (Ret.), *High Frontier: A Strategy for National Survival* (Washington, DC: High Frontier, 1983).

ORGANIZATION OF THE REPORT

The main body of this report begins with the discussion in **chapter 3** of the military utility of satellites and ASAT weapons. This chapter provides the conceptual framework necessary to understand how these various space systems contribute to or threaten U.S. national security. Current and projected Soviet and U.S. military satellite capabilities are examined, as are a variety of responses to such capabilities.

Chapter 4 provides a detailed technical look at the existing and projected ASAT capabilities of the United States and the Soviet Union. This chapter discusses both existing technologies and the possibilities for more advanced kinetic-energy, nuclear, and directed-energy ASAT weapons. It also considers the wide range of technical and political responses available to the United States to counter or compensate for Soviet ASAT capabilities.

Chapter 5 reviews the history of arms control related to ASAT weapons. This chapter describes the constraints imposed by treaties and agreements in force and discusses the international political barriers to ASAT development. The 1978-79 ASAT negotiations between the United States and the Soviet Union are examined, along with subsequent draft treaties proposed by the Soviet Union. Recent legislative and executive branch activities are also summarized.

Chapter 6 describes a number of different ASAT arms control provisions that might be sought by the United States. Restrictions on testing, possession, deployment, and use are all examined to determine whether they might contribute to U.S. national security. Provisions restricting spacecraft operation and orbits—so-called "rules of the road"—are also examined.

Finally, **chapter 7** provides a comparative evaluation of seven hypothetical legal/technical regimes. Each regime combines examples of technical measures and countermeasures discussed in chapter 4 with examples of arms control as discussed in chapter 6. Each of these hypothetical regimes describes the advantages and disadvantages of different combinations of ASAT weapons development, employment policies, defensive countermeasures, and arms control.

Chapter 3

MILSATs, ASATs, and National Security

Contents

Chapter 3

MILSATs, ASATs, and National Security

THE ROLE AND VALUE OF MILITARY SATELLITES

The Role of Military Satellites

Force Support and Force Enhancement

Satellites are used for a variety of military applications by the United states, the U.S.S.R., and—in smaller numbers—by several other nations. Most military satellites (MILSATs) perform nondestructive functions. For example, Soviet military satellites are used for meteorological surveillance, surveillance of ballistic missile launch areas to provide rapid warning of possible missile attack, relaying of radio communications to distant force elements, optical and radar reconnaissance of foreign force dispositions on land and at sea, interception of foreign radio communications and radar signals, transmission of radionavigation signals, and logistic support for space systems.[1] Even though these functions are nondestructive and the satellites which perform them are not considered weapons, they support force elements which would engage in direct combat and enhance the combat effectiveness of those force elements.

The value, or *utility*, of military satellites is very real, but it is extremely difficult to quantify.[2] The timeliness of information or the speed of communications may make the difference between winning a battle and losing one—or it may greatly affect the number of casualties suffered in a battle without deciding victory. In some cases satellites provide capabilities that could not be obtained in any other way—e.g., surveillance of areas which would otherwise be closed to our observation, or providing very early warning of enemy missile launches. In other cases, satellites provide a cheaper and easier way of of doing something that could be accomplished by other means—e.g., trans-Atlantic communications. Then there are the navigation satellites, which provide an added degree of precision which may be critical in some applications and only of marginal utility in others.

In a few special cases, satellites contribute to a military mission the objectives or requirements of which can be quantified, as can, therefore, the value of the support provided by satellites. For example, the use of navigation satellites may improve the accuracy with which certain munitions are delivered, thereby reducing wastage of munitions. If so, then the effectiveness of the munitions used would be "multiplied" by satellite support—they would be as effective as a larger number of munitions delivered without the assistance of navigation satellites. The effectiveness of munitions delivery systems would be similarly multiplied. For missions such as this, it is reasonable to think of satellites as "force multipliers," and the factor by which the forces are multiplied can in principle be used to assess the value of the satellite. This has led some analysts to assess the significance of anti-satellite weapons in terms of the additional forces which the United States would have to procure to maintain military capability if it could not use MILSATs—or could not rely on their availability—in a conflict severe enough to justify Soviet ASAT use.

[1] E.g., maintenance and retrieval.

[2] The term *utility* is used here in the sense defined by John von Neumann and Oskar Morgenstern, *The Theory of Games and Economic Behavior* (Princeton, NJ: Princeton University Press, 1953), pp. 26-27; i.e., as a numerical index of the relative preferability of an outcome [e.g., occurrence of nonnuclear war and survival of all high-altitude satellites] which could result from a decision [e.g., an agreement to ban ASAT weapon testing]. Of any two possible outcomes, the one having a higher utility would be preferred over the other. Practical methods of assessing the utilities of a decisionmaker for possible outcomes have been described and reviewed by M.W. Merkhofer, *Comparative Evaluation of Quantitative Decision-Making Approaches*, National Science Foundation report NSF/PRA-83014, April 1983.

Other notions of utility have been used in the classified literature on satellite utility. Some of these notions are vaguely defined, while others—e.g., force multiplication factors—are precisely defined and, in principle, objectively calculable, although less clearly related to national interests than are Von Neumann-Morgenstern utilities.

However, in assessing the importance of ASATs, it may be more important to consider the *dependence* of military capabilities on space systems. If space system support suddenly becomes unavailable to a force element which has become accustomed to it, the combat effectiveness of that force element may be reduced to lower than it was before it began using space system support. Its effectiveness will be reduced to a fraction of what it was with space system support; the smaller this fraction, the greater the force element's dependence on space support.

There is a trend in both the United States and the U.S.S.R. to use increasingly sophisticated satellites to perform more functions and to do so more capably. It is generally believed that because of its sophisticated and still advancing space technology and because of the global distribution of its interests, commitments, and forces, the United States derives considerable utility from space system support: without satellites, performance of many military missions would become impossible, and performance of others would require large increases in the unit strengths of various U.S. force elements. Other force elements probably derive negligible force multiplication from space support. In general, however, the utility of military satellites to both the United States and the Soviet Union is probably increasing.[3] It is also generally believed that because of the expense of other means of providing comparable support to these forces, the United States has not vigorously developed alternative means of support and has consequently become highly dependent on space system support.[4]

Whether the United States derives more military utility from space system support than does the Soviet Union probably cannot be answered in general terms,[5] although force multiplication of particular types of force elements has been estimated in several studies.[6] Insofar as such estimates are comparable, judgments of *comparative* space support differ. Such differences may attributable, in part, to exclusion from the scope of some studies, for reasons of security classification, of consideration of some types of satellites which are of great value to the United States but for which the U.S.S.R. has no counterpart or from which the U.S.S.R. might derive much less utility. The utility of some functions of such satellites may be unquantifiable in any case, and this may lead to their neglect in quantitative assessments of utility.

It is easier to argue that the United States is more dependent on space system support for performing important military functions than is the Soviet Union,[7] because the Soviet Union has less need for some types of space system support and more alternative terrestrial means of providing similar support [see table 3-1]. Moreover, the Soviet Union has

[3]See, e.g., Stephen M. Meyer, "Soviet Military Programs and the 'New High Ground'," *Survival*, Sept./Oct. 1983, pp. 204-215.

[4]Ibid.

[5]According to accepted (e.g., Von Neumann-Morgernstern) axioms of utility theory, utility cannot be assessed in absolute terms but only to within an affine transformation; hence interpersonal or international comparisons of utility are not justifiable in general. See, e.g., Thomas Schelling, *The Strategy of Conflict* (New York: Oxford University Press, 1960), p. 288n, and John Rawls, *A Theory of Justice* (Cambridge, MA: Harvard University Press, 1971), p. 90.

[6]Force multiplication, in those cases where it is meaningful and can be assessed, can be assessed in absolute terms and compared between nations.

[7]Just as international comparison of utility is unjustifiable in principle, international comparison of dependence on space systems is also unjustifiable, if dependence is defined as the loss in utility which it would suffer if its satellites were suddenly incapacitated. Nevertheless, it is apparent that the United States is more dependent on MILSATs *to perform important military functions* than is the Soviet Union, even though the value of these functions cannot be easily quantified.

Table 3-1.—Asymmetries in U.S. and Soviet Space System Need and Use

Asymmetry	United States	Soviet Union
MILSAT reliability	(+) High	(−) Lower
MILSAT endurance	(+) Long	(−) Shorter
Launch rate	(−) Low	(+) High
Stockpile of spare MILSATs	(−) Low	(+) Higher
C^3 requirements	(−) Global and oceanic	(+) Continental and littoral
Terrestrial C^3 alternatives (relative to requirements)	(−) Few	(+) More
Terrestrial alternatives for information collection (relative to requirements)	(−) Few	(+) More
Operational ASAT capability	(−) No	(+) Yes
ASAT altitude reach	(−) Low	(+) Higher
ASAT responsiveness	(+) High	(−) Low

greater capability to reconstitute satellites which provide such support in case they are lost in action, hence even to the extent the Soviet Union is dependent on space system support, it is less dependent on individual satellites for some functions.

Force Application

Satellites have also been used to provide destructive capabilities. For example, since 1968 the U.S.S.R. has tested a coorbital interceptor—a satellite which could be used to intercept and destroy other satellites. The U.S. Department of Defense estimates that the Soviet Union attained an operational anti-satellite capability with this weapon in 1971.[8] Although to date there has been little testing and apparently no long-term basing or actual use of weapons in orbit, there is increasing technological potential to do so and, in the United States, increasing overt interest in doing so. In particular, there is strong interest in the United States in using space-based (i.e., satellite) weapons for defensive missions, especially ballistic missile defense and air defense.

Satellites could also provide destructive capabilities in support of other missions. Public Soviet statements have indicated a decreasing interest—indeed, growing opposition—to space-based weapons, although such statements have been interpreted by some in the West as disingenuous and propagandistic, intended for political gain and strategic deception.[9]

Nonmilitary Functions Contributing to National Security

Satellites are also used for nonmilitary applications which contribute to national security, such as monitoring compliance with arms control agreements[10] and collecting data for scientific research which could improve future military capabilities.

Current and Projected MILSAT Capabilities

Important asymmetries exist between the space systems of the United States and the Soviet Union and between the ways in which these systems would be employed [see table 3-1]. However, simple comparisons of U.S. and Soviet space systems can be misleading. The Soviet Union has an operational ASAT weapon, the United States does not. Yet, the U.S. ASAT, if developed, will be more capable and versatile than the Soviet ASAT. U.S. satellites are more sophisticated, more reliable, and capable of performing more functions than their Soviet counterparts. Yet, the Soviet's rapid launch capability and policy of maintaining spares would allow them to reconstitute some space assets during a conflict.[11]

These factors are further modified by the different roles that satellites or ASAT weapons would play in different theaters of war at different levels of conflict. Soviet forces are deployed largely on the Eurasion land mass, and would, in many scenarios, be able to rely on terrestrial communication and information links. In many of the same scenarios, satellite communications would be critical to globally deployed U.S. forces which might lack the same terrestrial communication links. Even in peacetime, the United States relies heavily on surveillance satellites to monitor Soviet compliance with arms control agreements by monitoring Soviet activities which could not be monitored by other lawful means, although similar activities in the United States would be readily observable by Soviet personnel legally in the country.

---

[8]U.S. Department of Defense, *Soviet Military Power*, 4th ed., Apr. 1985, p. 55. According to analysts of the Congressional Research Service, the U.S.S.R. has also tested a fractional-orbit bombardment system (FOBS), and possibly also a multiple-orbit bombardment system (MOBS), which could employ satellites to bombard terrestrial targets with nuclear warheads [U.S. Congress, Library of Congress, Congressional Research Service, Science Policy Research Division, *Soviet Space Programs, 1971-1975*, Committee Print prepared for the U.S. Senate, Committee on Aeronautical and Space Sciences, Aug. 30, 1976].

[9]U.S. Department of Defense, Defense Intelligence Agency, *Soviet Military Space Doctrine*, report DDB-1400-16-84, Aug. 1, 1984 [UNCLASSIFIED].

[10]Les Aspin, "The Verification of the SALT II Agreement," *Scientific American*, vol. 240, No. 2, February 1979, pp. 38-45.

[11]President Ronald Reagan, *Report to the Congress: U.S. Policy on ASAT Arms Control*, Mar. 31, 1984 [UNCLASSIFIED].

Soviet MILSAT Capabilities

The Soviet Union currently uses satellites to perform missile launch detection, communications relaying, radionavigation, meteorological surveillance, photographic and radar reconnaissance, collection of electronic intelligence (ELINT: e.g., radar emissions), and other functions.[12] These functions support a variety of military applications. Of particular concern to the Administration and in Congress[13] are:

> . . . present and projected Soviet space systems which, while not weapons themselves, are designed to support directly the U.S.S.R.'s terrestrial forces in the event of a conflict. These include ocean reconnaissance satellites which use radar and electronic intelligence in efforts to provide targeting data to Soviet weapon[14] platforms which can quickly attack U.S. and allied surface fleets. In view of the fundamental importance of U.S. and Allied access to the seas in wartime, including for Allied reinforcement by sea, the protection of U.S. and allied navies against such targeting is critical.[15]

Soviet ELINT ocean reconnaissance satellites (EORSATs) attempt to detect, localize, and classify ships by detecting the radio signals emitted by their communications and radar systems, while Soviet radar ocean reconnaissance satellites (RORSATs) attempt to detect, localize, and classify ships by detecting radar "echoes" reflected by the ships. RORSATs and EORSATs are typically deployed at altitudes of about 250 and 425 kilometers, respectively, in nearly circular orbits inclined about 65° with respect to the equatorial plane [see figure 3-1 and table 3-2] From these altitudes, these satellites can observe shipping over a limited range. The observation "swath" of each satellite will eventually cover the entire earth and ocean surface between latitudes of about 65° north and south. If only one or two RORSATs or EORSATs were operational at one time—as has been customary in peacetime—then a ship in this latitude band would be exposed to observation only intermittently.[16] However, larger numbers of RORSATs or EORSATs could be operational during wartime.

Even if only intermittent, surveillance by RORSATs or EORSATs could assist Soviet forces in targeting Allied shipping to an extent dependent on details of satellite capabilities, such as resolution. For example, Soviet oceanographic radar satellites of the Kosmos-1500 class can obtain radar imagery with a resolution of only 1.5 to 2 kilometers [see figure 3-2], which is inadequate to distinguish an aircraft carrier from a tanker, for example.[17] Various countermeasures could be used by ships to evade observation by these satellites or to reduce the value of their observations to the enemy;[18] some of these are discussed below, in the section entitled "The Role and Value of Anti-Satellite Capabilities."

U.S. MILSAT Capabilities

The United States uses MILSATs to perform most of the functions performed by Soviet satellites, as well as some other functions.

---

[12]Ibid., pp. 7 and 12; and U.S. Navy, Office of the Chief of Naval Operations, *Understanding Soviet Naval Developments*, 4th ed., NAVSO P-3560 (Rev. 1/81), Jan. 1981 [UNCLASSIFIED], p. 46.

[13]*Hearings before the Subcommittee on Strategic and Theater Nuclear Forces of the Committee on Armed Services United States Senate, Testimony on Space Defense Matters in Review of the FY1985 Defense Authorization Bill*, S. Hrg 98-724, Pt. 7, pp. 3568-3569.

[14]Specifically, anti-ship cruise missiles launched against U.S. surface ships: see U.S. Department of Defense, *Soviet Military Power*, 3d ed., 1984; p. 41.

[15]Reagan, op. cit.

[16]Compare the discussion of radar search satellites in *MX Missile Basing*, OTA report OTA-ISC-140, September 1981.

[17]In 1983 the Soviet Union launched a civil ocean-surveillance satellite (Kosmos 1500) equipped with a side-looking radar and in the same year placed two satellites (Venera 15 and Venera 16) equipped with similar radar systems into orbits around Venus. The U.S.S.R. has exhibited radar imagery obtained by these satellites. Although the resolution of this radar imagery is poor compared to the 25-meter resolution imagery obtained by NASA's Seasat-A radar satellite in 1978 [see Figure 76B of U.S. Congress, Office of Technology Assessment, *MX Missile Basing*, OTA-ISC-140, Sept. 1981], the 40-meter resolution imagery obtained by NASA's experimental Shuttle Imaging Radar (SIR-A) device in 1982 [see figure 3-3], or the 25-meter resolution imagery obtained by NASA's experimental Shuttle Imaging Radar (SIR-A) device in 1984, Soviet satellite radar technology can be expected to improve, and synthetic aperture radar could be used on future satellites.

[18]Reagan, op. cit., p. 13. See also testimony of VADM Gordon R. Nagler, USN, before the HAC Subcommittee on Defense, Mar. 23, 1983.

Figure 3-1.—Ground Track of Soviet Radar Ocean Reconnaissance Satellite

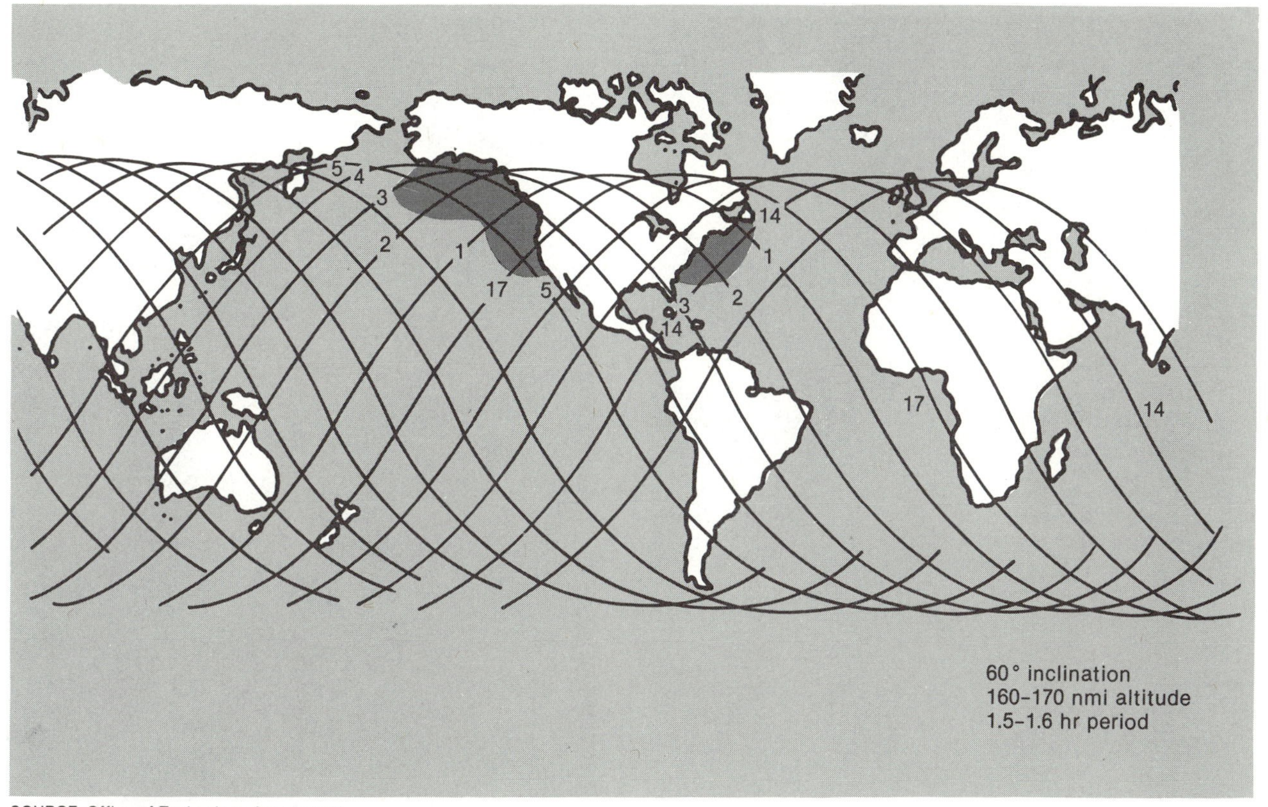

SOURCE: Office of Technology Assessment.

Table 3-2.—Orbits of Some Soviet Military Satellites

Perigee-Apogee (km) (km)	Inclination	Apparent mission*
35,785-35,785	0°	communications
19,000-19,200	65°	navigation
965-1,020	47°	communications
940-960	83°	meteorological
855-895	81°	meteorological
790-810	74°	communications
620-660	81°	electronic intelligence (ELINT)
425-445	65°	ELINT ocean reconnaissance (RORSAT)
760-40,000	63°	launch detection
400-40,000	63°	communications
356-415	73°	photo-reconnaissance[a]
250-265	65°	radar ocean reconnaissance (RORSAT)
172-351	67°	photo-reconnaissance[b]

[a]Maneuverable; initial parameters for Kosmos 1499 given.
[b]Maneuverable; initial parameters for Kosmos 1454 given.

SOURCES: NASA, *Satellite Situation Report,* vol. 24, No. 5, Dec. 31, 1984; and *Nicholas L. Johnson, *The Soviet Year in Space: 1983* (Colorado Springs, CO: Teledyne-Brown Engineering, 1984).

U.S. satellites are designed to have longer operational lifetimes in orbit than Soviet satellites, hence fewer satellite launches per year are required to perform similar functions. The complexity of a U.S. satellite may differ from that of a Soviet satellite performing the same function, and the value of this function to the United States may differ from its value to the U.S.S.R.

Attack Warning.—The United States uses infrared sensors aboard satellites in geostationary orbit to detect and promptly report ICBM and SLBM launches; these reports would provide early warning of a missile attack.[19]

[19]U.S. Department of Defense, *Report of the Secretary of Defense Caspar W. Weinberger to the Congress on the FY 1985 Budget, FY 1986 Authorization Request, and FY 1985-89 Defense Programs,* Feb. 1, 1984 [UNCLASSIFIED], p. 196.

Figure 3-2.—Radar Imagery From Kosmos 1500

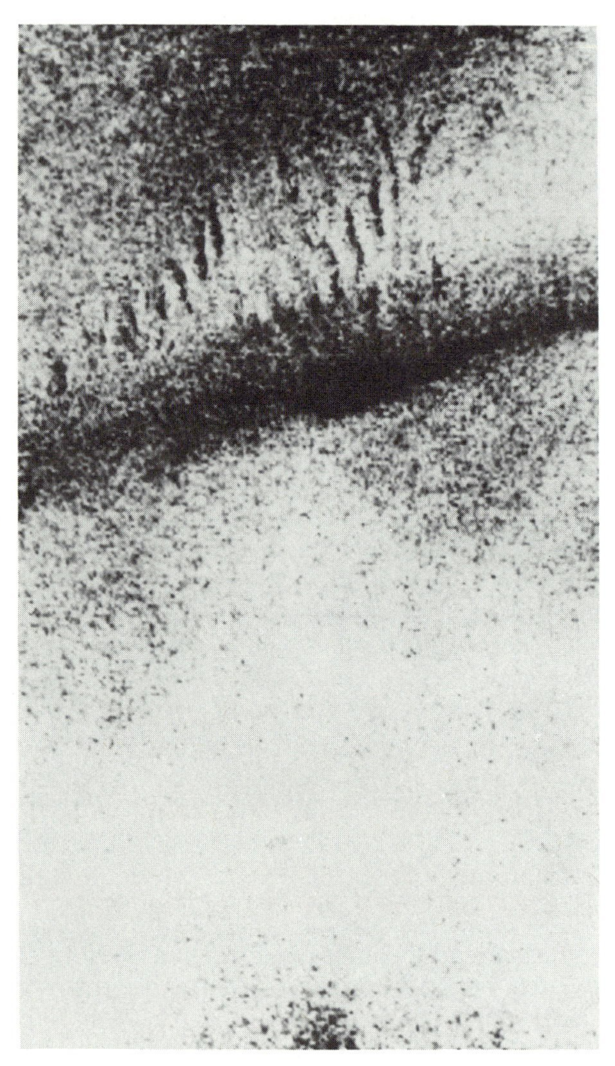

Left, imagery of Atlantic Ocean waves obtained by the Soviet Kosmos 1500 Oceanographic Research Satellite in 1983.

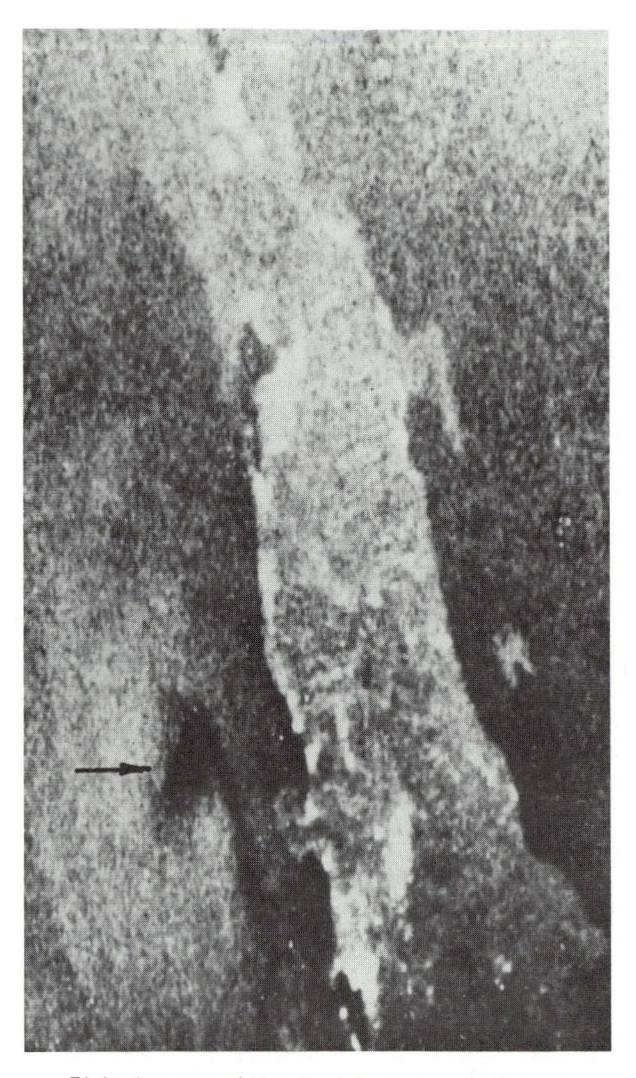

Right, imagery of Honshu Island, Japan obtained by same satellite.

SOURCE: Photographs courtesy of *Aviation Week and Space Technology* (reprinted by permission).

Navigation and Detection of Nuclear Detonations.—U.S. NAVSTAR satellites carry radionavigation beacons for use by Global Positioning System (GPS) receivers on aircraft, ships, and land vehicles. They also carry Integrated Operational Nuclear Detonation Detection System (IONDS) sensors designed to detect nuclear detonations in order to monitor compliance with the Limited Test Ban Treaty and other treaties in peacetime. In wartime they could be used to confirm a nuclear attack on the United States or its allies in order to support a decision by the National Command Authorities to retaliate, and to assess the success of a retaliatory strike.[20]

[20]Ibid.

Command and Control Communications.—The United States also uses several different military and commercial communications satellites to provide command and control communications among its globally distributed forces. An advanced satellite communications system called MILSTAR has been designed to replace these and is intended to provide survivable and enduring command and control communications to all four services at all levels of conflict, including general nuclear war.[21]

Meteorological Surveillance.—Defense Meteorological Satellite Program (DMSP) meteorological surveillance satellites provide timely information about weather conditions worldwide. This information is of considerable value in planning military operations, especially flight operations.

Compliance Monitoring.—Photoreconnaissance satellites are used to monitor compliance with arms control treaties and agreements; this function could not be performed as well by any alternative means which is politically acceptable in peacetime, and this function would be unnecessary during war with another party to such treaties, which would be suspended during such a war. However, in wartime the United States would attempt to collect intelligence using satellites[22] and other means, such as aircraft overflight, which are acceptable during wartime.

[21]The Honorable Caspar W. Weinberger, Secretary of Defense, *Annual Report to the Congress on the FY 1984 Budget, FY 1985 Authorization Request, and FY 1984-88 Defense Programs*, Feb. 1, 1983 [UNCLASSIFIED].

[22]The Honorable Richard Perle, Assistant Secretary of Defense for International Security Policy, has stated:

> We believe that this Soviet anti-satellite capability is effective against critical U.S. satellites in relatively low orbit, that in wartime we would have to face the possibility, indeed the likelihood, that critical assets of the United States would be destroyed by Soviet antisatellite systems. . . . If, in wartime, the Soviet Union were to attack critical satellites upon which our knowledge of the unfolding conventional war depended, . . . we would have little choice but . . . to deter continuing attacks on our eyes and ears, without which we could not hope to prosecute successfully a conventional war.

[*Hearings before the Subcommittee on Strategic and Theater Nuclear Forces of the Committee on Armed Services United States Senate, Testimony on Space Defense Matters in Review of the FY1985 Defense Authorization Bill*, S.Hrg. 98-724, Pt. 7., Mar. 15, 1984, p. 3452.]

Possible Advanced-Technology MILSAT Capabilities

Recent and prospective technological advances could be exploited by both the United States and the U.S.S.R.—at different rates—to develop MILSATs which could perform those functions now performed by MILSATs more effectively or economically. Such improvements are possible in the performance of each of the functions mentioned. Of particular interest and concern are possible marked improvements in ocean surveillance, logistic support of space systems, and anti-satellite capability which appear technologically feasible.

For example, radar ocean surveillance satellites using synthetic aperture radar techniques could provide radar imagery of sufficient resolution to permit classification of ships. An example of the potential quality of radar imagery is provided by the radar imagery obtained by the Shuttle Imaging Radar system SIR-A in 1981 [see figure 3-3]. The synthetic aperture radar carried by SIR-A distinguished features as small as 40 meters across. Earlier, in 1978, NASA's Seasat-A demonstrated a resolution of 25 meters; more recently, in 1984, SIR-B demonstrated a comparable resolution. Even finer resolution is possible, and several satellites could be deployed at once to provide frequent opportunities to observe each point on the ocean surface. [Deploying such satellites at higher altitudes for greater coverage would greatly increase the power required and hence also satellite cost.] Both the United States and the U.S.S.R. could deploy radar ocean surveillance satellites using high-resolution synthetic aperture radar technology in the future. Soviet deployment of radar ocean surveillance satellites with improved performance would threaten U.S. and allied shipping to a greater extent than does the existing RORSAT and would provide the United States with a greater incentive to maintain, or, if necessary, develop an ASAT capability to destroy such satellites or otherwise interfere with their performance. Future EORSAT performance could also be improved. With regard to the threat posed by

Figure 3-3.—Imagery Obtained by Synthetic-Aperture Satellite Radar

Image of cross-shaped array of radar reflectors near Lake Henshaw, California, obtained by SIR-A in 1981, demonstrating an image resolution of 40 meters. Adjacent reflectors are separated by 300 meters.

SOURCE: U.S. National Aeronautics and Space Administration, Jet Propulsion Laboratory.

such Soviet MILSATs to U.S. and allied shipping, the Administration has expressed concern that "as Soviet military space technology improves, the capabilities of Soviet satellites that can be used for targeting are likely to be enhanced and represent a greater threat to U.S. and allied security."[23]

The logistic support and anti-satellite capabilities of space systems could also be enhanced greatly in the future. Potential future ASAT capabilities are discussed in chapter 4 of this report.

Recent and prospective technological advances could also be exploited to enable future MILSATs to perform functions not now performed by MILSATs. For example, the United States could develop and deploy space surveillance satellites to detect and track foreign satellites. This capability could be used to detect impending attacks on U.S. or allied satellites in time to permit countermeasures to be used; it could confirm success of such attacks in order to support a decision to retaliate (not necessarily in kind); it could monitor compliance with possible future ASAT arms control or "Rules of the Road in Space" agreements; and it could provide targeting information for U.S. anti-satellite weapons. Space surveillance satellites could provide continuous space object detection and tracking capabilities which cannot be duplicated by ground-based radars (which have limited search range) and photographic or electro-optical sensors (which cannot be used in daytime or through overcast).

The United States and the U.S.S.R. could also develop satellite (i.e., space-based) weapons capable of attacking a variety of targets other than satellites—e.g., ballistic missiles, aircraft, ships, and fixed or mobile targets on land.[24] Assessment of the military capabilities which such weapons might eventually be able to provide requires an understanding of the feasibility of making them survivable at affordable costs and is beyond the scope of this report, which is limited to survivability issues. The feasibility and value of space-based ballistic missile defense system components—including weapons—is discussed in a companion OTA report, *Ballistic Missile Defense Technologies*.

The Value of Military Satellites

Estimation of the force multiplication effects which satellites might provide *under specific circumstances* might be done objectively, by means of combat modeling and simulation,

[23]Reagan, op. cit., p. 7. See also testimony of VADM Gordon R. Nagler, USN, before the HAC Subcommittee on Defense, Mar. 23, 1983.

[24]LTC David Lupton, USAF (Ret.), op. cit., pp. 36-47.

and analysis of historical combat data. The costs of the additional "equivalent forces" which the satellite services would, in effect, replace might be used as an upper bound on the value of the satellites under those circumstances. However, the expected utility, or worth, of such force multiplication cannot determined by such an analysis without assuming probabilities that the forces supported will be involved in various conflict situations, estimating the force multiplication expected in each such situation and the additional costs averted by such force multiplication, and averaging these averted costs over all such situations. This would be a demanding, and probably infeasible, analytical task, because assessment of such values and probabilities is necessarily subjective and hence subject to dispute. However, some judgments of value are conventionally accepted and can be rationalized to a considerable extent.

For example, the missile attack warning function performed by U.S. MILSATs has been described as being of "vital" importance to national security.[25] The navigation and nuclear detonation detection functions performed by the GPS and IONDS mission packages aboard NAVSTAR satellites have been described as "critical," as has the communications function to be performed by MILSTAR and currently performed by a variety of satellites [see box entitled "The Dependence of U.S. Command and Control Performance on Satellite Communications Systems"].[26] Surveillance satellites have been described as "critical assets,"[27] and the surveillance function which they perform has been called "vital."[28]

Other satellites are seldom judged to be of importance comparable to those aforementioned, although some are of considerable value. For example, after DMSP cloud imagery and DMSP-derived forecasts were made available to aircraft carriers operating in Southeast Asia, sorties of carrier-based aircraft scheduled for strike and reconnaissance missions which required clear weather in the target area were canceled when meteorological data from DMSP satellites showed overcast in the target area, and the aircraft which had been assigned to the canceled sorties were reassigned to other missions. This use of DMSP data decreased the number of aircraft required to perform a number of assigned missions in a given time, or, equivalently, increased the number of aircraft available to fly such missions.

As another example, the navigational accuracy of missiles and aircraft which rely on very accurate, unaided inertial navigation systems depends on geopotential anomaly data collected by geophysical research satellites.

Some of the functions performed by satellites, and the satellites which perform them, are most valuable in peacetime, while others would be more important in crisis, conventional war, or nuclear war. However, determination of the nature of the dependence of satellite value on the level of conflict is complicated by the fact that in many cases the

---

[25]The Honorable Hans Mark, then Secretary of the Air Force, in testimony in 1980 hearings before the U.S. Congress, House of Representatives, Committee on Science and Technology, stated that "two missions [strategic missile warning and surveillance] above all others stand out as being of vital importance to national security." [*United States Civilian Space Policy*, House Document 153, 96th Cong., 2d sess., July 23-24, 1980; pp. 93-94].

[26]Responses of LTG Robert D. Russ, USAF, Deputy Chief of Staff for Research, Development, and Acquisition, to questions submitted by The Honorable Norman Dicks in FY1985 Defense Appropriations hearings before the U.S. House of Representatives, Committee on Appropriations, Subcommittee on Defense, vol. V, 1984.

[27]The Honorable Richard Perle, Assistant Secretary of Defense (International Security Policy), has stated that "We believe that this Soviet anti-satellite capability is effective against critical U.S. satellites in relatively low orbit, that in wartime we would have to face the possibility, indeed the likelihood, that critical assets of the United States would be destroyed by Soviet anti-satellite systems."—Statement of The Honorable Richard Perle, Assistant Secretary of Defense (International Security Policy), *Hearings before the Subcommittee on Strategic and Theater Nuclear Forces of the Committee on Armed Services United States Senate, Testimony on Space Defense Matters in Review of the FY1985 Defense Authorization Bill*, Mar. 15, 1984, S. Hrg. 98-724, Pt. 7, p. 3452.

[28]The Honorable Hans Mark, then Secretary of the Air Force, in testimony in 1980 hearings before the U.S. Congress, House of Representatives, Committee on Science and Technology, stated that "two missions [strategic missile warning and surveillance] above all others stand out as being of vital importance to national security." [*United States Civilian Space Policy*, House Document 153, 96th Cong., 2d sess., July 23-24, 1980; pp. 93-94].

The Dependence of U.S. Command and Control Performance on Satellite Communications Systems

It has been estimated that about 70 percent of all U.S. military electronic communications are routed through communications satellites.[1] In peacetime, much of such traffic consists of administrative messages which would be inessential to the conduct of war. Therefore, the dependence of U.S. military capabilities on satellite communication systems, although undoubtedly substantial, is not necessarily represented accurately by the fraction of message traffic routed through satellites during peacetime. The dependence of U.S. military capabilities in several theaters on satellite communication systems is reflected qualitatively in the following views of military commanders and staff analysts.

Command and Control in the Southwest Asian Theater.—The dependence of command and control within the Southwest Asian theater and between that theater and the continental United States on satellite communications systems has been noted by Lieutenant General Robert C. Kingston, USA (Commander, Rapid Deployment Joint Task Force):

> . . . the challenge is to establish and maintain strategic communications upward, necessary linkages laterally, and tactical communications downward. Strategic connectivity in the [Southwest Asian] region is limited today. The backbone of the Defense Communications System cannot be accessed directly except by satellite or HF over long distances. The FLTSAT and DSCS II systems support this need for all the services. DSCS III, the follow-on satellite, will soon be launched with a new booster. The HF programs, however, especially in the anti-jamming arena, need better interoperability. At present, limited HF links must transmit beyond optimum distances to reach DCA entry points and are subject to frequent atmospheric interruptions.[2]

Command and Control in the Pacific Theater.—Similar concern has been expressed by Lieutenant General Joseph T. Palastra, Jr., USA (Deputy Commander-in-Chief, Pacific) about the dependence of command and control within the Pacific theater and between that theater and the continental United States on satellite communications systems:

> A critical problem is the Pacific area's heavy reliance on satellite and undersea cable systems. The Soviet Union's demonstrated ability to destroy satellites has made it urgent to implement countermeasures and modernize backup high frequency systems so we can deploy at least minimum essential communications.[3]

Command and Control in the European Theater.—Command and control between the European theater and the continental United States depends on similar means which are similarly vulnerable. Command and control within the European theater depends less on satellite communications, for reasons noted by Major General Robert A. Rosenberg, USAF (then Assistant Chief of Staff, Studies and Analyses, Headquarters, U.S. Air Force):

> It is interesting to watch a simulated war-game exercise in the central region of Europe when the communications circuits are removed to simulate loss or destruction. The German Post Office telephone system is used to contact another command center when the military primary lines are down.[4]

[1]Colonel Robert B. Giffen, *US Space System Survivability: Strategic Alternatives for the 1990s*, National Security Affairs Monograph Series 82-4 (Washington, D.C.: National Defense University Press, Fort Leslie J. McNair, 1982), pp. 22-23.

[2]Lieutenant General Robert C. Kingston, USA, "C[3]I and the Rapid Deployment Force," *Worldwide Deployment of Tactical Forces and the C[3]I Connection*, Proceedings of the National Security Issues 1982 Symposium, Oct. 4-5, 1982, MITRE Corp. document M82-64, 1982.

[3]Lieutenant General Joseph T. Palastra, Jr., USA, "Pacific Command Perspectives," *Worldwide Deployment of Tactical Forces and the C[3]I Connection*, Proceedings of the National Security Issues 1982 Symposium, Oct. 4-5, 1982, MITRE Corp. document M82-64, 1982.

[4]Major General Robert A. Rosenberg, USAF, "Satisfying C[3]I Requirements for Deployed Air Forces," *Worldwide Deployment of Tactical Forces and the C[3]I Connection*, Proceedings of the National Security Issues 1982 Symposium, Oct. 4-5, 1982, MITRE Corp. document M82-64, 1982.

value of a function performed by a satellite may be realized at a later time and at a different—probably higher—level of conflict than that at which the function was performed.

For example, missile attack warning data would be most valuable during a nonnuclear war, when anticipation of such an attack would be most intense and when it would be most important to be reassured that the nation is *not* under attack when such is indeed the case. Once the United States has confirmed that it has been so attacked, or is under attack, or decides to retaliate for a nonnuclear attack by Warsaw Pact forces against NATO allies, further warning information would be of value only if additional salvos were expected, as in "nuclear war-fighting" scenarios.

As another example, the value of geopotential anomaly data collected by satellite might be greatest during nuclear war, although to be of value then such data must have been collected and analyzed—a lengthy process—during peacetime.

The problem of assessing the values of MILSAT capabilities and ASAT capabilities will be discussed in greater detail in appendix D to this report.

The Vulnerability and Protection of Military Space Systems

The value of MILSAT functions performed during wartime can be realized only if MILSATs survive long enough to perform them. Most current MILSATs are vulnerable to deliberate nuclear or nonnuclear attack, and some are vulnerable to nondestructive electronic countermeasures and electro-optical countermeasures. However, only a few U.S. MILSATs are potentially vulnerable to current Soviet nonnuclear ASAT weapons, and these could be made less so.

Satellites operate as components of space systems, which include terrestrial components such as satellite control facilities and user terminals (the "ground segment") as well as the satellites themselves (the "space segment"). Attacks against either segment can disrupt the functioning of a military space system and negate or reverse the force enhancement it provides. Hence both MILSATs and their associated ground equipment must be effectively protected against attack if the value of their wartime functions is to be realized. Technical measures—i.e., active and passive countermeasures—can provide some protection by reducing the MILSAT vulnerability, and legal measures—i.e., arms control treaties treaty or customary law banning threatening activities in space—can provide some protection by constraining ASAT capabilities which could be used against MILSATs. Arms control measures could be used in combination with passive countermeasures to constrain potential ASAT threats and reduce vulnerability to those which would remain. However, some arms control measures would be incompatible with some active countermeasures, such as shoot-back with ASAT weapons.

U.S. responses to current and future ASAT threats need not be limited to legally constraining such threats and protecting and defending satellites from those which remain. Other possible responses include deterrence of ASAT attack by maintaining a capability to retaliate (not necessarily in kind), using nondestructive electronic countermeasures and electro-optical countermeasures against the space and ground segments of ASAT systems, attacking the ground segment of ASAT systems, augmenting other forces to compensate for MILSAT vulnerability, and reducing dependence on MILSATs.

[Technical countermeasures are discussed in greater detail in chapter 4 of this report, and arms control and other legal measures are discussed in chapters 5 and 6. In practice, passive countermeasures, and probably active countermeasures as well, would be used in conjunction with arms control; such combinations of arms control and technical countermeasures are discussed in greater detail in chapter 7 of this report.]

THE ROLE AND VALUE OF ANTI-SATELLITE CAPABILITIES

To the extent that enemy MILSATs could increase the effectiveness of enemy forces, capabilities to damage such MILSATs or to degrade their functioning would be valuable in wartime. In addition to such destructive and nondestructive ASAT capabilities, other options are available to reduce the utility of MILSATs to an enemy or to compensate for their disutility to the United States.

Electronic and electro-optical countermeasures can be used to provide nondestructive ASAT capabilities which could be used at any level of conflict. They would be particularly valuable during crises short of declared war, in which satellite destruction would be escalatory and hence in many cases undesirable. "Active" electronic countermeasures such as "jamming" (i.e., overloading enemy receivers with strong signals) and "spoofing" (i.e., sending deceptive signals) could be used to interfere with satellite functioning, as could such active electro-optical countermeasures as temporary blinding (or "dazzling": the optical counterpart to jamming) and spoofing. Such nondestructive ASAT measures are discussed in chapter 4 of this report.

At higher levels of conflict, ASAT weapons could be used to destroy enemy satellites. The inherent ASAT capabilities of nuclear weapons such as ICBMs and SLBMs could be used for negation of low-altitude MILSATs at nuclear levels of conflict, while at nonnuclear levels of conflict the inherent ASAT capabilities of the experimental nonnuclear ABM technologies demonstrated in Homing Overlay Experiment (HOE) tests could be used for low-altitude MILSAT negation, as could the U.S. Air Force's Miniature Vehicle ASAT weapon, when operational. Negation of satellites at higher altitudes or more rapid negation of low-altitude satellites would require more capable ASAT weapons such as are discussed in chapter 4 of this report.

Alternatively, or supplementally, various passive measures could be used in peacetime, crisis, and war *not* to interfere with the functioning of enemy MILSATs but rather to decrease the value of their functions to the enemy and, more importantly, to mitigate or compensate for the harm such satellites could cause the United States. Intelligence-gathering satellites may be particularly susceptible to such measures. For example, terrestrial force elements might employ maneuver to avoid observation by enemy imaging satellites, and they could use camouflage and concealment to prevent recognition if observed. Decoys which would also be "recognized" erroneously could be used to thwart image interpretation by causing confusion as to the numbers and locations of the assets simulated by the decoys. Ships, aircraft, and other assets might be designed to reflect radar signals only weakly in order to evade detection by radar satellites, and radio silence might be practiced, or covert signaling techniques used, in order to prevent detection of radio emissions by satellites. Of course, these passive measures would impose either operational constraints or financial costs or both.

Still other options are available to compensate for, rather than mitigate, the harm which foreign MILSATs could cause to U.S. and allied security. For example, to the extent that foreign MILSATs provide a force multiplier effect to foreign force elements which engage in direct combat, U.S. or allied force elements which might oppose these foreign force elements might be augmented in order to maintain relative combat strength. That is, force augmentation could offset the force multiplication provided to enemy forces by space support. Force augmentation would be costly, but not necessarily more costly than an ASAT capability of comparable security benefit.

These possible responses to threatening MILSAT capabilities are summarized in table 3-3. These response options would be more effective used in combination rather than individually.

Assessment of the value of ASAT capability is subject to the same methodological difficulties which confound assessment of the value of MILSAT capability.[29] These prob-

[29]The value of a U.S. ASAT capability to the United States in a conflict could be assumed to be equal in magnitude to the

Table 3-3.—Possible Responses to Enemy MILSAT Capabilities

- Force augmentation
  —offsets force multiplication by MILSATs
- Passive measures against satellite reconnaisance and targeting
  —Evasion
  —Concealment
  —Camouflage
  —Decoys
  —Radio/radar silence
  —Covert communications techniques
- Nondestructive anti-satellite measures
  —Electronic countermeasures
    - Jamming
    - Spoofing
  —Electro-optical countermeasures
    - Jamming
    - Spoofing
- Destructive anti-satellite measures
  —Use inherent ASAT capabilities of ICBMs, SLBMs, and ABMs
  —Develop deliberate ASAT weapons

lems, noted above, are discussed in greater detail in appendix D to this report. Although quantification of the value of ASAT capability is necessarily subjective or complicated, or both, it is clear that operational ASAT capabilities would have some value to the extent that they would be inexpensive, stabilizing, and compatible with other (e.g., diplomatic) options for enhancing security.

[negative] total value *to the United States* of those enemy MILSATs which it would be expected to destroy. No analytically sophisticated studies known to OTA have attempted to systematically assess and compare the utility of U.S. MILSATs to the United States with the disutility of Soviet MILSATs *to the United States* [but compare Stephen M. Meyer, op. cit., pp. 204-215]. It is these utilities which should be compared to determine whether the United States would fare better in a regime in which U.S. and Soviet MILSATs were mutually vulnerable to the other's ASAT weapons or in a regime in which U.S. and Soviet MILSATs were protected from ASAT threats by active or passive countermeasures or arms control.

MILITARY SPACE POLICY

Military Space Policy Issues

The fundamental task of military space policy formulation is deciding how much proposed military space programs are worth, both in terms of the resources for which they would compete with other proposed national programs, and in terms of opportunity costs which might be incurred by failing to take other actions (e.g., arms control) with which such programs are incompatible for nonbudgetary (e.g., legal) reasons. Because the costs, risks, and benefits differ in character, a political determination of levels at which military space efforts should be pursued is required.[30]

Prerequisite for this is the task of deciding the *relative* values of proposed military space systems, anti-satellite systems, and arms control agreements; judgment of these values also requires political choice.[31]

[30]Representative of such determinations are the annual budgets prepared by the executive branch and funds authorized and appropriated by Congress.

[31]Representative of such determinations are President Carter's Presidential Directive 37 (PD-37) issued in June 1978, President Reagan's July 4, 1982, statement on National Space Policy, and President Reagan's *Report to the Congress: U.S. Policy on ASAT Arms Control*, Mar. 31, 1984.

ASAT Policy Issues

Particularly apparent is the incompatibility between ASAT capabilities and ASAT arms control agreements; these would have different kinds of benefits and risks, and a decision to pursue one or the other must be based on a political judgement of their respective expected net benefits. A national *ASAT policy* reflects such a judgement and should attempt to establish goals for national efforts to enhance security by the chosen approach. In particular, a national ASAT policy should:

1. describe *military posture objectives*; it should attempt to answer the question: "What ASAT capabilities do we need, and why?"
2. indicate the extent to which pursuit of security by the chosen approach would probably benefit from increased spending in various areas of *research and technology development*, so that national research and development policy might be formulated cognizant of these potential benefits.
3. establish *ASAT arms control policy*; i.e., it should indicate types of arms control

provisions or agreed confidence-building measures which are judged to be in the national interest, in order to coordinate formulation of negotiating postures, in particular, and foreign policy in general;[32] and

4. specify U.S. *ASAT employment policy*; i.e., it should specify conditions under which U.S. use of ASAT capabilities—especially in conflicts short of declared war—would be deemed justifiable and in the national interest. In addition, if we desire to deter Soviet ASAT attacks, we must arrive at and announce a public policy which we believe will make this deterrence as effective as possible. Whether this policy should be explicit or instead one of calculated ambiguity is a matter of political judgment.

Ballistic Missile Defense as an ASAT Policy Issue

The incompatibility between ballistic missile defense capabilities and ASAT arms control agreements is apparent. Ballistic missile defense weapons which would be capable of attacking ballistic missile components *in space* generally would have some inherent capability to attack satellites at altitudes or ranges comparable to those at which ballistic missile components would be engaged, depending on satellite "hardness." The inherent ASAT capabilities of some possible advanced-technology BMD weapons would be considerable, and any arms control agreement which would attempt to limit the threat of such weapons to satellites must ban or limit such weapons. Hence a restrictive ASAT arms control agreement would also restrict BMD capabilities, just as the ABM Treaty—which limits, by intent, weapons capable of attacking ballistic missiles—also limits ASAT weapons with inherent BMD capability. Of all the opportunity costs which might be imposed on the United States by an agreement to limit ASAT capabilities, restrictions on the development and deployment of BMD capabilities beyond those already imposed by the ABM Treaty are considered most costly by those who believe that exploitation of advanced technology may make possible BMD weapons of great effectiveness.

A more fundamental incompatibility between ASAT and BMD capabilities is physical rather than legal: the most capable BMD systems now envisioned would use space-based sensors, and perhaps also space-based weapons, which would be subject to attack by deliberate ASAT weapons or by BMD or other weapons with inherent ASAT capabilities. Such BMD systems would be effective only if their sensors and weapons could be protected from such ASAT threats at reasonable cost by passive countermeasures, active countermeasures, arms control measures, or a combination of these. If so, perhaps such measures could be used to protect other satellites, or else the functions now performed by other satellites could be performed by secondary mission payloads "piggybacking" on BMD satellites. If not, such BMD capabilities would be of little value, if any, and the opportunity costs which would be incurred by restricting them would be small.

[Chapter 5 of this report contains a discussion of past arms control efforts which had the intent or effect of constraining ASAT capabilites. Other ASAT arms control options are described and evaluated on a provision-by-provision basis in chapter 6 and as packages of provisions in chapter 7.]

[32]Representative of such a policy is President Reagan's *Report to the Congress: U.S. Policy on ASAT Arms Control*, delivered Mar. 31, 1984.

Chapter 4

ASAT Capabilities and Countermeasures

Contents

Chapter 4

ASAT Capabilities and Countermeasures

ORGANIZATION OF THIS CHAPTER

A variety of technological options are available for space surveillance systems, stand-off weapons, and weapon and sensor platforms for anti-satellite uses. Current and projected U.S. and Soviet ASAT capabilities, including space surveillance capabilities, are described in the first section of this chapter. More advanced ASAT capabilities which could be deployed by the United States or the U.S.S.R. are described in the second section of this chapter. Some possible U.S. responses to Soviet development of such capabilities are described in the third section of this chapter. The principal conclusions about ASAT capabilities and countermeasures are summarized in the final section of this chapter. The actual military utility of these capabilities will be discussed in appendix D to this report, which is classified.

CURRENT AND PROJECTED ASAT CAPABILITIES

Generic ASAT System Components

A space defense system could include both passive countermeasures for protecting satellites and an ASAT system for interfering with enemy satellites in time of war. An ASAT system, whether deliberate or expedient, must be controlled by an associated command, control, communications, and intelligence (C^3I) system, which itself will have three types of subsystems, as illustrated abstractly in figure 4-1. First is the intelligence collection part—the space surveillance system—which would detect electromagnetic radiation emitted or reflected by a satellite and, using these measurements, attempt to track the satellite and determine its orbit. Careful interpretation of this information may allow characterization of the satellite—i.e., determination of its mass, shape, and other features—and even determination of the function of the satellite. On the basis of this interpretation, information would be communicated over command, control, and communications (C^3) links to command and control (C^2) centers, where authorities would consider the information in the context of other relevant information and possibly issue orders to *negate* the satellite or to interfere with its functioning nondestructively (e.g., using electronic countermeasures). If so, other C^2 elements would generate detailed instructions for an attack, and these would be communicated by C^3 links to an ASAT weapon system.

An ASAT system would have either nondestructive ASAT devices such as jammers or other electronic or electro-optical countermeasures, or ASAT weapons capable of damaging satellites (in which case it would be an ASAT *weapon* system), or both. In general, each weapon would consist of a *stand-off weapon* capable of damaging a satellite at a distance and either carried by a *platform* such as a satellite, rocket, airplane, or land vehicle, or else based on the ground at a fixed site. The stand-off weapon could be a kinetic-energy weapon (KEW) such as a gun or a fragmentation warhead, a directed-energy weapon (DEW) such as a laser or particle accelerator, or an ordinary "isotropic" nuclear warhead (so called because it would release roughly equal amounts of energy in all directions).

The weapon platform (if any) would carry the stand-off weapon to within lethal range of a targeted satellite. A highly maneuverable platform could pursue and collide with a targeted satellite; such a vehicle would be a "hit-to-kill" kinetic-energy weapon which would

Figure 4-1.—Generic Components of an ASAT System

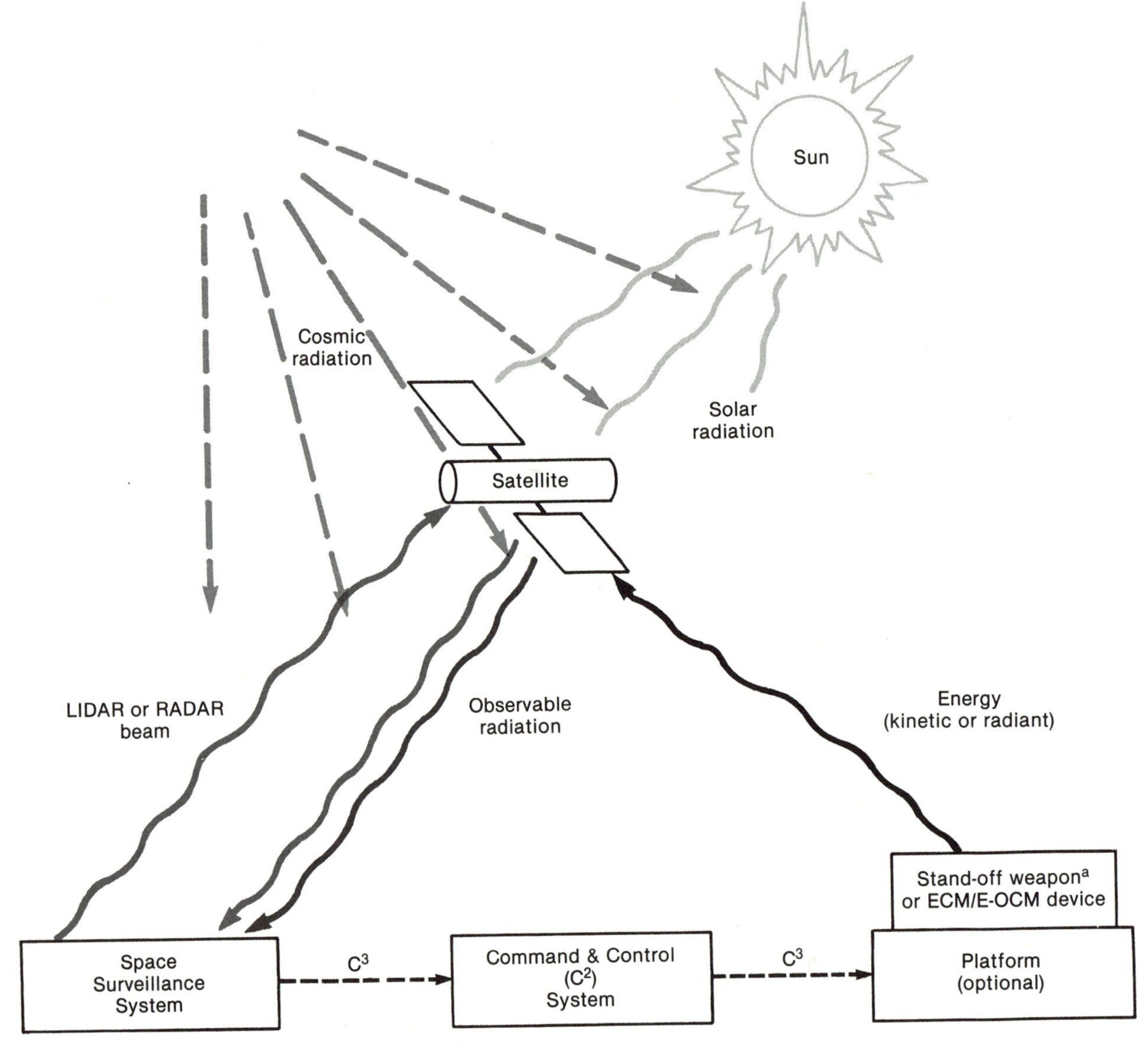

[a]Directed-energy weapon, kinetic-energy weapon, or isotropic nuclear weapon.

not need to carry a stand-off weapon. Alternatively, if the stand-off weapon had sufficient lethal range, it would not need to be maneuvered toward a targeted satellite but could instead be based on the ground or on a non-maneuvering platform.

Figure 4-2 illustrates a portion of the U.S. Space Defense System as it will appear when the planned anti-satellite weapon system is added, showing examples of its space surveillance, command and control, and ASAT weapon systems.

Soviet ASAT Capabilities

Space Surveillance

The ASAT capabilities of the Soviet Union depend on Soviet space surveillance systems.

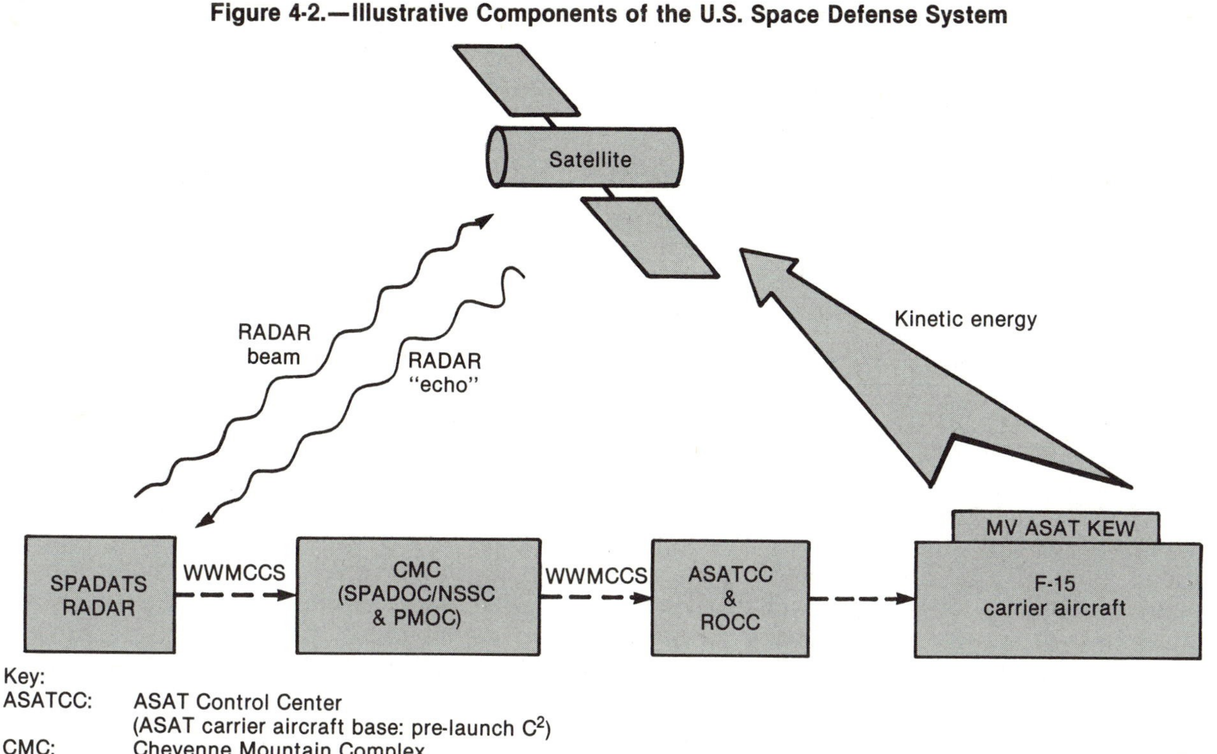

Figure 4-2.—Illustrative Components of the U.S. Space Defense System

Key:

ASATCC:	ASAT Control Center (ASAT carrier aircraft base: pre-launch C^2)
CMC:	Cheyenne Mountain Complex
MV:	Miniature Vehicle (satellite interceptor warhead)
NSSC:	National Space Surveillance Center
PMOC:	Prototype Mission Operations Center
ROCC:	Regional Operations [military air traffic] Control Center (post-launch C^2)
SPADATS:	SPAce Detection and Tracking System
SPADOC:	SPAce Defense Operations Center
WWMCCS:	World-Wide Military Command and Control System

The Soviet Union operates extensive military and civil networks of radar, LIDAR[1], and photographic space surveillance sensors linked together by satellite and terrestrial communications systems.[2] Soviet missile early warning radars and satellites can detect foreign satellite launches. Soviet radio/radar tracking ground stations can presumably detect and track satellites in low-Earth orbit and track satellites in higher orbits. In addition, the radars used by the Soviet ABM system and radio telescopes can be used to detect, track, and characterize satellites. The U.S.S.R. also uses ships for satellite tracking and communications and operates some tracking stations in foreign territory.

Soviet deep-space detection capabilities necessarily rely upon passive sensors, primarily telescopic cameras similar to the Baker-Nunn cameras formerly used by the United States and upon radio telescopes and ground-based military signal intelligence collection[3] systems. Passive optical sensors, whether photographic

[1]LIDAR is an acronym for LIght Detection And Ranging; it refers to a radar-like sensor system which transmits pulses of light, typically produced by a laser, and looks for reflections from objects. Ranges to objects can be inferred from the time delay of pulses reflected from it.

[2]*Soviet Space Programs: 1976-80*, Part 1, Committee Print, Committee on Commerce, Science, and Transportation, Senate, 97th Cong., 2d sess., December 1982. [USGPO 98-515 O]

[3]Such systems, which perform direction-finding or signal intercept functions, are called electronic support measure (ESM) systems.

or electro-optical, can detect sunlight reflected by a high-orbit satellite; the power of the reflected sunlight received by a distant optical sensor decreases only as the square of the range to the target, so passive optical sensors are more useful than radar for detecting high-orbit satellites.[4] Similar considerations make passive radio systems—i.e., radio telescopes or military electronic support measure systems—useful for detection and tracking at long range, provided the target is emitting a radio signal of some kind.

It is possible that the U.S.S.R. may develop electro-optical tracking sensors in the future; such sensors could provide surveillance information more quickly than can camera systems, which require development of photographic film or plates. Neither photographic nor electro-optical telescopes can detect or track satellites from the ground in daytime or through overcast, as radar can.

Weapons

The Soviet Union has been conducting a series of tests of coorbital satellite interceptors ("killer satellites") since 1968.[5] An artist's conception of such an interceptor is shown in figure 4-3. The U.S. Department of Defense estimates that these anti-satellite weapons became operational in 1971.[6] These weapons are believed to be capable of attacking satellites at altitudes up to 5,000 kilometers or even higher,[7] depending on their orbital inclinations; presumably they would be unable to attack satellites in orbits with unfavorable inclinations, i.e., very different from the latitude of the interceptor launch site. As of 1984 there appeared to be only two launch pads for Soviet coorbital interceptors, both located at the Tyuratam launch complex.[8] Several interceptors could be launched per day from this complex.[9]

In 1971 the testing of satellite interceptors was apparently suspended, then resumed in 1976, then again suspended in 1978, just before the U.S.-U.S.S.R. ASAT negotiations began, then again resumed in 1980 after suspension of those talks, then again suspended in August 1983 when Soviet President Yuri Andropov announced a unilateral moratorium on ASAT testing, stating that the U.S.S.R. would not test ASAT weapons if the United States did not. The U.S.S.R. continues to observe this unilaterally declared moratorium. The U.S.S.R. has never officially and publicly admitted developing or testing weapons of this type. However, on 29 May 1985, in an interview by a West German reporter in Geneva, Col. Gen. Nikolai Chervov, a senior department head on the Soviet General Staff, claimed that the U.S.S.R. had successfully developed a direct-ascent satellite interceptor similar to that tested by the United States in the early 1960s and operational until the mid-1970s.

[4]This is because the energy of a radar return, or "echo," from a satellite decreases as the fourth power of range to the target. Radar returns from satellites in high orbit are generally so weak that they cannot be detected by a radar rapidly scanning the sky; they can only be detected if the approximate position of the satellite is already known so that the radar can scan slowly for signals from that general direction, accumulating signal energy for a prolonged period of time. Hence ground-based radar can measure small changes of a high-altitude satellite's orbit but could not easily find a satellite which had maneuvered energetically since its last observation. Similar considerations limit the effective search range of LIDAR systems.

The energy of a radar echo also depends on the radar wavelength used. For example, at wavelengths much longer than a satellite's diameter, the echo energy decreases rapidly with increasing wavelength—as the inverse fourth power, according to Rayleigh's law. However, the echo also becomes more omnidirectional (as a consequence of Babinet's principle and—in the special case of spherical satellites—the Mie effect) and less dependent upon details of satellite shape and composition other than the electric susceptibility of the satellite, in accordance with Rayleigh's law.

[5]U.S. Department of Defense, *Soviet Military Power*, 4th ed., April 1985.

[6]Ibid., p. 55.

[7]Ibid., p. 56.

[8]U.S. Department of Defense, *Soviet Military Power*, 3rd ed., April 1984, p. 34.

[9]U.S. Department of Defense, *Soviet Military Power*, 4th ed., April 1985, p. 56.

Figure 4-3.—The Soviet Coorbital Satellite Interceptor

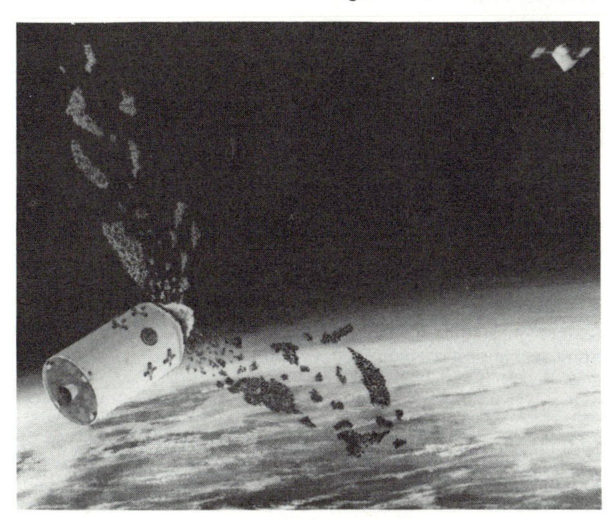

Artist's conception of a Soviet coorbital satellite interceptor attacking a satellite.

Launchpad and storage facility for satellite interceptors at Tyuratam

SOURCE: U.S. Department of Defense.

The Soviet Union has also been testing ground-based lasers which could have some ASAT capability [see figure 4-4]. The Department of Defense has stated that the Soviets "already have ground-based lasers that could be used to interfere with U.S. satellites." As of 1985, there are two experimental Soviet ground-based lasers with some ASAT capability,[10] both at Sary Shagan.[11]

In addition to weapons designed specifically for ASAT use, the U.S.S.R. could attack low-altitude satellites with its ABM interceptor missiles [illustrated in figure 4-4],[12] and presumably with ICBMs and SLBMs, although these might require some modification for ASAT use. All of these weapons are presumably armed with nuclear warheads, and their use in a nonnuclear conflict would be viewed as escalatory by the United States and presumably by the U.S.S.R. as well.

In addition to these destructive ASAT capabilities, the U.S.S.R. has a technological capability to jam satellite uplinks or downlinks with some effectiveness, and could use ground-based lasers for electro-optical countermeasures with some effectiveness.[13]

[10]Ibid., p. 56.
[11]Ibid., p. 58.
[12]Ibid., p. 56.

Operational Capabilities

Existing Soviet ASAT capability could be potentially effective for negating low-altitude U.S. MILSATs, such as those used for navigation (Transit) and meteorological surveillance (Defense Meteorological Satellite Program satellites). Commenting on this capability, the Honorable Richard Perle, Assistant Secretary of Defense (International Security Policy), has stated that:

> We believe that this Soviet anti-satellite capability is effective against critical U.S. satellites in relatively low orbit, that in wartime we would have to face the possibility, indeed the likelihood, that critical assets of the United States would be destroyed by Soviet anti-satellite systems.[14]

[13]Ibid., p. 56.
[14]Statement of The Honorable Richard Perle, Assistant Secretary of Defense (International Security Policy), in *Hearings Before the Subcommittee on Strategic and Theater Nuclear Forces of the Committee on Armed Services United States Senate, Testimony on Space Defense Matters in Review of the FY1985 Defense Authorization Bill* [S. Hrg. 98-724], March 15, 1984, Pt. 7, p. 3452.

Figure 4-4.—Soviet Anit-Satellite Capabilities

Artist's conception of a Soviet ABM interceptor missile, named GALOSH by Western analysts. GALOSH missiles might have capabilities to attack satellites at low altitudes.

Soviet high-energy laser facility at Sary Shagan. Two lasers there may be capable of damaging unprotected satellites at low altitudes.

SOURCE: U.S. Department of Defense.

The utility of these U.S. satellites in various types of conflicts is discussed in Appendix D to this report, which is a separate, classified document.

Projected Capabilities

The Soviet Union could develop weapons capable of attacking U.S. satellites at higher altitudes than can be reached by the current Soviet coorbital interceptor. Laser weapons, among other types, could be be used for this purpose. The U.S. Department of Defense estimates that:

> The Soviets are working on technologies or have specific weapons-related programs underway for more advanced anti-satellite systems. These include space-based kinetic-energy, ground- and space-based laser, particle beam, and radiofrequency weapons. The Soviets apparently believe that these technologies offer greater promise for future anti-satellite application than continued development of ground-based orbital interceptors equipped with conventional warheads.
>
> . . . In the late 1980s, they could have prototype space-based laser weapons for use against satellites. In addition, ongoing Soviet programs have progressed to the point where they could include construction of ground-based laser anti-satellite (ASAT) facilities at operational sites. These could be available by the end of the 1980s and would greatly increase the Soviets' laser ASAT capability beyond that currently at their test site at Sary Shagan. They may deploy operational systems of space-based lasers for anti-satellite purposes in the 1990s, if their technology developments prove successful.[15]

The Soviet Union also has the basic technology required to build space-based neutral particle beam weapons. The U.S. Department of Defense estimates that:

> A prototype space-based particle beam weapon intended only to disrupt satellite electronic equipment could be tested in the early 1990s. One designed to destroy the satellites could be tested in space in the mid-1990s.[16] [17]

[15] U.S. Department of Defense, *Soviet Military Power*, 4th ed., 1985.

[16] Ibid.

[17] However, as recently as 1984, while projecting this capability, the Administration noted that *We have, as yet, no evidence of Soviet programs based on particle-beam technology* [President Ronald Reagan, "Report to the Congress: U.S. Policy on ASAT Arms Control," 31 March 1984 (UNCLASSIFIED)].

Future manned Soviet spacecraft, such as the expected Soviet "space shuttle"—and, especially, the "space plane"—could have greater maneuverability and possibly non-cooperative rendezvous capability and, if so, some inherent ASAT capability.[18]

U.S. ASAT Capabilities

The U.S. Space Defense System is a network of systems used for space surveillance and for command and control. An anti-satellite weapon system will be added to it in the near future. Figure 4-2 illustrates a portion of the prospective U.S. Space Defense System as it will appear the planned anti-satellite weapon system is added. The figure shows examples of the space surveillance, command and control, and ASAT weapon systems which will be part of the U.S. Space Defense System.

Space Surveillance

U.S. ASAT capabilities, like those of the Soviet Union, depend on space surveillance capabilities. Like the U.S.S.R., the United States can use its missile attack warning radars and satellites to detect satellite launches and can track satellites after launch using ground-based and shipboard radar, LIDAR, passive optical, and passive radio sensors. The Space Detection and Tracking System (SPADATS) acquires, processes, stores, and transmits data from such sensors, including Naval Space Surveillance (NAVSPASUR) radar interferometers and Air Force Spacetrack radars and Ground-based Electro-Optical Deep-space Surveillance System (GEODSS) sensors [see figure 4-5]. The United States operates more foreign-based space surveillance facilities than does the U.S.S.R., and consequently relies less on shipboard systems, although such systems are used.

The effective search range of U.S. ground-based radar systems is limited to low-Earth orbit, although some radars can track a satellite out to geosynchronous altitude if the satellite's approximate position is already known.[19] The range at which a ground-based radar can track low-altitude satellites is limited by the requirement for an unobscured line of sight to the satellite. For example, a satellite at an altitude of 185 kilometers (100 nautical miles) would be below the horizon if farther away than 1,590 kilometers slant range.[20]

For detection of satellites in deep space the United States relies on a system of telescopic electro-optical sensors called GEODSS, for Ground-based Electro-Optical Deep-space Surveillance System [see figure 4-6].[21] The GEODSS network, when completed, will provide world-wide deep-space surveillance coverage using sensors at five sites[22]:

- Socorro, New Mexico (Site I);
- Taegu, South Korea (Site II);
- Maui, Hawaii (Site III);
- Diego Garcia, in the Indian Ocean (Site IV); and
- Portugal (Site V).

The main telescopes at each GEODSS facility are designed to detect objects as dim as a star of visual magnitude 16.5, or a reflective sphere about the size of a soccer ball in geosynchronous orbit.

Command and Control

Under present policy, the U.S. National Command Authorities (NCA) would have to authorize satellite negation. Actual operational control of a negation mission would be exercised by the USAF Space Command

[18]Current manned Soviet spacecraft (Soyuz, Salyut) do not have a significant inherent ASAT capability: they have little maneuver capability and have not demonstrated coorbital rendezvous with non-cooperative spacecraft. Rendezvous with co-operative spacecraft is typically performed by an automatic system which relies on a transponder on the passive spacecraft.

[19]See note 4, supra.

[20]I.e., 1,560 kilometers as projected on the Earth's surface. A closer satellite at this altitude might be below the horizon, if mountains or other terrain features obscured the view. In addition, the azimuthal coverage of some radars is limited.

[21]D.D. Otten, E.I. Bailis, and J.G. Klayman, "GEODSS: Heavenly Chronicler," *Quest* (Redondo Beach, CA: TRW, Inc., August 1980), pp. 3-23.

[22]Sites I, II, and III are presently operational, and site IV should become operational this year. Sites I, II, and III each have two main 40-inch telescopes and one 15-inch auxiliary telescope, while sites IV and V will have three main telescopes each. Site equipment is designed to be relocatable within 2 weeks.

Figure 4-5.—Space Detection and Tracking System

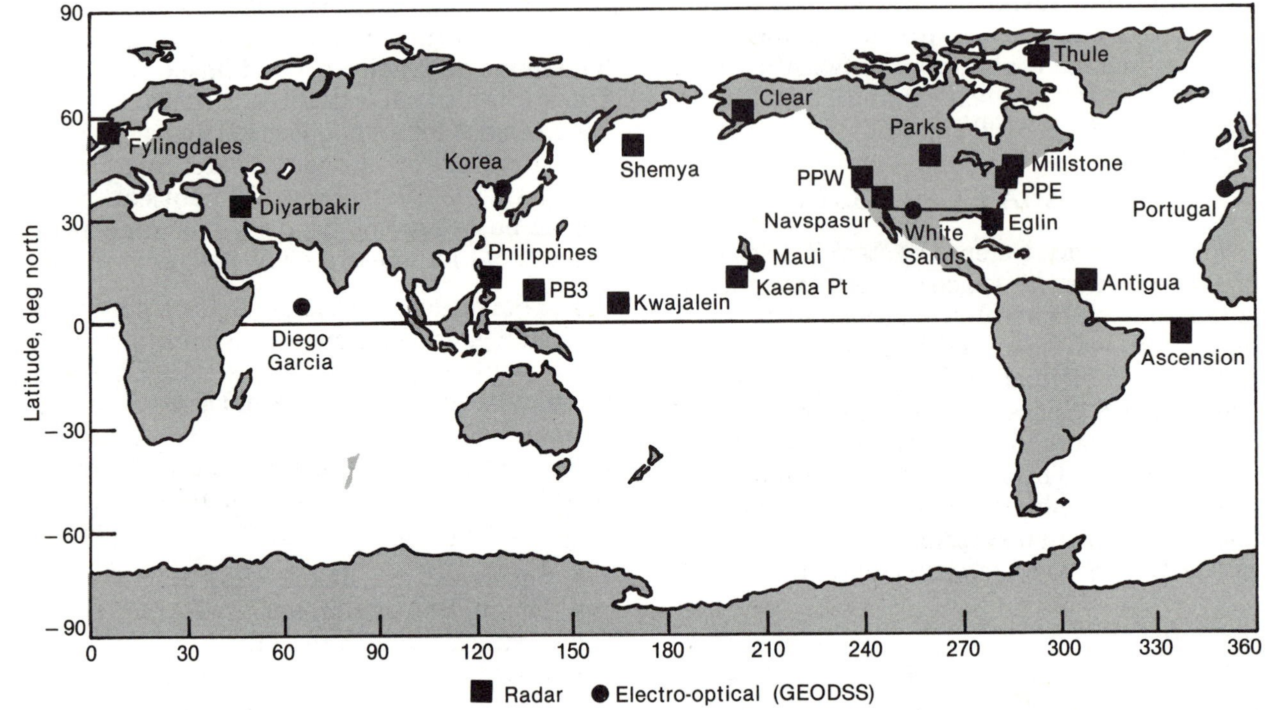

SOURCE: The Aerospace Corp.

Space Defense Operations Center (SPADOC) in the Cheyenne Mountain Complex using other assets of the Space Defense Command and Control System [see Figures 4.7 and 4.8].[23]

On October 1, 1985, satellite negation will become the responsibility of a new unified space command which will also exercise operational command over U.S. military space systems which provide support to the combatant forces of other unified and specified commands. Creation of a unified space command was proposed in order to increase the effectiveness and responsiveness of U.S. space systems and to ensure a clear chain of command from the NCA to combatant forces.[24] Creation of the U.S. Space Command was authorized by the President on November 30, 1984.

[23]R.S. Cooper, Director, Defense Advanced Research Projects Agency, *The U.S. Anti-satellite (ASAT) Program*, statement before the Senate Foreign Relations Committee, 25 April 1984.

[24]Hon. Verne Orr, Secretary of the Air Force, *USAF FY85 Report to the 98th Congress of the United States of America*, 8 February 1984; reprinted in S. Hrg 98-724.

Weapons

In 1959 the United States successfully intercepted the Explorer VI satellite using an air-launched ballistic missile developed for other purposes and, the following year, began to develop—but abandoned before testing—a coorbital SAtellite INTerceptor system (SAINT) designed specifically for the purpose of inspecting and destroying satellites. The United States also maintained an operational direct-ascent satellite interceptor capability from 1963 until 1975, using nuclear-armed Nike-Zeus missiles (Project Mudflap, 1963-1964) and Thor missiles (Project 437, 1964-1975).[25]

The United States has no deliberate operational ASAT capability at the present time, although its nuclear-armed ICBMs and SLBMs have some inherent ASAT capabilities, as was demonstrated by the nonnuclear exoatmospheric ABM Homing Overlay Experiment

[25]M. Smith, "Anti-satellites (Killer Satellites)," CRS Issue Brief IB81123, 22 August 1983.

Figure 4-6.—Deep Space Surveillance Systems

Left: Today, surveillance of deep space is performed by ground-based electro-optical surveillance systems such as the Ground-based Electro-Optical Deep Space Surveillance (GEODSS) system, which is operated by the U.S. Air Force Space Command as part of the Space Detection and Tracking System (SPADATS). Here, an operator at a GEODSS control and display console studies a 2.1° field of view of deep space. On a clear night, a GEODSS main telescope can automatically survey over 500 such fields per hour and detect reflective satellites as small as soccer balls at geosynchronous altitudes.

SOURCE: U.S. Air Force.

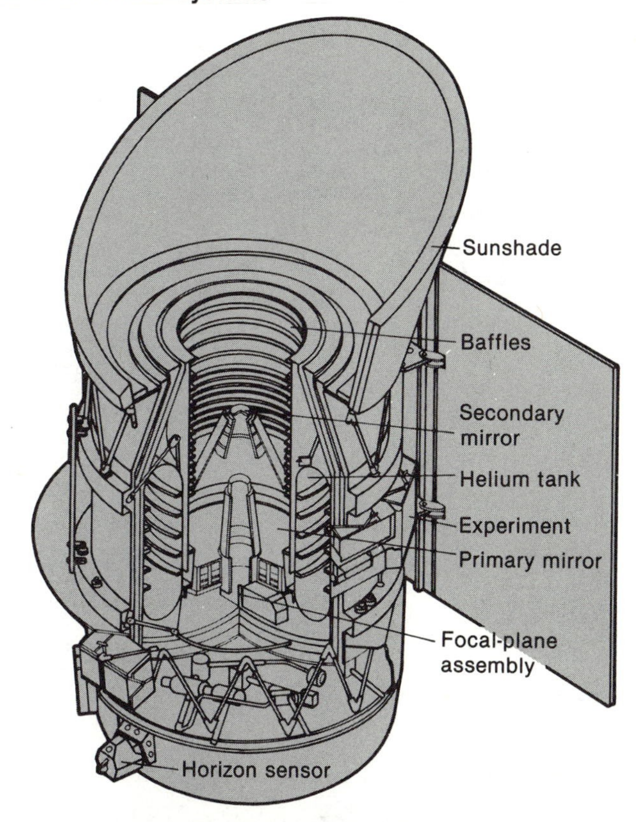

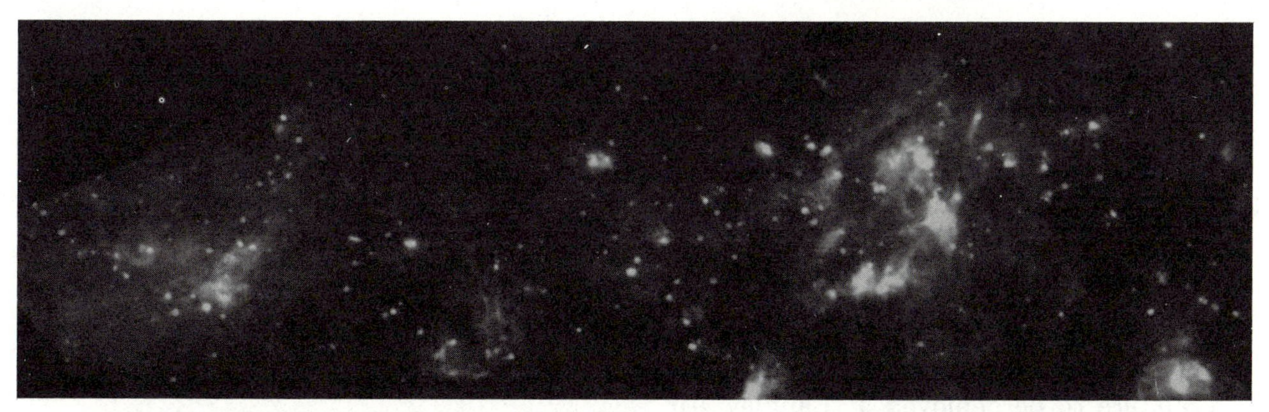

In the future, surveillance of deep space could be performed both day and night, regardless of weather conditions, by infrared telescopes on satellites. Such satellites could use technologies similar to those used in the U.S./British/Dutch InfraRed Astronomical Satellite (IRAS, above right) but would have to be much more capable in order to scan the sky more quickly: IRAS required most of a year to survey most of the sky, including the 28x11° view (below) of a portion of one of the Milky Way's spiral arms.

SOURCE: U.S. National Aeronautics and Space Administration, Jet Propulsion Laboratory.

Figure 4-7.—ASAT Mission Operations Concept

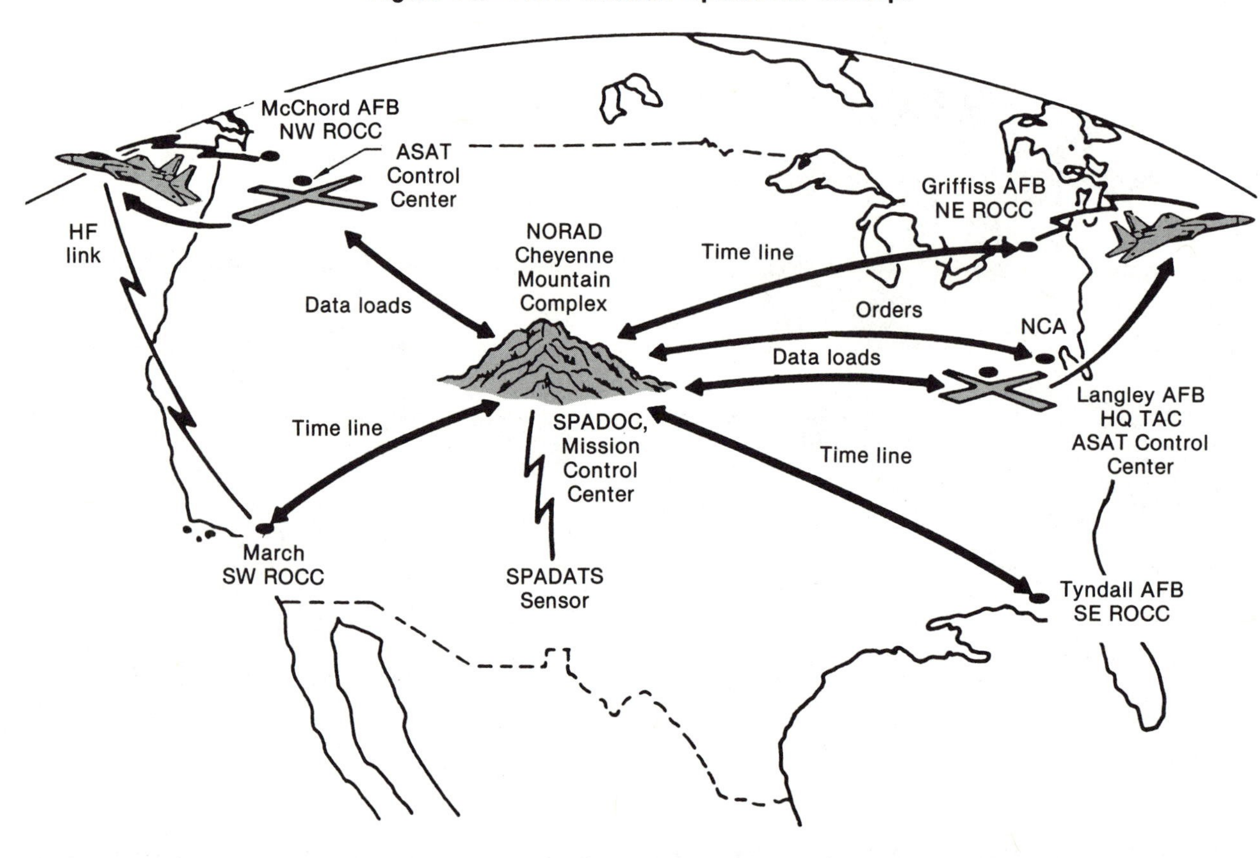

This illustration shows communication links which would be used for a satellite negation operation.

SOURCE: The Aerospace Corp.

(HOE) test of 10 June 1984. However, the U.S. Air Force is flight-testing an air-launched, direct-ascent anti-satellite weapon called the Miniature Vehicle ("MV"). This infrared-guided nonnuclear kinetic-energy weapon [see figure 4-9] is mounted on a two-stage SRAM/ ALTAIR booster which is carried aloft and launched from an F-15 aircraft. F-15 carrier aircraft are to be deployed at Langley Air Force Base in Virginia and at McChord Air Force Base in Washington state [see figure 4-10]. When operational, it will be able to attack low-altitude Soviet satellites which perform reconnaissance and targeting functions; these are viewed as most threatening by the United States.[26] If based as planned, these weapons will not be able to attack Soviet satellites which provide missile attack warning, navigation, and advanced communications functions.[27]

The Air Force plans to hold 12 flight tests of the ASAT system. Two of the twelve tests have been held—the first in January 1984 and the second in November 1984. In the January test, the ASAT missile was targeted at a point in space to determine whether the two-stage SRAM/ALTAIR booster could deliver the miniature homing vehicle to the vicinity of the target point. The Air Force considered the test a success: the proper functioning of the first and second stage propulsion systems and the

[26]President Ronald Reagan, *Report to the Congress: U.S. Policy on ASAT Arms Control*, 31 March 1984 [UNCLASSIFIED].

[27]R.S. Cooper, Director, Defense Advanced Research Projects Agency, *The U.S. Anti-satellite (ASAT) Program*, statement before the Senate Foreign Relations Committee, 25 April 1984.

Figure 4-8.—Mission Control Events

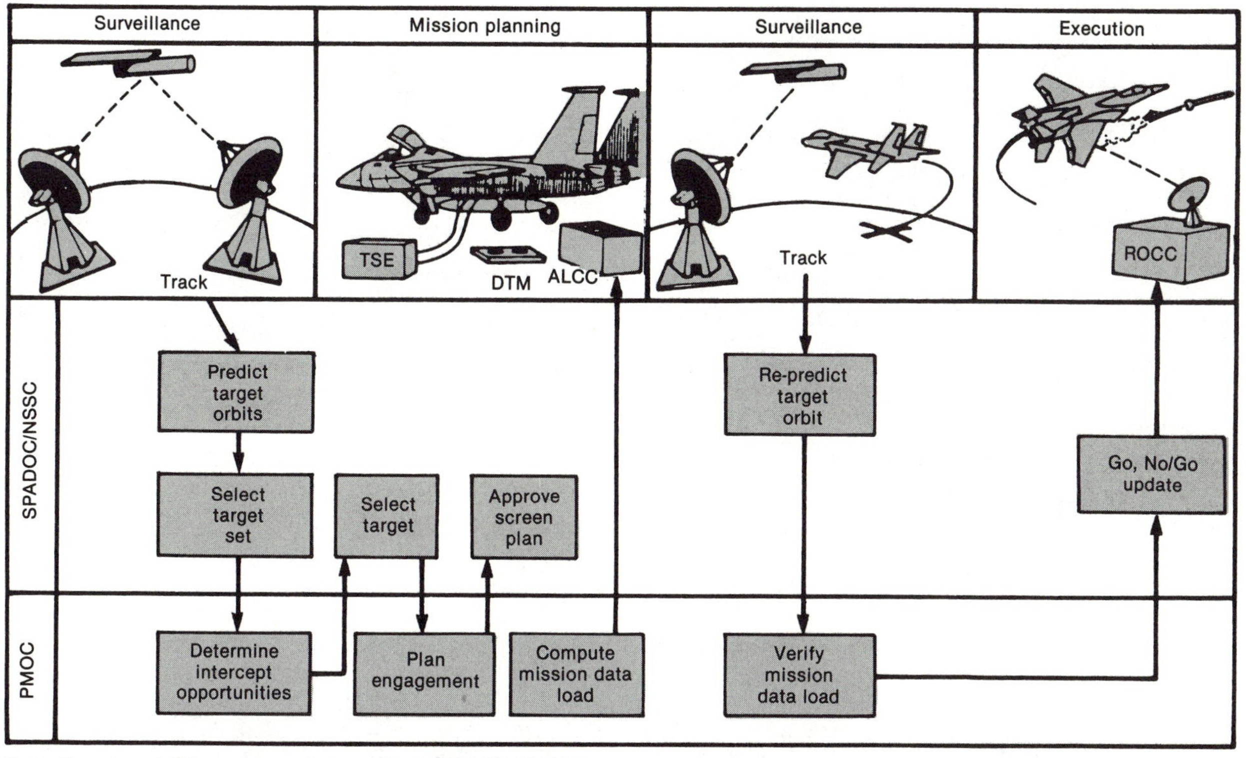

Execution of a satellite negation mission using the USAF MV ASAT weapon would begin with an alerting order. The target would then be tracked by SPADATS and its emphemeris would be updated by the National Space Surveillance Center (NSSC) for use in generating mission tapes to be loaded on the carrier aircraft for targeting of the ASAT weapon. After an execution order is issued from the Prototype Mission Operations Center (PMOC) in the Cheyenne Mountain Complex (CMC), the carrier aircraft would take off and the missile would have to be launched before sensor coolant exhaustion. Before missile launch, the carrier aircraft would receive updated information and instructions by high-frequency radio data link from a Regional Operations Control Center (ROCC), which is an element of the current military air traffic control center.

SOURCE: The Aerospace Corp.

missile guidance system was successfully demonstrated.[28]

The objective of the November 1984 test was to reaffirm the performance of the missile and to demonstrate the capability of a miniature homing vehicle to acquire and track an infrared-emitting body (in this test, a star) against the radiant background of deep space.

[28]U.S. Congress, General Accounting Office, "Report to the Honorable George E. Brown, Jr., House of Representatives: Status of the U.S. Antisatellite Program," report GAO/NSIAD-85-104, June 14, 1985.

The Air Force considered the test a partial success.[29]

Successful completion of U.S. ASAT tests would provide confidence that the weapon could perform as specified if actually used. Funds for completion of the planned flight test program have been appropriated by Congress subject to certain limitations, partly because of concern that once the weapons are proven effective, the Soviet Union would cease to observe its self-imposed moratorium on the testing of ASAT weapons and might be reluc-

[29]Ibid.

Figure 4-9.—The USAF Miniature Vehicle

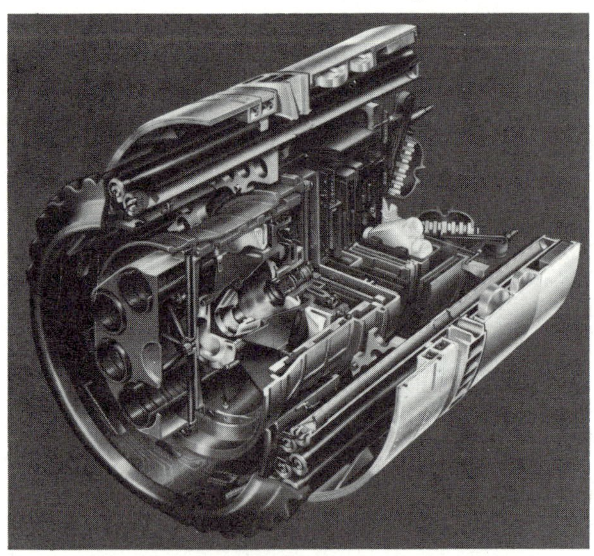

Cutaway view of the ASAT missile warhead (the Miniature Vehicle).

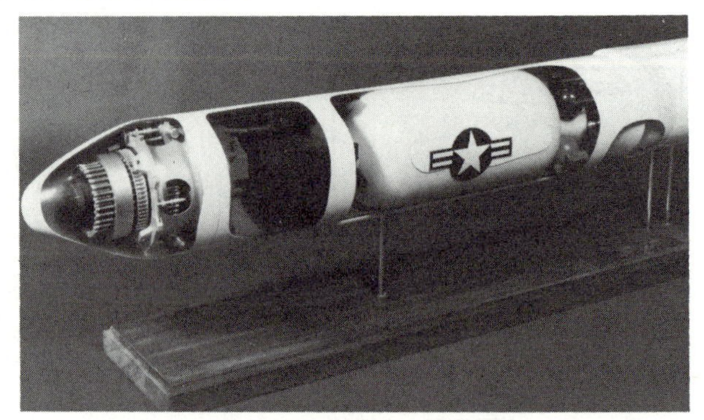

Cutaway view of the ASAT missile, showing MV in nose cone atop ALTAIR upper stage on SRAM (Short Range Attach Missile) booster.

ASAT missile carried by F-15 fighter during refueling operation.

SOURCE: U.S. Air Force.

Figure 4-10.—ASAT Mission Profile

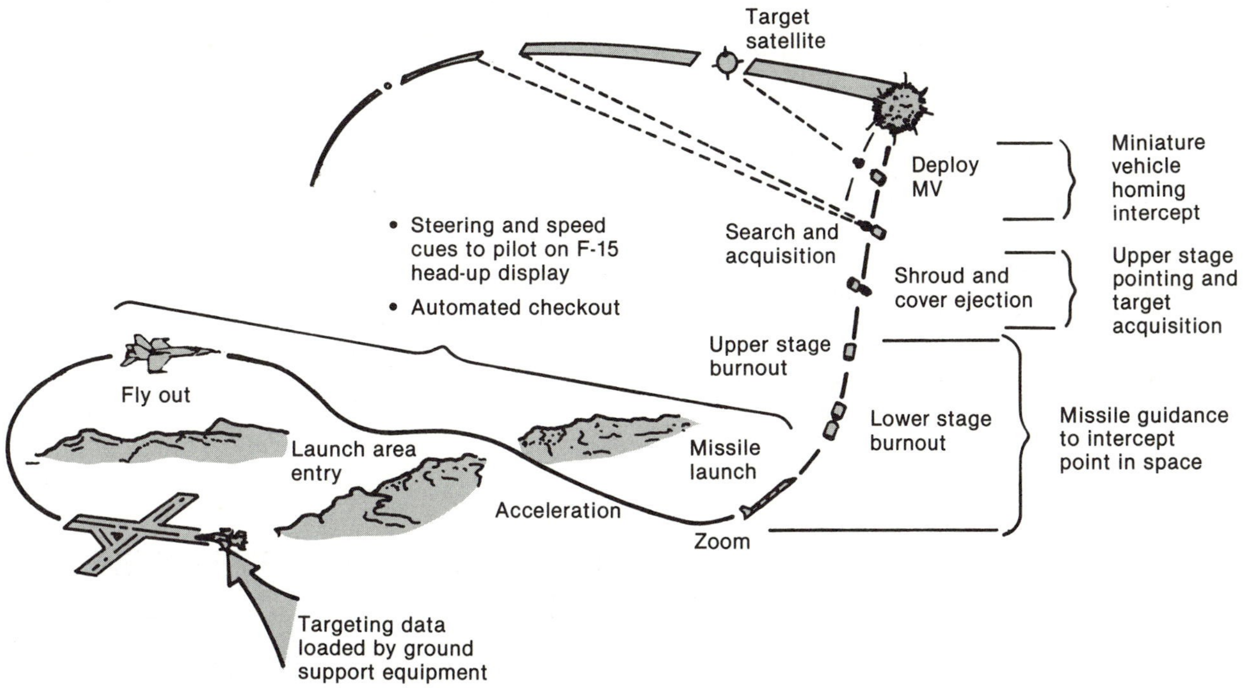

SOURCE: The Aerospace Corp.

tant to agree to arms control measures which would ban them. Because the small weapons could be easily concealed, if they are proven effective and then banned by agreement, Soviet authorities might suspect that the United States retained a capability to use hidden ASAT weapons.

The estimated costs of the U.S. ASAT program are listed in table 4-1.

Table 4-1.—Estimated Costs of the U.S. Air Force Space Defense and Operations (ASAT) Program

Activity	Cost	Reference
Research, development, testing and evaluation (Program element 64406F)	$1.40 B	1)
Procurement (missiles and aircraft)	$2.64 B	1)
Military construction (Program element 12450F)	$0.04 B	1)
Operation and maintenance, FY 1986	$0.009 B	2)
Operation and maintenance, FY 1987	$0.0129 B	2)
Operation and maintenance, FY 1988	$0.0283 B	2)
Operation and maintenance, FY 1989	$0.0453 B	2)
Operation and maintenance, FY 1990	$0.0461 B	2)
Total	$4.2416 B	

SOURCES: 1) U.S. Congress, General Accounting Office, "Report to the Honorable George E. Brown, Jr., House of Representatives: Status of the U.S. Antisatellite Program," report GAO/NSIAD-85-104, June 14, 1985.
2) U.S. Air Force.

POSSIBLE ADVANCED-TECHNOLOGY ASAT WEAPON SYSTEMS

It will be possible in a few years to build space surveillance systems, ASAT weapon systems, and ASAT countermeasure systems much more capable than those used by the United States or expected to become operational soon. These advanced systems could use technologies expected to be available eventually in both the United States and the Soviet Union. The variety of possible systems is so great that this report will discuss only those which seem most promising or threatening with respect to criteria of responsiveness, survivability, altitude reach, economy, early availability, controllable lethality (for destructive applications), and usefulness at nonnuclear levels of conflict. The most promising ASAT technologies will be discussed in this section as possible U.S. options. Space surveillance systems, although essential components of space defense systems, will not be discussed in this section, because the most promising space surveillance technologies would be used to best advantage in space-based space surveillance systems; these were discussed in chapter 3 as possible advanced-technology MILSAT capabilities.

Table 4-2 lists the major categories of ASAT weapons, organized, according to physical means of causing damage, into three categories: isotropic (nondirectional) nuclear weapons, kinetic-energy weapons (projectiles), and directed-energy weapons (particle beam weapons, radio-frequency weapons, and laser weapons). Because the boundary between destructive directed-energy devices (weapons) and nondestructive directed-energy devices (e.g., radio jammers, lasers used to overload optical sensors, or particle-beam generators used to upset the functioning of electronic systems) is blurred, being one of power or mode of use rather than kind, nondestructive directed-energy devices will not be distinguished from directed-energy weapons except where necessary.

Isotropic Nuclear Weapons

Ordinary nuclear weapons, when detonated in space, radiate energy and disperse debris more or less uniformly in all directions. Hence they will be called isotropic nuclear weapons (INW) to distinguish them from nuclear explo-

Table 4-2.—Types of ASAT Weapons

Isotropic nuclear weapons (INW):
Ground-based
—Coorbital interceptor
—Direct-ascent ("pop-up") interceptor
Space-based
—Coorbital interceptor (nuclear "space mines")[a b]
Kinetic-energy weapons (KEW):
Ground-based
—Coorbital interceptor
—Direct-ascent ("pop-up") interceptor
Space-based
—Coorbital interceptor ("space mines")[a]
—Noncoorbital interceptor[c]
Directed-energy weapons (DEW):
Ground-based
—High-power radio-frequency (HPRF) and active electronic countermeasures (ECM)
—High-energy laser (HEL) and active electro-optical countermeasures (E-OCM)
Space-based
—High-power radio-frequency (HPRF) and active electronic countermeasures (ECM)
—High-energy laser (HEL)[d] and active electro-optical countermeasures (E-OCM)
—Neutral-particle beam (NPB)

[a]A "space mine" is an expendable ASAT weapon predeployed in space so as to be capable of destroying enemy satellites almost instantly. They could be armed with INW, KEW, or DEW. If armed with a short-range weapon, a space mine must be coorbital.
[b]Prohibited by the 1967 Outer Space Treaty.
[c]Intercepts targets at high velocity from parking orbit.
[d]Including nuclear explosive powered X-ray lasers (XRLs), which are nuclear directed-energy weapons (NDEW).

sive-powered directed-energy weapons (NDEW) which will be discussed subsequently. INW could be carried to within lethal range of a satellite or satellites by a rocket and detonated. Early U.S. ASAT weapons were of this type, and the United States or the U.S.S.R. could use nuclear-armed ICBMs, SLBMs, and ABM interceptors in this manner. Alternatively, INW could be used as nuclear space mines: they could be concealed aboard satellites which are continuously or occasionally within lethal range of enemy satellites and detonated on command.

As ASAT weapons, nuclear weapons have several legal, political, and strategic disadvantages: they can only be used at the nuclear level of conflict—or in any case their use would escalate a conflict to the nuclear level—and when used they may upset or damage unhardened friendly and neutral satellites at ranges which depend on weapon yield but which can be very large. In addition, they cannot legally be based in orbit, this being prohibited by the Outer Space Treaty of 1967. Moreover, existing U.S. procedures for safeguarding nuclear weapons and for preventing their unauthorized use are expensive and time-consuming, and the Soviet Union may have similar safeguards now and incentives to retain them in the future. On the other hand, the advantages of isotropic nuclear weapons are their present availability, their economy (relative to other weapons of comparable range), their concealability (from present surveillance systems), their great lethal range (as compared to kinetic-energy weapons) against unhardened satellites, the difficulty of hardening satellites against nuclear detonations at close range, and their adaptability for delivery by a variety of launch vehicles and orbital platforms, including those with poor guidance accuracy and no pointing capability.

By comparison, nuclear directed-energy weapons are not now available and, if developed, would require platforms with moderately accurate pointing capability. However, it is possible in principle to build nuclear directed-energy weapons, such as X-ray lasers, which could have far greater lethal range than the nuclear explosive devices which power them and which could be feasibly and economically delivered by platforms with adequate pointing capability. The theoretical potential of such weapons,[34] if realized in practice, would make them superior to isotropic nuclear weapons for ASAT applications. The *potential* ASAT capabilities of NDEW, therefore, deserve greater concern than do those of isotropic nuclear weapons which are, however, of more immediate concern because of their existence and *demonstrated* capability.

Existing nuclear-armed Soviet ABM missiles could be used against low-altitude satellies. Prior testing of such weapons against satellites would not be required to demonstrate the reliability of subsystem operation.

[34]See, e.g., F.V. Bunkin, V.I. Derzhiev and S.I. Yakovlenko, "Specification for pumping X-ray laser with ionizing radiation," *Soviet Journal of Quantum Electronics*, vol. 11, No. 7, July 1981, pp. 971-972, and note that many such lasers could be powered by one "exotic" source of pulsed x-radiation.

Isotropic nuclear weapons concealed aboard satellites used as "space mines" could attack without warning and would pose a greater threat, because reactive countermeasures could not be used for protection. Nuclear space mines could be lethal against satellites as hard as any now operational at such range that the testing of their trailing capability—e.g., in geostationary orbit—although observable, might not be interpreted as such. Protection against such weapons would require evading them during placement (which would be relatively economical only for a small, cheap satellite), opposing their placement by defending an agreed or unilaterally declared "keep-out zone" around satellites from penetration by any spacecraft which *might* contain a nuclear weapon, or possibly, after penetration and trailing by such a spacecraft, evading (e.g., under cover of smoke and chaff) it until out of its probable lethal range, at which time it could be attacked by ASAT weapons of greater range, if available.

Development and, when necessary, operation of close-look inspection satellites equipped with gamma-ray spectrometers or other instruments capable of detecting materials used in nuclear explosives would afford additional protection: if inspection by such satellites were anticipated, the designer of a nuclear space mine would have to shield its nuclear explosive device with tons of material[35] in order to prevent collection of *prima facie* evidence of having placed a nuclear weapon in orbit in violation of the Outer Space Treaty of 1967. Placing so massive a space mine into orbit would be more costly, and it could be evaded at less relative cost than a smaller, unshielded space mine could be.

[35]U.S. and Soviet scientific spacecraft, landers as well as orbiters, have carried gamma-ray spectrometers capable of detecting low concentrations of fissionable materials as deep as half a meter below lunar and planetary surfaces [U.S. National Aeronautics and Space Administration, *A Forecast of Space Technology 1980-2000*, NASA Special Report SP-387, 1976; and cf. the recent Soviet "VEGA" Venus/Halley's Comet probe.]. The United States could develop such sensors for use on satellites for purposes of monitoring compliance with the Outer Space Treaty.

The risk which such weapons may pose to U.S. spacecraft now or in the future is mitigated to some extent by the fact that they would be useful only in a nuclear war, in which the ground segments of space systems would also be significantly vulnerable and might be attacked by preference, and more rapidly. This and the other aforementioned disadvantages of ASAT INW pose disincentives against developing them, attempting to base them in orbit covertly, and using them at nonnuclear levels of conflict

Several conclusions follow from these considerations:

1. It is feasible to build direct-ascent and coorbital isotropic nuclear weapons.
2. Weapons of this type could be as small and inexpensive as many existing satellites.
3. Such weapons could be developed and tested covertly.
4. Soviet nuclear weapons could threaten U.S. satellites at low and high altitudes now and in the future but only at nuclear levels of conflict.
5. Protection of U.S. satellites from coorbital INW may require defense of a keep-out zone around low-altitude satellites or designing future low-altitude satellites to be at least as small and inexpensive as the nuclear space mines which threaten them. Many more options are available for defending high-altitude satellites from direct-ascent nuclear interceptors.

Kinetic-Energy Weapons

Anti-satellite kinetic-energy weapons pursue satellites and destroy them by direct impact or at close range using a gun or a fragmentation warhead. One example of an ASAT KEW is a coorbital interceptor which approaches its target at a low closing velocity before destroying it. Another example is a direct-ascent ("pop-up") interceptor which is launched from the Earth's surface or from an airplane and which approaches its target at a high closing velocity. Such an interceptor could also be based in space in a parking orbit, from which it could enter a transfer orbit in which it would close with its target at high velocity, just as a pop-up interceptor would. "Pop-up" interception requires less energy per unit interceptor payload than does coorbital interception, particularly at low altitudes. A coorbital interceptor, on the other hand, could be used as a "space mine," continuously observing and trailing its target, prepared to destroy it almost instantly on command or (if salvage-fused) when attacked or disturbed in specified ways. Used in this way, a coorbital ASAT KEW could take advantage of the element of surprise in an attack, leaving an enemy no time to react with active defenses or reactive passive countermeasures such as evasion or deployment of decoys or smoke.

A coorbital interceptor could pursue a target (its "quarry") indefinitely if it had as much velocity change capability ("delta-V") as its quarry. If it also had as much acceleration capability as its quarry, it could, with tracking capability, pursue its quarry continuously, otherwise its quarry would be able to maneuver out of lethal range temporarily. Under some conditions, an interceptor may be able to trail its quarry using less delta-V than its target uses for evasion; however, calculation of required velocity change is complicated,[36] particularly in the case of many-against-one interception. In any case, if comparable propulsive technologies are available to both pursuer and evader, pursuit can be successful if the interceptor's mission payload—its armament (if any) and guidance and fuzing systems—is lighter than that of its quarry.

It should be possible to build space mines with very lightweight armament. For example, a directional fragmentation warhead similar to that of a Claymore mine could project 100,000 one-gram pellets in a pattern which would cover a 100 x 100 meter area with 10 pellets per square meter at a range of 1 kilometer. This would be adequate to destroy unarmored satellites as small as about a meter in diameter with high probability. The pellets for such a warhead would weigh 100 kilograms, and its explosive charge could weigh less than that. Even lighter warheads of this type may be possible, possibly having dispersion angles as small as those of shotguns or even high-power rifles (e.g., 10 centimeters dispersion at a range of 1 kilometer).

Guns and rockets could also be used as armament by a KEW space mine. For example, a single unguided rocket which deploys a 10 x 10 meter net weighing about 1 kilogram and having a 1-gram weight at each of its 100 nodes (knots) could achieve the same kill probability as the Claymore-type warhead if its aiming error were less than 5 milliradians (5 meters at 1 kilometer) after rocket burnout and net deployment. A space mine using such a rocket as armament might weigh as little as 50 kilograms but could destroy much heavier unhardened satellites.

It would not be relatively economical to use such mines to attack smaller, cheaper satellites, which could be useful for some military applications, such as communications.[37] However, it would be economical to use space mines to attack larger satellites. If such satellites are armored as a countermeasure, the space mine could also be made more lethal, and this would probably require less mass than would be required for armor against it at the assumed lethal range[38]. There appears to be no relatively economical means of protecting large satellites against a surprise attack by such mines, once they are emplaced. Safety from such attacks would require opposing their placement by defending an agreed or unilaterally declared "keep-out zone" around satellites, or else, once they are emplaced, evading them (e.g., under cover of smoke and chaff) until out of their range, at which time they could be attacked by ASAT weapons of greater range.

[36]G.M. Anderson, "Differential Dynamic Programming Feedback Control for a Pursuing Spacecraft with Limited Fuel,"(New York, NY: American Institute of Aeronautics and Astronautics, AIAA Paper A78-31925, 1978).

[37]For example, the U.S.S.R. operates a constellation of lightweight satellites in low orbit, and as early as 1961 the United States operated SYNCOM communications satellites weighing only 31 kilograms in geosynchronous orbit.

[38]See the discussion of hardening in the discussion of countermeasures, below.

Space mines the size of those assumed (2 meters in diameter) could be detected by ground-based radars if at low altitude or, if at high altitude, by electro-optical sensors (e.g., GEODSS) if they do not employ measures (e.g., black paint) to evade detection or by space-based long-wavelength infrared space surveillance sensors even if they do employ such measures. In order to demonstrate reliability, they would have to be tested in space by trailing satellites or space debris. This activity could be observed and would be noticeable.

The U.S. Department of Defense has estimated that the U.S.S.R. views development of directed-energy weapons as a more promising approach for improving future antisatellite capabilities than further development of ground-based kinetic-energy ASAT weapons.[39] However, Soviet development and testing of interceptors of the type which is currently operational have demonstrated an interest, and some capability, in coorbital, nonnuclear kinetic-kill approaches to ASAT capability. Given sufficient incentive, the U.S.S.R. might choose to continue these efforts and, if so, could eventually produce smaller, cheaper weapons of longer endurance. Once testing of such weapons in space is observed, there might be insufficient time for the United States to react by developing new generations of small, inexpensive satellites against which use of small space mines would be uneconomical.[40]

It can be concluded from these considerations that it is feasible to build coorbital kinetic-energy ASAT weapons. Weapons of this type could be smaller and cheaper than most satellites as presently designed. If such weapons are deployed by the U.S.S.R., protecting future U.S. satellites from them may require defense of a keep-out zone around U.S. satellites or designing future satellites to be at least as small and inexpensive as the space mines which threaten them.

Directed-Energy Weapons

Several types of directed-energy weapons could be used for ASAT purposes, including ground- and space-based systems powered by nuclear explosives, nuclear reactors, or non-nuclear energy sources. They include high-power radio-frequency (HPRF) generators, high-energy laser (HEL) weapons, and neutral particle beam (NPB) weapons. They could also include non-laser sources of short-wavelength radiation.

High-Power Radio-Frequency Weapons and Electronic Countermeasures

HPRF weapons—which include high-power microwave (HPM) weapons—are devices capable of producing intense, damaging beams of radio-frequency radiation.[41] HPRF generators could be used to overload and damage satellite electronic equipment at high power levels or, at lower power levels, merely to temporarily overload satellite electronic systems (i.e., for "jamming").[42] Radio-frequency (RF) generators could therefore be useful at all levels of conflict.

HPRF weapons could be ground-based or based in space. Ground-based HPRF weapons, unlike ground-based laser weapons, could operate through cloud cover. However, the maximum pulse energy per unit area which can be beamed through the atmosphere is limited.[43] Space-based HPRF weapons would not be so limited, but, like ground-based HPRF weap-

[39] U.S. Department of Defense, *Soviet Military Power*, 4th ed., April 1985, p. 44.

[40] For example, Soviet testing of interceptors of the type now operational was first observed in 1968, and the U.S. Department of Defense has judged, in retrospect, that an operational capability with these interceptors was achieved three years later, in 1971. [Ibid., p. 55.]

[41] I.e., electromagnetic radiation at wavelengths of 1 millimeter or longer.

[42] This continuum of HPRF effects makes distinction between "weapons" and "jammers" difficult. For arms control purposes, however, a distinction could be made based on power-aperture product, as is done in the ABM Treaty. Wavelength-dependence should be considered as well, because at a given power-aperture product, shorter wavelengths can be radiated with greater brightness and deliver more power per unit area at long range.

[43] The maximum pulse energy per unit area which can be beamed through the atmosphere is limited to about 1 joule per square meter by the phenomenon of dielectric breakdown of the atmosphere, which occurs at higher energy fluence levels.

ons, would have to be very large in order to concentrate their HPRF energy into a narrow beam. Even a relatively wide beam might be able to damage satellites of existing designs at considerable range, but this is uncertain, and hardening satellites against HPRF radiation is possible.

The lethality of HPRF weapons would be less certain than the lethality of other DEW because of uncertainties about target vulnerability, i.e., about the beam energy per unit area required to damage a particular enemy satellite. This will depend on details of the design of the satellite and would be difficult to predict with accuracy even if those details were known. Moreover, even though it is possible in principle to harden satellites to withstand intense HPRF radiation, it is difficult to verify by modeling and simulation, and exorbitantly expensive to verify by testing, the actual degree of hardness achieved. Hence, although many concepts for HPRF weapons have been studied, none of them are as resistant to countermeasures as are some HEL and neutral particle beam concepts.

Radio-frequency generators of lower power will continue to be valuable for providing capabilities to jam and spoof (i.e., deceive) satellite radio systems.[44] The vulnerability of satellites to jamming and, especially, spoofing, is also uncertain and probably varies greatly among satellites. However, low-power RF generators useful for jamming and spoofing would be much cheaper than HPRF generators; indeed, some existing electronic countermeasure systems could be used against satellite links. Because of the prevalence and ambiguity of these capabilities, it would be difficult or impossible to eliminate the non-destructive ASAT capabilities of ECM systems by means of arms control agreements. Use of passive electronic counter-countermeasures, when necessary, would be preferred.

High-Energy Laser Weapons and Electro-Optical Countermeasures (Ground-Based)

High-energy laser weapons are devices capable of producing intense, damaging beams of optical radiation[45] by means of the phenomenon of stimulated emission of radiation. High-energy lasers could be used to permanently damage satellites, or, at lower power levels, to jam optical communication systems and to "dazzle" optical sensor systems (i.e., to overload them, temporarily blinding them). This continuum of effects makes the distinction between "laser weapons" and "active electro-optical countermeasures" one of degree rather than kind and therefore difficult.[46] Lasers of several types, employing different materials and physical processes and operating at different wavelengths, might be suitable for use as weapons; some of these types are described in the box entitled "Types of Lasers."

HEL weapons could be space-, ground-, air-, or sea-based. Of the many possible types of lasers useful as ground-based ASAT weapons, continuous-wave deuterium-fluoride chemical lasers have been of particular interest because of their simplicity, the maturity of their technology, and the possibility of focusing their infrared beams using mirrors of relatively rough surface quality (compared to that which would be required at shorter visible and ultraviolet wavelengths).

Repetitively-pulsed free-electron lasers and electric-discharge excimer lasers have also been of interest because they would operate at short wavelengths at which small beam divergence angles could be achieved using smaller mirrors than would be required for infrared beams, although they would have to be

[44] Even at very low power levels radio-frequency generators might be able to confuse ("spoof") satellites by beaming deceptive signals at them. Some possible spoofing techniques have been described by Col. Robert B. Giffin, USAF, in *US Space System Survivability: Strategic Alternatives for the 1990s*, National Security Affairs Monograph Series 82-4, National Defense University Press, Fort Leslie J. McNair, Washington, DC, 1982; p. 26. Electronic counter-countermeasures can be used to reduce the susceptibility of MILSATs to spoofing.

[45] I.e., electromagnetic radiation at wavelengths shorter than 1 millimeter.

[46] For arms control purposes, however, a distinction could be made based on power-aperture product, as is done in the ABM Treaty. Wavelength-dependence should be considered as well, because at a given power-aperture product, shorter wavelengths can be radiated with greater brightness and deliver more power per unit area at long range.

Types of Lasers

A **laser** is described by the **lasant** material it uses as a source of coherent optical radiation (e.g., deuterium fluoride), the characteristic **wavelength** or wavelengths at which the lasant emits such radiation, the method of **pumping** (i.e., energizing) the lasant (e.g., chemical lasers, electric-discharge lasers, optically pumped lasers, gasdynamic lasers, etc.), the **source of energy** used for pumping (e.g., chemical electrical, nuclear reactor, nuclear explosive, etc.), and its **pulse waveform** (e.g., single-pulse, repetitively pulsed, or continuous-wave).

Many materials can be used as lasants; these can be in solid, liquid, or gaseous form (consisting of molecules or atoms) or in the form of a plasma (consisting of ions and electrons). Lasant materials useful in high-energy lasers include carbon dioxide, carbon monoxide, deuterium fluoride, hydrogen fluoride, iodine, xenon chloride, krypton fluoride and selenium, to mention but a few. Some of these—e.g., xenon chloride and krypton fluoride—are molecules which cannot exist under ordinary conditions of approximate thermal equilibrium but must be created by special "pumping" processes in a laser. Such molecules are called **excimers**, a contraction for "excited dimer" (a dimer is a molecule consisting of two atoms).

Free-electron lasers do not use lasants but instead generate radiation by the interaction of an electron beam with a static magnetic or electric field. Loosely speaking, free-electron laser technology resembles and evolved from that used by particle accelerators ("atom smashers"), while other (i.e., **bound-electron**) lasers use lasants heated by electrical current or chemical combustion and resemble fluorescent lamps or rocket engines in some respects. Some lasers use mirrors, lenses, diffraction gratings, or other optical elements to recirculate the laser beam through the lasant in order to achieve adequate amplification; such lasers are called **cavity lasers**, and the arrangement of optical elements used to recirculate the beam is called the laser cavity or resonator. In other lasers, the beam passes through the lasant only once; such cavity-less lasers are called **superradiant lasers**, and the process of single-pass beam generation they use is called **superradiance, superfluorescence**, or **amplified spontaneous emission**.

of higher optical quality. Free-electron lasers can probably be made more efficient than other short-wavelength lasers and could be ground- or space-based. Electric-discharge excimer lasers might be available sooner, but probably cannot be made as efficient as free-electron lasers can be, nor can they be tuned, as free-electron lasers can be, to wavelengths which would minimize beam degradation by atmospheric effects (in the case of ground-based lasers) and optimize target damage.[47]

Beams from ground-based HEL weapons would be subject to a variety of phenomena which would disturb their propagation through the atmosphere. These phenomena include absorption, scattering, thermal blooming, dielectric breakdown, and the refractive effects of atmospheric turbulence. Most serious is beam absorption and scattering by clouds: ground-based HEL weapons, unlike ground-based HPRF weapons, could not operate through cloud cover.[48]

[47]Soviet laser development has emphasized other technologies than these; the "best" laser technologies for near-term Soviet weapons may differ from those which would be "best" for near-term U.S. weapons. Cf. N.N. Sobolev and V.V. Sokivikov, "The Carbon Monoxide Laser: Review of Experimental Results," *Soviet Journal of Quantum Electronics*, vol. 2, Jan.-Feb. 1973, p. 305 ff.; M.M. Mann, "CO Electric Discharge Lasers," *AIAA Journal*, vol. 14, No. 5, May 1976, pp. 549-567; A.A. Stepanov and V.A. Schlegov, "Continuous-Wave Reaction-Product Chemical Lasers (Review)," *Soviet Journal of Quantum Electronics*, vol. 12, No. 6, June 1982, pp. 681-707; and S. Kassel, *Soviet Free-Electron Laser Research* (Santa Monica, CA: The Rand Corporation, report R-3259-ARPA, May 1985);

[48]Dielectric breakdown is not as serious at optical wavelengths as it is at radio wavelengths. Scattering causes beam degradation, especially at short wavelengths, but is not insurmountably problematic in clear weather. Thermal blooming of a beam focused on a distant satellite can be controlled by phase compensation techniques, or by using a laser which operates at a wavelength at which atmospheric absorption is not severe. A deuterium fluoride laser, for example, operates at such a wavelength and future free-electron lasers could also operate at such wavelengths.

The effects of atmospheric turbulence also pose a serious problem for ground-based lasers: without compensation, the beam of a ground-based laser would diverge at so great an angle that it would be unable to damage anything other than the most sensitive earth-pointed optical sensors on satellites at geostationary altitude. Such compensation appears possible, in principle.

Another serious problem for ground-based lasers is the infrequency with which a low-altitude satellite would pass within view of a ground-based laser site. The interval between such passes might be days or weeks, and, until it exhausted its maneuver fuel, a maneuvering satellite could completely avoid coming within range. Deployment of a large number of ground-based lasers would provide more frequent opportunities to engage satellites and would increase the difficulty and expense of evasive maneuver and of attempts to attack the laser sites. It would also increase the probability that a number of the lasers would not be overcast by impenetrable cloud cover. Of course, such proliferation would be expensive.

An alternative approach would be to base ASAT lasers on ships, which would have some flexibility to remain in clear weather near the ground tracks of high-priority targets and distant from most anti-shipping threats. Another solution to the coverage problem would be to deploy steerable reflectors on satellites to relay the beams from ground-based lasers to targets or to other relay satellites. The operational capabilities provided by such a system would be similar to those provided by space-based laser weapons with an unlimited power supply, except that beam availability would be contingent on the absence of overcast at at least one ground-based laser site and on the survival of both a ground-based laser and an orbital reflector.

A high-altitude satellite, on the other hand, would be within view of a ground-based laser site for a prolonged or indefinite period. For example, a ground-based laser could irradiate a satellite in geostationary orbit continuously, weather permitting.

High-Energy Laser Weapons and Electro-Optical Countermeasures (Space-Based)

The beams of space-based laser weapons would not have to pass through the atmosphere and could damage unhardened satellites at great range. A much smaller force of such lasers than would be required for effective ballistic missile defense could pose a threat to a nation's most critical satellites. However, space-based laser weapons, like other satellites, would be subject to attack by ASAT weapons.

Several types of lasers could be used as space-based laser weapons, each having some particular advantage. For example, of those lasers which could damage satellites by overheating them, hydrogen-fluoride chemical lasers are particularly attractive because of their simplicity, while carbon-monoxide electric-discharge lasers are attractive because of their potentially high electrical efficiency. Free-electron lasers are attractive because they can operate at short wavelengths at which small beam divergence angles can be achieved using small mirrors. Excimer lasers would be bulky and less suitable for space basing.

Space-based lasers of very low power, if of an appropriate wavelength, could dazzle or permanently blind optical sensors used by other weapons for homing guidance or for beam pointing. More powerful lasers could be used to attack and damage a satellite by overheating it and possibly melting its "skin," or tearing its "skin" as a result of the hammer-like mechanical impulse which pulsed laser radiation can generate on a target surface.[49]

[49]The amount of mechanical impulse which a given amount of beam energy can generate can be estimated on the basis of published impulse-coupling models, such as that of P.E. Nielson, "High-Energy Laser-Matter Coupling in a Vacuum," *Journal of Applied Physics*, vol. 50, No. 6, June 1979, pp. 3938-3943, modified to account for the depth to which the beam radiation will penetrate before surface vaporization begins. In *Directed Energy Missile Defense in Space* [OTA Background Paper OTA-BP-ISC-26, April 1984], Dr. Ashton Carter estimated that a laser beam fluence of 20 kilojoules per square centimeter would produce an impulse intensity of 10 kilotaps [10,000 dyne-seconds per square centimeter], i.e., an impulse coupling of 0.5 dyne-seconds per joule. This estimate assumes that the laser pulse is so brief that little heat is conducted to a depth greater than

Space-based reflectors could relay beams from ground-based lasers to a target, or another reflector, beyond the horizon. These could be used in much the same manner as space-based lasers and would be, in effect, space-based lasers with an unlimited fuel supply.

All of these weapons have certain disadvantages, however. They would use large, expensive, power-handling mirrors; they would engage targets sequentially, thus giving other enemy satellites time to use reactive passive countermeasures (e.g., smoke); and they would be subject to attack by expendable, single-shot weapons (INW, KEW, or DEW) against which reactive countermeasures and shoot-back would be ineffective. Hence space-based lasers which can damage one or several targets instantly, without warning, using a single pulse and which are cheap enough to be considered expendable, if feasible, would be most attractive. Because they would exhaust their fuel or destroy themselves as soon as they were used, they would be invulnerable to shoot-back although subject, and possibly vulnerable, to preemptive attack.

One type of laser which might be useful as an expendable, single-shot, space-based weapon is the X-ray laser. X-ray lasers are presently only in the earliest stages of development.[50] X-ray lasers (also called "x-rasers" or XRLs) could be very simple in design; they might be thin fibers of lasing material powered ("pumped") by intense, pulsed radiation from another laser, a nuclear explosion, or some other source. The beam from such a device would diverge at an angle roughly equal to the square root of its wavelength divided by the square root of the length of the fiber.[51]

The U.S. Department of Energy is investigating the feasibility of developing nuclear-pumped X-ray laser weapons. Nuclear explosive pumping is of interest because even if only a small fraction of the energy of a nuclear explosion could be converted into X-ray laser beams, it could still be lethal at great range. Most details of this research, except for the fact of its existence, are classified. However, the U.S. Department of Energy has stated that:

1. The U.S. Department of Energy is interested in and is conducting research on certain types of nuclear explosive powered directed-energy weapons (NDEW)—viz., X-ray lasers, visible-light weapons, microwave weapons, and charged-particle beam weapons—as well as on nuclear explosive powered kinetic-energy weapons (NKEW).
2. Underground nuclear tests at the Nevada Test Site have been and continue to be a part of this research.
3. NDEW could engage multiple targets using multiple beams, providing high leverage.
4. NDEW could damage targets at ranges of thousands of kilometers.
5. Nuclear explosive powered X-ray laser weapons would damage targets by means of ablative shock.

---

the depth to which the beam radiation will penetrate before surface vaporization begins. Longer pulses can produce a much greater impulse coupling. For example, an impulse coupling of 20 dyne-seconds per joule has been measured in experiments using infrared lasers [P. Bournot, et al., "Mesure de la Pression Induite sur une Cible Metallique par une Laser CO_2 Impulsionnel," *Journal de Physique*, Tome 41, Suppl. au No. 11, Colloque C9, Novembre 1980, pp. 81-86].

[50] Successful operation of an X-ray laser was claimed by a Lawrence Livermore National Laboratories group headed by Dennis Matthews at the meeting of the Division of Plasma Physics of the American Physical Society in Boston on October 29, 1984. Their laser operated at two wavelengths near 20 nanometers. The design of the laser is described by D.L. Matthews, et al., in " Demonstration of a Soft X-Ray Amplifier," *Physical Review Letters*, 54:110-113, 1985. More recently, the U.S. Department of Energy has stated that the nation's nuclear-weapons laboratories conducted an underground test of a nuclear explosive powered X-ray laser at the Nevada Test Site laser and achieved lasing [see note 52, infra].

[51] Provided the square of this angle is greater than twice the cross-sectional area of the fiber divided by the square of its length. For example, a thin, one meter long XRL operating at a wavelength of 20 nanometers would produce a beam with a divergence angle of 100 microradians. This divergence angle is large compared to those achievable by lasers operating at longer wavelengths at which mirrors can be used, and only a small fraction of the energy in a beam diverging at such an angle would would be intercepted by a satellite-sized target at a range of thousands of kilometers. X-rays cannot be focussed using conventional lenses and mirrors, but they can be focussed using diffraction gratings and other special optical elements [R.W. Waynant and R.C. Elton, *Proceedings of the IEEE*, vol. 64, No. 7, July 1976, pp. 1059-1092]. Such techniques may be useful for generating X-ray laser beams of very high brightness.

6. Lasing by a nuclear explosive powered X-ray laser has been demonstrated in an underground nuclear test at the Nevada Test Site.[52]

Weapons powered by nuclear explosions would have several disadvantages compared to nonnuclear weapons: nuclear explosives are banned from orbit by the Outer Space Treaty[53], they would require elaborate, costly, and time-consuming command and control arrangements under present U.S. policy, they would be useful only at the nuclear level of conflict or would signal escalation to that level, and they might disrupt radio propagation and damage allied and neutral satellites when used, depending on burst times, locations, and details of weapon design. For these reasons there is also interest in developing expendable, single-pulse non-nuclear laser weapons, if these should prove feasible and economical. Such lasers might operate at short X-ray and gamma-ray wavelengths[54] or at longer wavelengths (e.g., iodine lasers). Non-laser sources of single-pulse directional radiation may also be useful as weapons.

[52]On January 14, 1983, in the first official public discussion of U.S. research on nuclear explosive pumped X-ray lasers, the Presidential Science Advisor, Dr. George Keyworth, suggested that such lasers might eventually be of great military significance, and he called the "bomb-pumped X-ray laser" program "one of the most important programs that may seriously influence the nation's defense posture in the next decades." [Quoted by William J. Broad, in "Reagan's 'Star Wars' Bid: Many Ideas Converging," *New York Times*, Mar. 4, 1985, p. A1 ff.]. Subsequently, Major General William W. Hoover, USAF (Ret.), Assistant Secretary of Energy for Defense Programs, elaborated on the U.S. nuclear directed-energy research, saying that the nation's nuclear-weapons laboratories conducted an underground test at the Nevada Test Site of an X-ray laser and achieved lasing. More recently, the Department of Energy has released the other statements quoted here.

[53]Some have, however, questioned this interpretation, arguing that the intent of the Outer Space Treaty was to ban weapons of mass destruction from orbit, not ASAT or BMD weapons which would cause minor collateral damage, if any.

[54]One concept for a short-wavelength X-ray laser envisions using a very brief and intense laser pulse to stimulate coherent, collective radioactive decay of nuclei which have been energized by exposure to neutrons in a reactor [C.B. Collins, et al., "The Coherent and Incoherent Pumping of a Gamma Ray Laser with Intense Optical Radiation," *Journal of Applied Physics*, vol. 53, No. 7, July 1982, pp. 4645-4651]. Recent experimental results [B.D. DePaola and C.B. Collins, "Tunability of Radiation Generated at Wavelengths Below 1 Å by Anti-Stokes Scattering from Nuclear Levels," *Journal of the Optical Society of America B*, vol. 1 December 1984, pp. 812-817,; C.B. Collins and B.D. Paoli, "Observation of Coherent Multiphoton Processes in Nuclear States," *Optics Letters*, vol. 10, January 1985, pp. 25-27] have verified that some of the problems of developing such a laser can be solved. However, it is not yet known whether nuclei suitable for use in such a laser exist.

Neutral Particle Beam Weapons

Powerful particle accelerators similar to those used for scientific research, isotope production, and fusion power applications could be used as particle-beam weapons to attack satellites. Because electrically charged particles would travel along spiraling paths within the Earth's magnetic field, electrically neutral particles such as atoms of hydrogen, deuterium, tritium, or heavier elements would be used by such weapons. Because such atoms would become ionized and hence charged if they passed through matter as dense as the upper atmosphere, such weapons must be based in space and are useful only against targets in space, although relatively small weapons of this type could be kept on the ground ready for launch into orbit and, after some on-orbit testing and calibration, for use.

A neutral particle beam (NPB) weapon might consist of a negative ion source, a particle accelerator, beam focusing and pointing magnets, and a "stripping" device—e.g., a gas cell[55]—which strips the negative ions of their extra electrons, thereby neutralizing them, as well as a power source and other ancillary equipment,[56] as shown in figure 4-11. These components could resemble those presently in use for other purposes and need not be much larger to provide a modest ASAT capability. For example, the hydrogen atoms produced by an accelerator at the Los Alamos Meson Physics Facility (LAMPF) have energies of 800 million electron volts[57] (MeV) and could

[55]T.D. Hayward, et al., *Negative Ion Beam Processes*, Los Alamos National Laboratory, report UC-34c, January 1976, UNCLASSIFIED; J. H. Fink, "Photodetachment Technology," American Institute of Physics, Conference Proceedings No. 111, pp. 547-560, 1984; V. Vaņek, et al., "Technology for a Laser Resonator for the Photodetachment Neutralizer," American Institute of Physics, Conference Proceedings No. 111, pp. 568-584, 1984.

[56]K. Boyer, "Directed-Energy Beam Weapons," *Proceedings of the Society of Photo-Optical Instrumentation Engineers*, Volume 474, 1984, pp. 79-86.

[57]The electron volt (eV) is a unit of energy; about 6.25 quintillion electron volts equals one joule, the Systeme Internationale unit of energy.

Table 4-3.—A Comparison of Laser Weapons

	Space-based laser (single-pulse)	Space-based laser (repetitive pulse or continuous wave)	Ground-based laser	Ground-based laser and space-based reflectors
Subject to shoot-back by ASAT/DSAT	+ No	− Yes	+ No	− Yes (relay)
Subject to terrestrial attack	+ No	+ No	− Yes	− Yes (laser)
Subject to evasion and reactive deployment of decoys and shielding	+ No	− Yes	− Yes	− Yes
Cost per shot	− High	− High	+ Low	+ Low
Target availability	+ High	+ High	− Low	− Low
Requires survivable space surveillance for targeting	+ No	− Yes	− Yes	− Yes

Figure 4-11.—Artist's Concept of a Neutral Particle Beam Weapon

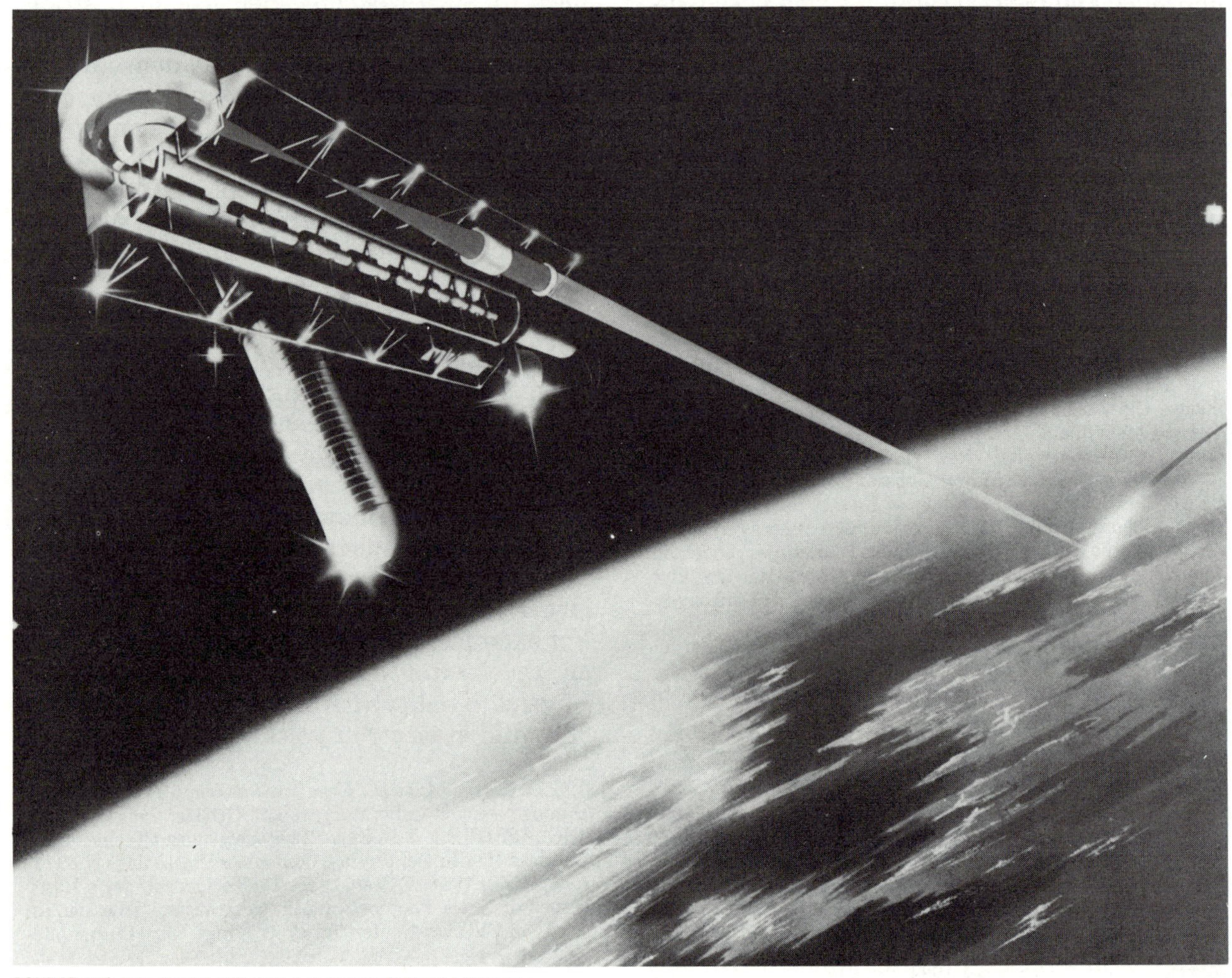

SOURCE: U.S. Department of Energy, Los Alamos National Laboratory.

penetrate aluminum shielding 1 meter thick. The hydrogen atom beam produced by this accelerator has a current of 1 milliampere; irradiation of an unhardened satellite at a range of 40,000 kilometers for several minutes by a beam of this current and particle energy could upset the functioning of its electronic circuits.

A turboalternator-powered NPB weapon might require about 25 tons of liquid hydrogen, liquid oxygen, tankage, and other "overhead,"[58] to deliver an absorbed dose of 10 kilograys[59] through shielding at a range of 40,000 kilometers, regardless of the thickness of the shielding, provided that maximum shield thickness is known or assumed in advance and that the weapon is designed to penetrate a shield of such thickness.[60] An absorbed radiation dose of about 10 kilograys would permanently damage most radiation-resistant high-density silicon integrated circuits. **Many existing and planned spacecraft use, or will use, high-density integrated circuitry with a radiation hardness three or four orders of magnitude lower.**[61] Ten times this mass—250 tons—would be required to damage circuits hardened to withstand 100 kilograys, at this range.[62] On the other hand, only 1.6 tons would be required to damage such circuits at a range of 1,000 kilometers, and perhaps as little as a kilogram would suffice to upset or damage integrated circuits of existing hardness levels at a range of 1,000 kilometers.

Although it may prove possible to harden high-density electronics to withstand 10 kilograys without suffering permanent damage, it is unrealistic to expect that all satellites will be hardened to that extent; transient upset of electronics, possibly causing memory loss in computers, could occur at doses several orders of magnitude lower. Hence a neutral particle beam weapon could attack not only satellites, but decoys for satellites presumed subject to upset, using very little fuel and overhead—e.g., 250 kilograms to deliver a dose of 10 grays at a range of 40,000 kilometers (the distance from low orbit to geosynchronous orbit).

According to some estimates, a NPB weapon with sufficient fuel to operate for 1,000 seconds must weigh about 4 tons per megawatt of beam power[63] The U.S. Space Shuttle, or its expected Soviet counterpart, could deploy into low orbit a NPB weapon weighing as much as 30 tons [see figure 4-12].[64] A heavier weapon could be launched into low orbit by the anticipated Soviet heavy-lift launch vehicle, which is expected to carry payloads as large as 150 tons. Even heavier weapons could be assembled in space by the United States or the Soviet Union.

A neutral-particle beam could be made to diverge at a small angle and would therefore

[58]Such as fuel for propulsion of the extra weapon fuel, oxidizer, coolant, and tankage.

[59]A gray (Gy) is the Systeme Internationale unit for absorbed energy dose. One gray is one joule per kilogram, or 100 rads. One kilogray is a tenth of a megarad.

[60]A hydrogen atom having a kinetic energy of 50 MeV could penetrate about a centimeter of aluminum shielding. If the thickness of the shield were increased, particle energy, and hence also weapon size, would have to be increased in order to penetrate the shielding, but the amount of beam energy and weapon fuel required need not increase. The reason for this is that as particle energy is increased, beam divergence can be decreased, and the same number of particles per unit area per second could be delivered (over a smaller area) with a lower total beam current (particles per second). Hence a high-energy, low-current weapon could penetrate thicker shielding and deliver the same radiation dose in the same time over a smaller cross-sectional area of a target than could a lower-energy, higher-current weapon of equal beam power (which equals particle energy times beam current).

[61]Doses of 100 grays would probably upset electronic circuits on most satellites. It would be possible to shield such circuits, but the shield mass required would increase more rapidly than would the mass of a weapon which could penetrate it. On the other hand, it is possible to fabricate integrated circuits capable of withstanding radiation doses as high as 100 kilograys.

[62]Low-density gallium arsenide circuits which can withstand 100 kilograys have been fabricated, as have higher-density silicon circuits. For example, the Sandia National Laboratories of the U.S. Department of Energy has fabricated a pin-for-pin equivalent of the Intel Corporation's 8085 8-bit microprocessor chip which can function after absorbing a 100-kilogray dose of gamma radiation and can withstand single-event upsets caused by 140 MeV particles. Sandia plans to fabricate a 32-bit silicon microprocessor chip hardened to withstand 10 kilograys.

[63]E.g., see K. Boyer, "Directed-Energy Beam Weapons," **Proceedings of the Society of Photo-Optical Instrumentation Engineers**, Volume 474, 1984, pp. 79-86.

[64]U.S. Department of Defense, *Soviet Military Power*, 3d ed., 1984, p. 44.

Figure 4-12.—Artist's Conception of the Space Shuttle Deploying a Neutral Particle Beam Weapon

SOURCE: U.S. Department of Energy, Los Alamos National Laboratory.

have a relatively small diameter even at great distances—e.g., 40 meters diameter at a range of 40,000 kilometers. A beam this small must be pointed at a target with an accuracy of about 1 microradian, and this presents a more difficult problem in the case of a neutral particle beam than in the case of a continuous-wave or repetitively pulsed laser. The tracking and pointing systems of such lasers can quickly sense reflected beam energy from targets and thereby determine whether their beams are actually on target. However, a target would emit very little radiation when irradiated by a neutral particle beam. Sensors within several hundred kilometers of a target might be able, during NPB irradiation, to detect enough x-radiation or gamma or other radiation from the target to determine that the target had been hit, but it could not detect such radiation fast enough to correct beam pointing errors based on its presence or absence.

A neutral particle beam weapon could acquire (i.e., detect) a target using a passive long-wavelength infrared (LWIR) sensor; it could then track the target using an active optical tracker (LIDAR) and use this tracking information to determine the angle at which its beam must be pointed at the target. However, this approach only guarantees that the optical tracker is pointed at the target and cannot directly sense whether the beam itself is pointed at the target. That is, **open-loop** pointing must be used for neutral particle beams, while more accurate **closed-loop** pointing can be used by lasers. Determining whether open-loop pointing of a neutral particle beam can be done with an accuracy comparable to beam divergence angles may require testing of neutral particle beam generators in space against instrumented targets.

Aside from these difficulties of pointing and kill assessment, neutral particle beam weapons are among the most promising near-term options for nonnuclear ASAT weapons because of the maturity and demonstrated performance of their component technologies and the relative diseconomy of hardening targets against neutral particle beams. However, neutral particle beam weapons, like other space-based sequential-fire weapons, would be subject to attack by single-shot weapons against which shoot-back and other reactive countermeasures would be ineffective. They would have little operational effectiveness unless their survivability can be assured.

It appears, then, that if accurate open-loop beam pointing can be demonstrated, it will be feasible to build neutral particle beam weapons which, if deployed in low orbit, would pose a serious threat to satellites in low and high orbit. Against this threat only shoot-back would be economical for protecting low satellites, while hardening and deception might protect high-altitude satellites at low relative cost. At close range (1,000 kilometers), neutral particle beam weapons could damage satellite electronics of current hardness levels and upset harder future satellite electronics using relatively little fuel and tankage (etc.)—probably about 1 kilogram per shot. The cost per shot at a satellite hardened to this level, or at a decoy simulating such a satellite, would be very cheap relative to the cost per satellite or decoy. Hardening satellite electronics would increase the cost per shot, although probably not to the cost of the smallest satellites or precision decoys: damaging high-density electronics hardened to the greatest extent now foreseeable would require about 160 kilograms per shot.

Such weapons, if uncountered, could place low-altitude satellites at risk at relatively low cost. They could also upset or damage unhardened electronics on satellites in synchronous orbit, from low orbit, using little mass per shot (e.g., 25 kilograms for a dose of 10 grays); however, hardening high-density electronics on such satellites to 10 kilograys and using upset-tolerant circuit design could increase the mass requirement for damage to perhaps 25 tons per shot and would increase required irradiation time enough to permit use of reactive passive countermeasures such as generation of smoke or deployment of reaction decoys. Shielding satellites—as distinct from hardening their electronics—would be economically unfavorable against larger weapons: a disproportionately small increase in weapon mass, with no increase in mass per shot, could compensate for an increase in shield mass.

POSSIBLE U.S. RESPONSES TO SOVIET ASAT CAPABILITIES

How might the United States respond to increasingly threatening Soviet ASAT capabilities? Several options are available. For example, an increase in U.S. ASAT capabilities might—but would not necessarily—be an appropriate response. Although useful for attacking Soviet MILSATs, they might be unable to protect U.S. MILSATs. Other possible responses include reduction of dependence on military satellites, augmenting U.S. combat forces (to offset possible loss of force enhancement as a result of an ASAT attack), use of passive or active countermeasures for defense or retaliation, and arms control efforts or other diplomatic initiatives intended to constrain foreign ASAT capabilities or reduce incentives to use them.

Reduction of Dependence on Military Satellites

The United States routinely uses satellites to support its forces deployed worldwide to reinforce allies, protect sea lines of communication, and pursue other national interests. Routine use of satellites for such purposes has engendered a considerable degree of dependence on them. In the past, when satellites were less expensive and less vulnerable than other means to these ends, dependence on satellites was not risky. In the future, satellites may become so vulnerable that use, if possible, of more expensive but less vulnerable means of performing functions now performed by space systems may be required for adequate security.

Most functions now performed by space systems could be performed by alternative terrestrial systems, although terrestrial systems providing comparable performance of some functions would be unaffordable or politically unacceptable. Missile launch detection and space surveillance could be performed by airborne optical sensors; navigation by advanced inertial navigation systems which can recognize local gravity gradient patterns, nuclear detonation detection by ground-based over-the-horizon electromagnetic pulse (OTH EMP) sensors, and radio communications relaying by MF ground-wave, HF ground-wave and sky-wave, VHF meteor-burst, UHF tropospheric scatter, and UHF/SHF/EHF and light-wave airborne repeaters. Reconnaissance could be performed in wartime by aircraft overflight at great expense and risk.

On the other hand, the kind of information presently collected by satellites in peacetime to monitor compliance with arms control agreements cannot be obtained by other means which are acceptable in peacetime. Unauthorized aircraft overflight, for example, would be unacceptable. However, arms control treaties—like other treaties except those defining "laws of war"—are suspended during war between parties, so a survivable space-based means of performing this function is not required.

Although alternative terrestrial systems for performing some functions may be infeasible or very expensive, alternative terrestrial systems for performing other functions may be only slightly more expensive than space systems. For example, providing intra-theater single-channel UHF communications to mobile ground elements in the 1980s by means of satellite communications transponders would be less expensive than use of remotely piloted vehicles (aircraft) carrying transponders, but only slightly so [see figure 4-13]; a slight increase in satellite vulnerability would make the costs favor use of RPVs. Whether alternative terrestrial systems are worth their cost depends on the function to be performed and is a matter of judgment deserving periodic reconsideration.

Force Augmentation

As satellites have been acquired and integrated into military systems, they have presumably increased the effectiveness of force elements which would engage in direct combat, so that each such force element can now fight as well as, say, one and a half could without MILSAT support. Having become dependent on satellite support, if suddenly deprived of that support the effectiveness of a force element would be reduced, not just to that of a force element unaccustomed to such support but probably more so because of the disorganizing effect of losing sources of information it had come to count on. Nevertheless, it is possible, in principle, to augment current forces so that in the event of sudden loss of MILSAT support, larger future forces, although impaired in effectiveness, would fight as well as or better than would current forces using satellite support. The force augmentation required by this criterion might, under some circumstances, be modest.

Figure 4-13.—Cost Comparison: Satellite and RPV Relays for Single-Channel UHF Tactical Theater Communications

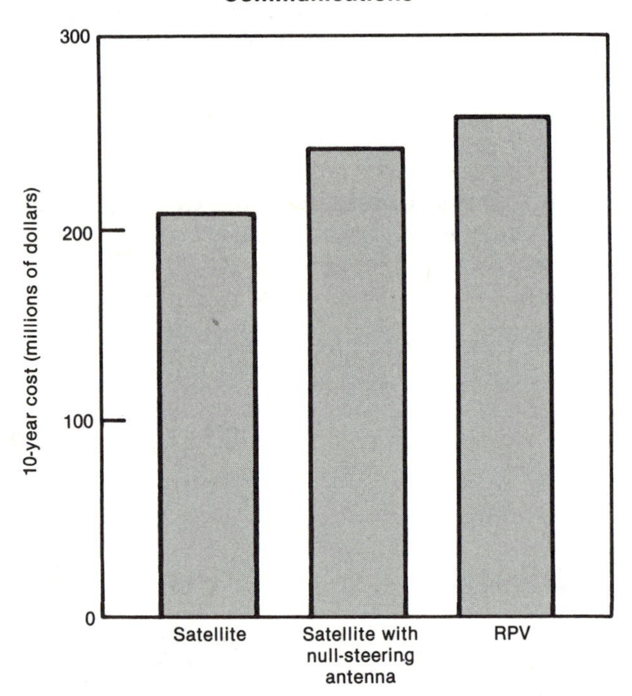

Passive Countermeasures

"Passive" countermeasures against ASAT capabilities include hiding, deception, maneuver, hardening, electronic countermeasures and electro-optical countermeasures, and proliferation, as well as combinations of these measures [see table 4-4]. Some of these countermeasures—e.g., hardening—are truly passive, requiring no satellite activity for their effectiveness, while others—e.g., evasion—require satellite activity—and hence attack warning—and are not truly passive, although they are nondestructive.

For expository purposes it is convenient to discuss each type of countermeasure in isola-

Table 4-4.—Passive Countermeasures Against ASAT Attacks

Hiding
  e.g., satellite miniaturization and orbit selection
Deception
  e.g., deploying lightweight decoys
Maneuver
  e.g., evasion
Hardening
  e.g., use of shielding
Electronic countermeasures and electro-optical countermeasures
  e.g., use of shorter wavelengths and more highly directional antennas
Proliferation
  e.g., of on-orbit spare satellites

tion from the others in the context of a one-against-one ASAT attack, as will be done here. However, it is anticipated that such countermeasures, if used, might be used in combination in the context of a many-against-many (i.e., force-against-force) space battle. It is in such a context that the potential effectiveness of a countermeasure should be assessed. However, the number and complexity of possible contexts precludes any attempt at assessment of countermeasure effectiveness from being exhaustive or conclusive.

Hiding

"Hiding" measures are measures taken to evade detection by surveillance systems. In some sense the effective use of hiding, if feasible, would be most desirable, because it would eliminate the need for other countermeasures, all of which require increases in on-orbit mass in the form of decoys for deception, fuel for evasion, shielding for hardening, or spares for proliferation.

Different hiding measures are required against different types of surveillance sensors. Space surveillance systems may be of two types: active and passive. Active sensors irradiate a target with electromagnetic radiation in order to "see" it, while passive sensors look for electromagnetic radiation emitted by the target or reflected by the target from natural sources, e.g., the Sun. At optical wavelengths (infrared, visible, and ultraviolet), both active sensors (LIDAR) and passive sensors may be used[65] In general, passive optical sensors on satellites in low Earth orbit may be able to detect satellites as small as a meter in diameter at altitudes between a few hundred kilometers and geosynchronous altitude.

Passive LWIR and visible light sensors can work better in combination than either can alone. For example, painting a satellite black would prevent it from reflecting sunlight and thereby make it invisible to passive visible light sensors. However, painting a satellite black would cause it to absorb more solar radiation and become hotter. In thermal equilibrium it would emit more LWIR radiation, making it detectable at greater range by a passive LWIR sensor [see box entitled "Long-wave Infrared (LWIR) Space Surveillance Sensors"].

Operating a satellite at very low altitude can make it difficult to detect using space-based infrared sensors which must view it against the radiant Earth or Earth limb background. More satellites would be required to perform a given function at lower altitude, and user equipment might also have to be more complex and expensive.

Deception

The use of decoys to induce an enemy to waste firepower on false targets—or to withhold fire for fear of doing so—can always be made effective, if the decoys are sufficiently realistic, i.e., "credible" to enemy space surveillance systems. A decoy can always be made credible at a cost less than or equal to that of the satellite it mimics, because a spare satellite could be used as a decoy, and it would be preferable to do so rather than spend as much on a nonfunctional decoy. **The critical question is whether a decoy can be made credible at a much lower cost than that of the satel-**

[65] In most cases passive sensors are preferable to active sensors because they do not require a high-power radiation source for irradiating targets, they are themselves consequently more difficult to detect, and their effective range can be increased more economically, because a twofold increase in range to a target decreases the irradiance received by a passive sensor only fourfold as compared to sixteenfold in the case of an active sensor.

Long-Wavelength Infrared (LWIR) Space Surveillance Sensors

The following discussion describes the physical basis for the estimated detection capabilities of passive LWIR sensors and for the ineffectiveness of hiding measures against them.

Every object emits electromagnetic radiation as a result of thermal processes; the amount of power emitted by an object increases in proportion to its surface area and would increase sixteenfold if its temperature were doubled. A satellite at a typical operating temperature—about 300° K—emits most of its thermal radiation at a wavelength of about 10 microns [10 millionths of a meter], which is in the LWIR portion of the electromagnetic spectrum. The wavelength at which thermal radiation is most intense varies in inverse proportion to the temperature of its source; thus the surface of the Sun—which has a temperature of about 6,000° K—emits thermal radiation (sunlight) which is most intense at a wavelength of 0.5 microns, a wavelength in the visible portion of the electromagnetic spectrum. Passive optical sensors which can detect LWIR radiation can detect the thermal radiation emitted by satellites, while those sensitive to visible radiation are best suited for detecting sunlight reflected by satellites. Visible light sensors are preferred for ground-based space surveillance systems because LWIR thermal radiation is absorbed strongly by the lower atmosphere, although LWIR telescopes have been operated on mountain peaks and on aircraft.

Passive LWIR and visible light sensors can work better in combination than either can alone. For example, painting a satellite black would prevent it from reflecting sunlight and thereby make it invisible to passive visible light sensors. However, painting a satellite black would cause it to absorb more solar radiation and become hotter.[1] In thermal equilibrium it would emit more LWIR radiation, making it detectable at greater range by a passive LWIR sensor. Conversely, making the satellite surface highly reflective would reduce its absorptivity and hence also its emissivity, which equals the absorptivity at each wavelength; this would make the satellite less visible to passive LWIR sensors.

Passive optical sensors may be designed to detect targets within view above the horizon (ATH) or below the horizon (BTH). BTH detection is more difficult than ATH detection, and the image processing required by BTH sensors is more complicated than that required by ATH sensors. ATH sensors are therefore preferred for passive optical space-based surveillance systems.

Space-based LWIR ATH sensors could not easily detect satellites which orbit at altitudes so low that they are actually within the upper atmosphere. Satellites at such low altitudes would be in the "Earth limb" as viewed by a space-based LWIR ATH sensor, which would have to view the satellite just above the horizon through a thick layer of air which would absorb much LWIR radiation and which would emit, reflect, and scatter LWIR background radiation against which the satellite would have to be viewed. Satellites at such low altitudes would experience considerable atmospheric drag and would slow down and reenter the atmosphere sooner unless they could maneuver and carried enough fuel to compensate for the drag. Operating in such an altitude regime to evade detection by space-based LWIR ATH sensors would also impose operational penalties in some cases. For example, surveillance satellites could not "see" as far from lower altitudes.

Required image processing sophistication would be more difficult for general surveillance than for warning of interception of the satellite carrying the sensor. This is because the spot of focused radiation from an interceptor approaching a satellite-borne sensor "staring" at the celestial background in the direction of the approaching interceptor would not move on the sensor's focal plane, and a single detector element of the sensor could accumulate thermal radiation from the interceptor until sufficient energy for detection has been accumulated. By contrast, the spot of focused radiation from an interceptor traveling in some arbitrary direction would move on the sensor's focal plane, limiting the time available to a detector element for accumulating image energy.

[1] S. Sternberg and V.P. Landon, "Satellite Systems" (ch. 17), R.E. Machol (ed.), *Systems Engineering Handbook* (New York: McGraw-Hill, 1965).

If the target's angular position and velocity is approximately known by other means, the number of photons received by each detector element over which the target's image is expected to move could be added in order to accumulate image photons and average out the noise photons. However, if the target's angular position and velocity is not known accurately by other means, the number of possible averages which must be calculated in this way to detect the target reliably increases very rapidly with increasing uncertainty about target position and velocity, and the required mass of image-processing hardware required will increase correspondingly.

A low-orbit LWIR space surveillance satellite with a 1 meter nonemissive primary mirror and a focal plane detector array cooled to 77° K could detect a black-painted satellite 1.5 meters in diameter at a range of 35,000 kilometers using a 1 millisecond integration (energy accumulation) time. In one millisecond the image of a small satellite at geosynchronous altitude would move across the face of a detector element the size of the telescope's "spot size," if the telescope were on a satellite at 1,000 kilometers altitude and "stared" continuously toward the zenith. It would be feasible and affordable, but costly (several billion dollars), to deploy a constellation of satellites of this performance which stare continuously in all directions above the horizon. Even smaller objects could be detected at greater range using a larger primary mirror, or—less economically in the limit—more image-processing hardware.

lite it mimics, as well as cheaper than an enemy's cost to identify it (e.g., by dispatching a coorbital interceptor to observe it at close range) **or to attack it** in a manner which would negate the satellite it simulates. This critical question remains unanswered.

Before any of these costs can be estimated, the designs of decoys and enemy surveillance and weapon systems must be specified. Many design choices are available. For example, for the same amount of money, a few highly realistic decoys (called "precision decoys" or "replica decoys") could be built, or a much larger number of less realistic but less expensive decoys could be built. A "precision" decoy designed to simulate an on-line satellite would probably have to simulate attitude control, stationkeeping, signal transmission, power generation and heat dissipation, and other observable functions and properties. The subsystems required to do this would be relatively expensive. If such decoys could not be distinguished from actual satellites, all such decoys would have to be attacked in a manner which would damage the satellites they simulate.

Alternatively, inexpensive "traffic" decoys could be made to simulate only those features of a capital satellite which might be measurable cheaply, quickly, and remotely. Reflective balloons or clouds of smoke and chaff could be used as "reaction decoys," i.e., they could be deployed in reaction to warning of an impending attack.[66] Even if such decoys could deceive an enemy for only a limited period of time, they could be effective in some situations. However, reaction decoys offer no protection against single-shot ASAT weapons (e.g., "space mines") which can destroy satellites almost instantly before reaction decoys can be dispensed. Ingeniously designed lightweight decoys might be both inexpensive and highly credible to passive remote sensors; whether they will be is uncertain.

[66]For example, a satellite under attack by a pop-up infrared-homing interceptor could dispense several lightweight decoys which resemble it, from a distance, in infrared brightness temperature and color temperature. If the interceptor cannot distinguish such decoys from the capital satellite until it has flown past, then the decoys would be adequately "credible." Because satellite mass cannot be measured both quickly and inexpensively, cheap, lightweight decoys could be effective as reaction decoys.

Even if lightweight decoys cannot be recognized as such after prolonged remote passive observation, it might be possible to recognize them using directed-energy devices.[67] However, use of directed-energy devices in such a manner might be provocative in peacetime and possibly more expensive than the cost of lightweight decoys.

If lightweight decoys cannot be distinguished from actual satellites, all such decoys would have to be attacked, although not necessarily in a manner which would damage the satellites they simulate, because the cost of attacking and observably damaging a lightweight decoy with some types of ASAT weapons could be comparable to, or smaller than, the cost of attacking a decoy in a manner which would damage a satellite which it simulates. For example, ground-based lasers might be able to attack lightweight decoys inexpensively. Neutral-particle beam weapons could also be used to attack decoys at long range, as could single-pulse lasers, although less economically.

The large number of possible designs for decoys and enemy surveillance and weapon systems renders assessment of the cost-exchange ratios[68] of future systems infeasible at this time. Futhermore, even if the decoy design were specified, it would be difficult to estimate decoy costs accurately. Rough preliminary estimates of satellite cost and of uncertainty in satellite cost are sometimes derived from estimates of satellite subsystem mass and complexity. However, analysis of historical cost data reveals considerable variation in satellite cost for satellites of comparable small mass.[69]

Deception is more advantageous when used in combination with other passive measures such as hardening and proliferation. For example, dormant spare satellites can be hardened and made to resemble cheaper decoys.

These considerations lead OTA to conclude that **the question of whether decoys can be made credible at a much lower cost than that of the satellites they mimic or than an enemy's cost to identify them or to attack them in a manner which would negate the satellites they simulate remains unanswered,** and that answering this question is essential in any attempt to assess prospects for making future satellites adequately survivable. An affirmative answer will probably require detailed designs for decoys which are inexpensive and lightweight as well as credible to possible future Soviet surveillance systems.

Maneuver

Satellites may maneuver in order to complicate enemy surveillance and targeting and to evade enemy fire. Satellites which do not *maneuver* are nevertheless unavoidably *mobile*, although in fixed orbits. Because of this property, the relation of maneuver to attrition is different in space than on land or at sea, and proximity in terms of orbital elements (e.g., apogee, perigee, inclination, etc.) has as much tactical significance as does momentary proximity in space. A maneuver, loosely speaking, is an action which changes a satellite's Keplerian orbital elements. Pursuit of another satellite and evasion of an interceptor are examples of maneuvers.

In order to continuously evade an interceptor—whether pop-up or coorbital—a satellite must have an acceleration capability and a velocity change ("delta-V") capability about as

[67]Referring to the possibility of active discrimination of decoys from ballistic missile reentry vehicles, Dr. Gerold Yonas, Chief Scientist of the Strategic Defense Initiative Organization, has written that "directed energy, even in a very early period, could be used in an interactive mode to assist in midcourse discrimination [Gerold Yonas, "Strategic Defense Initiative: The Politics and Science of Weapons in Space," *Physics Today*, June 1985, pp. 24-32; cf. remarks attributed to Dr. Yonas in E.J. Lerner, "Star Wars: Part II—Survivability and Stability," *Aerospace America*, vol. 23, No. 9, Sept. 1985, pp. 80-84].

[68]The relevant cost-exchange ratio is the **minimax** cost-exchange ratio, i.e., the ratio of decoy costs to Soviet ASAT sensor and weapon costs which would be incurred if decoys were designed to **minimize** the **maximum** cost-exchange ratio which the U.S.S.R. could subsequently force on the United States by judicious choice of ASAT sensor and weapon designs. The cost-exchange ratio of a specific future system is of questionable relevance, unless the system can be shown to have a cost-exchange ratio close to the minimax cost-exchange ratio.

[69]There has been less variation in cost per kilogram among satellites of large mass.

great as those of the interceptor, but somewhat more or less, depending on initial positions and velocities.[70] Acceleration and delta-V can be maximized by minimizing the mission payload, so that a large fraction of the spacecraft's initial mass is contributed by its engines (for acceleration) and fuel (for delta-V). Because an interceptor's payload can be quite small—perhaps comparable to that of a shoulder-fired anti-tank missile—an interceptor might have acceleration and delta-V capabilities which would be much more costly to provide to satellites with large mission payloads such as long-range directed-energy weapons. If so, it would be difficult for such satellites to evade small but sophisticated interceptors.

Hardening

For each type of ASAT weapon, there exist hardening techniques which can reduce the range at which the weapon would be effective. For example, satellites may be hardened to withstand the effects of ordinary, isotropic nuclear weapons by avoiding reliance on photovoltaic cells—which are vulnerable to weapon X-rays—for power, by using massive shielding to block gamma radiation, and by using Faraday shielding, magnetic shielding, and fault-tolerant electronic design to reduce vulnerability to system-generated electromagnetic pulse. Of course, such practices cannot protect a satellite from a nearby nuclear explosion, but they can force an attacker to expend at least one nuclear warhead per satellite and credible decoy to destroy them with confidence.

Shielding, or armor, of different types can offer protection against some types of projectiles, pulsed or continuous lasers, and neutral particle beams. Different types of shields would be required for protection against different types of ASAT threats. For example, shields could be used against projectiles, pulsed lasers, and neutral particle beams, respectively. For example, NASA developed shields to protect a Halley's Comet probe craft from 0.1 g meteoroids impacting at 70 kilometers per second.[71] Such shields could be all-aspect shields which completely surround a satellite, or they could be "shadow shields" deployed between the defended satellite and a weapon which poses a threat to it. Shadow shields could be deployed on a boom or they could be independent, "free-flying" satellites.

Shadow shields could be lighter than all-aspect shields, but a separate shadow shield might be required for each known or suspected threatening weapon. All-aspect shields would be superior to shadow shields in that they could defend a satellite from multiple sequential or simultaneous attack from any direction or all directions and from covert weapons and they would require no power or warning information for their operation.

Massive shields could also protect satellites from laser radiation and from neutral particle beams. Relatively little shield mass would be required to protect a satellite from beam of low-energy particles (i.e., those having energies of less than 50 to 100 MeV), but the shield mass required would increase sharply if particles of higher energy, produced by larger NPB weapons, were used.

Semiconductor microelectronic circuits inside satellites could also be made more resistant to ionizing radiation such as would be produced by a neutral particle beam. For example, the Sandia National Laboratories of the U.S. Department of Energy have fabricated a pin-for-pin equivalent of the Intel Corp.'s 8085 8-bit microprocessor chip which can function after absorbing a 100-kilogray dose of gamma radiation and can withstand single-event upsets caused by particles with energies as great as 140 MeV.

The use of asteroidal materials such as nickel for large, massive, all-aspect shields has been proposed. Possible advances in space mining, manufacturing, and transportation—which would require large investments—might

[70] G.M. Anderson, op. cit.

[71] J.P.D. Wilkinson, "A Penetration Criterion for Double-Walled Structures Subject to Meteoroid Impact," *AIAA Journal*, vol. 7, No. 10, October 1969, pp. 1937-1943.

someday make use of asteroidal material for such purposes cheaper than use of terrestrial material.[72]

These considerations suggest that, in general, shielding against weapons of relatively low capability is feasible and in many cases may be less expensive than the weapons against which it can offer protection. However, as weapons are made larger, more capable, and more numerous, the cost of protection against such weapons generally increases more rapidly than the cost of the weapons and and begins to exceed the cost of the weapons at some point.

Electronic Countermeasures and Electro-Optical Countermeasures

Passive electronic and electro-optical countermeasures can provide protection—analogous to "hardening"—against nondestructive ASAT measures. For example, communication links can be made increasingly resistant to jamming by using more transmitter power (which ultimately becomes uneconomical) or signal bandwidth (which is limited except at extremely high radio frequencies and optical frequencies), or—in some applications—by using larger antennas or shorter wavelengths for greater directionality of transmission and reception, or by transmitting at a lower data rate. Command encryption can prevent spoofing, and use of spread-spectrum modulation and time-division multiplexing techniques can provide significant resistance against uplink and downlink jamming and against downlink exploitation (e.g., by anti-radiation missiles).

Proliferation—Replenishment

Another countermeasure against ASAT attack is proliferation of satellites, so that even if a large fraction of the satellites were damaged by hostile action, enough undamaged satellites would remain to perform their assigned functions. The number of additional satellites needed to assure survivability of a required number of them would depend on enemy ASAT capabilities. The extra satellites could be placed in orbit or else kept on Earth to be launched into orbit after an ASAT attack to replenish those satellites destroyed by the attack.

Unless on-orbit spare satellites are also deployed, replenishment could not be relied on to maintain uninterrupted performance of satellite function, which is essential for such applications as early warning of missile attack and other strategic command and control functions. If it is cost-effective for an enemy to negate an operational satellite, it would probably be cost-effective for the enemy to negate replacements, if enemy ASAT capability survives. It would also be cost-effective for an enemy to maintain enough ASAT weapons or fuel to avoid exhaustion of ASAT capability before replenished satellites can be negated. Hence replenishment appears unattractive as a countermeasure unless enemy ASAT capability can be destroyed before replenishment is attempted, and unless the satellites to be replaced need not function without interruption.

Proliferation—On-orbit Spares

Spare satellites could also be pre-deployed in orbit, where they could remain dormant until needed or else be used routinely to provide redundant capability in peacetime. Dormant satellites would need to listen for radio commands to activate and might need to report their status occasionally but in general would require little power generation, cooling, attitude control, or exposure of antennas or other sensors while dormant and could be made harder than operational satellites. Their armor could have a simple shape easily mimicked by inexpensive decoys; hence proliferation of on-orbit spares would work more effectively in conjunction with hiding, deception, and hardening measures. However, an enemy which can negate an operating satellite might be able, by the same means, to negate an on-orbit spare once it became operational. Proliferating and simulating dormant spare satellites will not preserve the functioning of a constellation of

[72] C. Meinel, "Near-Earth Asteroids: Potential Bonanza for Ambitious Military Space Projects," *Defense Science 2003 +*, February-March 1985, p. 40 ff.

satellites if the spares can be identified and negated quickly and cheaply after being brought "on-line." Hence the use of on-orbit spares would be most attractive if enemy ASAT weapons could be themselves negated soon after space combat begins.

Proliferation—Modularization and Segregation

Another form of proliferation is the partitioning of satellite subsystems into modules which can be segregated and deployed on different satellites. For example, the function of a high-capacity comsat could be performed by several small comsats which pass message "packets" to one another over radio or laser crosslinks.[73] Functions such as stationkeeping might be performed by maneuverable satellite "tenders," each of which could visit one satellite after another, adjusting their positions and velocities as needed. Segregation of subsystems would require forgoing economies of scale in peacetime in order to reduce vulnerability.

Combined Passive Countermeasures

Passive countermeasures work better in combination than individually. For example, use of decoys for deception would confer little protection against some (e.g., nuclear) ASAT weapons unless maneuver were used to disperse the decoys. It is therefore important to consider the effectiveness of "packages" of passive countermeasures against various ASAT capabilities, which can also supplement and complement one another and which should also be considered packages, or postures.

Active countermeasures could, and probably would, be used to complement passive countermeasures unless prohibited by a comprehensive ban on possession of ASAT weapons. Hence in hypothesizing ASAT threats to be countered by passive measures alone, it is appropriate to consider as threats only those capabilities which are unlikely to be banned or those which might be developed and deployed (or retained) covertly. The former include nondestructive ASAT capabilities (e.g., ECM and E-OCM), and inherent ASAT capabilities of allowed weapons (e.g., ICBMs); the latter include the existing Soviet coorbital interceptor, direct-ascent and coorbital nuclear interceptors, space-based or pop-up X-ray laser weapons, and possibly ground-based lasers. Of these, the nuclear weapons might be based in space disguised as, or aboard, a different type of satellite, the presence of which would be ob-

Figure 4-14.—Low-Cost Packet-Switching Communications Satellite

The Global Low-Orbiting Message Relay Satellite (GLOMR), shown here, is designed to receive messages sent to it, store them, and relay them to ground facilities. The GLOMR program of the Defense Advanced Research Projects Agency (DARPA) is intended to demonstrate that communications satellites operating in this manner can be produced at relatively low cost. If sufficiently inexpensive, such satellites could be deployed at lower cost than the cost of attacking them with some types of ASAT weapons, and so many could be deployed that other weapons might require a long time to destroy most of them.

SOURCE: U.S. Department of Defense.

[73] The Defense Advanced Research Projects Agency (DARPA) is investigating the feasibility of a packet-switched network of transponders on low-altitude satellites, or on Earth.

servable but the nature of which might be impossible to ascertain except by prolonged close observation or invasive sensing techniques.[74] In general, nonnuclear space-based weapons could not be expected, with confidence, to perform well, unless they had been previously—and observably—tested in space.

Active Measures

Passive countermeasures against ASAT attacks may be supplemented by active measures intended to deter ASAT attack or to defend satellites if deterrence should fail. Active measures can therefore be used for either defensive or retaliatory purposes. Defensive active measures are *active countermeasures* against ASAT attacks. Retaliatory active measures do not counter ASAT attacks but instead fulfill explicit or implied threats of retaliation which were intended to deter ASAT attacks. Active measures used for either purpose can be either nondestructive (e.g., electronic countermeasures and electro-optical countermeasures) or destructive (e.g., shoot-back), as shown in table 4-5.

Defensive Countermeasures

Shoot-Back.—"Shoot-back" usually refers to counter-attacking space-based ASAT weapons, but can also denote counter-attacks against the ground segment of ASAT weapon systems (e.g., satellite control facilities). Many weapons capable of shoot-back would themselves be subject to shoot-back, making the effectiveness of shoot-back highly dependent on the types and numbers of ASAT and other weapons deployed and on the incentives for preemptive attack which ASAT weapon vulnerabilities, if any, could create. Analysis of the effectiveness of shoot-back is therefore very complicated in general, although simple in certain important cases.

For example, **shoot-back would be ineffective against expendable, single-shot space mines** employing kinetic-energy, directed-energy, or nuclear destructive mechanisms. Such weapons would damage their targets almost instantly, if at all, and destroy themselves in the process, leaving nothing of value to shoot back at. Moreover, **shoot-back using sequential-fire weapons which are vulnerable to attack by expendable, single-shot weapons would be ineffective**, because they could be damaged by single-shot weapons after attacking only one target. However, space-based single-shot ASAT weapons would themselves be subject to **preemptive** attack—i.e., "shoot-first" instead of "shoot-back." If such weapons were mutually deployed in space, if each such weapon could instantly destroy several similar weapons and if such weapons were not salvage-fused to fire if disturbed, the resulting preemptive advantage could cause a condition of "crisis instability," in which each nation, desiring peace but fearing (perhaps mistakenly) an imminent attack by the other, would have reason to initiate hostilities.[75]

However, it is conceivable that even if such incentives should induce escalation from peace or low-level conflict to war in space, the preemptive ASAT attack which would begin such a war might reduce incentives for further escalation, either "vertically" (to higher levels of conflict) or horizontally (to other theaters of conflict, e.g., Earth). For example, if the United States and the U.S.S.R. continue to possess strategic offensive missile forces of considerable counterforce capability, and if each were to deploy a large BMD system

[74]E.g., see U.S. Patent 4,320,298.

[75]See M.B. Callaham and F.M. Scibilia, *Proceedings of the Society of Photo-Optical Instrumentation Engineers*, vol. 474, 1984, pp. 107-114.

Table 4-5.—Active Measures Against ASAT Attack

Active measures
Defensive measures:
Nondestructive
e.g., jamming
Destructive
shoot-back
attack on ground-based ASAT command and control facilities
Retaliatory measures:
ASAT counterattack (retaliation in kind)
Horizontal escalation (to terrestrial theaters)

which relied on vulnerable space-based components, then each nation might fear that the other nation (also fearing a preemptive attack) might attack these BMD components preemptively with some confidence that its BMD system could limit damage from a retaliatory missile attack, if any. Each nation would therefore have an incentive to attack the other's space-based BMD components preemptively. However, after having done so, the attacker would have no motive to launch a preemptive, damage-limiting *missile* attack, because it could assume that the other nation—now highly vulnerable to retaliation—would not seriously consider such an option. Hence, under these assumptions, escalation instability would exist during crises in peacetime (thus "crisis instability") but not at the level of war confined to space.

It might be supposed that the crisis instability which would accompany mutual deployment of such weapons could be eliminated by salvage-fusing them to fire if disturbed in certain ways (presumably indicative of an attack). If salvage-fusing were feasible and actually used (or believed to be used), there might be little incentive to fire first even if an attack were expected. However, salvage-fusing some types of weapons against some other types may be inordinately difficult or infeasible. Moreover, even though an "intelligent" salvage-fusing system might be able to distinguish among different types of disturbances, it could not be made completely reliable or infallible in discrimination, so there would be a risk that some natural disturbance (e.g., a meteoroid impact) might trigger such a weapon to fire, possibly at several similar enemy weapons, possibly triggering them to fire, etc. Similar consequences could follow an accidental attack or a "catalytic" attack by a third party. Moreover, if salvage-fused space-based weapons only held an enemy's space-based assets at risk, the prospect of losing such assets in retaliation for an attack might be considered an acceptable or favorable trade by a nation less dependent on space assets.

Regardless of whether salvage-fusing were employed, mutual deployment of single-pulse weapons would not be expected to create strong proliferation incentives: with mutual salvage-fusing each side could plausibly lose all its important space assets in such an exchange regardless of whether it had many such weapons or only a few deployed; it would therefore have no incentive to deploy more weapons than would be required to negate all threatening satellites *except* single-pulse ASAT weapons, against which neither preemptive attack nor shoot-back would be effective. Without salvage-fusing, the side which failed to preempt could plausibly lose all its important space assets in such an exchange regardless of whether it had many such weapons or only a few deployed, and would therefore gain nothing by deploying many weapons.

Electronic Countermeasures and Electro-Optical Countermeasures.—Active electronic and electro-optical countermeasures (jamming, blinding, and spoofing) could be used against some near-term ASAT command uplink systems, KEW homing systems, and DEW acquisition, tracking, and pointing systems which have inadequate counter-countermeasures.

Attack on Ground-Based ASAT Weapons or Support Systems.—At present there appear to be only two launch pads for Soviet coorbital interceptors, both at Tyuratam, and only two Soviet ground-based lasers of significant ASAT capability, both at Sary Shagan. Hence attacking such ground-based facilities with conventional or nuclear weapons could be very effective, especially if preemptive, but would be viewed by some in the United States as escalatory with respect to attacking or defending satellites using nonnuclear weapons.

Retaliatory Measures

The ability to respond to ASAT attack by active measures could be maintained and publicized in an attempt to deter such an attack in the first place. Postures and policies intended to enhance deterrence could act as adjuncts to or substitutes for active and passive countermeasures. Even in the event of deployment of advanced ASAT weapons such as expendable single-pulse lasers against which "shoot-back" and passive measures might be ineffective, postures and policies intended to enhance deterrence could enhance security, although they cannot guarantee security.

In pursuing security through deterrence, it is appropriate to develop retaliatory capabilities which place at risk targets of sufficient value to deter attack and which do not exacerbate crisis instability. The first of these considerations implies that to deter ASAT attack, retaliation need not necessarily be "in kind"—i.e., against satellites. In fact, an ability to retaliate in kind, however thoroughly or swiftly, would be inadequate to deter an ASAT attack if the attacking nation valued destruction of enemy satellites more than survival of its own. For example, if the U.S.S.R. developed a capability to quickly destroy all on-orbit U.S. satellites, then even if the United States could destroy all on-orbit Soviet satellites in retaliation, the U.S. capability to mount such a retaliatory response —although valuable in the event—might not deter a Soviet first use of ASAT weapons. Soviet leaders might judge the continued deployment of U.S. MILSATs to be more detrimental to Soviet interests than survival of Soviet MILSATs is valuable to Soviet interests. If this were the case, an ability to retaliate against [more valuable] terrestrial assets would be required to successfully deter an ASAT attack. Such retaliatory capabilities might be provided by terrestrial or space-based weapons. [A separate, classified appendix to this report (Appendix D) contains a more detailed discussion of the utility of military satellites to the United States and to the U.S.S.R.]

The second consideration—avoidance of crisis instability—precludes reliance on destabilizing weapons to provide retaliatory capabilities. Space-based ASAT weapons capable of instantly destroying several satellites, including similar ASAT weapons, would be most destabilizing **unless** salvage-fused but would be prone to accidental firing **if** salvage-fused. By comparison, an ideal weapon for deterring ASAT attack would be nonnuclear and hence usable at all levels of conflict without escalating the level of conflict. It could survive a preemptive attack and destroy enemy assets of sufficient value to deter an attack while causing little collateral damage.

Diplomatic Measures

In addition to the military measures discussed above, diplomatic measures such as arms control initiatives and negotiation of "rules of the road" for space operations could be useful responses to foreign development of threatening ASAT capabilities. The variety of possible measures is great, and assessment of their advantages is complicated; this topic is discussed in detail in chapter 6.

SUMMARY OF FINDINGS REGARDING ASAT CAPABILITIES AND COUNTERMEASURES

The most important conclusions which may be drawn from the preceding discussion of ASAT capabilities and countermeasures are:

1. **Nonnuclear** ASAT weapons which are now deployed or being tested by the United States and the U.S.S.R. are limited in altitude capability and responsiveness and can attack only a subset of currently deployed opposing MILSATs, although this subset includes important MILSATs.
2. The inherent ASAT capabilities of existing nuclear weapons such as U.S. and Soviet ICBMs and Soviet ABM interceptor missiles are substantial. Such weapons could pose a threat even to satellites in synchronous orbit, but are useful only at the highest levels of conflict.
3. Technologies applicable to future ASAT weapons are so varied, and many so promising, that future ASAT weapons, if developed, would be able to attack and dis-

able virtually all MILSATs of current types as currently deployed. Hence to maintain the survivability of constellations of future MILSATs, it will be necessary that the development and deployment of such weapons be constrained by arms control or that future satellites be protected from them by passive or active countermeasures, or that a combination of these approaches be pursued.

4. Of individual passive countermeasures which might be used against advanced ASAT weapons, only deception (use of decoys) is likely to be effective against all types of ASAT weapons, and deception is likely to be economical (relative to the offense) only if the decoys, and the satellites they mimic, are lightweight and inexpensive. Use of deception in combination with maneuver, hardening, and proliferation might offer economical protection for lightweight satellites.
5. Active countermeasures—electronic countermeasures, electro-optical countermeasures, and shoot-back—would be ineffective against an aggressive or preemptive surprise attack using expendable, single-shot ASAT weapons (e.g., kinetic-energy, directed-energy, and nuclear "space mines"). Actively defending keep-out zones around critical satellites might be able to protect such satellites against emplacement of short-range space mines but not against advanced, long-range space mines.
6. Of future ASAT weapons now foreseeable, those which would be most effective if used in a preemptive or aggressive surprise attack—i.e., expendable, single-shot ASAT weapons—would be space-based and therefore subject to such attacks by similar weapons. The cost of protecting them from such attacks—which must necessarily be by passive means—would exceed the cost of attacking them. Such weapons, if mutually deployed, would provide or increase incentives to attack preemptively in crises in which similar attacks are anticipated. Salvage-fusing such weapons to fire if disturbed would reduce but not necessarily eliminate incentives to preempt but would increase risks of accidental attack.
7. A capability to confirm the occurrence and identify the perpetrator of an ASAT attack and to retaliate in proportion, but not necessarily in kind, might deter ASAT attacks. A capability to retaliate in kind—i.e., against the attacker's satellites—could contribute to deterrence, if this capability were survivable, but if this capability were vulnerable to ASAT attack, it could undermine deterrence by posing an opponent an incentive to attack preemptively. However, a capability to retaliate in kind would be inadequate to deter an ASAT attack by an adversary nation which values destruction of U.S. satellites more highly than survival of its own.
8. Strict arms control measures could not be expected to eliminate the inherent ASAT capabilities of weapons such as ICBMs nor provide complete confidence that no ASAT weapons have been developed and deployed covertly. In an arms control regime which bans ASAT weapons, use of passive countermeasures would be required to reduce the residual risk posed by weapons such as ICBMs. However, prohibiting the testing of ASAT capabilities of weapons would preclude the attainment of confidence that certain types of advanced, nonnuclear ASAT weapons would perform reliably if used and would therefore also reduce incentives to develop such weapons or to attempt to deploy them covertly. A ban on testing would also render more difficult, costly, and risky any attempt to attain confidence, by covert testing, that other types of advanced, nonnuclear ASAT weapons (e.g., ground-based lasers) would perform reliably if used and would therefore also reduce incentives to develop such weapons or to attempt to deploy them covertly.

Prohibiting the basing in space of weapons with ASAT capabilities, to the extent that compliance with such a ban could be verified, would forestall the creation of strong incentives to attack such weapons preemptively when a similar attack is feared. Even in the absence of such strict restraints, if ASAT weapons are based in space, an agreement banning unauthorized close approach to foreign spacecraft could reduce the ambiguity of such provocative acts and thereby reduce the risk of ASAT attack resulting from misunderstanding, while providing a legal basis for anticipatory self-defense against ASAT weapons of short effective range.

Chapter 5

ASAT Arms Control: History

Contents

Chapter 5

ASAT Arms Control: History

INTRODUCTION

This chapter discusses the constraints on ASAT development imposed by the treaties and agreements currently in force. It also briefly examines the history of ASAT weapons development and deployment, and describes the previous attempt by the United States and the Soviet Union to conclude a treaty further restricting such weapons. The issue of ASAT weapons and ASAT arms control, a politically volatile topic, has stimulated considerable interest in the U.S. Congress over the last several years; this chapter also discusses the history of the major pieces of legislation in the 97th, 98th, and 99th Congresses (1981-85) which concerned ASAT negotiations and weapons development.

Chapter 4 examined how certain passive and active ASAT countermeasures might contribute to U.S. national security and provide protection for critical space assets. Building on the historical background presented in this chapter, chapter 6 will examine the contribution that ASAT arms control might make to these same goals, analyzing a number of potential ASAT arms control regimes and identifying those which might be appropriate for the United States to pursue. The interaction between technical countermeasures and arms control is examined in chapter 7.

CONSTRAINTS IMPOSED BY TREATIES AND AGREEMENTS IN FORCE

To evaluate future space arms control measures it is first necessary to understand the constraints that existing treaties and other international agreements place on military space activities. No single treaty fully specifies which space activities are allowed and which prohibited, and existing agreements do not apply uniformly to all countries. All nations are presumably bound by the provisions of the Charter of the United Nations,[1] customary international law, and the "general principles of law recognized by civilized nations."[2] States party to the 1967 Outer Space Treaty[3] and the Limited Test Ban Treaty[4] accept additional restrictions on their space activities. The United States and the Soviet Union agreed bilaterally in the context of SALT I (the ABM Treaty[5] and the Interim Agreement to limit offensive arms) not to disturb the function of satellites used to verify compliance with those treaties and to forgo the development of space weapons to counter ballistic missiles. The relevant provisions of these instruments are discussed below.

[1]As a general rule, only states party to a treaty are bound by its terms. An exception to this rule appears in Article 2 (6) of the U.N. Charter which provides: "The Organization shall insure that states which are not members of the United Nations act in accordance the Principles (of the Charter) so far as may be necessary for the maintenance of international peace and security." Charter of the United Nations, *1970 Yearbook of the United Nations*, p. 1001. See also: Ian Brownlie, *Principles of Public International Law* (3d ed., 1979).

[2]Statute of the International Court of Justice, Art. 38, *1970 Yearbook of the United Nations*, p. 1013.

[3]"Treaty on Principles Governing the Activities of States in the Exploration and Use of Outer Space, Including the Moon and Other Celestial Bodies," 18 U.S.T. 2410; T.I.A.S. 6347.

[4]"Treaty Banning Nuclear Weapon Tests in the Atmosphere, in Outer Space and Under Water,"14 U.S.T. 1313, T.I.A.S. 5433.

[5]"Treaty Between the United States and the U.S.S.R. on the Limitation of Anti-Ballistic Missile Systems," Oct. 3, 1972, 23 U.S.T. 3435, T.I.A.S. 7503.

Charter of the United Nations[6]

Article 2(3) of the U.N. Charter directs nations to "settle their international disputes by peaceful means in such a manner that international peace and security, and justice, are not endangered." Article 2(4) requires that nations "refrain . . . from the threat or use of force . . . in any . . . manner inconsistent with the purposes of the United Nations." It could be argued that these statements and other general principles of customary international law in some ways inhibit the *use* of ASATs.[7]

It is important to note that the responsibilities imposed by Article 2 of the U.N. Charter are modified by Article 51, which states, "Nothing in the present charter shall impair the inherent right of individual or collective self-defense." Taken together, Articles 2 and 51 do indicate general international censure of the use of force, but do not limit specific weapon systems.

Limited Test Ban Treaty[8]

The Limited Test Ban Treaty of 1963 prohibits nuclear weapons tests "or any other nuclear explosion" in outer space, as well as in the atmosphere or under water. The treaty therefore prohibits the testing, in space, of exotic ASAT weapons that would derive their power from a nuclear explosion[9]—a consequence probably not anticipated by the treaty's drafters. The Limited Test Ban Treaty would not limit the development or testing, on Earth or in space, of other nonnuclear components for such weapon systems. The power source could be tested underground on Earth,[10] as are other nuclear weapons, and the nonnuclear components could be tested separately in space.

The 1967 Outer Space Treaty[11]

Article III of the Outer Space Treaty states that space activities shall be carried out in accordance with international law "in the interest of maintaining international peace and security and promoting international co-operation and understanding." This Article expresses the sentiment of the drafters that space be used to benefit mankind and contribute to peace.

In contrast to the general language of Article III, Article IV of the Outer Space Treaty establishes a clear prohibition against placing "in orbit around Earth any objects carrying nuclear weapons or any other kinds of weapons of mass destruction."[12] Orbiting weapons using nuclear explosions for power would presumably be included. This provision does not limit ground-based ASATs or ASATs which use conventional explosives or other means to destroy a target. Neither does it ban nuclear-armed "pop up" ASAT interceptors that ascend directly to their targets without entering into orbit.[13]

---

[6]Supra, note 1.

[7]Terrestrial international law is explicitly extended to space by Article III of the 1967 Outer Space Treaty, which states that the exploration and use of outer space shall be conducted "in accordance with international law, including the Charter of the United Nations."

[8]Supra, note 4.

[9]An additional limitation on nuclear-pumped space weapons can be found in the "Threshold Test Ban Treaty" of 1974. Article 1 prohibits tests of nuclear weapons greater than 150 kilotons in yield, banning even underground testing of any nuclear-driven weapon requiring an explosion larger than that. The Threshold Test Ban Treaty was signed by the United States but has yet to be ratified. "Treaty Between the United States of America and the Union of Soviet Socialist Republics on the Limitation of Underground Nuclear Weapon Tests," reprinted in, *Arms Control and Disarmament Agreements,* U.S. Arms Control and Disarmament Agency (1982 ed.), p. 167.

[10]Subject to other treaty limitations—see previous note.

[11]Supra, note 3.

[12]According to Ambassador Arthur Goldberg, chief U.S. negotiator of the Outer Space Treaty, weapons of mass destruction include "any type of weapon which could lead to the same type of catastrophe that a nuclear weapon could lead to" (Hearings Before the Senate Foreign Relations Committee on Executive D, 90th Cong., 1st sess., p. 23.) In 1948, the U.N. Commission for Conventional Armaments advised the Security Council that the term "weapon of mass destruction" would include, "atomic weapons, radio-active material weapons, lethal chemical and biological weapons, and any weapons developed in the future which have characteristics comparable in destructive effect to those of the atomic bomb or other weapons mentioned above." (Resolution adopted by the Commission for Conventional Armaments at its 13th meeting, Aug. 12, 1948. U.N. Security Council, S/C.3/32/Rev. 1, Aug. 18, 1948.)

[13]Testing of such weapons which involved detonating nuclear warheads in space would be banned by the Limited Test Ban Treaty.

Article IX of the Outer Space Treaty directs nations to "undertake appropriate international consultations" before proceeding with any activity that might cause "potentially harmful interference with the activities of other states in the peaceful exploration and use of outer space." It is possible to argue that states developing ASATs (weapons *intended* to cause "harmful interference") should do so only after "appropriate international consultations." Nonetheless, the vague wording of Article IX and the forced nature of such an interpretation reduce the Article's value as an arms control provision.[14]

Taken together, the provisions of the Outer Space Treaty afford satellites some measure of legal protection against attack. The precise nature of this protection is unclear since the treaty was not drafted for the specific purpose of limiting deliberate hostile activities. The treaty clearly does not limit the development, testing, or deployment of nonnuclear weapons capable of interfering with satellites of other nations; moreover, the U.N. Charter provision for self-defense might be taken to permit such interference in some cases.

Strategic Arms Limitation Talks (SALT I & II)[15]

The verification provisions of SALT I and II state that the parties shall use "national technical means" (NTM) of verification to monitor adherence to the Agreements. NTM is understood, though not explicitly specified, to include certain reconnaissance satellite systems. The SALT Agreements further state that "Each Party undertakes not to interfere with the [NTM] of the other Party" as long as these assets are operated "in a manner consistent with generally recognized principles of international law." The SALT Agreements implicitly sanction the use of satellites for verification of treaty compliance and provide some measure of protection against peacetime attack on these assets. These Agreements do not, however, restrict the development, testing, or deployment of ASAT systems capable of attacking NTM. In addition, whatever legal protection these Agreements provide is limited to systems used to verify the SALT Agreements. Other space systems used for combat support during hostilities would not be protected under the SALT provisions.

Article IX of SALT II prohibits the development, testing, or deployment of "systems for placing into Earth orbit nuclear weapons or any other kind of weapons of mass destruction, including fractional orbital missiles." This provision was included to limit the development of Fractional Orbital Bombardment Systems (FOBS), in which missiles enter partial Earth orbit and fly the long way around the Earth rather than taking the much more direct trajectory of normal ICBMs. However, this provision could also be read as expanding the prohibition of Article IV of the Outer Space Treaty. Whereas Article IV prohibits only the act of orbiting nuclear weapons, Article IX of SALT II would seem to prohibit in addition the development, testing and deployment of systems (e.g., launchers) to accomplish the orbiting of these weapons. So interpreted, Article IX could create an additional legal barrier to the development of

[14]The "Accidental Measures" Agreement of 1971 requires the United States and the Soviet Union to "notify each other immediately in the event of . . . signs of interference with [missile warning systems] or with related communication facilities." However, this agreement places no limitations on the development or use of ASAT capabilities. "Agreement on Measures to Reduce the Risk of Outbreak of Nuclear War Between the United States of America and the Union of Soviet Socialist Republics," reprinted in U.S. Arms Control and Disarmament Agency, *Arms Control and Disarmament Agreements* (Washington, D.C.: U.S. Government Printing Office, 1982), p. 159.

[15]"Interim Agreement Between the United States of America and the Union of Soviet Socialist Republics on Certain Measures with Respect to the Limitation of Strategic Offensive Arms" (SALT I) reprinted in, *Arms Control and Disarmament Agreements,* U.S. Arms Control and Disarmament Agency (1982 ed.), p. 139.; "Treaty Between the United States of America and the Union of Soviet Socialist Republics on the Limitation of Strategic Offensive Arms" (SALT II), reprinted in U.S. Arms Control and Disarmament Agency *Arms Control and Disarmament Agreements* (Washington, DC: U.S. Government Printing Office, 1982), p. 246.

The completed SALT II agreement was signed by President Carter and General Secretary Brezhnev on June 18, 1979. Although Senate consent to ratification has not been given, Presidents Carter and Reagan both declared that they would do nothing to jeopardize the treaty as long as the Soviet Union abided by it.

orbital, nuclear-pumped, directed-energy weapons.

Anti-Ballistic Missile (ABM) Treaty[16]

In the ABM Treaty, the United States and the Soviet Union agreed not to deploy anti-ballistic missiles except under the very limited conditions set forth in the treaty.[17] Each party also undertook not to "develop, test, or deploy ABM systems or components which are sea-based, air-based, space-based, or mobile land-based."

The distinction between advanced ASAT and BMD technologies is not always clear. As noted by Secretary of Defense Weinberger in a report to Congress,[18] directed-energy weapons "could perform a variety of missions, such as antisatellite or ballistic missile defense." For the purposes of the ABM Treaty, "an ABM system is a system to counter strategic ballistic missiles or their elements in flight trajectory." Therefore, ASAT weapons would be prohibited by the ABM Treaty if they were capable of countering strategic ballistic missiles.* Such systems are banned unless they are fixed, land-based and deployed at permitted sites. The testing of ASAT weapons of lesser capability would not be inhibited by the treaty. If an ASAT weapon became capable of intercepting missiles, it would fall within the terms of the ABM treaty. This capability test includes future systems having components "based on other physical principles" than those of the ABM system components (interceptors, launchers, and radars) described in Article II of the treaty. However, the ABM Treaty does not control highly capable ASAT systems lacking ABM capability, and it does not clearly indicate how such capability is to be inferred.

[16]"Treaty Between the United States of America and the Union of Soviet Socialist Republics on the Limitation of Anti-Ballistic Missile Systems," supra, note 5.

[17]Article III limits each side to two fixed, ground-based ABM deployment areas (later reduced to one), each of which has limitations on radar facilities and interceptors and launchers. Agreed Statement D provides that should ABM systems based on other physical principles than missile interceptors be developed in the future, these would be subject to discussion and agreement. Unless such systems were explicitly permitted by future agreement, they would continue to be banned.

[18]C. Weinberger, *Fiscal Year 1985 Annual Report to Congress* (Washington, DC: U.S. Government Printing Office, February 1984), p. 263.

*The ABM Treaty is discussed in grater detail in OTA's report, *Ballistic Missile Defense Technologies,* OTA-ISC-281, app. A.

INTERNATIONAL POLITICAL BARRIERS TO ASAT DEVELOPMENT

Although there are few clear international legal barriers to ASAT *development*, the desire of the United States to remain in some way responsive to international opinion creates certain inhibitions to the unrestrained pursuit of weapons that are based or operate in space. In the United Nations and other international fora, the United States and, to a lesser extent the Soviet Union, have been criticized for their military activities in space. Some view the "militarization" or "weaponization" of space as breaking a de facto political taboo; others see it as a violation of customary international law. Some of our allies, responding to strong domestic political pressures to limit the arms race, see U.S. and Soviet cooperation in controlling space weapons as one means to reduce international tension. It is important to note, therefore, that there may be a significant political or diplomatic cost to developing space weapons.

Opposition to "space weapons" derives, in part, from the belief that space is a unique environment which must be preserved for "peaceful" activities and should be responsive to international controls. The "uniqueness" of space is seen as deriving from the fact that some space activities, such as remote sensing and satellite communications are inherently global in effect; i.e., they pass over the territory of other countries and may require international coordination, such as frequency allocation. These characteristics have resulted in the development of a number of successful international institutions such as the International Telecommunication Union (ITU), the International Telecommunication Satellite Organization (INTELSAT), and the International Maritime Satellite Organization (INMARSAT).

The fact that certain space activities have been the subject of international controls has fostered a belief among some that all space activities should somehow require international consent. In this view, an unrestrained arms race between the United States and the Soviet Union in space is seen as threatening the interests of nations not having strong space programs as well as being a threat to peace.

The United Nations, and in particular its Committee on the Peaceful Uses of Outer Space (COPUOS), has been responsible for five international treaties dealing with space. Each of these treaties emphasizes to some degree the necessity that the exploration and use of space be for "peaceful purposes." In addition, many world leaders (including every President of the United States since Eisenhower), scholars, and jurists have, since the beginning of the space age, emphasized the unique nature of space and its ability to contribute to peace and the common good.

Having been nurtured for over 25 years, the idea that space is a unique environment which should be used for peaceful purposes has come to be considered by some to be a principle of customary international law. As a result, the development of weapons which would operate in or through space has met with strong opposition in international fora.

The 1982 General Assembly Resolution on the "Prevention of an Arms Race in Outer Space" reflects this international concern.[19] The Resolution reaffirms the belief of the General Assembly that "space [activities] should be for peaceful purposes and carried on for the benefit of all peoples," and notes "the important and growing contribution of satellites for . . . the verification of disarmament agreements and . . . their use to promote peace, stability and international cooperation." Pointing out the "threat posed by anti-satellite systems and their destabilizing effect for international peace and security," the Resolution urges all states "to contribute actively to the goal of preventing an arms race in outer space and to refrain from any action contrary to that aim." Finally, it requests the U.N. Committee on Disarmament to consider "the question of negotiating effective and verifiable agreements aimed at preventing an arms race in space."

[19]U.N.G.A. Doc. A/36/192, Aug. 11, 1982.

ASAT NEGOTIATIONS—PAST AND PRESENT

Background

The first test of a weapon against a satellite was conducted by the United States in 1959 when a Bold Orion missile launched from a B-47 aircraft successfully passed within 20 miles of the U.S. Explorer VI satellite as it passed over Cape Canaveral. In 1963 and 1964, the U.S. Army operated a system of nuclear-armed direct-ascent ASAT interceptors on Kwajalein Atoll in the Pacific Ocean. From 1964 until 1970, another such system was maintained on Johnston Island by the Air Force; this system was formally decommissioned in 1975.[20]

[20]Marcia Smith, " 'Star Wars': Antisatellites and Space-Based BMD" (Washington, DC: Library of Congress, Congressional Research Service, Issue Brief IB81123, Nov. 26, 1984); Paul Stares, "Deja Vu: The ASAT Debate in Historical Context," *Arms Control Today*, December 1983, p. 2.

The Soviet Union initiated a series of ASAT tests in 1968 which continued through 1971.[21] Although the United States suspected that the Soviets were developing an "inspect and destroy" capability, this system was not seen as posing so significant a threat to U.S. assets that a response was necessary. However, when the Soviets conducted another series of ASAT tests between 1976 and 1978, U.S. officials began to express concern.

The Carter Administration adopted a "two-track" policy. In March 1977, President Carter announced that he had suggested to the Soviets that "we forgo the opportunity to arm satellite bodies and also to forgo the opportunity to destroy observation satellites."[22] Also in that month, the Department of Defense announced that U.S. military space programs were being accelerated.[23]

1978-79 Negotiations

The Soviets responded positively to Carter's proposal for ASAT negotiations and in March 1978, agreement was reached on an exploratory meeting. Three rounds of the ASAT limitation talks were held: June 8-16, 1978, in Helsinki; January 23-February 16, 1979, in Bern; and April 23-June 15, 1979, in Vienna. The third round of talks ended when the two sides felt they had gone as far as they could without further consultation and study in their respective countries.[24] According to Ambassador Robert W. Buchheim, head of the U.S. delegation for most of the 1978-79 ASAT talks, the two delegations agreed that when either decided that it was ready to resume active negotiations, the other party would be so notified through diplomatic channels.[25] Although neither side formally withdrew from the negotiations, after the Soviet invasion of Afghanistan, the U.S. refusal to ratify the SALT II Treaty, and a general deterioration of U.S.-Soviet relations, discussions never resumed.

The 1978-79 talks did not result in an ASAT arms control agreement; however, they did clarify some of the concerns of the two parties. The talks focused on two main topics: limits on ASAT use, and limits on the development of ASAT capabilities.[26]

During the 1978-79 negotiations, as now, the different status of the U.S. and Soviet ASAT programs was a substantial impediment to progress. The Soviet ASAT system was considered by the United States to be "operational," whereas the potentially more capable American system had yet to be tested. A limited moratorium or treaty would have given the United States time to conduct ground-based research and development while inhibiting Soviet ASAT weapons tests in space. An indefinite test ban, on the other hand, might have locked the United States into a position of ASAT inferiority.

Soviet Draft Treaties

In 1981 and again in 1983, the Soviets submitted draft space weapon treaties to the United Nations.[27] U.S. experts disagree as to why the Soviets have continued to advocate space weapon arms control. One theory holds that the Soviet interest is not in arms control, but rather in propaganda. Since the Reagan Administration was not actively seeking limitations on space weapons, the Soviets could portray the United States as being responsible for the escalation of the arms race and the

[21]*Soviet Space Programs: 1976-80*, Committee Print, Senate Committee on Commerce Science and Transportation, 97th Cong., 2d sess., December 1982, p. 184.

[22]*The Washington Post*, Mar. 10, 1977, p. A4.

[23]Ibid.

[24]Lynn F. Rusten, "Soviet Policy on Antisatellite (ASAT) Arms Control" (Washington, DC: Library of Congress, Congressional Research Service, 84-670-S, June 22, 1984), p. 1.

[25]"Arms Control and the Militarization of Space," Hearings Before the Subcommittee on Arms Control, Oceans, International Operations and Environment of the Committee on Foreign Relations on S.J. Res. 129, U.S. Senate, 97th Cong., 2d sess., Sept. 20, 1982, pp. 54-55.

[26]For a more detailed discussion of the 1978-79 talks, see: Walter Slocombe, "Approaches to an ASAT Treaty," *Space Weapons—The Arms Control Dilemma*, Bhupendra Jasani (ed.) (London: Taylor and Francis, 1984), p. 149; and Lynn F. Rusten, op. cit.

[27]"Draft Treaty on the Prohibition of the Stationing of Weapons of Any Kind in Outer Space," U.N. General Assembly, Doc. A/36/192, August 1981; "Treaty on the Prohibition of the Use of Force in Outer Space and From Space Against the Earth," U.N. Doc. A/38/194, Aug. 26, 1983.

"militarization" of space. Another hypothesis is that the Soviets have a genuine interest in limiting ASAT technology because this is an area where the United States would be able to excel. Since the Soviets have clearly stated their opposition to the Reagan Administration's plans to develop space-based BMD technologies, their interest in arms control in space—especially after March 1983—could be intended to inhibit the progress of this program.

The United States refused to participate in multilateral negotiation with the Soviets on either the 1981 or 1983 draft treaties. Whatever the true reason or combination of reasons for Soviet interest in ASAT arms control, the Soviets have used this issue—and the U.S. refusal to negotiate—effectively in their political propaganda. The Soviet position has been, until recently, that their space program has been purely peaceful in nature. Since 1958, according to Soviet Foreign Minister Gromyko, the Soviet Union "invariably stated and continues to state that space should be a sphere of exclusively peaceful cooperation."[28]

From an American point of view, the Soviet propaganda seems absurd since the Soviet Union has an "operational" ASAT and a very active military space program. From the point of view of many nonaligned governments, as well as important segments of the populations of our allies, the fact that the Soviet Union was as responsible for the "militarization of space" as the United States, or more so, did not lessen the culpability of the United States for refusing to negotiate. As a result, the Soviet propaganda on the "militarization" of space was initially successful in enhancing the international image of the Soviets while fostering criticism of the United States. More recently, the inability of the United States and the Soviet Union, in the summer of 1984, to come to an agreement regarding ASAT weapon and other arms control negotiations (discussed in detail below) served to shift some of the burden of the "militarization" issue back to the Soviets.

Since their introduction at the United Nations, a good deal of attention has been given to the language of the two Soviet draft treaties. It is useful to examine these drafts since they provide valuable insights into how the Soviets have been thinking about arms control in space.

1981 Soviet Draft Treaty

The provisions of the 1981 and 1983 Soviet draft treaties reflect the major issues raised in the 1978-79 negotiations. Articles I and III, the operative provisions of the 1981 Soviet draft treaty, state:

> I. The member states undertake not to put into orbit . . . objects with weapons of any kind, . . . and not to deploy such weapons in outer space in any other way, including also on piloted space vessels of multiple use . . .
>
> III. Each member shall . . . not destroy, damage, or disturb the normal functioning and not to alter the flight trajectory of space vehicles of other member states where the latter have . . . been put into orbit in strict accordance with . . . Article I.

Because it prohibited only weapons stationed in orbit, the 1981 draft would not have restricted the testing, development, and deployment of ground-based or air-launched ASATs. Accordingly, the United States and the Soviet Union could have kept their current ASAT systems and also pursued future technologies such as ground-based or air-borne directed-energy weapons. The 1981 draft treaty would, however, have prohibited the development of space-based BMD systems.

According to Article III of the 1981 draft, parties would agree not to "destroy, damage, or disturb the normal functioning and not to alter the flight trajectory of space vehicles." Presumably, signatories to such a treaty could agree as to the meaning of the words "destroy," "damage," and "alter the flight trajectory." It is less clear that a quick consensus

[28]Ibid. However, on May 29, 1985, in an interview by a West German reporter in Geneva, Col. Gen. Nikolai Chervov, a senior department head on the Soviet General Staff, claimed that the U.S.S.R. had successfully developed a direct-ascent satellite interceptor similar to that tested by the United States in the early 1960s and operational until the mid-1970s.

could be reached on what would be inhibited under the injunction against "disturbing the normal functioning." Would this prohibit interference with ground stations or the use of electronic countermeasures such as jamming or spoofing? Had the treaty been negotiated, these issues would have certainly been the subject of great attention and possible compromise.

Article II of the 1981 proposed treaty states that space vehicles shall be used in "strict accordance with international law." This language seems to reflect the often stated Soviet belief that certain space activities—e.g., the operation of direct-broadcast satellites—are a violation of national sovereignty. However, under the terms of Article III, the only satellites that would be denied the treaty's protection would be objects carrying "weapons of any kind."

1983 Soviet Draft Treaty

In August 1983, when then Soviet Chairman Andropov met with several U.S. Senators he made the following statement:

> . . . (T)he Soviet Union considers it necessary to come to an agreement on a complete ban of tests and of deployment of any space-based weapons for striking targets on Earth, in the air and in space.
>
> Furthermore, we are ready, in the most radical way, to resolve the issue of anti-satellite weapons—to agree to eliminate anti-satellite systems already in existence and to ban creation of new ones.
>
> At the forthcoming session of the General Assembly of the United Nations, we will introduce proposals developed in detail on all these issues.[29]

As indicated by Chairman Andropov, on August 22, 1983, Soviet Foreign Minister Gromyko submitted a new draft treaty to the U.N. General Assembly.[30] The new draft was more comprehensive than the 1981 draft, and in particular went beyond it in calling for a ban on all testing of ASAT systems and the elimination of all existing ASAT systems (see appendix A). However, it also repeats many of the themes of the 1981 draft and of the 1978-79 negotiations.

Article 1 prohibits the "use or threat of force in outer space and the atmosphere and on the Earth through the utilization of . . . space objects" and the "use or threat of force against space objects." The "use or threat of force" language echoes the language of Article 2 of the U.N. Charter. Since by the terms of Article III of the Outer Space Treaty, the U.N. Charter already applies to space, it is unclear what this provision would add to existing international law. Article I does make it clear that: 1) space objects are not to be used to threaten objects in "outer space and the atmosphere and on the Earth"; and 2) space objects themselves are not to be threatened. This article would prohibit threats from space-based assets—e.g., ASAT or BMD weapons—and threats to space-based assets, whether from ground-, air-, sea-, or space-based systems.

Article 2 has five sections. Section 1 prohibits testing and deploying space-based weapons; this goes well beyond the simple "no-use" provision of the 1981 draft, which is repeated in section 2. Section 3 repeats the prohibition of the 1981 draft against destroying, damaging, disturbing the normal function or changing the flight trajectory of space objects of other states.

Under section 4 of Article 2, parties agree not to "test or create new anti-satellite systems and to destroy any anti-satellite systems that they may already have." There is no attempt in the treaty to define what constitutes an "anti-satellite system." Presumably, it would include both the proposed U.S. and current Soviet orbital interceptors. It is unclear how systems, such as the Soviet GALOSH ABM, which might have some ASAT capability, would be dealt with under the draft treaty.

Section 5 of Article 2 prohibits the "test or use of manned spacecraft for military, includ-

[29] "Dangerous Stalemate: Superpower Relations in Autumn 1983, A Report of a Delegation of Eight Senators to the Soviet Union," Senate Doc. 98-16, Sept. 22, 1983, p. 28.

[30] "Treaty on the Prohibition of the Use of Force in Outer Space and From Space Against the Earth," U.N. Doc. A/38/194, Aug. 22, 1983.

ing antisatellite, purposes." Because of the limitations that this would place on the U.S. Space Shuttle, it is unlikely that the United States would agree to such a provision. In any case, since the SALT agreements allow verification by "national technical means" (NTM) and the Shuttle is the launch vehicle for Government payloads—including satellites used for NTM—this provision would seem to conflict with current Soviet and U.S. agreements.

Congressional Interest in ASAT Arms Control and Executive Response

Following the introduction of the two Soviet draft treaties, the Reagan Administration expressed no interest in negotiating these or any other limitations on ASAT weapons. As time passed, Members of Congress in both Houses began to apply pressure on the Administration to halt ASAT testing and to begin negotiations with the Soviets. This pressure was applied most effectively in amendments to the Department of Defense authorization and appropriations bills.

The following resolutions concerning space weapons were introduced in the 97th Congress (1981-82).[31] None of them were reported out of committee or passed by either House:

- **Senate Resolution 129** (introduced by Pressler, R-S.Dak.) calling for resumption of ASAT limitations talks.
- **Senate Executive Resolution 7** (Pressler), calling for negotiation of a protocol to the 1967 Outer Space Treaty that would provide a complete and verifiable ban on ASAT development, testing, deployment, and use.
- **Senate Resolution 488** (Matsunaga, D-Hawaii), calling for talks with the Soviet Union concerning the possibility of establishing a weapons-free international space station.
- **House Joint Resolution 607** (Moakley, D-Mass. and 29 cosponsors), calling for the immediate negotiations for a ban on space weapons of any kind.

The number of bills and resolutions on space weapons introduced in the 98th Congress (1983-84) rose dramatically, with all but one dying in committee.[32] The exception was S.J.Res. 129 (Pressler and 28 others), which was reported favorably out of the Senate Foreign Relations Committee and significantly modified before being introduced, and later withdrawn, as an amendment to the fiscal year 1985 DOD authorization bill. A resolution suggesting that international cooperation in space be pursued as an alternative to the arms race was passed by Congress and signed into law (S.J.Res. 236; Public Law 98-562), but only after most of the language concerning the arms race had been deleted. The most important actions of the 98th Congress resulted from amendments to the DOD authorization and appropriation bills.

The Fiscal Year 1984 DOD Authorization Bill

While the House of Representatives was debating the fiscal year 1984 DOD authorization bill (H.R. 2969), two amendments concerning ASAT weapons were introduced. The first, introduced by Representative George Brown (D-Calif.), would have denied procurement funding for the ASAT weapon; the second, introduced by Representative Seiberling (D-Ohio), would have prohibited the flight testing of the ASAT until authorized by Congress. Both amendments were defeated.

In the Senate, an amendment introduced by Senator Tsongas and unanimously passed prohibited the expenditure of funds for tests of

[31]For a more detailed history of congressional activity during the 97th and 98th Congresses, see Marcia S. Smith, " 'Star Wars': Antisatellites and Space-Based BMD" (Washington, DC: Library of Congress, Congressional Research Service, Issue Brief IB81123, Nov. 26, 1984).

[32]Legislation not reported from committee in either House or Senate:

Legislation Opposed to Space Weapons: H.J. Res. 87 (Kastenmeier, D-Wis.); H.J. Res. 120 (Moakley, D-Mass. and 130 others); S.J.Res. 28 (Tsongas, D-Mass. and 8 others); H.J.Res. 523 (Dicks, D-Wash. and 58 others) and 524 (Dicks and 55 others); H.J.Res. 531 (Brown, D-Calif. and 96 others).

Legislation in Favor of Space Weapons: S.Res. 100 (Wallop, R-Wyo. and 14 others); S. 2021 (Armstrong, R-Colo.); H.Res. 215 (Whitehurst, R-Va.); H.Res. 259 (Bennett, D-Fla. and 17 others:); H.R. 3073 (Kramer, R-Colo. and 13 others)

Source: Smith, op. cit.; and Library of Congress SCORPIO database.

explosive or inert ASAT weapons (i.e., exempting directed-energy weapons) against objects in space, unless the President determined and certified to Congress that: 1) the United States was endeavoring in good faith to negotiate a treaty with the Soviet Union for a mutual, verifiable, and comprehensive ban on ASATs; and 2) that pending such an agreement, such tests were necessary for the national security.

The Fiscal Year 1984 DOD Appropriation Bill

Following a proposal by Representative McHugh, the House Appropriations Committee deleted the fiscal year 1984 ASAT procurement funds pending a report from the President on his policies regarding arms control in space. The Senate Appropriations Committee took no similar action, but during floor debate, the Senate adopted an amendment introduced by Senator Tsongas requiring the President to submit a report on the national security implications of the Strategic Defense Initiative.

In the course of the House and Senate conference on the appropriations bill, the conferees agree to provide $19.4 million for advance procurement for the ASAT program as proposed by the Senate, instead of no funds as proposed by the House. However, the conferees direct that these funds not be obligated or expended until 45 days following submission to Congress of a comprehensive report on U.S. policy on arms control. The appropriations bill, as amended, was passed by both Houses and signed into law (Public Law 98-212).

President Reagan's March 1984 Report on ASAT Arms Control

On March 31, 1984, the Reagan Administration issued its "Report to the Congress on U.S. Policy on ASAT Arms Control," thus satisfying the requirements of the fiscal year 1984 DOD appropriation bill. The report stated that the Administration was "studying a range of possible options for space arms control with a view to possible negotiations with the Soviets." However, it concluded that "no arrangements or agreements beyond those already governing military activities in outer space have been found to date that are judged to be in the overall interest of the United States and its Allies." The report stated that the search for effective ASAT arms control was impeded by the "difficulties of verification, diverse sources of threats to U.S. and Allied satellites and threats posed by Soviet targeting and reconnaissance satellites which undermine conventional and nuclear deterrence," and it emphasized the necessity for the development of a U.S. ASAT weapon.

The Fiscal Year 1985 DOD Authorization Bill

When considering the fiscal year 1985 DOD authorization bill, the House approved an amendment introduced by Representative George Brown. The amendment prohibited the use of funds for ASAT testing against objects in space until the President certified to Congress that the Soviet Union had conducted an ASAT test after the enactment of the bill. The House later accepted an amendment (offered by Representative Gore) to the Brown amendment which limited testing until the President certified that either the Soviet Union *or another* foreign power had conducted such a test.

The Senate Armed Services Committee recommended that the fiscal year 1984 authorization language restricting ASAT tests be relaxed to permit ASAT tests against objects in space provided only that the President certified such tests to be essential for pursuing arms control arrangements. During floor debate, the Senate adopted a compromise amendment offered by Senators Warner and Tsongas that prohibited spending funds for testing ASAT weapons against objects in space until the President certified to Congress:

- that the United States was endeavoring in good faith to negotiate a mutual and verifiable agreement with the strictest possible limitations on ASATs consistent with the national security interests of the United States;
- that pending agreement on such a ban, tests against objects in space were necessary to avert clear and irrevocable harm to the national security;

- that such testing will not constitute an irreversible step which will gravely impair prospects for negotiations; and
- that testing is fully consistent with U.S. obligations under the ABM Treaty.

With some minor changes, the Warner-Tsongas amendment was adopted in the House and Senate conference report.

The Fiscal Year 1985 DOD Appropriation Bill

Fiscal year 1985 appropriations for the Department of Defense were included in the Continuing Appropriation Bill (Public Law 98-473). The appropriation bill, as enacted, reflects the compromise reached on the DOD authorization bill. The only differences are that no tests against an object in space are permitted before March 1, 1985, or 15 days after the President submits the required certifications, whichever is later, and no more than three tests against objects in space are permitted in fiscal year 1985.

Current Activities in ASAT Arms Control

On June 29, 1984, about 3 months after the President's March 31 report had been released, the official Soviet news agency Tass announced that the Soviet Government had offered to start talks "to prevent the militarization of outer space."[33] "To provide favorable conditions for the achievement of agreement," Tass reported that the Soviet Union was prepared, "to impose on a reciprocal basis a moratorium on the tests and deployment of these weapons, starting with the date of the opening of the talks."[34] The Soviets suggested that such meetings should take place in Vienna in September 1984.

In response, the Reagan Administration stated that it was now ready to "discuss and seek agreement on feasible negotiating approaches which could lead to verifiable and effective limitations on antisatellite weapons."[35]

[33]"Soviet and U.S. Statements on Space Weapons Negotiations," *New York Times*, June 30, 1984, p. 4.
[34]Ibid.
[35]Ibid.

The Administration also announced that in addition to discussing space weapons it intended "to discuss and define mutually agreeable arrangements under which negotiations on the reduction of strategic and intermediate-range nuclear weapons can be resumed."[36] However, the Administration stressed that there were "no preconditions on the U.S. willingness" to discuss the entire range of arms control issues.

The Soviets objected to discussing strategic and intermediate-range missiles at the same time as space weapons. The Soviets proposed that the parties publish a joint public announcement that would define the purposes of the talks as being limited to the subject of space weapons and would endorse the concept of a moratorium on testing. The United States responded that it was prepared to talk about space weapons but that it was not prepared to agree to a moratorium.[37] The Soviets rejected the U.S. position and declared that it made the talks "impossible."[38] Although the U.S. Administration sent new messages modifying and "clarifying" its initial stand, these too were spurned by the Kremlin.

In the weeks following the initial exchanges there was little communication between the parties. The real argument seemed to be over which side would take the blame for refusing to negotiate. No meeting was held in September although both Washington and Moscow continued to express interest in arms control in space.

Six months later, on January 8, 1985, U.S. Secretary of State George Schultz and Soviet Foreign Minister Andrei Gromyko concluded 2 days of talks concerning the structure of future arms control negotiations. They jointly released a communique indicating that planning would commence on "the forthcoming U.S.-Soviet negotiations on nuclear and space arms" on "a complex of questions concerning space and nuclear arms, both strategic and in-

[36]Ibid.
[37]"Soviets Say U.S. Makes Talks on Space 'Impossible'," *The Washington Post*, July 28, 1984, p. A1.
[38]Ibid.

termediate range . . . The objective of the negotiations will be to work out effective agreements aimed at preventing an arms race in space" along with constraining terrestrial arms and increasing strategic stability.[39]

Negotiations between the United States and the Soviet Union began in Geneva in March 1985. Throughout these negotiations, the Soviet delegation has insisted that the termination of President Reagan's Strategic Defense Initiative is a necessary first step to any reduction in offensive arms. U.S. negotiators have, for their part, argued that advanced ballistic missile defense systems could provide a means by which both parties could safely negotiate deep reductions in their nuclear arsenals. As a result of this deadlock, both sides appear to remain far from agreement on antisatellite limitations.

On August 20, 1985, pursuant to the Fiscal Year 1985 DOD Authorization Act (discussed above), the President certified that the four requirements set out by Congress had been fulfilled.[40] President Reagan's decision to test the U.S. MV ASAT weapon against an object in space has reinvigorated congressional debate on the ASAT issue.

The Soviet response to the U.S. ASAT program has fluctuated. In 1983, President Andropov implied that the U.S.S.R. would rescind its self-imposed moratorium on ASAT testing if the United States *began* its ASAT test program. Then, in May 1985, in an interview with a West German reporter, Col. Gen. Nikolai Chervov, a senior department head of the Soviet General Staff, stated that the U.S.S.R. would rescind its moratorium if the United States *completed* testing the F-15 launched ASAT weapon. Most recently, the official Soviet news agency Tass, said that if the United States "holds tests of antisatellite weapons against a target in outer space," the Soviet Union "will consider itself free of its unilateral commitment not to place antisatellite weapons in space."[41]

[39]Statement text from *The Washington Post*, Jan. 9, 1985, p. A14.

[40]Presidential Determination No. 85-19 of August 20, 1985, *Federal Register*, vol. 50, No. 165, Aug. 26, 1985, pp. 34441-34443.

[41]*Washington Post,* Sept. 5, 1985, p. A-17.

Chapter 6

ASAT Arms Control: Options

Contents

Chapter 6
ASAT Arms Control: Options

INTRODUCTION

This chapter explores how various ASAT arms control provisions might affect the long-term national security interests of the United States. The interaction between these arms control provisions and the unilateral satellite survivability measures that the United States might pursue is discussed in greater detail in chapter 7. Four types of arms control are presented below: restrictions on ASAT testing, possession, use, and "rules of the road" for space. Each of these provisions is described and an assessment is given of its ability to protect U.S. space assets and contribute to other long-term U.S. goals. Potential conflicts between ASAT arms control and the development of military capabilities (e.g., the U.S. MV ASAT program and the Strategic Defense Initiative) are also examined.

The development of anti-satellite weapons poses a significant threat to the military satellites of both the United States and the Soviet Union. These military satellites, in turn, provide information and services which can be threatening to either side. Here lies the inherent difficulty in arriving at acceptable ASAT arms control agreements—given the choice, the United States would like to protect its own satellites while eliminating any military threat posed by Soviet satellites.[1] Since such a one-sided advantage is not possible, it is reasonable to examine whether there are mutual restraints that would contribute to national security and protect U.S. satellites, yet allow an adequate response to the threat posed by Soviet satellites.

The debate over ASAT arms control contains many familiar themes: To what extent can the the United States monitor Soviet compliance? Do the Soviets intend to cheat, and if so, can they? What recourse would the United States have if faced with either clear or ambiguous Soviet violations? Is the United States better off pursuing arms control, technological superiority, or some combination of both? What will be the response of our allies to new development programs or arms control proposals? These issues are discussed below, both in the context of specific arms control provisions and in a more general discussion of monitoring treaty compliance.

Most of the provisions discussed in this chapter would require the United States and the Soviet Union to enter into a bilateral agreement to limit ASAT weapons development. As a result of the technologies involved, their theater of operation, and the closed nature of Soviet society, it is unlikely that the United States could monitor Soviet compliance with complete certainty. The United States can know only part of what the Soviet Union does and little or nothing about what it intends; therefore, any arms control agreement involves some degree of risk. **For the purposes of this discussion, the value or danger of a particular arms control provision is measured by its likely impact on U.S. national security after allowance is made for possible covert Soviet violations. In other words, given the risks of entering into an agreement with the Soviets, are we better off with or without a particular provision?**[2]

This chapter focuses primarily on bilateral treaties of unlimited duration. Other arrangements for ASAT constraints, such as multi-

[1]Not all Soviet military satellites threaten the United States. Presumably, the United States would like the Soviet Union to retain some reconnaissance and early warning satellites since these satellites contribute to stability by allowing verification of arms control agreements and by assuring the Soviets that they are not under nuclear attack.

[2]It is important to note that "risk," as it is used here, does not imply merely the probability that the Soviets can or would violate a particular provision of an ASAT agreement. Rather, risk signifies both the probability that the Soviets would violate the agreement and the threat to U.S. national security that would likely result from such a Soviet violation.

lateral agreements, joint declarations, executive agreements or even unilateral declarations, might also be in the national interest. However, this report is limited exclusively to bilateral agreements for the sake of simplicity and primarily to formal treaties of unlimited duration because they are the hardest to obtain and have the most lasting effect on national policy and programs.

PROVISIONS RESTRICTING ASAT TESTING

An agreement that established limits on testing could be a useful means by which to prevent the development of reliable, dedicated ASAT systems. The effectiveness of test restrictions is assumed to derive from the naturally conservative nature of military planners. Many informed observers believe that, except when forced by necessity, Soviet military planners would be reluctant to rely on systems that have not been tested near their full capabilities, particularly in situations where the stakes are high, second chances may not come, and the penalties for failure could be severe.

A test ban would prevent the testing which would increase confidence in new ASAT weapons or, at minimum, would force testing to be done covertly, under less than optimal conditions. In either case, the result would be to erode the confidence that an ASAT system would work as planned, when needed.

There are several ways to frame a test ban. The most comprehensive would be a ban on all "testing in an ASAT mode." For the purposes of this discussion, "testing in the ASAT mode" would include tests of ground-, sea-, air-, or space-based systems against targets in space or against points in space. Testing of ASAT systems or components on the ground would not be prohibited. Such an approach would avoid both the necessity of defining an ASAT weapon and of restricting systems that, although not designed as ASATs, might have some inherent ASAT capability. For example, it is possible that the Soviet GALOSH ABM system might have some ASAT capability. An agreement that banned all "testing in an ASAT mode" would not require the United States and the Soviet Union to agree whether GALOSH *was* an ASAT or if it *could function as* an ASAT, but would simply ban the testing of this system *as* an ASAT.

More limited "no-test" agreements could also be used to inhibit the development of specific types of ASATs or to place restrictions on certain types of testing. Such a treaty might be used to ban the testing of only ASAT weapons that would be based in space. Alternatively, it might be used to ban the testing of specific space-based ASAT weapons (e.g., directed-energy weapons or space mines) thought to be particularly destabilizing. Such a ban might also limit ASAT testing to low altitudes to protect critical early warning and communication satellites that are in higher orbits.

All of these examples of limited test bans could be further modified by agreed limitations on allowable numbers of tests. For example, the United States and the Soviet Union might agree to limit themselves to only 10 tests over the next 5 years, or to a set number (either constant or declining) of tests per year for the duration of the agreement.

Monitoring Compliance With a Test Ban

In past bilateral agreements between the United States and the Soviet Union, the Soviet Union has tended to take advantage of treaty ambiguities and to engaged in activities that—although sometimes difficult to characterize—bordered on treaty violation.[3] It is prudent, therefore, to assume that should

[3]There is reason to believe that the Krasnoyarsk radar, when complete and ready for operation, will violate the ABM Treaty.

Photo credit: U.S. Air Force

F-15 launched, MV anti-satellite weapon currently under development in the United States.

an ASAT test ban be negotiated, the Soviets would comply only to the extent that the United States was able to verify its compliance. This being the case, it is important to examine some of the problems associated with monitoring the wide range of Soviet activities that might be related to ASAT weapon development.

Scope of Monitoring Task

One barrier to verifying compliance with a test ban is the enormous volume of space where illicit activities might be conducted. Verification of compliance with a SALT or START arms control agreement involves inspection of a number of areas in the Soviet Union or its immediate airspace. This area, although large, is relatively well determined and is amenable to close inspection by space-based photographic reconnaissance satellites. The region where space activities must be monitored starts at altitudes of about 100 km and can range well past geosynchronous orbit at 36,000 km. In addition, advanced ground-based ASATs could be located anywhere in the Soviet Union and air-based ASATs might even operate from non-Soviet airfields.

Although the volume of space is indeed large, space-based ASAT activities must start on the ground. Relevant ground sites, including launch facilities, can be observed by an extensive array of U.S. monitoring facilities; launches of ICBMs and similar vehicles from Soviet territory can be detected. To some extent, the problems created by the large volume of space are offset by the fact that space is transparent and accessible to monitoring. Current weaknesses in ground-based surveillance systems can be mitigated by putting surveillance systems into space.

If the Soviets were to develop air-based, ground-based, or "pop-up" directed-energy weapons, these would require extensive testing. It is likely that some portion of this testing could be conducted out of the sight of (e.g., indoors or underground) U.S. monitoring assets. However, full development would probably require some in-space testing against targets. Possible targets could, in principle, be monitored to see if they are being illuminated by strong lasers, are giving off gases, are being unexpectedly accelerated, or are emitting unusual signals. Air- and ground-based systems might be detectable by national technical means. Nonnuclear, space-based systems would be quite large and might emit detectable amounts of hydrogen fluoride or other gases.

Problems of Discrimination

Verifying treaty compliance is complicated by the growing number and variety of Soviet space launches. Although the launch rate may decrease in the future as the Soviets develop longer lived satellites, space surveillance requires a body of experience with each additional type of satellite in order to classify its function and discriminate between unusual activity and routine behavior. The functional characteristics distinguishing ASAT weapons, such as space mines, from other satellites may not be readily observable. Some occurrences might have multiple interpretations. For example, a satellite fragmenting in orbit could be accidental, the test of a self-destruct mechanism (either to avoid capture or to prevent large components from falling back to Earth), or the test of a space mine.[4] All national technical means have imperfect discrimination, and the physical differences between permitted and prohibited satellites may be small.

Although the annual number of Soviet launches is large, the number of new satellites or satellites engaged in "unusual activities" is relatively small. Even if U.S. national technical means of verification could not *by direct observation* distinguish between space mines and normal satellites, other indicators, such as orbital parameters, proximity to other—particularly U.S.—satellites, and other sources of intelligence might supply the needed information. If in addition to a test ban the treaty also included some mechanism for resolving ambiguities—e.g., the Standing Consultative Committee established in SALT I—the problem might be further resolved.

Assuming that the difficulties associated with deliberate ASAT systems were resolved, it would still be necessary to reach some agreement concerning tests of advanced, ground-based, BMD systems. Should conventional ground-launched BMD systems be developed —similar to the system recently demonstated in the U.S. Homing Overlay Experiment (HOE) —they may have some limited ASAT capability.

Covert Development

There are numerous ways for the Soviets to engage in covert ASAT development. It is possible that space mine or orbital interceptor tests could be masked as legitimate rendezvous operations or satellite repair missions. ASAT weapons might be directed against points in space or space debris, thereby obviating the need for recognizable target satellites. ASAT vehicles or their targets could be instrumented to store test data for broadcast over the Soviet Union or for deorbit in a reentry capsule, thereby preventing the United States from intercepting test information. Nuclear-armed ICBMs or ABM launchers such as the Soviet GALOSH might also be tested (though not detonated) in a manner which would be difficult to characterize. Relatively low-powered lasers capable of blinding satellite sensors are already available and

[4]One could, of course, ban all deliberate explosions in space. If such a ban were made part of a more general test ban, there would be less ambiguity to resolve.

Photo credit: U.S. Department of Defense

Artist's conception of the nonnuclear ABM inteceptor recently tested in the Homing Overlay Experiment (HOE). The current Soviet GALOSH nuclear ABM interceptor or future, nonnuclear systems based on HOE technology could complicate the process of monitoring an ASAT weapon test ban.

might be tested without being clearly identified as being ASATs.

It is, on the other hand, possible to exaggerate the threat posed by covert development. The United States is sufficiently familiar with the operational characteristics of the current generation of Soviet ASAT interceptors to make its covert testing unlikely. The development of a new system would require an extensive testing program, some portion of which we would almost certainly identify. New or unusual orbiting vehicles would be noticed, especially maneuvering ones. Monitoring equipment could be developed that would detect the laser illumination of Soviet satellites, and which could aid in monitoring Soviet directed-energy facilities. (See table 6-1, below). Soviet efforts to hide covert testing might serve to narrow down the regions where the United States needs to concentrate its verification efforts. In any case, it is likely that an ASAT test limitation agreement would provide the means by which parties could inquire about suspicious activities.

Utility of an ASAT Test Ban

Considering both the limitations of U.S. monitoring capabilities and the possible ill-intentions of the Soviet Union, what then is the value of an ASAT test ban? To answer this question completely, one must examine the specific test bans being considered in combination with possible technical countermeasures (this is done in chapter 7). However, some preliminary generalizations can be helpful.

Some of the satellites that the United States relies on for critical information are now vulnerable and few in number. With respect to these specific systems, a small degree of Soviet cheating under a test ban agreement might have a significant effect on U.S. security. On the other hand, the United States has been quite successful at monitoring past Soviet space activities and the deployment of more capable monitoring assets—e.g., space-based surveillance systems—could substantially aid the process of treaty monitoring.

It is important to note that modest satellite survivability measures would reduce the risk posed by current ASAT weapons and could do much to reduce the risk posed by covert weapons development. In the absence of an agreement limiting ASAT weapon development, the United States must still monitor Soviet activities but modest survivability measures might not be effective. Without limitations, advanced ASATs would pose a greater risk to a larger number of satellites and failure to effectively monitor these advanced ASATs could create a significant danger to U.S. national security.

Comprehensive Test Ban

A ban which prohibited all testing "in the ASAT mode" would severely reduce the likelihood that the Soviet Union could successfully develop advanced, highly capable ASAT weap-

Table 6-1.—Sensor Technology for Compliance Monitoring

Prohibitable action	Observables	Sensors
ASAT attack:		Attack sensors:
KEW[a] impact	acceleration	accelerometers
Pulsed HEL[b] irradiation	acceleration	accelerometers
Continuous HEL irradiation	heating	thermistors
NPB[c] irradiation	ionization	ionization detectors
Keep-out zone penetration	position of thermal radiation source (ASAT)	space-based LWIR[d] thermal imager[e f]
Interception test	positions of thermal radiation sources (ASAT and target)	space-based LWIR thermal imager[e f]
NPB ASAT operation	thermal radiation from ASAT	space-based LWIR thermal imager[e f]
HEL ASAT operation	thermal radiation from ASAT	space-based LWIR thermal imager[e f]
Irradiation of target with NPB	gamma radiation from target	gamma-ray spectrometer[g]
Irradiation of target with pulsed HEL	thermal radiation from target	space-based LWIR thermal imager
Irradiation of target with pulsed HEL	reflected radiation from target	space-based multispectral imager
Irradiation of target with continuous HEL	position of thermal radiation source (target)	space-based LWIR thermal imager
Irradiation of target with continuous HEL	reflected radiation from target	space-based multispectral imager
Nuclear explosive aboard satellite	gamma radiation from fissile or fusile nuclei activated by cosmic radiation or by particle beams	gamma-ray spectrometer (and optional particle beam generator)

[a]Kinetic-energy weapon.
[b]High-energy laser.
[c]Neutral particle beam.
[d]Long-wavelength infrared.
[e]The LWIR telescope on the Infrared Astronomical Satellite (IRAS) exemplifies demonstrated space-based thermal imager technology; this instrument is described in *Astrophysical Journal,* 278 (1, Pt. 2); L1-L85, Mar. 1, 1984 (Special Issue on the Infrared Astronomical Satellite).
[f]Radar and passive radio direction-finding methods could also be useful for tracking, if hiding measures are not employed by the penetrating spacecraft. LWIR tracking is emphasized here because it is difficult to counter by such measures.
[g]A target irradiated by a high-energy neutral particle beam will emit gamma rays, neutrons, and other observable particles, just as it will, at a slower rate, when bombarded by natural cosmic rays. These gamma rays could be detected by a gamma-ray spectrometer such as those which have been carried by Soviet Venusian and lunar landers and by U.S. NASA Ranger and Apollo spacecraft. (NASA report SP-387, pp. 3-20.)

ons. The categories of weapons eliminated might included space mines capable of "shadowing" valuable military assets in any orbit, or directed-energy weapons with kill radii of hundreds to thousands of kilometers. In the absence of an agreement limiting the development of these weapons, each side might seek continually more effective means to attack threatening satellites and to defend valuable assets. This could result in a potentially destabilizing arms race in space. The "instantaneous kill" ability of the most advanced ASATs would be destabilizing in a crisis, since each side would have the incentive to "shoot first" or else risk the loss of its space assets.

A comprehensive test ban would require both the United States and the Soviet Union to cease testing their current generation of ASAT weapons. The Soviet ASAT is already considered operational. Assuming the United States also had an operational ASAT when the agreement entered into force, each side's existing system would pose some threat to the other side. Over time, a comprehensive test ban would gradually erode each side's confidence in its respective weapons, thereby reducing the possibility of their use. If a test ban were combined with additional restrictions on possession or deployment, this might result in somewhat greater security.

A comprehensive test ban would be less effective at reducing the threat posed by weapon systems with "inherent" ASAT capability. ICBMs, SLBMs, and ABM interceptors with nuclear payloads are examples of systems with inherent ASAT capability. Although these systems lack the kind of guidance necessary to intercept a satellite with great precision, the long-range destructiveness of their nuclear payloads makes them potentially effective ASATs. However, some of the ASAT threat posed by nuclear weapons is offset by their very nature. The collateral physical, political, and military consequences of using nuclear

ICBMs or ABMs as ASATs could well deter their use in most conflicts short of a terrestrial nuclear war.

The Shuttle's recent success at retrieving and refurbishing satellites strongly suggests the ASAT potential of future maneuverable spacecraft. However, the range, effectiveness and reaction time of even advanced maneuverable systems would be substantially less than that of future dedicated ASATs. Although the development of maneuverable spacecraft would not be inhibited by most ASAT testing limitations, some limits could be placed on operating them in an ASAT mode.

The Soviet draft treaties and the 1983 unilateral Soviet moratorium on ASAT testing suggest that the Soviets would be willing to negotiate about a comprehensive ASAT test ban. To date, the U.S. response to Soviet suggestions has been to point out that since the Soviets have an "operational" ASAT and the U.S. testing program has just begun, a comprehensive test ban would prevent the United States from ever having a reliable interceptor ASAT and would increase the threat posed by a Soviet "breakout." Nonetheless, the United States has continued to express interest in ASAT negotiations and has not ruled out the possibility that it would agree to some kind of test limitations.

Limited Test Bans

Should a comprehensive test ban be considered undesirable or nonnegotiable, it might still be worthwhile to limit testing to the current generation of ASATs or to ASATs only capable of attacking satellites in low-Earth orbit. A ban which limited each side to testing its current ASAT would have three advantages: 1) a ban on testing new types of ASATs would reduce the likelihood that advanced ASATs, such as space mines or space-based directed-energy weapons, would be developed; 2) the threat to critical early warning and communication satellites would be diminished; and 3) the United States would retain the ability to negate Soviet low-orbiting, targeting, and data collection satellites judged to pose a threat to U.S. surface forces.

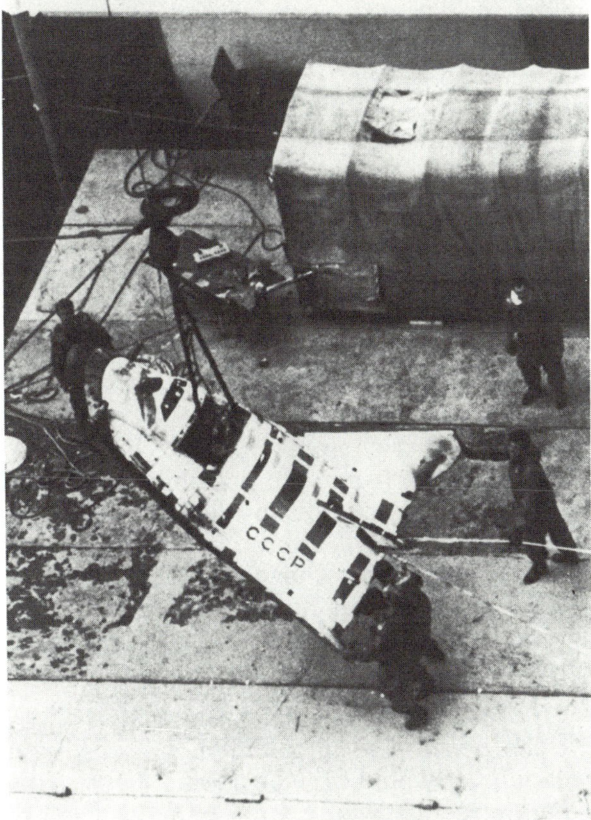

Photo credit: U.S. Department of Defense

Ocean recovery of what is believed to be an unmanned scale model of a new Soviet space plane. The development of maneuverable spacecraft would not be inhibited by most ASAT testing limitations, but some restrictions might be placed on operating such spacecraft "in an ASAT mode."

If a limited test ban could restrict each side to its current, low-orbit, ASAT capability, it would, in effect, create high-altitude "no attack zones."[5] This might encourage adversaries to move some Earth monitoring space assets into those zones. The development of high-altitude data collection systems would re-

[5] If advanced directed-energy weapons with kill radii of thousands of kilometers are developed, such "no attack zones" might be meaningless.

quire considerable time and expense. For the next decade and perhaps beyond nations would probably be forced to operate their current low-altitude systems.

Although it is possible that some reconnaissance satellites might be able to function from higher orbits with some degradation of performance, radar satellites—useful in tracking surface ships—would have substantially greater difficulty. Current systems employ active radar, which means that the strength of the return signal decreases as the fourth power of the range to the target. The substantial increase in range necessary to take advantage of a high-altitude "no-attack" zone would severely degrade the performance of current systems. It is possible that, over time, improvements in technology could solve the problems created by the increase in range. Nonetheless, by the time this occurred new ECM and EOCM capabilities might also be developed that could help to negate systems taking advantage of high-altitude "sanctuaries."

PROVISIONS RESTRICTING ASAT POSSESSION OR DEPLOYMENT

An agreement which sought to restrict the possession or deployment of ASAT weapons could be either comprehensive or limited. A *comprehensive ban* might prohibit the possession or deployment of any deliberate "anti-satellite system." A *limited ban*, on the other hand, might allow the possession of some ASAT weapons but not others, or establish limitations on allowable ASAT capabilities or on the number and kind of deployments.

In order to establish a comprehensive ban on the possession or deployment of ASAT weapons, it would first be necessary to come to an agreement as to what exactly was being banned. As explained above, the existence of systems that have an inherent ability to attack satellites complicates the process of eliminating all ASAT *capability*. A ban on all systems with ASAT capabilities would be so broad as to be unworkable since it would include ICBMs, SLBMs, ABMs, and maneuverable spacecraft such as the Shuttle. On the other hand, a ban on deliberate ASAT systems alone might allow the development of non-ASAT systems having sophisticated ASAT capabilities. For this reason, the most effective comprehensive ban on possession and deployment would probably be one which was also accompanied by a prohibition on testing non-ASAT systems in an ASAT mode.

Many types of limited-possession regimes can be imagined. The United States and the Soviet Union might decide to keep the ASATs they are currently testing, but prohibit the possession or deployment of more advanced systems. Alternatively, each side might be allowed to have one designated system in addition to the one they are currently testing; the capabilities of this additional system might or might not be limited (e.g., low-Earth orbit capability only). Still another regime might limit the parties to ground-based ASAT weapons and ban possession of weapons that would be based in space.

In all of these limited-possession regimes additional restrictions on the number and location of allowable ASAT deployments could be added.

Monitoring Compliance With Limitations on Possession and Deployment

A comprehensive ban on the possession or deployment of the existing Soviet ASAT weapon would raise some important monitoring problems. The launch vehicle for the Soviet ASAT is used in several other non-ASAT roles. These launchers will remain available even if the Soviet ASAT weapon is banned.

Since the ASAT weapon itself is small, it would be difficult for the United States to verify with high confidence that the Soviets had not clandestinely retained a stockpile.[6]

A limited possession ban that granted either side the right to possess and deploy the ASAT weapon it was currently testing would raise fewer verification problems. Significant Soviet cheating would involve covertly testing and developing a new and unproven advanced ASAT weapon, rather than simply hiding an existing system. As discussed above, it is likely that such a development program would include some testing requirements that were observable.

Given the small size of the current Soviet ASAT weapon, restrictions on the number of ASAT weapons that could be deployed at each launch site would be difficult to monitor in the absence of onsite inspection. Even onsite inspection would not provide complete security, since ASATs could be covertly stored and easily transferred to the launch area when needed. The United States would have higher confidence at monitoring restrictions on the allowable number of launch sites. Restriction on launch facilities would increase the time between ASAT launches and decrease the probability of sudden, multiple kills. A combination of restrictions on both the allowable number of launch sites and on the number of ASAT weapons that could be stored at each site could reduce the likelihood of a surprise attack or, at minimum, reduce the effect of such an attack.

Utility of Limitations on Possession and Deployment

A comprehensive ban on ASAT possession and deployment is complicated by: 1) the existence of the Soviet, and, in the near future, the U.S. ASAT weapons; 2) the fact that the lack of possession or deployment could not be monitored with high confidence; and 3) the fear that a ban on possession and deployment—even if monitored with high confidence—would not eliminate the knowledge of how to build these systems, and that the forces might be reconstituted at some time in the future.

Balancing these three concerns is the understanding that for the Soviets to retain some ASAT weapons in violation of a possession or deployment ban would not in itself be threatening—they must also be able to use these ASAT weapons in a way that is militarily significant. Differences of opinion exist as to the military significance of minor violations of a ban on possession and deployment. Some argue that the Soviets must be able to launch a sufficient number of ASAT weapons with sufficient rapidity to gain an important military advantage. To do this, the ASAT weapons and launch vehicles would have to be pre-mated and held in readiness, activities that would probably be observable. Others believe that the Soviets would not have to launch a mass ASAT attack in order to gain important military advantages. They point out that in some limited war scenarios, destroying a very small number of critical satellites could have grave consequences. Therefore, there might not be a need for a large number of observable, pre-mated ASAT weapons and launchers.

Assuming the United States did have advance notice of Soviet ASAT activities, it could respond through diplomatic channels or through a Standing Consultative Committee, if established. Even short-term notice of intent to use ASAT weapons would allow the United States time to maneuver its satellites or take other appropriate action.

The fact that every element of an agreement cannot be monitored with high confidence does not necessarily mean it has no value. It is extremely difficult to monitor the "no nuclear weapons in space" provision of the Outer Space Treaty and yet the United States continues to adhere to it. Presumably, this is because the benefits of the treaty outweigh the risk posed by potential Soviet cheating.

[6]The U.S. ASAT weapon currently under development is quite small. However, the Soviet monitoring task is easier because the U.S. ASAT weapon requires large and distinctive support equipment and because significant expenditures for military facilities, personnel, and weapons procurement would be revealed in the annual authorization and appropriation process of Congress or by the popular press.

Even if a ban on possession and deployment could not be monitored with high confidence it would, at minimum, oblige the Soviets to conduct future ASAT weapons tests covertly. This would complicate maintenance of the current system, make upgrades difficult and advanced ASAT development less likely. The combination of these effects would make U.S. satellite survivability programs more effective and might discourage the use of ASAT weapons.

Space systems with inherent rather than intentional ASAT capabilities would be difficult to restrict by a comprehensive ban on possession or deployment. Nonetheless, such systems pose only a modest threat to critical U.S. assets. Those systems which employ nuclear warheads (e.g., ICBMs, SLBMs, ABMs) might only be used in a terrestrial nuclear war or at the risk of precipitating one. They would also risk damage to the attacker's own satellites. Future maneuverable spacecraft, although capable of some limited ASAT activity, would not be able to provide the rapid, multiple-kill capability likely to be obtained from future dedicated ASAT systems, and are therefore a considerably lesser threat.

A regime which banned only *deliberate ASAT systems* and disregarded systems with some *inherent ASAT capability* would still be useful in as much as it would reduce the threat of the most highly capable and destabilizing future ASATs systems. Nonetheless, a ban on the possession and deployment of ASAT systems would probably be most valuable if accompanied by a prohibition against the testing of non-ASAT systems in an ASAT mode.

The 1983 Soviet draft treaty contained an example of a comprehensive ban on possession. Article 2(4) of the draft treaty would have required that parties undertake, "Not to test or create new anti-satellite systems and to destroy any anti-satellite systems that they may already have." Given the past statements of Soviet officials and the draft treaties proposed by the Soviet Union, it is likely that the Soviets would be willing to negotiate a comprehensive ban on *possession*. It is less clear whether they would be willing to negotiate some form of limited ban. If their primary concern is protecting all of their space assets, then a limited ban might not be acceptable. If, on the other hand, their purpose in negotiating any ban is to limit the development of more effective ASATs or space-based BMD technologies, then there might be some partial bans that they would find acceptable.

PROVISIONS RESTRICTING ASAT USE

Perhaps the least complicated ASAT agreement would be one that prohibited hostile acts against satellites. Such an agreement would probably not attempt to limit specific ASAT systems, but would instead prohibit the *use* of all ASAT *capabilities.* Although a "no use" agreement could not strictly be considered "arms control," in as much as both parties would be free to develop and deploy any number and kind of advanced ASAT system, it might usefully define what constituted a "hostile act" against a satellite. This agreed definition of "hostile act" might serve to avoid some future conflict brought about by a confusion of intentions. It would also establish a satellite attack as an unambiguous warning of further aggressive intent.

Although a "no use" agreement might not substantially reduce the threat of ASAT attack, it could serve as a useful component of other, broader arms control agreements. The definition of prohibited acts that might reasonably result from the negotiation of a "no use" agreement could lead to a clearer understanding of the systems capable of performing those acts. This, in turn, might assist in the negotiation of agreements that prohibited the testing, possession, or deployment of ASAT weapons.

Monitoring Compliance With a "No Use" Agreement

Compliance with a "no-use" agreement would be relatively easy to monitor. This is particularly true for the current generation of ASAT weapons. The monitoring task would become slightly more difficult if the Soviets were to follow the U.S. example and develop an air-launched interceptor. This would allow them to launch an ASAT attack outside of the Soviet Union—perhaps even from the western hemisphere if the appropriate facilities were installed in Cuba.

If it were possible to covertly develop ground-based directed-energy facilities or more flexible air-based facilities, these might be used to damage the sensors of a U.S. satellite in such a manner as to mimic an equipment malfunction. This is particularly true when the object of an attack is not to destroy the satellite, but rather to "blind" or "dazzle" delicate sensors.

On the other hand, the effective use of on-board monitoring equipment could substantially reduce this threat. To a limited degree, satellites now have some on-board "state-of-health" monitoring equipment. It is possible to augment these sensors to determine whether a failure is due to an internal flaw or whether it has been externally induced. These sensors might, for example, measure incident laser light, rises in temperature, or sudden accelerations. The inclusion of "state-of-health" monitoring equipment on satellites combined with a future space-based surveillance system could provide the necessary ingredients to verify a "no use" treaty with high confidence.

Utility of a "No Use" Agreement

In order to judge the utility of a "no use" agreement it is first necessary to understand what such an agreement could and could not accomplish. Even if a "no use" agreement could be monitored with very high confidence, in an environment of unconstrained ASAT development, a "no use" treaty might make only a small contribution to protecting U.S. space assets. Should nations eventually possess directed-energy weapons or space mines with an instantaneous and multiple kill capability, there will be significant advantages to being a first user of these weapons. Nations may find themselves in the position of having to use or lose their offensive space-based assets. If the measure of effectiveness of a "no use" agreement is how well it protects U.S. satellites in an otherwise unconstrained environment, then one would have to conclude such an agreement was of limited value.

Although the U.N. Charter and the Outer Space Treaty both *implicitly* prohibit hostile acts against the satellites of other countries, there may be some value to obtaining a formal agreement that such hostile interference is a violation of international law and potentially a cause of war. A "no use" ASAT treaty would, like the Geneva protocol on use of poisonous gases, establish more clearly the "law of civilized nations." Codifying what is already implicit in international law might serve to inhibit the willingness of nations to attack satellites in a crisis before hostilities have broken out on Earth and perhaps even for some period of low intensity conflict.

Although a "no use" agreement would not, in itself, substantially reduce the risk or the effect of an ASAT attack, it would serve as a useful addition to other, more comprehensive, ASAT limitations. For example, an agreement that restricted ASAT testing would benefit from the clear statement that hostile acts against satellites were forbidden. Such an agreement would assist in developing the principle that the goal of ASAT limitations was to protect space assets and to keep space from becoming an area of unrestrained conflict and not simply to control the development of one or another class of offensive weapons.

It is likely that some type of "no use" agreement would be acceptable to the Soviet Union. It would, of course, be necessary to clearly define what constituted "use" under the agreement. In their 1983 draft treaty, the Soviets defined "use" as meaning "to destroy, damage, disturb the normal functioning or change the flight trajectory." Although the United States might agree in principle with the intent of such a provision, it is unlikely that it would accept this exact language. The Soviet phrase "disturb the normal functioning" might be interpreted as prohibiting the use of electronic countermeasures, and this interpretation would probably be unacceptable to the United States.

The 1983 draft treaty of the Union of Concerned Scientists (UCS) is similar to the Soviet draft except the phrase "disturb the normal functioning" is replaced by "render inoperable." The UCS language is, from a U.S. perspective, probably more acceptable, since it would seem to cover only actions that harm the satellite and not those that make its job harder. In the absence of formal negotiations, it is impossible to assess Soviet intentions or willingness to compromise on this point.

PROVISIONS RESTRICTING SPACECRAFT OPERATION AND ORBITS

Whether or not the United States and the Soviet Union agree to restrict ASAT weapons or capabilities it might be useful to negotiate a set of "rules of the road" for military space operations. These rules could serve the general purpose of reducing confusion and encouraging the orderly use of space, or they could be designed specifically to aid in the defense of space assets. Examples of general rules might include agreed limits on minimum separation distance between satellites or restrictions on very low-orbit overflight by manned or unmanned spacecraft. These general rules might also be used to establish new, stringent requirements for advance notice of launch activities. Specific rules for space defense might include declared and possibly defended "keep-out zones," grants or restrictions on the rights of inspection, and limitations on high-velocity fly-bys or trailing. It might also be desirable to establish a means by which to obtain timely information and consult concerning ambiguous or threatening activities.

Precedents can be found for each of the general rules suggested above. The clearest example of international acceptance of "rules of the road" is the 1960 multilateral agreement on "International Regulations for Preventing Collisions at Sea."[7] This agreement established the rule of international conduct on the high seas and provided the basis for the 1972 Soviet-U.S. treaty on the "Prevention of Incidents On and Over the High Seas."[8] In this latter agreement, the United States and the Soviet Union established more specific rules for the operation of their respective warships.

In the civilian communications field, nations have agreed to work with the International Telecommunication Union (ITU) to develop rules to insure orderly use of the geostationary orbit and the radiofrequency spectrum. Those nations possessing military satellites might wish to establish an organization, or more limited working groups, to develop similar technical rules of conduct for military space activities.

The Chicago Convention of 1945 established the fundamental principle of state sovereignty over territorial airspace.[9] The 1967 Outer Space Treaty established the equally important principle that space should be freely available for the use and exploitation of all nations.[10] Since the beginning of the space age, nations have wrestled with, but failed to re-

[7]33 U.S.C. 1051; T.I.A.S. 5813

[8]23 U.S.T. 1168; T.I.A.S. 7379.

[9]"Convention on International Civil Aviation" (Chicago 1947), 61 Stat. 1180, 15 U.N.T.S. 295, T.I.A.S. 1591

[10]"Treaty on Principles Governing the Activities of States in the Exploration and Use of Outer Space, Including the Moon and Other Celestial Bodies, Jan. 27, 1967 (Article I) 18 U.S.T. 2410, T.I.A.S. 6347.

solve the question of how to characterize the boundary between airspace and outer space.[11] As one author has observed, the very important difference between these two regimes is that "states shoot at aircraft not authorized to be in their airspace; they do not shoot at satellites passing over that airspace."[12] This distinction will become increasingly harder to make as maneuverable space vehicles become more capable. Will nations continue to allow overflight of their territory by military spacecraft which are also capable of aerodynamic flight? Practical and internationally consistent "rules of the road" may be necessary to resolve this problem.

Article IV of the "Convention on Registration of Objects Launched into Outer Space" currently requires signatories to supply the Secretary-General of the U.N. with information concerning its space objects and launches. However, since the Convention requires only that the signatories supply this information "as soon as practicable," it is of little use in clarifying ambiguous activities in a timely manner. Article 4 of the "Agreement on Measures to Reduce the Risk of Outbreak of Nuclear War Between the United States of America and the Union of Soviet Socialist Republics" also requires that "each Party . . . notify the other Party in advance of any planned missile launches if such launches will extend beyond its national territory in the direction of the other Party." Unfortunately, since this article does not apply to space launch vehicles it is of little use as a means to protect space assets. As space launches become more numerous and varied, an agreement providing for *timely* notification of launch and information on the characteristics of the vehicle may be essential to avoid crisis through confusion.

International law recognizes that the concept of sovereignty extends to more than a nation's land mass. For example, a country's territorial waters and contiguous airspace are considered to be sovereign and defendable elements of that country. Extrapolating from this concept, one method for protecting satellites would be to negotiate or declare "keep-out zones" around the most critical space assets. The agreement or declaration of these "keep-out zones" might also include the right to defend these zones once declared. Precedent for the concept of "keep-out zones" can be found in the history of the SALT negotiations pertaining to submarines. During the course of these negotiations a number of proposals were discussed such as, "no-submarine zones" which would have prohibited missile-carrying submarines from operating in certain parts of the ocean, and "no-ASW zones" (anti-submarine warfare) that would have allowed the unhindered operation of submarines in select areas to ensure that reliable retaliatory forces would exist to deter a possible first strike.

Photo credit: National Aeronautics and Space Administration

Artist's conception of the U.S. Space Shuttle servicing a Space Station. As commercial and scientific space activities increase, internationally accepted "rules of the road" may be necessary to ensure that both military and nonmilitary space activities are conducted in a safe and orderly manner.

[11] This question has been considered almost annually in the U.N. Committee on the Peaceful Uses of Outer Space but has yet to be resolved. The position of the United States has been that such a delimitation has not been necessary and, indeed, might impede beneficial space activities.

[12] "Anti-Satellite Weapons, Arms Control Options, and the Military Use of Space," William J. Durch, U.S. Arms Control and Disarmament Agency, contract No. AC3PC103, July 1984, p. 3.

Negotiated or declared "keep-out zones" would have to be reconciled with Article II of the 1967 Outer Space Treaty which states, "outer Space . . . is not subject to national appropriation by claim of sovereignty, by means of use or occupation, or by any other means." "Keep-out zones" might be considered by some nations to be contrary to the Outer Space Treaty's ban on "national appropriation." A counter argument might hold that current international practice with respect to communication satellites in geosynchronous orbit already incorporates a variation of the "keep-out zone" principle. Current geosynchronous orbit must be space several degrees apart in order to avoid frequency interference. Therefore, such a satellite precludes the placement of other satellites near its position in the orbital arc.

In order to reduce uncertainty regarding the purpose of certain satellites and the tension likely to result from unauthorized close approach, it might be useful to establish rules regarding inspection, high-velocity fly-by and trailing. Such agreements might allow close-approach and inspection under certain circumstances (e.g., prior consent) but might otherwise ban high-velocity fly-by and trailing—either of which could be a prelude to satellite attack.

One of the functions of a regime of rules in space would be to reduce instances where provocative or threatening activities are observed but not explained. To resolve this problem, a forum or a "hot line" might be established through which questionable space activities could be discussed in a timely manner. Precedent exists for this in the 1971 "Agreement on Measures to Reduce the Risk of Outbreak of Nuclear War," which requires the United States and the Soviet Union to notify each other "in the event of signs of interference with [early warning systems] or with related communication facilities, if such occurrences could create a risk of outbreak of nuclear war." The 1971 agreement might be strengthened to require consultation regarding activities that might threaten satellites and not just activities which create a risk of nuclear war.

Monitoring Compliance With a "Rules of the Road" Agreement

The ability to monitor individual "rules of the road" with high confidence would vary directly with the specific measures adopted. As a general rule, however, monitoring "rules of the road" would be easier than monitoring other "arms control" regimes. The primary purpose of such rules would not be to restrict substantially the activities of the parties, but rather, to make the intentions behind these activities more transparent. Although the degree of protection for U.S. space assets to be gained from a "rules of the road" agreement would be less than from other arms limitation regimes, the costs are also correspondingly less for failure to completely verify compliance. One must assume that in the absence of ASAT arms control, both ASAT development and satellite survivability programs will be given high priority. This being the case, offensive and defensive measures would be available to respond to violations of "rules of the road."

Utility of a "Rules of the Road" Agreement

The "rules of the road" discussed above—if implemented in the absence of restrictions on ASAT weapon development—would not remove the threat of ASAT attack. If they were defended, "keep-out zones" would probably offer the closest thing to security in such a regime. Space mines designed to shadow satellites and detonate on command would lose a great deal of their utility if held at bay by a defended keep-out zone. If these zones were sufficiently large, or if satellites were appropriately shielded, they might even be effective against nuclear space mines. Keep-out zones would be less effective against advanced directed-energy weapons with kill radii of thousands of kilometers. However, these might be controlled by other arms control measures.

"Keep-out zones" combined with defensive satellites (DSATs) would offer substantial—though still incomplete—protection but would likely be extremely expensive. As an alternative to defended "keep-out" zones, the United States might wish to develop redundant systems and an ability to rapidly reconstitute lost assets.

"Rules of the road" would be substantially more effective at encouraging the orderly use of space by the military and at reducing the chances of escalation or misunderstanding in a crisis. Even in the absence of controls on ASAT weapons it would be valuable to have a multinational consensus concerning ambiguous activities such as close-approach, very low-orbit overpass, and high-velocity fly-by. If the "rules of the road" were part of other limitations on ASAT weapons and capabilities they would likely contribute to the effectiveness of these agreements and make their implementation more manageable.

Whether "rules of the road" were negotiable would depend on the specific provisions chosen. The negotiations pertaining to such rules might require the United States, the Soviet Union, and perhaps others, to sit down and discuss secret and extraordinarily sensitive issues relating to the operation of military space assets. Some rules, such as very low-orbit overflight by manned, reusable vehicles, may be so politically sensitive as to not be amenable to discussion. Other rules, such as "keep-out zones" and minimum separation distance for satellites, may not be desirable because they are not technically possible at altitudes where the majority of current U.S. and Soviet satellites are located. On the other hand, some rules, such as high-velocity fly-by, or close inspection might lend themselves to discussion and agreement.

The United States and the Soviet Union may wish to adopt "rules of the road" as a result of their increased use of space for military—including ASAT—purposes, or because they are engaged in negotiations designed to limit the arms race in space. "Rules of the road" might be an attractive companion agreement to far-reaching limits on ASAT weapon development. On the other hand, in the total absence of ASAT weapon limitations, there would be a need to clarify ambiguous activities before it became necessary to "use or lose" offensive space weapons. The negotiability—or lack thereof—of "rules of the road" can only be discovered as a result of serious negotiation between interested parties.

BMD AND ASAT TREATIES OF LIMITED DURATION

Each of the regimes examined above could be negotiated as a treaty of indefinite or limited duration or, alternatively, as one which remains in force as long as periodic reviews are favorable. Each of these alternatives has its advantages and disadvantages. Treaties of indefinite duration are more effective at discouraging the pursuit of banned activities, yet require a greater degree of foresight regarding the long-term interests of the signatories and can foreclose technological options for the indefinite future.[13] Treaties of limited duration allow parties to take advantage of future technological options, yet can encourage aggressive development programs designed to reach fruition at the termination of the designated period. Treaties which call for a periodic reassessment of agreed limitations in theory have great flexibility, yet, in practice, often result in a strong presumption that they should be continued.

[13] Treaties of unlimited duration usually contain a clause which allows the signatories to withdraw from the treaty if their "supreme national interests" are threatened. In addition to "supreme national interest clauses," treaties may also contain specific unilateral or agreed statements regarding specific understandings about related events. For example, The 1972 ABM Treaty contains a unilateral statement by the United States which links the continued viability of the treaty to "more complete limitations on strategic arms."

The United States might, for example, enter into a treaty limiting ASATs with the explicit and public reservation that we would withdraw from this treaty if and when we were ready to test and deploy a ballistic missile defense system in ways that the ASAT Treaty would forbid. Alternatively, we might take the public position that we intended to restrict our BMD activities so as to remain within the limits of an ASAT Treaty. While the former position would suggest a treaty of limited duration and the latter a treaty of unlimited duration, this need not be the case. It would be perfectly possible to sign a treaty of unlimited duration, with the standard provision allowing for withdrawal, accompanied by a clear statement of some of the conditions under which we intended to withdraw.

From one point of view, the exact language in a treaty regarding its duration is less important than the intentions of the parties. After all, there have been numerous examples of treaties of unlimited duration that were violated soon after they were signed and examples of treaties of limited duration that continued in force after they had expired (e.g., the "Interim Offensive Agreement" signed at SALT I). The real issue is whether the parties believe that adherence to the treaty in question continues to be in their national security interest.

The Reagan Administration has recently indicated that it intends to conduct ASAT tests to gather information useful in advanced BMD research.[14] Given the close connection between these two technologies, an ASAT treaty of even limited duration would require modification of current SDI program plans. Thus, to the extent that the United States wishes to maintain the most rapid pace of advanced BMD research within the bounds of the ABM Treaty, such a treaty would not be desirable. Conversely, to the extent that the United States wishes to slow the pace of Soviet BMD research and is willing to defer decisions regarding the testing of space-based or space-directed weapons, an ASAT treaty of limited duration could contribute to that result.

[14]The purpose of tests "in an ASAT mode" would be to investigate advanced technologies without violating the ABM Treaty. The Department of Defense recently told Congress that, "To ensure compliance with the ABM Treaty the performance of the demonstration hardware will be limited to the satellite defense mission. Intercepts of certain orbital targets simulating anti-satellite weapons can clearly be compatible with this criteria." "Report to the Congress on the Strategic Defense Initiative," Department of Defense, 1985, app. B, p. 8.

COMPLIANCE MONITORING, VERIFICATION, AND RECOURSE

Verification of compliance with an arms control treaty provision involves three distinct processes: monitoring the activities of other parties to the treaty, interpretation of the information obtained by monitoring, and, assessment of the risk which such activities pose to U.S. security. Each of these processes presents a different set of problems and opportunities to the intelligence community. Should violations or potential violations of treaty obligations be discovered during the verification process, then it becomes necessary to decide what, if any, action is to be taken in response. Verification of compliance and recourse are discussed in greater detail below.

Monitoring

When discussing the ability of the United States to monitor Soviet treaty compliance, it is important to distinguish existing and planned capabilities from potential capabilities. Existing and planned monitoring capabilities are described in chapter 4. Some of the existing systems used to monitor compliance with SALT and other arms control agreements would be useful for monitoring compliance with possible ASAT arms control provisions.[15]

[15]Some of these capabilities have been described in general terms by Congressman Les Aspin in "The Verification of the SALT II Agreement," *Scientific American*, vol. 240, No. 3, February 1979, pp. 38-45.

For example, capabilities to monitor the construction and dismantling of ICBM launchers —the number of which is constrained by the SALT II agreement—could also be used to monitor the construction and dismantling of launchers for boosters used for ASAT weapons.

By investing in new monitoring systems and personnel, future monitoring capabilities can be made more comprehensive than existing capabilities. To a limited extent, one can actually "buy" more monitoring capability. (See table 6-1). However, such additional capabilities would, in most cases, require years of work and substantial expenditures of funds. As in weapon system procurement, it will be necessary to judge "how much is enough"—i.e., to determine the level of investment above which the value of monitoring capability improvement obtainable per dollar ceases to be worth a dollar.

The fact that future monitoring systems could be more capable than current systems does not mean that all monitoring problems can be solved by spending more money on advanced technologies. Some activities will always be unmonitorable (e.g., some forms of underground testing), other dual-purpose activities (e.g., manned spaceflight) will often be difficult to characterize. Although future technologies will increase our ability to monitor the activities of other countries, similar technologies may make the job of treaty verification more difficult. Specific examples of these problems are presented above in the discussions of specific treaty provisions.

Interpretation

Once indications of a potentially prohibited activity have been detected by monitoring systems, the data must be further interpreted to determine the intent of the activity and how the activity affects specific treaty agreements. For example, suppose that while a space-weapon ban is in effect, the deployment or construction of a large mirror is observed in space. In this case, the monitoring data might be scrutinized to determine whether the mirror was capable of reflecting intense laser beams and changing its pointing direction quickly—as a prohibited weapon system might—or whether, instead, it was only capable of reflecting low-intensity radiation and changing its pointing direction slowly, as communication system or telescope components might.[16] The ability to make such a determination would depend both on the sophistication of the monitoring system employed and prior knowledge regarding similar activities.

Even if the monitoring system provides data sufficient to clearly identify the nature of a questioned activity, it still remains to be determined whether that activity is prohibited by the language of the relevant treaty. In the example of a mirror deployed in space, there would remain the question of whether deployment in space of any large mirror capable of reflecting intense laser beams would be a violation. Since similar mirrors have been proposed for peaceful purposes (e.g., propulsion of laser-powered rockets[17]), even if the relevant agreement defined weapons in terms of their capabilities rather than intended uses, there could be ambiguity as to the legality of deploying such mirrors.

When ambiguities are foreseen, treaty language can be worded to avoid them. However, history has demonstrated that it is extremely difficult to foresee all the significant ambiguities that could arise in an arms control agreement.

Assessment

If monitoring data are interpreted to indicate that an activity prohibited by a treaty (or possibly inadvertently allowed by ambiguity of the treaty) is taking place (or about to take place), the risk which the activity poses to U.S. security must be assessed. This assessment must take into consideration at least three factors: 1) the threat to U.S. national security posed by the specific violation; 2) assuming the

[16]The large deployable reflector (LDR) under development by NASA is an example of such a component.

[17]R.R. Berggren and G.E. Lenertz, "Feasibility of a 30-Meter Space Based Laser Transmitter," NASA-CR-134903, 1975 [NTIS accession number N-7611421].

violation, the extent to which the relevant treaty still contributes to U.S. national security; and 3) the ability of the United States to take actions which will prevent, mitigate, or compensate for damage that might be caused by the violation. The result of such an assessment will often imply the appropriate nature of the recourse to be pursued.

Recourse

Given the many different activities that ASAT arms control could restrict and the numerous ways that such agreements could be violated, it is difficult to make generalizations about how the United States might or should respond. Faced with a clear violation of a major treaty provision that seriously jeopardized U.S. national security, the United States would be wise to withdraw from the treaty in question. If, on the other hand, the existence of a violation was uncertain and it pertained only to a subsidiary portion of an otherwise valuable treaty, then it might be appropriate to seek consultation to resolve this particular activity while leaving the treaty otherwise intact.[18] Alternatively, unilateral defensive countermeasures or R&D on treaty compliant offensive measures might be pursued to hedge against breakout. The hardest questions are those that arise somewhere between these two examples.

ASAT arms control raises a number of questions common to all high-technology treaty restrictions. For example, if one party violates a test ban on advanced directed-energy ASATs and then, when confronted with the violation, declares its intent not to repeat this violation, what is the appropriate response? Some would argue that the damage has been done. One side has had the opportunity to verify a technology which it may have been developing covertly over a period of years. The side which remained in compliance has lost not only the information it could have gotten from similar tests, but potentially, years of research experience. Others might argue that limited testing or minor ambiguities offer no real and enduring military advantage.

Other responses to clear or uncertain treaty violations include negotiating modifications to the agreement or matching cheating with identical or equivalent conduct. Negotiating modifications can be a long and contentious process, particularly if the negotiations require one party to admit to treaty violations or ambiguous conduct. Given the differences between Soviet and U.S. force structure and technology base, matching cheating with identical conduct is often not a useful alternative. For example, the United States may not desire to build a Krasnoyarsk-style radar. On the other hand, matching cheating with equivalent conduct (the so-called "parallel interpretation" alternative) runs counter to notion that a treaty should have one common understanding which is accepted by both parties.

---

[18]Others have argued that the mere fact that a treaty has been violated is as important as the national security impact of the violation. For example, Colin Gray writes:

> The Soviet noncompliance issue is not important as a matter of ethics or because the sanctity of international legal norms must be upheld . . . Nor is Soviet cheating primarily important in terms of military advantage and disadvantage . . . (T)he greatest danger . . . results from the loss of U.S. credibility. . . . (W)ar is more likely to explode out of a mutual diplomatic miscalculation (than a military imbalance). That miscalculation could be rooted . . . in a Soviet lack of respect for the quality of determination in U.S. policy.

Colin Gray, "Moscow is Cheating," *Foreign Policy*, No. 56, fall 1984, pp. 141-152.

Chapter 7

Comparative Evaluation of ASAT Policy Options

Contents

Table

Chapter 7

Comparative Evaluation of ASAT Policy Options

POLICY OVERVIEW

ASAT Policy Choices

Over the next 5 years, the United States will have to make key decisions regarding research and development programs for anti-satellite weapons and countermeasures and for ballistic missile defense (BMD) systems. In addition, the United States must also consider whether it wishes to seek agreement with the Soviet Union to halt or limit the development of certain weapons that would operate from space or against space objects. This chapter analyzes the relationships between offensive and defensive weapons programs and arms control. In so doing, it utilizes the technology discussions contained in chapters 3 and 4 and the discussions of arms control found in chapters 5 and 6.

As discussed in chapter 6, those regimes which require negotiated arms control agreements could be either of limited or unlimited duration. Opponents of developing BMD systems might prefer an agreement of unlimited duration. Agreements of limited duration—perhaps 5-10 years—might be attractive to proponents of advanced BMD research if they could be fashioned so as not to interfere with plans to develop and test prototype BMD weapons. Such agreements would have the added benefit of temporarily constraining the development or testing of advanced ASAT weapons which could attack space-based BMD system components.

Alternative Legal/Technical Regimes

This chapter considers possible arms control provisions, ASAT postures, and countermeasures together as packages in order to examine their interaction. Since there are many conceivable packages, it is necessary to select a limited number for analysis. These packages have been constructed so that each will have at least one advantage over the others considered and so that each contains elements which might reasonably be expected to coexist in the same proposal. Consideration of these regimes is intended to facilitate assessment of the effectiveness and desirability of different combinations of ASAT and BMD technology development, satellite survivability, and arms control.

The seven regimes considered in the remaining sections of this chapter are[1]:

1. **Existing Constraints.** The first regime is defined by treaties and agreements presently in force. The ways in which this legal regime would affect technology developments designed to protect U.S. satellites or to place Soviet satellites at risk will be examined.
2. **A Comprehensive Anti-Satellite and Space-Based Weapon Ban.** Regime two could be established by adhering to treaties and agreements presently in force and, in addition, agreeing to forgo the possession of deliberate anti-satellite weapons, the testing—on Earth or in space—of any deliberate ASAT capability, the testing in an "ASAT mode"[2] of systems with inherent ASAT capabilities, and deployment—on Earth or in space—of any ASAT weapon.
3. **An ASAT Weapon Test Ban and a Space-Based Weapon Deployment Ban.** The third

[1]These regimes might usefully include elements not discussed here. For example, regimes 2, 3, 4, and 5 might also include a "no-use" provision which would prohibit the parties from destroying or "rendering inoperable" each others satellites.

[2]Testing in an "ASAT mode" would include tests of land-, sea-, air-, or space-based systems against targets in space or against points in space.

regime could be created by adhering to treaties and agreements presently in force and, in addition, agreeing to forgo testing in an "ASAT mode" and the deployment of any weapon in space. This regime differs from regime 2 most importantly in that it would *not* ban possession or testing—*on Earth*—of deliberate ASAT weapons.

4. **A "One Each/No New Types" Regime.** Regime 4 includes arms limitation provisions which would permit the United States and the Soviet Union to test and deploy their current ASATs but would prohibit testing of more advanced systems. Advanced systems prohibited would include those capable of operating or attacking targets at higher altitudes and those that would be deployed in space. For the purposes of this assessment, the U.S. MV will be considered to be the only deliberate "current" U.S. ASAT.
5. **Rules of the Road.** The fifth regime illustrates the advantages and disadvantages of establishing "keep-out zones" around individual, high-value satellites.
6. **Space Sanctuaries.** Regime 6 would provide high-altitude sanctuaries where satellites could operate but where the testing or deployment of weapons would be forbidden.
7. **A Space-Based BMD Regime.** The seventh regime might result from U.S. or Soviet withdrawal from the ABM Treaty followed by the deployment of space-based BMD systems.

As table 7-1 demonstrates, the regimes discussed here can be characterized both by the extent to which they rely on negotiated arms controls and by the extent to which they allow or encourage ASAT development. With the exception of the "Existing Constraints" and the "Space-Based BMD" regimes, all other regimes involve some type of arms control. With the exception of the "Comprehensive Anti-satellite and Space-Based Weapon Ban," and perhaps, the "ASAT Weapon Test Ban and Space-Based Weapon Deployment Ban," all other regimes assume some level of ASAT development. These regimes demonstrate that although anti-ASAT arms control arguments and pro-ASAT weapon arguments are related, there are many distinguishing features. ASAT arms control proponents believe that an ASAT treaty is in the national interest; those who support ASAT weapon development believe that this also is in the national interest. However, ASAT arms control proponents do not necessarily oppose all types of ASAT development and ASAT weapon proponents do not necessarily oppose all types of ASAT arms control.

Table 7-1.—Effect of Regimes on ASAT Development and Arms Control

	Restrict with arms control	Develop ASAT weapons
Existing constraints	No	Yes
Comprehensive ASAT and space-based weapon ban	Yes	No
Test ban and space-based weapon ban	Yes	Yes/No[a]
One each/no new types	Yes	Yes[b]
Rules of the road	Yes	Yes[c]
Space sanctuary	Yes	Yes[c]
Ballistic missile defense	No	Yes

[a]In this regime ASAT weapons could be developed, tested, and deployed on Earth but not in space. The United States could pursue ASAT development within the bounds of the treaty, or it could forego ASAT development entirely.
[b]All ASAT weapons other than "current types" could not be tested or deployed in space.
[c]Development and deployment optional but strongly supported by advocates of this regime.

Although the individual regimes vary considerably, all of them should be assessed with two important considerations in mind:

1. **First, if we wish to continue to use space for military purposes, a commitment to satellite survivability is essential whether or not any arms limitation agreements are in force.** The existence of space systems with some inherent ASAT capability makes it impossible to ban the ability to attack satellites. Therefore, even under the most restrictive ASAT arms control regime, programs for satellite survivability and countermeasures must be pursued. In the absence of arms control limitations on

ASATs, ensuring satellite survivability will be a more demanding task.

2. **Second, the United States should exercise caution in its reliance on space assets to perform tasks essential to the national security.** No matter what arms control or satellite survivability measures are taken, there will always be some risk that critical satellites can be destroyed or rendered inoperable. The value of continued and future reliance on space systems must be balanced against the probability that such assets may not be available in a conflict situation.

REGIME 1: EXISTING CONSTRAINTS

Legal Regime

The United States could decide that there are no additional arms control limitations relating to space weapons that are in its national security interest. If so, development of anti-satellite and space-based weapons by the United States and the U.S.S.R. could continue unrestrained except by the Limited Test Ban Treaty, the Outer Space Treaty, and the ABM Treaty.[3]

Even in the absence of new arms control limitations there are restrictions on what the United States and the Soviet Union can do in space. As discussed in chapter 5, under existing international law and the treaties to which the United States is a party, the following activities are already banned:

- **Unprovoked Attack on Another Country's Satellite:** Subject to the right of individual or collective self-defense, Article 2 of the U.N. Charter prohibits the use or threat of force. A similar sentiment is to be found in Article III of the 1967 Outer Space Treaty. The SALT and ABM Treaties also prohibit interference by either state with space assets used by the other to monitor those treaties.
- **Placement of Nuclear Weapons in Orbit:** Article IV of the 1967 Outer Space Treaty (OST) prohibits orbiting nuclear weapons. This would include nuclear "space mines" and, presumably, ASATs that used a nuclear explosion as a power source.
- **Detonation of Nuclear Weapons in Space:** The 1963 Limited Test Ban Treaty (LTBT) prohibits nuclear weapons tests or other nuclear explosions in space. This would prohibit the full testing of ASATs that use nuclear explosions for destruction or as a power source.
- **Development, Testing, or Deployment of Weapons Capable of Countering Strategic Ballistic Missiles, or Their Elements in Flight:** Space-based weapons sophisticated enough to "counter strategic ballistic missiles or their elements in flight" are banned under the terms of the 1972 ABM Treaty. This establishes a somewhat vague upper limit on the capabilities of advanced ASAT weapons.

To summarize, the existing international legal regime prohibits the use of ASAT capability except in national or collective self-defense, the testing or deployment of space-based weapons with strategic BMD capability, and the testing in space or deployment in orbit of nuclear space mines or ASATs that

[3]The unratified Threshold Test Ban Treaty, and the unratified SALT II Treaty, if adhered to, would supply additional restrictions. The Threshold Test Ban Treaty, which was signed in 1974, prohibits testing *on Earth* of nuclear weapons with a yield greater than 150 kilotons. Should a nuclear ASAT weapon require a nuclear explosive of greater yield than this, it could not be fully tested without violating the Threshold Test Ban Treaty. Under Article IX of the Salt II Treaty, the parties agreed not to develop, test, or deploy "systems for placing into Earth orbit nuclear weapons." This might be interpreted to include nuclear ASAT weapons.

would require a nuclear detonation as a power source. The existing regime places few restrictions on the current ASAT research and development programs of either the United States or the Soviet Union.

Offensive Posture

In the absence of further restrictions, the following weapons could be *developed*, *tested*, and *deployed* as deliberate ASAT weapons by either the United States or the Soviet Union, *if deployed in compliance with the ABM Treaty* (i.e., so as not to be capable of countering strategic ballistic missiles or their elements in flight)[4]:

- **Coorbital Interceptors:** Ground-launched, nonnuclear coorbital interceptors—e.g., the current Soviet ASAT—are allowable under the existing regime. Ground-based nuclear systems could be developed and deployed but not tested in space. There are no restrictions on nonnuclear coorbital interceptors predeployed as space mines.
- **Direct-Ascent Interceptors:** Ground-launched or air-launched direct-ascent interceptors—e.g., the U.S. ASAT being developed—are allowable. Direct-ascent interceptors carrying nuclear weapons could be developed and deployed but not tested in space.
- **Ground-Based or Airborne Lasers:** There are no restrictions on nuclear or nonnuclear ground based lasers, or on airborne lasers that would not require a nuclear explosion in the atmosphere.
- **Space-Based Lasers:** Nonnuclear, space-based lasers are allowable.
- **Space-Based Neutral Particle Beam Weapons:** There are no restrictions on space-based neutral particle beam weapons.
- **Maneuverable Spacecraft:** Although not necessarily "deliberate" ASAT systems, maneuverable spacecraft could be given substantial ASAT capabilities under the existing regime.

In addition to these deliberate ASAT systems, other weapon systems such as ICBMs or ABMs that have some ASAT capability could be developed and deployed, but could not be completely tested as ASAT weapons. Such systems could be tested in space as long as they were not detonated. The SALT agreements and the ABM Treaty do place other restrictions on ICBMs and ABMs.

Defensive Posture

The United States and the Soviet Union could develop, test, deploy, and use defensive measures such as hiding, deception, evasion, hardening, and proliferation without legal restraint in the existing regime. In addition to such passive countermeasures, nondestructive active countermeasures such as electronic countermeasures (ECM) and electro-optical countermeasures (E-OCM) could also be used. ECM and E-OCM are likely to be available and inexpensive and are unlikely to by restricted by arms control agreements; however, these countermeasures could be defeated at a reasonable cost.

Many destructive active countermeasures would also be allowed under the present regime. Satellites could be given a self-defense capability (shoot-back) or provided with an escort defense (DSAT). The current ASAT interceptors being developed by the United States and the Soviet Union (respectively, the U.S. Air Force Miniature Vehicle and Soviet coorbital interceptor) are not capable of attacking each other. However, many advanced ASAT weapons that could be built in the current regime would have *some* effectiveness against *some* types of ASATs. For example,

[4]The constraint that ASAT weapons not be deployed so as to be capable of countering strategic ballistic missiles or their elements in flight is restrictive, but several deployment schemes can be conceived which would be both lawful and useful. For example, a neutral particle beam weapon of relatively low power might be deployed in geosynchronous orbit for ASAT or DSAT purposes. It might be capable of damaging an enemy satellite or ASAT several hundred kilometers away within several seconds, but incapable of damaging a distant ballistic missile during its flight time of a few minutes. Deployment of such weapons might also be allowed in low orbit, if the U.S.-Soviet Standing Consultative Commission—which was established by the ABM Treaty to consider allegations of treaty violations—should agree that such weapons, if never tested as BMD systems, could not reasonably be expected to have a significant BMD capability.

a space-based neutral particle beam weapon, in addition to its ASAT role, could also be used as a DSAT to provide "enclave defense"—i.e., to defend a number of distant satellites from other weapons such as coorbital or direct-ascent interceptors or continuous-wave lasers. However, neutral particle beam weapons deployed as DSATs could not shoot back effectively at larger neutral particle beam ASATs, nor could they shoot back effectively at expendable single-pulse weapons such as predeployed nuclear "space mines" or some nuclear or nonnuclear directed-energy weapons.

Moreover, if shoot-back is to be effective, space objects with known or suspected ASAT capabilities would have to be fired upon while still some distance from U.S. satellites believed to be in danger. As discussed above, attacking an approaching spacecraft is prohibited by international law except in self-defense and one could not be certain that the approaching spacecraft had a hostile intent until it was too late. Hence, active defense against suspected "space mines" might be considered to be unlawful in the existing regime, although deployment of means for such defense may not be.

Neither passive nor active countermeasures could guarantee the survival of satellites attacked by some advanced directed-energy weapons. Although, as discussed in chapter 4, the cost of destroying small, inexpensive satellites and decoys with advanced directed-energy weapons might exceed the cost of building such satellites and decoys. Security for large and expensive satellites might ultimately have to rely on an attempt to deter ASAT attacks by credibly threatening retaliation against enemy space-based or terrestrial assets. A credible retaliatory capability would require a means of discovering that U.S. satellites had been attacked and identifying the attacker. This would probably require attack sensors mounted on satellites and a space-based surveillance system to track and distinguish ASATs from meteorites or space debris. The latter could also be used to verify compliance with future ASAT arms control agreements, if any, or for targeting future ASAT (or DSAT) weapons, if any.

Net Assessment

Treaties and agreements presently in force create no significant barrier to the development, testing, and deployment of very capable, nonnuclear ASAT weapons.[5] The current regime also allows a wide range of active and passive countermeasures, including the development of satellites capable of defending themselves by striking at attacking ASAT weapons.

The primary advantage of the current regime is that it allows the almost unrestrained application of U.S. technology to the related problems of protecting U.S. satellites and placing threatening Soviet satellites at risk. Under this regime, the United States would be free to use its comparative advantage in advanced technology to keep pace with expected developments in Soviet ASATs and other military satellites. Advanced U.S. ASATs might discourage the development of more capable Soviet military satellites designed to place U.S. terrestrial assets at risk. In addition, the United States would be free to respond to Soviet ASAT weapons with increasingly sophisticated defensive weapons and countermeasures, thereby reducing the probability that the Soviets could successfully use their intentional or inherent ASAT capabilities. Effective ASAT capability could also give the United States a powerful countermeasure against potential Soviet space-based BMD systems.

In addition, research and development on new ballistic missile defense technologies can also proceed without the constraints that might be imposed by certain ASAT arms control regimes. Testing of advanced ASATs could provide valuable information that would contribute to the development of very capable BMD systems. Such testing in the "ASAT

[5]ASAT weapons capable of operating in an "ABM mode" are, or course, limited by the ABM treaty. See discussion, supra, p. 127.

mode'' could allow some research to go forward that, if designated as BMD research, might be considered to be inhibited by the ABM Treaty.

The primary disadvantage of the current regime is that it might lead to an expensive and potentially destabilizing arms race in space. Rather than protecting satellites, a competition in space weapons might severely reduce their military utility. Under conditions of unrestrained competition, security might be purchased only at the price of a substantial and sustained commitment to the development of increasingly sophisticated offensive and defensive space weapons. In such an environment, ensuring the survivability of satellites would require more than simple hardening or evasion. Costly measures might have to be taken such as the deployment of precision decoys, pre-deployed spares, or the ability to quickly reconstitute ones space assets. Satellites capable of defending themselves or a companion satellite might also have to be developed and deployed.

Should space mines or directed-energy weapons be deployed, they might be capable of the almost instantaneous destruction of a large number of critical satellites and ASATs. This could force nations into a situation in which they must "use or lose" their own pre-deployed space weapons. This might supply the incentive to escalate an otherwise manageable crisis. If missile early warning and communication satellites were highly vulnerable, crisis stability might be lessened. The malfunction of such satellites could be misinterpreted as a sign of imminent attack, since potential nuclear aggressors would find such satellites to be attractive targets.

Another potentially destabilizing factor is that some satellites (particularly communication satellites) play a dual role—they are intended to be force multipliers in a conventional war, yet they are to play a key role in managing a conflict so as to avoid unwarranted escalation. In the event of a conventional war, the possessor of a capable ASAT system would have a strong incentive to attack satellites that were providing support to conventional enemy forces. Destruction of these satellites, however, might contribute to escalation from conventional to nuclear war.

An unrestrained competition in ASAT weapons would also increase the risk posed to space-based ballistic missile defense systems. Such systems are likely to have many critical assets based in low-Earth orbit. So situated, extensive precautions would have to be taken to protect them from even modest ASAT weapons.

It is possible that an ASAT weapon competition could also inhibit the use of space for commercial and scientific purposes. Manned space stations would be quite vulnerable to ASAT attack. Should considerable ASAT testing take place, the resulting debris could prove harmful to scientific and commercial satellites.

REGIME 2: A COMPREHENSIVE ANTI-SATELLITE AND SPACE-BASED WEAPON BAN

Legal Regime

This regime could be established by adhering to treaties and agreements presently in force and, in addition, agreeing to forego the possession of deliberate anti-satellite weapons, the testing—on Earth or in space—of any deliberate ASAT weapon, the testing in an "ASAT mode" of systems with inherent ASAT capabilities, and the deployment—on Earth or in space—of any ASAT weapon.[6] In addition, the U.S.S.R. would be required to destroy all its

[6]Such an agreement might resemble the draft treaty proposed to the United Nations by the U.S.S.R. in August of 1983, except the testing or use of manned spacecraft for military purposes would not, in general, be banned as proposed in Article 2 of the 1983 Soviet draft treaty. (U.N. Document A/38/194, Aug. 23, 1983). The fifth provision of Article 2 of this proposed

coorbital interceptors and the United States would be required to destroy the direct-ascent interceptor it is currently developing.

Offensive Posture

In this regime, the United States could not maintain any deliberate ASAT weapons, whether dedicated or multi-role, nor would the U.S.S.R. be allowed to do so. Space systems with inherent ASAT capabilities such as ICBMs, ABMs, and maneuverable spacecraft would still be allowed, but they could not be tested in an "ASAT mode."

Defensive Posture

Under a comprehensive ASAT ban the United States would retain the right to deploy and use passive countermeasures such as hiding, deception, evasion, hardening, and proliferation. The United States would not be allowed to develop, possess, test, or deploy weapons for satellite self-defense, defensive satellites (DSATs), or other systems intended to have anti-satellite capabilities, even for defensive purposes.

If the U.S.S.R. complied fully with the letter of such a comprehensive ASAT ban, the risk posed to U.S. satellites would be limited to the risk posed by possible Soviet use of ICBMs, SLBMs, ABM interceptors, and possible future highly maneuverable spacecraft. If U.S. satellites were hardened against the effects of nuclear explosions to a modest degree, only low-altitude U.S. satellites would be at significant risk of damage by such inherent ASAT capabilities, and then primarily at the nuclear level of conflict. Assuming Soviet compliance, U.S. warning and communications satellites in high-altitude orbits would enjoy a high degree of security in this regime.

Net Assessment

Although this regime would contain the most far-reaching arms control provisions and therefore might be most effective at preventing the development of new and more threatening ASAT weapons, it would have the disadvantage of being the most difficult to verify. Unlike an ASAT Test/Space-based Weapon Deployment Ban (regime 3), a comprehensive ban would prohibit *possession* of ASAT weapons *on Earth*. Because it is difficult to obtain information about Soviet military affairs, the United States would have to assume that the Soviet Union could possess some number of their current ASAT weapon.

The current Soviet coorbital interceptor is a relatively small spacecraft launched on much larger, general-purpose boosters. Maintaining such boosters and their launchpads would be allowed, and it would have to be assumed that the U.S.S.R. would continue such activities. Construction of additional boosters and launchpads would also be allowed by an ASAT ban of the type considered here. Hence the U.S.S.R. could maintain and even expand its ASAT force with some confidence that the United States could not gain *unambiguous evidence of a violation* of an ASAT possession ban. However, even if the U.S.S.R. maintained some coorbital interceptors, it could not test them without risking almost certain detection, and in time the confidence of Soviets in a long-untested and never perfected ASAT weapon might erode.

There would always be the possibility that the Soviets might develop a new type of ASAT weapon with the intention of using it, without prior testing, *in extremis* (e.g., if anticipating an imminent attack). For example, the U.S.S.R. might equip an existing booster or satellite vehicle with a nuclear explosive—either an isotropic nuclear weapon or possibly a nuclear directed-energy weapon—and maintain it in readiness for launch or actually launch it into space. The military utility of

treaty would obligate parties "not to test or use manned spacecraft for military, including anti-satellite, purposes." If this provision were stricken or changed to read "not to test or use manned spacecraft for anti-satellite purposes," the resulting draft treaty, if acceded to by the United States and the U.S.S.R., would establish a regime of the type considered in this section. The fifth provision of Article 2 of the proposed Soviet draft treaty would obligate parties "Not to test or create new anti-satellite systems and to destroy any anti-satellite systems they may already have."

such untested systems would be questionable, particularly if the United States aggressively pursued available satellites survivability measures.

Since the United States might agree to a comprehensive ASAT ban only after considerable political friction over question of compliance and verification, it would be important to consider how such a ban might make a greater contribution to U.S. national security than a ban on ASAT testing and space-based weapon deployment (regime 3). The purpose of both bans would be to prevent the use of ASATs, or, at minimum, to reduce the probability that an ASAT attack would be effective. An ASAT test ban would primarily affect weapons reliability, while an ASAT possession ban, if observed, would affect both availability and reliability. It is conceivable that the risk posed by possible illegal Soviet use of ASAT weapons might be somewhat lower in a regime in which the Soviets could not lawfully possess ASAT weapons. Presumably, the inability to overtly possess ASAT weapons would diminish one's ability to use them effectively. Furthermore, an absolute ban on possession might make it less likely that the current generation of ASAT weapons could be upgraded and held in readiness in significant numbers.

However, if the United States could only be confident that the Soviets were complying with a treaty to the extent we could verify compliance, then the United States would not have confidence that this regime offered any greater protection to our satellites than does regime 3 (test ban and space-based weapons ban).

REGIME 3: AN ASAT WEAPON TEST BAN AND SPACE-BASED WEAPON DEPLOYMENT BAN

Legal Regime

This regime would ban what can be monitored with greater confidence—testing in an "ASAT mode"[7] and ASAT deployment in space. Everything that is prohibited under the current regime would continue to be prohibited. In addition, further testing—in space—of the current Soviet coorbital interceptor and the U.S. direct-ascent interceptor would be prohibited, as would the placement of any weapons in space. Unlike regime 2, this regime would not attempt to ban testing, possession, or deployment of ASAT weapons *on Earth*.

Offensive Posture

Although they could not be tested *overtly* in an "ASAT mode," a number of weapons which have some limited ASAT capability already exist or could be developed. ICBMs, ABMs, and maneuverable spacecraft already exist and have inherent ASAT capabilities which pose some threat to satellites. It might be possible to increase the ASAT potential of these systems without violating a ban on the testing of ASAT weapons. In addition, upon entry into force of a ban on ASAT testing, the United States and the U.S.S.R. would possess deliberate ASAT weapons which would have undergone some developmental testing, although possibly not enough to perfect their designs. Such weapons could be maintained in partial readiness. However, without operational testing for reliability evaluation and training purposes, confidence in the effectiveness of such weapons would probably degrade in time.

Advanced ASAT weapons such as neutral particle beam weapons or x-ray lasers could be developed and maintained in partial readiness, but could not be completely tested. Confidence that such weapons would perform adequately if used might be so low that one would not rely on them in an aggressive first strike nor find it cost-effective to develop them for that purpose. On the other hand, one might

[7]Testing in an "ASAT mode" would include tests of ground-, air-, sea-, or space-based systems against targets in space or against points in space. Testing on the ground of ASAT systems or components would not be prohibited.

use them, if attacked, to degrade enemy capabilities supported by satellites, and might find it cost-effective to develop them for that purpose. That is, the discrepancy between offense conservatism and defense conservatism might decrease the risk which untested weapons could pose if possessed by an aggressive nation.

Defensive Posture

In this regime, testing and deployment in space of advanced ASAT weapons would be prohibited but might be attempted by the U.S.S.R. covertly or after a breakout.[8] Hence the choice of passive countermeasures in this regime would be influenced by the same considerations which favor deception and modest nuclear hardening in the existing regime. Such measures would be more effective, however, in a test-ban regime because it could be assumed that the ASAT threat would be reduced to some degree by the arms control provision. Passive countermeasures would also be more important in this regime, because destructive active security measures—e.g., shoot-back with reliable, tested DSAT weapons—would not be an option. Deep-space surveillance would be even more desirable in this regime than in the existing regime, because of the need to monitor compliance as well as for its role in providing attack assessment information. Hence, in a test-ban regime, attack sensors, space-based LWIR sensors, satellite decoys, and modest nuclear hardening would be at least as desirable, as in the existing regime, if not more so.

Nondestructive active countermeasures such as ECM and E-OCM would be desirable, if not inherent, in a test-ban regime, just as in the existing regime. Destructive active countermeasures, on the other hand, would be severely constrained: new ASAT weapons useful as DSATs could be developed but could not be tested nor deployed in space. An untested NPB or XRL built and readied for quick launch and use as a DSAT could not be responsive enough to use for defensive shoot-back against expedient ASAT weapons such as ICBMs but might have value if maintained for retaliatory shoot-back.

[8]Although deployment of an NPB in space—a prerequisite for testing—would probably be observable, maintaining an untested NPB weapon on Earth in readiness for quick launch might not be, and would be allowed. Maintaining an untested XRL weapon on Earth in readiness for quick launch might also be difficult to detect and would also be allowed under the terms of an ASAT test and SBW deployment ban. Illegal deployment of an untested XRL in space would be difficult and costly to observe. However, an enemy could have little confidence in the reliability and performance of untested NPB or XRL weapons, so such weapons would not be as threatening as in the existing regime in which NPB weapons could be legally tested in space.

Net Assessment

A negotiated ban on the testing of weapons in space or against space objects would limit the nature and extent of U.S. and Soviet arms competition in space. Advanced ASAT directed-energy weapons which could threaten high-altitude satellites with prompt destruction could not be lawfully tested and attempts to extensively test such weapons covertly would probably be detectable. Although such a ban could not eliminate all threats to satellites, it would substantially reduce the cost and complexity of ensuring a reasonable level of satellite survivability. The United States would still benefit from hardening its satellites to some extent and deploying spares and decoys, but the more elaborate, expensive, and possibly ineffective precaution of developing and deploying DSATs would be prohibited and, indeed, less attractive. In the absence of reliable, effective ASATs, satellites would be of greater utility since the United States might have higher confidence that they would be available when needed.

Relative to the existing regime, the primary advantage of a regime banning testing of ASAT capabilities and deployment of space-based weapons would be that highly valued U.S. satellites in higher orbits—e.g., the future MILSTAR system—could be protected with some confidence from advanced ASAT weapons, especially if protected as well by passive countermeasures. The fact that advanced ASATs could not be overtly tested would reduce the probability that they would be developed and deployed. If they were developed and used without prior or complete testing, the improbability of their success compounded with the improbability of their attacking an operational satellite rather than a decoy (if such are deployed) would afford such satellites considerable protection and would, at least, disproportionately increase an enemy's cost for an effective ASAT capability. In addition, a ban on testing advanced ASAT weapons and deploying them in space would plausibly inhibit future competition in developing space-based weapons and would discourage development and covert testing and deployment of ASAT weapons of types which would pose the strongest incentives for preemptive ASAT attack. These benefits might be deemed advantageous by both the United States and the U.S.S.R.

As in the existing regime, the United States could retain a capability to attempt to negate low-altitude Soviet satellites (e.g., RORSAT) with its MV ASAT in the event of war and to respond in kind to a Soviet ASAT attack. However, confidence in the operational capability of this system might degrade over time without continued operational testing.

From the point of view of those interested in preserving the present agreement beween the United States and the Soviet Union limiting ballistic missile defenses, another advantage of an ASAT test ban would be its prevention of tests of ASAT technologies with potential BMD applications.

On the other hand, from the point of view of those favoring intensive BMD research, a primary disadvantage of this regime, relative to the existing regime, is that the testing of some types of advanced BMD weapons might be prohibited. Such limitations could be slightly more restrictive than those of the ABM Treaty, and would be very restrictive compared to a regime in which the ABM Treaty was no longer in force [regime 7]. Finally, it must be recognized that a ban on testing ASAT capabilities and deploying space-based weapons would not offer absolute protection for satellites; there would remain some possibility that an untested or partially tested ASAT, if suddenly deployed and used, might actually work well enough to overcome passive countermeasures.

REGIME 4: A "ONE EACH/NO NEW TYPES" REGIME

Legal Regime

A "one each/no new types" regime might be established by adhering to agreements currently in force and further agreeing to ban the deployment in orbit of any weapon and the testing in space, "in an ASAT mode" of any system *except* the currently operational type of Soviet coorbital interceptor and the U.S. MV direct-ascent interceptor.[9] Research on advanced systems and testing of these systems *on Earth* would not be prohibited.

Offensive Posture

Offensive postures in a "no new types" regime would be as in an ASAT test ban and space-based weapon deployment ban regime (regime 3), except ASAT weapons of the single allowed type would almost surely be maintained for offensive ASAT missions in wartime.[10] It is possible that each side would be satisfied with the capabilities such fully tested weapons could provide and would be less tempted than it would be in a test ban regime to covertly develop advanced ASAT weapons.

[9]Although the U.S. Department of Defense has stated its belief that the Soviets have two ground-based lasers which could be used against satellites [U.S. Department of Defense, *Soviet Military Power*, 1984, p. 35], testing of such lasers as ASAT weapons would be prohibited. If these lasers had already been tested as ASATs by the time a "no new types agreement" could enter into force then this regime might have to be appropriately modified.

[10]It is possible, of course, that one or both nations would decide—as the United States did after ratifying the ABM Treaty—that its allowed system was not worth maintaining.

Defensive Posture

Passive countermeasures appropriate in a test-ban regime would also be appropriate in this regime, and for the same reasons. In addition, the unambiguous, if limited, threat posed by the one allowed ASAT weapon would provide an additional incentive to deploy passive countermeasures tailored to that weapon. For example, evasion might effectively counter coorbital interceptors such as those tested by the U.S.S.R., and maneuver—although not literally "evasion"—could complicate targeting of the U.S. MV. These countermeasures would probably be developed and employed even though they would not be effective against more capable weapons which might be developed but not tested nor deployed in space.

ECM and E-OCM would be allowed in this regime as in a test-ban/space-based weapon ban regime. Current U.S. and Soviet ASAT weapons would be insufficiently responsive to be effective for defensive shoot-back; however, they could be used in retaliation.

Net Assessment

The primary advantage of a "no new types" regime, relative to the existing regime, would be that critical U.S. satellites in higher orbits could be protected with some confidence from advanced ASAT weapons. If developed and used without prior testing, it is possible that such advanced ASAT weapons would not work properly. If they did work, it would not be clear that they could overcome the survivability measures that could be given satellites in this regime. More generally, a ban on testing advanced ASAT weapons would inhibit to some extent future arms competition in space.

Assuming the United States had successfully developed its MV ASAT, a "no new types" regime might be particularly desirable. Such an agreement could prohibit the testing of Soviet ground-based lasers or MV-type ASAT weapons and limit them to their current, unsophisticated ASAT weapon. Of course, this would make such an agreement less acceptable to the Soviet Union. Should the Soviets test advanced ASAT weapons before such an agreement can enter into force, such an agreement would be less advantageous to the United States. However, since such an agreement might avert the risks posed by even more advanced—particularly directed-energy ASATs—a "no new types" agreement might still be considered valuable and negotiable by both the United States and the Soveit Union.

As in the existing regime, the United States could retain a capability to negate low-altitude Soviet satellites (e.g., RORSAT) in the event of war and to respond in kind to a Soviet ASAT attack. A primary disadvantage of a "no new types" regime, relative to the existing regime, would be that allowed U.S. ASAT capabilities would be inadequate to negate threatening Soviet satellites if such satellites were moved to higher orbits—a feasible but difficult and costly Soviet countermeasure. As in the test ban and space-based weapon ban regime, the testing of some types of advanced BMD weapons which would be allowed in the existing regime would be limited in this regime. Such limitations could be slightly more restrictive than those of the ABM Treaty and would be very restrictive compared to a regime in which the ABM Treaty was no longer in force.

Finally, it must be recognized that the reliability of protection afforded high-altitude satellites by a ban on testing "new types" would be uncertain; there would remain some probability that an untested advanced ASAT, if suddenly deployed and used, might actually work.

REGIME 5: RULES OF THE ROAD

Legal Regime

A legal regime providing for "keep-out zones" around satellites could be established by a "rules of the road" agreement similar to the "Rules of the Road at Sea" Treaty.[11] As discussed in chapter 6, such an agreement would not prohibit development, testing, or deployment in space of advanced ASAT weapons but would, instead, attempt to enhance security by establishing rules regarding space activities such as close approach of foreign satellites, advance notice of launch activities, high-velocity fly-bys, minimum separation distance between satellites, low-altitude overflight, and "keep-out zones."[12]

"Keep-out zones" would probably offer the closest thing to security in a "rules of the road" regime. The following "rules of the road" are illustrative of those which might be agreed should it be decided that "keep-out zones" are in the U.S. national security interest:

- Keep 100 kilometers and three degrees out-of-plane from foreign satellites below 5,000 km.
- Keep 500 km from foreign satellites above 5,000 km except those within 500 km of geosynchronous altitude.
- One pre-announced close approach at a time is allowed.
- In the event of a violation of the rules above, the nation of registry of the satellite which most recently initiated a maneuver "burn" is at fault and guilty of trespass.
- Satellites trespassing upon keep-out zones may be forcibly prevented from continued trespass.

The rationale for these rules is as follows:

- ASAT weapons such as nuclear interceptors would have to be kept at a range of several hundred kilometers from moderately hardened satellites in order to protect such satellites; advanced ASAT directed-energy weapons might have to be kept much farther away.[13]
- Satellites in geostationary orbit are already so closely spaced that a keep-out zone sufficiently large to protect satellites from nuclear attack could not be established around such satellites without displacing satellites already there and reducing the number of geostationary orbital slots available to other nations in the future.
- There are now very few satellites in supersynchronous orbits,[14] but critical strategic warning and communications functions could be performed by satellites in such orbits. Should space systems be developed to operate in this region, there would be adequate room to accommodate large keep-out zones.
- There are presently few satellite *orbits* in deep space[15] but below geosynchronous orbital altitude. The most notable exceptions are the orbits of various Soviet satellites in highly elliptical, semi-synchronous "Molniya-type" orbits, U.S. Air Force Satellite Data System (SDS) satellites in similar highly elliptical orbits, and U.S. (NAVSTAR) and Soviet (GLONASS) navigation satellites in semi-synchronous circular orbits. Although there are, or soon will be, many such satellites de-

[11] 16 UST 794, TIAS 5813.

[12] In addition to agreeing to such "rules of the road," the United States and the Soviet Union might have to modify their commitment to the 1967 Outer Space Treaty (18 U.S.T. 2410; T.I.A.S. 6347). Since Article II of the Outer Space Treaty states that "outer space . . . is not subject to national appropriation by claim of sovereignty, by means of use or occupation, or by any other means," this could be interpreted as prohibiting establishment of such keep-out zones. A contrary argument maintains that a precedent for "keep-out zones" can be found in the international acceptance of the principle that a satellite should not be placed in geostationary orbit if it will interfere with a satellite already in that orbit.

[13] *In re*: NDEW, see, e.g., L.A. Wojcik, "Separation Requirements for Protection of High-Altitude Satellites from Coorbital Anti-Satellite Weapons," (Pittsburgh, Pa.: Carnegie-Mellon University, Department of Engineering and Public Policy, dissertation, March 1985); *in re*: NPB weapons, see ch. 4 of this report.

[14] I.e., higher than geosynchronous orbital altitude.

[15] I.e., higher than 3,000 nautical miles, or about 5,600 kilometers.

ployed, several satellites will (or could) occupy the same orbit. For example, 24 NAVSTAR satellites will occupy only three orbits, with eight satellites following one another around each of the three orbits. Hence there would be enough room in this region of space to accommodate keep-out zones of several hundred kilometers radius around the satellites presently deployed there.

- There are too many satellites in low-Earth orbit—particularly below the inner Van Allen radiation belt which extends from about 1,800 km (1,000 nmi) to about 5,600 km (3,000 nmi)—to accommodate keep-out zones of several hundred kilometers radius around the satellites presently deployed there. Indeed, many satellites have perigees within several hundred kilometers of the Earth's surface. Requiring keep-out zones of several hundred kilometers radius around low-altitude satellites would therefore be impractical.
- However, it would be feasible to establish smaller keep-out zones around satellites in low orbit and, in addition, to prohibit satellites from entering an orbital plane inclined less than, say, three degrees from the orbital plane of a foreign satellite at such altitudes. Specifying a minimum angular separation between orbital planes would prevent continuous trailing; for example, two satellites in 1,000 km circular orbits with orbital planes separated by three degrees would approach each other closely every 53 minutes, if properly "phased," but would separate by as much as about 400 km at intermediate times and would be separated by at least 200 km about half the time. If, in addition, such satellites were phased so as to not approach one another more closely than 100 km at any time, their separation would vary between 100 km and more than 400 km, at minimum. Under such rules, although satellites would occasionally approach one another so closely as to be mutually vulnerable to, for example, covert on-board nuclear weapons, such approaches would not all occur simultaneously. Therefore, adequately hardened, low-altitude satellites could not be instantly and simultaneously destroyed by relatively primitive ASAT weapons.
- There would be some value in allowing one pre-announced close approach at a time as an exception to the rules above. Such an exception would, for example, permit an inspection satellite carrying a gamma-ray spectrometer to trail a foreign satellite while trying to determine whether the foreign satellite carried fissionable material, possibly in violation of Article IV of the 1967 Outer Space Treaty. A disadvantage of such an exception would be that a trailing "inspection" satellite could carry a weapon and destroy the trailed satellite at close range. However, deployment of one on-orbit spare for any truly essential satellite would eliminate this risk.
- Although, given adequate space surveillance, it could be verified that two foreign satellites approached one another more closely than would be allowed by these rules, there could be a problem in determining which nation or other party would be guilty of a violation. It is difficult to predict, to within an accuracy of 100 km, where a satellite will be in several months as the result of an orbital transfer or stationkeeping maneuver. This is particularly true if the satellite is at very low altitude where it would be subject to atmospheric drag or at very high altitude where it would be subject to the lunar gravitational field. Hence, inadvertent close approach might be possible. Legal allocation of responsibilities in such a regime might follow precedents established in maritime and, especially, aeronautical law, which specifies minimum separation distances between aircraft and gives right-of-way to relatively unmaneuverable aircraft such as aerostats (balloons) and gliders. One possibility would be to give right-of-way to satellites already in orbit and, by implication, to assign fault to whichever spacecraft most recently initiated or continued a maneuver "burn."

The rules suggested above are intended to be illustrative rather than precise. Careful framing of an agreement would be required in order to prohibit unintended abuses such as establishment of a *de facto* barrier to deep space by deploying many small satellites in low orbit in order to fill an altitude band with keep-out zones. Rationing keep-out zones—e.g., 10 per nation—could solve this problem, but careful study may be required to foresee other possible abuses. In addition to its technical problems, this regime is likely to have a significant political dimensions inasmuch as it will affect the rights of all present and future spacefaring nations.

Offensive Posture

A "keep-out zone" agreement would not constrain offensive postures, and these could be as in the existing regime. The protection afforded by defended keep-out zones would diminish the effectiveness of some types of weapons such as coorbital interceptors and thereby diminish incentives to include them in a space order of battle. However, the effectiveness of advanced ASAT weapons—e.g., directed-energy weapons—would not be significantly reduced by keep-out zones of the size considered here.

Defensive Posture

A "keep-out zone" agreement would not constrain defensive postures, and these could be as in the existing regime. Decoys might be an attractive defensive measure in this regime, because "keep-out zones" would inhibit or preclude certain types of close inspection which might otherwise be able to distinguish decoys from valuable satellites (see discussion in chapter 4). The deployment of DSATs or self-defense weapons would also be attractive, because such weapons could be used to enforce *agreed* keep-out zones. In the existing regime, attempts to enforce a declared keep-out zone by firing upon a "violating" suspected (but not proven) ASAT would probably be considered unlawful unless lethal capability and hostile intent of such spacecraft could be established.

Net Assessment

An agreement establishing minimum satellite separation rules could establish important legal rights to actively defend satellites, and would be an improvement over the existing regime if an active defense posture were desired. Enforcing agreed keep-out zones using DSATs would provide protection against relatively primitive ASAT weapons such as the current Soviet coorbital interceptor. However, keep-out zones large enough to protect satellites from advanced directed-energy weapons could be accommodated only beyond geosynchronous altitude.

A "keep-out zone" regime would have the advantage of not limiting research, development, and deployment of ASAT, DSAT, and BMD technologies. On the other hand, since a defended "keep-out zone" would provide significant protection against current ASAT weapons, it would encourage the development of more advanced systems. Such systems would likely increase in sophistication until the more advanced directed-energy technolgies reduced the effectiveness of "keep-out zones."

REGIME 6: SPACE SANCTUARIES

Legal Regime

A legal regime prohibiting the deployment of weapons in deep space (i.e., at altitudes greater than 3,000 nmi (5600 km)) or the testing of any weapons against instrumented targets or other objects in deep space could be established by a "Deep-Space Sanctuary." Such an agreement would be similar in some respects to the Antarctic Treaty,[16] the Outer Space Treaty,[17] the Treaty for the Prohibition

[16]The text of the Antarctic Treaty is reprinted in U.S. Arms Control and Disarmament Agency, *Arms Control and Disarmament Agreements* (Washington, D.C.: U.S. Government Printing Office, 1982), pp. 22-26.

[17]Ibid., pp. 51-55.

of Nuclear Weapons in Latin America,[18] the so-called Seabed Arms Control Treaty,[19] and other treaties and agreements which establish demilitarized or de-weaponized zones. Such an agreement would not prohibit development, testing, or deployment in space of ASAT weapons but would attempt to enhance security by banning the testing and deployment of weapons in deep space where critical strategic satellites are presently based. At present, such systems are invulnerable to currently operational tested ASAT weapons.

In addition to such an agreement, other relevant agreements currently in force (Limited Test Ban Treay, Outer Space Treaty, ABM Treaty) could remain in force in a "deep-space sanctuary" regime. Amendment of the Outer Space Treaty would not be an issue, since, unlike the "keep-out zone" regime, the "space sanctuary" regime could not be considered as a national appropriation of space.

Offensive Posture

Offensive postures appropriate in a "keep-out zones" regime [regime 5] would also be appropriate in a deep-space sanctuary regime, and for the same reasons. However, nuclear or kinetic-energy weapons—which would require more time to reach a satellite in deep space than to reach a satellite inside a small keep-out zone—would be less attractive as ASAT weapons than in a "keep-out zones" regime. Advanced directed-energy weapons, when feasible, would be the most capable ASAT weapons allowed in this regime, as in a "keep-out zones" regime.

Defensive Posture

Passive countermeasures appropriate in a "keep-out zone" regime would also be appropriate in this regime, and for the same reasons. However, as in a "keep-out zone" regime, passive countermeasures could not economically protect large and expensive satellites as high as in geosynchronous orbit from advanced directed-energy weapons, which would be allowed in low orbit and which could be adequately tested against instrumented target satellites in low orbit. As in the existing regime, small, inexpensive satellites might be protected from such advanced weapons because they might cost more to attack than to build.

Active countermeasures appropriate in the existing regime would also be appropriate in this regime, and for the same reasons. As in the existing regime, attacking suspicious approaching ASAT weapons would be unlawful at low altitudes where such objects would have rights of innocent passage. Deployment in deep space of "shoot-back" capabilities or DSATs would probably be prohibited since it might be impossible to differentiate these weapons from offensive weapons.

Net Assessment

The primary advantage of this regime would be that it could protect satellites in high orbits from the current generation of ASAT weapons. In addition, a deep-space sanctuary regime would constrain ASAT development less than would a comprehensive test ban regime or a no-new-types regime. However, should the United States and the Soviet Union choose to pursue advanced ASAT weapons, a space sanctuary might offer only limited protection.

The greatest risks in a space sanctuary regime would be posed by advanced directed-energy weapons which could be tested and deployed at low altitudes. Such testing and deployment would probably be adequate to guarantee effectiveness against targets at higher altitudes. Satellites at very high, supersynchronous altitudes might still derive some protection from this regime, but violation of the sanctuary by highly maneuverable kinetic-energy weapons or by satellites covertly carrying powerful nuclear or directed-energy weapons would remain a risk.

[18] Ibid., pp. 64-75; the texts of Protocols I and II thereto are reprinted in ibid., pp. 76 and 77, respectively.

[19] Formally titled "Treaty on the Prohibition of the Emplacement of Nuclear Weapons and Other Weapons of Mass Destruction on the Seabed and the Ocean Floor and in the Subsoil Thereof," the text of which is reprinted in ibid., pp. 103-105.

REGIME 7: A SPACE-BASED BMD REGIME

Legal Regime

If the United States or the Soviet Union withdrew from the ABM Treaty, this would, in addition to allowing ballistic missile defense, eliminate constraints on ASAT capabilities now imposed by that Treaty. The resulting regime would allow both advanced ASAT and space-based BMD weapons. Withdrawal from the Limited Test Ban Treaty and the Outer Space Treaty would also be necessary if the United States or the Soviet Union desired to test and deploy space weapons that used nuclear explosives as a power source.

Offensive Posture

In a space-based BMD regime, ASAT options would be less constrained than in the existing regime and advanced ASAT weapons would be more essential for defeating space-based enemy BMD system. In such a regime, advanced space-based weapons could be deployed at low altitudes and used as ASAT or DSAT weapons as well as for BMD. Some space-based weapons which would be useful—but not preferred—for satellite negation might be deployed in this regime because of their usefulness as BMD weapons. For example, kinetic-energy weapons and continuous-wave lasers which could destroy fast-burn boosters deep within the atmosphere might be preferred as BMD weapons over neutral particle beam or X-ray laser directed-energy weapons. The latter, although more useful in an ASAT role, could not readily penetrate the atmosphere and therefore may have more limited value as BMD weapons.

Defensive Posture

In a space-based BMD regime, defensive measures would be less constrained and more essential than in the existing regime. Advanced space-based weapons could be deployed at low altitudes and then used as ASAT or DSAT weapons. In a DSAT role, these weapons could offer some protection to low-altitude satellites. However, such satellites would probably remain vulnerable to attack by larger weapons or by expendable single-shot weapons (e.g., single-pulse lasers) which could attack from great range unless held at bay by large "keep-out zones." As discussed in chapter 4, it is possible that future technological advances might allow decoys to be developed that were cost-effective when compared to future offensive weapons and discrimination capabilities.

In evaluating offensive and defensive postures in a space-based BMD regime, it is necessary to assume that future technology will confer an advantage to ASAT countermeasures vis-a-vis ASAT capabilities. Although such an assumption may be unjustified at present, if the United States is to deploy advanced space-based BMD weapons then it must also have developed highly effective countermeasures to ASAT weapons. It would be irrational for the United States to seek to establish a "space-based BMD" regime unless it judged that adequate numbers of the space-based BMD components would survive or unless it judged that non-space-based BMD components could provide an adequate defense without space-based components. Scenarios illustrating each of these conditions are imaginable; for example:

1. The United States may judge that BMD systems with space-based components *could not be destroyed* by the U.S.S.R.: For example, the United States might deploy, in addition to ground-based BMD components, space-based electromagnetic launchers for kinetic-energy weapons and defend them by hardening, deception, and shoot-back. Deceptive measures employed might include massive decoys made from asteroidal material such as nickel.[20] While

[20]It is speculated that the cost of transporting such material to low Earth orbit and refining and fabricating finished products with it there may eventually be several orders of magnitude lower than the cost of refining and forming such materials on Earth and transporting the products to space. Should this forecast prove accurate, deception may have a favorable cost-exchange ratio even against ASAT systems which can discriminate decoys on the basis of mass density.

the Soviets might be able to destroy some BMD components, the system as a whole would survive.

2. The United States may judge that space-based BMD components *would not be destroyed* by the U.S.S.R.: Even if future technology does not favor ASAT countermeasures to the extent assumed in (1), ASAT countermeasure technology could be so effective that the Soviet leadership would be unwilling to pay the costs of defeating the countermeasures.
3. The United States may desire an extensive BMD system *without space-based components:* For example, U.S. aspiration might be limited to defense of hardened facilities which house strategic retaliatory forces or command and control systems; this might be accomplished using ground-based radars and interceptors but would require deployment of more of these over larger areas than is allowed by the ABM Treaty. Alternatively, the United States might desire an extensive BMD system capable of defending industry and population using only ground-based weapons.

Net Assessment

Depending on one's viewpoint, the principal advantage, or disadvantage, of a space-based BMD regime would be that it would allow the United States and the Soviet Union to deploy highly capable weapons in space. Since even a limited BMD system would probably make a very good ASAT system decision to proceed with BMD deployment necessarily includes a decision not to proceed with certain types of ASAT arms control.[21]

On March 23, 1983, the President called for a vigorous research program to determine the feasibility of highly effective, advanced-technology BMD systems, suggesting that the deployment of such systems, if feasible, would be desirable. Before the United States deployed space-based BMD systems it would have to determine, first, that the contribution that such systems made to U.S. security was great enough to compensate for the threat which similar opposing systems would pose to U.S. satellites, and second, that space-based BMD components could be protected at competitive cost against advanced ASAT weapons.

The threat to satellites would be greater in a space-based BMD regime than in any other regime because the BMD weapons would likely have extensive ASAT capabilities. The expense of equipping all military satellites with countermeasures against such capabilities would be considerable, particularly if, as some fear, deployment of space-based BMD systems will lead to a major arms race in both offensive and defensive weapons. However, if, as some argue, space-based missile defenses can make us more secure and encourage the Soviets to make real reductions in offensive missiles, this would reduce the threat of U.S./Soviet conflict and to contribute to a mutual desire to protect space assets. In a world where conflict was less likely, satellite vulnerability would be less important.

ASAT countermeasures must prove to be effective for space-based BMD platforms if a decision to deploy them is to make sense. It is possible that large improvements in the effectiveness or economy of passive countermeasures such as combinations of hardening, deception, and proliferation might provide the needed protection. If such improvements occur, they might also be used effectively for satellites in the other regimes discussed above. Alternatively, the superior fire-power or massive shielding of BMD weapons might give them a degree of protection unattainable by smaller, less capable satellites.

[21]It is possible that in a space-based BMD regime one might also wish to negotiate "rules of the road" such as "keep-out zones," or perhaps even a deep-space sanctuary.

Appendix A

Soviet Draft Treaty on the Prohibition of the Use of Force in Outer Space and From Space Against the Earth

Appendix A

Soviet Draft Treaty on the Prohibition of the Use of Force in Outer Space and From Space Against the Earth

U.N. General Assembly document A/38/194, Aug. 22, 1983

The States Parties to this Treaty,

Guided by the principle whereby Members of the United Nations shall refrain in their international relations from the threat or use of force in any manner inconsistent with the purposes of the United Nations,

Seeking to avert an arms race in outer space and thus to lessen the danger to mankind of the threat of nuclear war,

Desiring to contribute towards attainment of the goal whereby the exploration and utilization of outer space, including the Moon and other celestial bodies, would be carried out exclusively for peaceful purposes,

Have agreed on the following:

Article 1

It is prohibited to resort to the use or threat of force in outer space and the atmosphere and on the Earth through the utilization, as instruments of destruction, of space objects in orbit around the Earth, on celestial bodies or stationed in space in any other manner.

It is further prohibited to resort to the use or threat of force against space objects in orbit around the Earth, on celestial bodies or stationed in outer space in any other manner.

Article 2

In accordance with the provisions of article 1, States Parties to this Treaty undertake:

1. Not to test or deploy by placing in orbit around the Earth or stationing on celestial bodies or in any other manner any space-based weapons for the destruction of objects on the Earth, in the atmosphere or in outer space.
2. Not to utilize space objects in orbit around the Earth, on celestial bodies or stationed in outer space in any other manner as means to destroy any targets on the Earth, in the atmosphere or in outer space.
3. Not to destroy, damage, disturb the normal functioning or change the flight trajectory of space objects of other States.
4. Not to test or create new anti-satellite systems and to destroy any anti-satellite systems that they may already have.
5. Not to test or use manned spacecraft for military, including anti-satellite, purposes.

Article 3

The State Parties to this Treaty agree not to assist, encourage or induce any State, group of States, international organization or natural or legal person to engage in activities prohibited by this Treaty.

Article 4

1. For the purposes of providing assurance of compliance with the provisions of this Treaty, each State Party shall use the national technical means of verification at its disposal in a manner consistent with generally recognized principles of international law.
2. Each State Party undertakes not to interfere with the national technical means of verification of other States Parties operating in accordance with paragraph 1 of this article.

Article 5

1. The States Parties to this Treaty undertake to consult and co-operate with each other in solving any problems that may arise in connection with the objectives of the Treaty or its implementation.
2. Consultations and co-operation as provided in paragraph 1 of this article may also be undertaken by having recourse to appropriate in-

ternational procedures within the United Nations and in accordance with its Charter. Such recourse may include utilization of the services of the Consultative Committee of States Parties to the Treaty.

3. The Consultative Committee of States Parties to the Treaty shall be convened by the depositary within one month after the receipt of a request from any State Party to this Treaty. Any State Party may nominate a representative to serve on the Committee.

Article 6

Each State Party to this Treaty undertakes to adopt such internal measures as it may deem necessary to fulfil its constitutional requirements in order to prohibit or prevent the carrying out of any activity contrary to the provisions of this Treaty in any place whatever under its jurisdiction or control.

Article 7

Nothing in this Treaty shall affect the rights and obligations of States under the Charter of the United Nations.

Article 8

Any dispute which may arise in connection with the implementation of this Treaty shall be settled exclusively by peaceful means through recourse to the procedures provided for in the Charter of the United Nations.

Article 9

This Treaty shall be of unlimited duration.

Article 10

1. This Treaty shall be open to all States for signature at United Nations Headquarters in New York. Any State which does not sign this treaty before its entry into force in accordance with paragraph 3 of this article may accede to it at any time.
2. This Treaty shall be subject to ratification by signatory States. Instruments of ratification and accession shall be deposited with the Secretary-General of the United Nations.
3. This Treaty shall enter into force between the States which have deposited instruments of ratification upon the deposit with the Secretary-General of the United Nations of the fifth instrument of ratification, provided that such instruments have been deposited by the Union of Soviet Socialist Republics and the United States of America.
4. For States whose instruments of ratification or accession are deposited after the entry into force of this Treaty, it shall enter into force on the date of the deposit of their instruments of ratification or accession.
5. The Secretary-General of the United Nations shall promptly inform all signatory and acceding States of the date of each signature, the date of deposit of each instrument of ratification or accession, the date of entry into force of this Treaty as well as other notices.

Article 11

This Treaty, of which the Arabic, Chinese, English, French, Russian and Spanish texts are equally authentic, shall be deposited with the Secretary-General of the United Nations, who shall send duly certified copies thereof to the Governments of the signatory and acceding States.